Understanding and optimising the nutraceutical properties of fruit and vegetables

It is widely recognised that agriculture is a significant contributor to global warming and climate change. Agriculture needs to reduce its environmental impact and adapt to current climate change whilst still feeding a growing population, i.e. become more 'climate-smart'. Burleigh Dodds Science Publishing is playing its part in achieving this by bringing together key research on making the production of the world's most important crops and livestock products more sustainable.

Based on extensive research, our publications specifically target the challenge of climate-smart agriculture. In this way we are using 'smart publishing' to help achieve climate-smart agriculture.

Burleigh Dodds Science Publishing is an independent and innovative publisher delivering high quality customer-focused agricultural science content in both print and online formats for the academic and research communities. Our aim is to build a foundation of knowledge on which researchers can build to meet the challenge of climate-smart agriculture.

For more information about Burleigh Dodds Science Publishing simply call us on +44 (0) 1223 839365, email info@bdspublishing.com or alternatively please visit our website at www.bdspublishing.com.

Related titles:

Achieving sustainable cultivation of temperate zone tree fruits and berries Volume 1: Physiology, genetics and cultivation
Print (ISBN 978-1-78676-208-5) Online (ISBN 978-1-78676-210-8; 978-1-78676-211-5)

Achieving sustainable cultivation of vegetables
Print (ISBN 978-1-78676-236-8) Online (ISBN 978-1-78676-238-2; 978-1-78676-239-9)

Achieving sustainable cultivation of tomatoes
Print (ISBN 978-1-78676-040-1) Online (ISBN 978-1-78676-042-5; 978-1-78676-043-2)

Achieving sustainable cultivation of mangoes
Print (ISBN 978-1-78676-132-3) Online (ISBN 978-1-78676-134-7; 978-1-78676-135-4)

Improving the nutritional and nutraceutical properties of wheat and other cereals
Print (ISBN 978-1-78676-479-9) Online (ISBN 978-1-78676-481-2; 978-1-78676-482-9)

Chapters are available individually from our online bookshop:
https://shop.bdspublishing.com

climate•SMART•publishing
IN AGRICULTURAL SCIENCE

BURLEIGH DODDS SERIES IN AGRICULTURAL SCIENCE
NUMBER 116

Understanding and optimising the nutraceutical properties of fruit and vegetables

Edited by Professor Victor R. Preedy, King's College London, UK and Dr Vinood B. Patel, University of Westminster, UK

burleigh dodds
SCIENCE PUBLISHING

Published by Burleigh Dodds Science Publishing Limited
82 High Street, Sawston, Cambridge CB22 3HJ, UK
www.bdspublishing.com

Burleigh Dodds Science Publishing, 1518 Walnut Street, Suite 900, Philadelphia, PA 19102-3406, USA

First published 2022 by Burleigh Dodds Science Publishing Limited

Library of Congress Control Number: 2022934577

British Library Cataloguing in Publication Data
A catalogue record for this book is available from the British Library

ISBN 978-1-78676-850-6 (print)
ISBN 978-1-78676-853-7 (PDF)
ISBN 978-1-78676-852-0 (ePub)
ISSN 2059-6936 (print)
ISSN 2059-6944 (online)

DOI 10.19103/AS.2022.0101

Typeset by Deanta Global Publishing Services, Dublin, Ireland

Contents

Part 2 Phytochemicals in fruits and vegetables: glucosinolates and organosulfur compounds

Series list

Title	Series number
Achieving sustainable cultivation of maize - Vol 1 From improved varieties to local applications Edited by: Dr Dave Watson, CGIAR Maize Research Program Manager, CIMMYT, Mexico	001
Achieving sustainable cultivation of maize - Vol 2 Cultivation techniques, pest and disease control Edited by: Dr Dave Watson, CGIAR Maize Research Program Manager, CIMMYT, Mexico	002
Achieving sustainable cultivation of rice - Vol 1 Breeding for higher yield and quality Edited by: Prof. Takuji Sasaki, Tokyo University of Agriculture, Japan	003
Achieving sustainable cultivation of rice - Vol 2 Cultivation, pest and disease management Edited by: Prof. Takuji Sasaki, Tokyo University of Agriculture, Japan	004
Achieving sustainable cultivation of wheat - Vol 1 Breeding, quality traits, pests and diseases Edited by: Prof. Peter Langridge, The University of Adelaide, Australia	005
Achieving sustainable cultivation of wheat - Vol 2 Cultivation techniques Edited by: Prof. Peter Langridge, The University of Adelaide, Australia	006
Achieving sustainable cultivation of tomatoes Edited by: Dr Autar Mattoo, USDA-ARS, USA and Prof. Avtar Handa, Purdue University, USA	007
Achieving sustainable production of milk - Vol 1 Milk composition, genetics and breeding Edited by: Dr Nico van Belzen, International Dairy Federation (IDF), Belgium	008
Achieving sustainable production of milk - Vol 2 Safety, quality and sustainability Edited by: Dr Nico van Belzen, International Dairy Federation (IDF), Belgium	009
Achieving sustainable production of milk - Vol 3 Dairy herd management and welfare Edited by: Prof. John Webster, University of Bristol, UK	010
Ensuring safety and quality in the production of beef - Vol 1 Safety Edited by: Prof. Gary Acuff, Texas A&M University, USA and Prof. James Dickson, Iowa State University, USA	011
Ensuring safety and quality in the production of beef - Vol 2 Quality Edited by: Prof. Michael Dikeman, Kansas State University, USA	012
Achieving sustainable production of poultry meat - Vol 1 Safety, quality and sustainability Edited by: Prof. Steven C. Ricke, University of Arkansas, USA	013
Achieving sustainable production of poultry meat - Vol 2 Breeding and nutrition Edited by: Prof. Todd Applegate, University of Georgia, USA	014

Introduction

As populations in many developed countries age and the burden of chronic disease increases, there remains a need to establish effective preventative measures. Fruit and vegetables are a natural source of vitamins and minerals which can contribute to good health.

This volume reviews the associated health benefits of key horticultural crops, including apples, broccoli and cranberries. The book is split into five parts: Part 1 chapters examine phytochemical compounds in fruits and vegetables, specifically focusing on polyphenols. Chapters in Part 2 focus on glucosinolates and organosulfur compounds in fruits and vegetables. Part 3 chapters discuss the role of phytochemicals in preventing diseases such as cancer and cardiovascular disease. Chapters in Part 4 highlight ways to analyse and optimise phytochemical compounds in fruits and vegetables. Part 5 chapters conclude the book by providing three case studies that focus specifically on improving the nutraceutical properties of cranberries, apples, broccoli and other brassicas.

Part 1 Phytochemical compounds in fruits and vegetables: polyphenols

The book opens with a chapter that examines the advances in understanding the nutraceutical properties of antioxidants in fruits and vegetables. Chapter 1 begins by highlighting the most common antioxidants in fruits and vegetables such as polyphenols, carotenoids, vitamins, selenium and zinc. The chapter then moves on to review the mechanisms in which these antioxidants function within the human body. A section on how these antioxidants can play a crucial role in the prevention of chronic diseases such as cancer and cardiovascular disease is also included. This is then followed by an analysis of their application as natural pigments, preservatives, edible films and coatings, natural emulsifiers, stabilisers and nutraceuticals. The chapter concludes by emphasising how critical antioxidants in fruits and vegetables are in the human diet.

Chapter 2 focuses on understanding phenolic compounds in fruits and vegetables. The chapter first examines the characteristics of phenolic compounds, focusing specifically on the two main subgroups they can be divided into, flavonoids and non-flavonoids. It then moves on to discuss the effects of cultivation and post-harvest operations on phenolic compounds, as well as how the COVID-19 pandemic highlighted the need to include phytochemicals such as polyphenols in diets to increase general immunity. The chapter also looks at ways to improve phenolic compounds in fruits and vegetables, before concluding with an overview of why it is crucial to develop our understanding of the importance of phenolic compounds to improve human health.

The subject of Chapter 3 is understanding the nutraceutical properties of flavonoids in fruits and vegetables, drawing attention to the chemical structure and groups flavonoids can be divided into. The chapter first provides an overview of the flavonoids present in fruits and vegetables, focusing on flavones, flavonols, flavanones, flavanols, isoflavones, neoflavonoids, flavanols, anthocyanidins and chalcones. Individual sections and subsections are provided for each group of flavonoids, focusing on their chemical structure and their role as nutraceuticals in human health.

Expanding on topics previously discussed in Chapter 3, Chapter 4 expands on the review of understanding the nutraceutical properties of flavonoids in fruits and vegetables by addressing the mechanisms of action for flavonoids. The chapter first looks at the antioxidant properties of flavonoids and their role in preventing auto-immune diseases. It then looks at the anti-microbial, anti-fungal and anti-viral activity of flavonoids and their role in treatment of diabetes. The chapter goes on to discuss the anti-cancer properties of flavonoids and their anti-neoplastic activity in tumour suppression. It also reviews their role in preventing cardiovascular disease and anti-thrombogenic activity of flavonoids, as well as their neuro-protective, anti-ulcerogenic, anti-inflammatory and hepato-protective activity. Finally, the chapter discusses biotechnological approaches for enhanced production of nutraceuticals in fruits and vegetables.

Part 2 Phytochemicals in fruits and vegetables: glucosinolates and organosulfur compounds

Part 2 begins with a chapter that focuses on the health promoting effects of glucosinolates and their breakdown products. Chapter 5 first reviews the natural sources of glucosinolates, highlighting one of the most common groups of plants – Brassicales – and how well-known species such as broccoli, cabbage, kale, cauliflower and turnip are some of the best natural sources of these phytochemicals. The chapter also examines the various agricultural and environmental factors that affect the composition and levels of glucosinolates in vegetables. Sections on the potential health effects and antinutritional properties of glucosinolates are also provided, which are then followed by a discussion on dietary intake, absorption and digestion of glucosinolates in the human diet.

Expanding on topics previously discussed in Chapter 5, Chapter 6 looks at the nutraceutical potential of glucosinolates. The chapter reviews the different classes of glucosinolates and their breakdown products, focusing specifically on aliphatic, indole and aromatic glucosinolates first, then goes on to examine isothiocyanates, thiocyanate, nitriles, oxazolidine-2-thione and epithionitriles. A section on the hydrolysis of glucosinolates and the glucosinolates-myrosinase system is also provided, which is then followed by a discussion on how glucosinolates can be analysed, and finally the mechanisms of action and how glucosinolates can be used as nutraceuticals.

The final chapter of Part 2 focuses on understanding the health benefits and nutraceutical properties of organosulphur compounds in vegetables. Chapter 7 first looks at the bioavailability of organosulphur compounds. It then moves on to discuss the health benefits of these compounds, drawing attention to anti-inflammatory activity, anticarcinogenic effects, antioxidant and antimicrobial activity, improving immune function and their use in neurogenerative disorders. A section on the nutraceutical applications of organosulphur compounds is also provided.

Part 3 Phytochemicals and the prevention of disease

The first chapter of Part 3 examines advances in understanding the role of plant phytochemicals in preventing cancer. Chapter 8 begins by emphasising the importance of plant-based foods as sources of cancer-preventative substances as they contain various types of phytochemicals. The chapter then goes on to examine the mechanisms of chemoprevention, which is then followed by an analysis of phytochemicals and their mode of action in the prevention and treatment of cancer. A section on the cancer-preventive effects of antioxidant and anti-inflammatory activities is also provided. The chapter moves on to review the angiogenesis suppression activities of phytochemicals, their function in cell death pathways and the combined use of phytochemicals with antineoplastic agents. The importance of delivery systems for phytochemicals is also discussed. Sections on routes of administration for phytochemicals and combining phytochemicals with other applications are also provided.

Chapter 9 discusses advances in understanding the role of plant phytochemicals in preventing cardiovascular disease. The chapter first reviews the complex nature of cardiovascular disease and its link to diet as well as the disease's comorbidities. It then assesses the current state of research on the protective and therapeutic effects of phytochemicals in relation to cardiovascular disease. A section on the types of phytochemical compounds is also provided, focusing specifically on flavonoids, phenols, organosulfur compounds, alkaloids, lignans, sterols, tannins and soluble fibre. The chapter moves on to examine how wild crop relatives and herbs and spices can be considered as sources of phytochemicals, before it addresses the range of initiatives that are currently in place in order to enhance phytochemicals in the diet to prevent or treat cardiovascular disease.

Part 4 Analysing and optimising phytochemical compounds in fruits and vegetables

Part 4 first begins with a chapter that analyses the advances in screening and analysing phytochemical compounds in fruits and vegetables. Chapter 10 reviews the various state of the art methods currently available for extracting phytochemicals before screening and analysis can be performed. The chapter then moves on to review the methods for phytochemical analysis, drawing attention to high-performance thin-layer chromatography, liquid chromatography, gas chromatography, mass spectrometry, infrared spectroscopy and nuclear magnetic resonance. A case study on using advanced methods for the analysis of phenolic compounds in apples is also included.

The next chapter of Part 4 examines the agronomic factors affecting phytochemical compounds in fruits and vegetables. Chapter 11 first reviews the various phytochemicals that are available in fruits and vegetables, drawing attention to those previously discussed in earlier chapters. It then addresses the environmental factors that affect phytochemicals, focusing specifically on lighting, temperature and relative humidity. A section on the current agronomic practices in use to improve the yield and quality of fruits and vegetables is also included, discussing practices such as variety selection, fertilisation, phytohormones and phytochemicals induced by abiotic stress. It also includes a review of the phytochemical changes during harvesting and different stages of maturation. A case study on the use of organic and inorganic nitrogen to influence phytochemical levels in lettuce is also provided.

Chapter 12 examines the metabolism of phytochemical compounds in fruits and vegetables in the gut. It first discusses how phytochemicals can be classified, indicates their dietary sources and intake and highlights their health effects. The chapter focuses on categories of phytochemicals previously highlighted such as phenolics and alkaloids but also draws attention to terpenoids and sulphur-containing compounds. The chapter then moves on to discuss the digestion of phytochemical compounds, focusing on oral and gastric digestion, small intestinal digestion and colonic digestion. A case study on a dynamic multistage gastrointestinal model for the study of phytochemical biotransformation by gut microbiota is also included. Factors affecting digestion such as the food matrix and background diet are also reviewed.

Part 5 Case studies

The first case study chapter in Part 5 looks at the advances in understanding and improving the nutraceutical properties of cranberries. Chapter 13 first reviews the clinical evidence supporting the health benefits of cranberry consumption such as the use of cranberries to protect urinary tract infections as well as their use to improve dental and gastrointestinal health, cardiometabolic and skin health, cognition and cancer prevention/treatment. It also discusses future directions in the elucidation of mechanism of actions for health benefits and approaches to maximising bioefficacy of cranberry-related food/products.

Chapter 14 discusses advances in understanding and improving the nutraceutical properties of apples. It begins by examining the health effects of apple in humans, then moves on to discuss apple processing and pomace generation. Individual sections on various nutraceutic compounds in apples are also provided, focusing specifically on phenolic compounds, dietary fibre and pectin. The chapter then moves on to review improving apple cultivation through research on food bioactives, before concluding with an overview of why apple and apple pomace are important sources of phenolic compounds, dietary fibre and pectin which in turn, have significant health benefits.

The final chapter of the book draws attention to advances in understanding and improving the nutraceutical properties of broccoli and other brassicas. Chapter 15 discusses the available evidence demonstrating the benefits of the *Brassica* family, particularly broccoli, and offers information to help ensure that public health messaging reflects current science. It demonstrates that *Brassica* vegetables have advantages beyond helping achieve basic nutritional requirements and could provide specific nutrients towards reducing the risk of cancer and other diseases. Knowledge of the bioavailability of *Brassica* bioactive compounds, particularly of phenolic compounds and glucosinolates, is critical to understanding such benefits. Finally, it discusses additional support and strategies to maintain *Brassica* nutritive value since the content of particular compounds varies significantly due to the different factors in the food supply chain.

Part 1

Phytochemical compounds in fruits and vegetables: polyphenols

Chapter 1

Advances in understanding the nutraceutical properties of antioxidants in fruits and vegetables

Ugunujhie Agbaje, Mallaidh Hyndman and Soraeya Kharaty, School of Food Science and Environmental Health, Technological University Dublin - City Campus, Ireland; and Swarna Jaiswal and Amit K. Jaiswal, School of Food Science and Environmental Health, Technological University Dublin - City Campus and Environmental Sustainability and Health Institute, Technological University Dublin - City Campus, Ireland

1 Introduction

Antioxidants are natural or synthetic substances that inhibit oxidation, a chemical reaction within the human body that can produce free radicals and chain reactions, resulting in cell damage. They play a vital role in growth, development, detoxification, and effective immune responses. Oxygen is known as a highly reactive atom which is commonly known to become part of damaging molecules within the body known as free radicals. These free radicals form during normal metabolism or through external factors such as X-rays, ultraviolet radiation, and exposure to pollution. Free radicals such as superoxide (O_2^-), hydrogen peroxide (H_2O_2), or peroxynitrite ($OONO^-$) have the capabilities of attacking cells and other structures like cellular proteins, lipids, and deoxyribonucleic acid (DNA) (Kazmierczak-Baranska et al., 2020) within the human body, resulting in a loss of structure and function of those cells. Free radicals have also been linked to the initiation and progression

http://dx.doi.org/10.19103/AS.2022.0101.03

of tumour cells, while also enhancing their metastatic potential (Singh et al., 2019). Oxidative stress is one of the reasons for promoting the ageing of the body and many diseases, and its mechanism is mostly related to the generation of reactive oxygen species (ROS).

Dieticians recommend a diet that promotes a high volume of fruits and vegetables, as these are universally regarded as healthy. They supply a wide range of vitamins and minerals to the human diet and are prominent sources of phytochemicals that function as antioxidants. Plants produce antioxidants as a form of protection. The antioxidants act as a barrier to the damage that may be caused when reactive species produced during photosynthesis are exposed to ultraviolet light, resulting in damage to their cellular structure (Laxa et al., 2019). Moderate consumption of approximately five servings of fruit and vegetables a day is recommended to adults, as this diet will aid in mitigating the development of chronic diseases such as cancer, diabetes, and cardiovascular diseases (CVD), with convincing evidence provided by Boeing et al. (2012) in a critical review of literature in 2012.

This chapter aims to provide recent advances in the understanding of various nutraceuticals properties of fruit and vegetables. The chapter also briefs on common types of antioxidants found in fruits and vegetables. The mechanisms in which these antioxidants function within the human body have been illustrated, with emphasis on their role in the prevention of chronic diseases such as cancer and CVDs. Furthermore, applications of nutraceuticals properties of fruit and vegetables have been discussed.

2 Antioxidants in fruits and vegetables

The consumption of fruits and vegetables can reduce the risk of oxidative damage to cells within the human body. They are considered excellent sources of antioxidants as they contain considerable levels of biologically active compounds that transmit health benefits that surpass basic nutrition (Oomah and Mazza, 2000). Common dietary antioxidants which are found in fruits and vegetables include polyphenols, carotenoids, vitamins such as vitamin A, C, and E, selenium, and zinc. Each of these offers an array of health benefits when consumed, with the shared objective of reducing the effect of free radicals on the body's immune system, gastrointestinal system, and overall health.

2.1 Polyphenols

Polyphenols are organic compounds commonly synthesized within plant species. They are mainly present in fruits, vegetables, whole grains, and green teas. They are a well-known group of phenolic systems and are categorized by a minimum of two phenyl rings and one or more hydroxyl substituents

(Singla et al., 2019). The term 'polyphenol' is commonly used today in relation to flavonoids, tannins, and phenolic acid, as well as the numerous chemically modified or polymerized derivatives. There are more than 8000 types of polyphenols that have been identified, which are split into four main classes: phenolic acids, flavonoids, stilbenes, and lignans. Figure 1 below includes the chemical structures of different classes of polyphenols.

Phenolic acids are a class of polyphenols that exhibit a variety of functions including plant growth, development, and defence (Kumar and Goel, 2019). They are a powerful group of compounds that are both water- and lipid-soluble and exhibit antioxidative activities through various mechanisms which include removing free radicals, binding metal ions, and inducing the expression of a range of genes that are responsible for synthesizing enzymes, with the function of reducing oxidative stress (Kaurinovic and Vastag, 2019). Common sources of phenolic acids within the human diet are apples, mangos, berries, plums, cherries, kiwis, citrus fruits, and onions, with a daily intake recommendation of 200 mg/day (Kumar and Goel, 2019; Scalbert and Williamson, 2000).

Flavonoids account for approximately two-thirds of dietary polyphenols (Koch, 2019). They are defined as a class of polyphenolic secondary metabolites and are found in the form of aglycones or glycosides in many fruits and vegetables (Hernández-Rodríguez et al., 2019). They consist of a basic structure of two aromatic rings which are bound together by three carbon atoms to form an oxygenated heterocycle (Pandey and Rizvi, 2009). There are more than 4000 varieties of flavonoids that have been identified. These are split into six

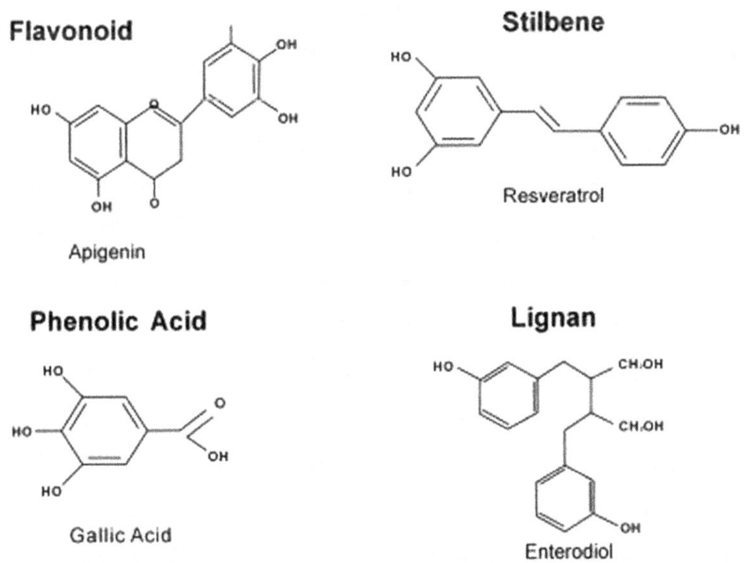

Figure 1 Chemical structures of different dietary polyphenols (Yoon and Baek, 2005).

sub-categories based upon the variation in the type of heterocycle involved and their extent of alkylation and/or glycosylation (Spencer et al., 2008). These sub-categories are flavonols, flavones, flavanones, flavanols, anthocyanins, and isoflavones. Leafy vegetables, onions, apples, berries, soybeans, and citrus fruits are classified as important sources of dietary flavonoids (Janabi et al., 2019).

Stilbenes contain two benzene rings joined by a molecule of ethanol or ethylene (Pandey and Rizvi, 2009). It is mainly derived from the skin of grapes in the form of the trans isomer of resveratrol. It is mainly extracted during the production of red wine, where it is found at concentrations of between a few tenths of a milligram and a few milligrams per litre (Moreno and Peinado, 2012).

Lignans are chemical compounds, which contain a 2,3-dibenzylbutane structure, formed by the dimerization of two cinnamic acid residues (Pandey and Rizvi, 2009). They are found in plants and act as dietary antioxidants within the human body, presenting a range of health benefits. They are known as phytoestrogens and can bind to oestrogen receptors in the breast tissue (Wcislo and Szarlej-Wcislo, 2014). Studies have shown that an increase in the consumption of dietary lignans had exhibited a significant reduction of approximately 40–53% in mortality and a 33–70% reduction in mortality by breast cancer (Calado et al., 2018). The richest sources of lignans in the diet are flaxseeds, where the principal lignan precursor, secoisolariciresinol diglucoside, is found (Simpson and Amos, 2017). Cruciferous vegetables, such as broccoli and cabbage, strawberries, and apricots are also considered excellent sources of these polyphenols.

The health benefits of polyphenols have been studied for decades. They have been known to prevent blood clots (Behl et al., 2020), reduce blood sugar levels, and lower the risk of heart disease (Sakaki et al., 2019). They are best known for their capabilities of reducing inflammation (Cory et al., 2018) which is why they are commonly prescribed to patients suffering from chronic illnesses, as they increase the production of anti-inflammatory molecules like IL-10, IL-4, IL-13, and adiponectin (Lee at al., 2021). In addition to the above, polyphenols have also been known to inhibit the absorption of already oxidized products, such as lipid hydroperoxides.

Polyphenols, such as polyphonic acid, are typically absorbed through the gut barrier (Kawabata et al., 2019). However, larger polyphenols are poorly absorbed. Once they are absorbed, they are linked to glucuronide, sulphate, and methyl groups in the gut mucosa and inner tissues.

2.2 Carotenoids

Carotenoids are lipid-soluble pigments which are found in algae, plants, and photosynthetic bacteria. They are known as very efficient physical scavengers of ROS. These pigments typically have a 40-carbon chain backbone composed of

8 isoprene molecules (Fig. 2) (Scott and Stuart, 2020). They are responsible for producing bright yellow, red, and orange pigmentations in fruits, vegetables, and other plants. There have been more than 600 different carotenoids identified and characterized to date (Young and Lowe, 2018). They are commonly split into two classifications of carotenoids: carotenes and xanthophylls. Carotenes consist of only hydrogen and carbon atoms, with beta-carotene being the most common form. Xanthophylls consist of one or more oxygen atoms, with lutein being the most common form.

Important members of oxygenated carotenoids are lutein, zeaxanthin, β-cryptoxanthin, capsanthin, astaxanthin, and fucoxanthin (Abdel-Aal et al., 2013). Astaxanthin is a natural antioxidative pigment, which has been found to prevent degenerative diseases (Gao et al., 2021) and cancer (McCall et al., 2018) and stimulate the immune system to detoxify free radicals within the human body. Carotenoids are usually absorbed intact with approximately 80% absorbed. However, they have a lower absorption rate compared to other

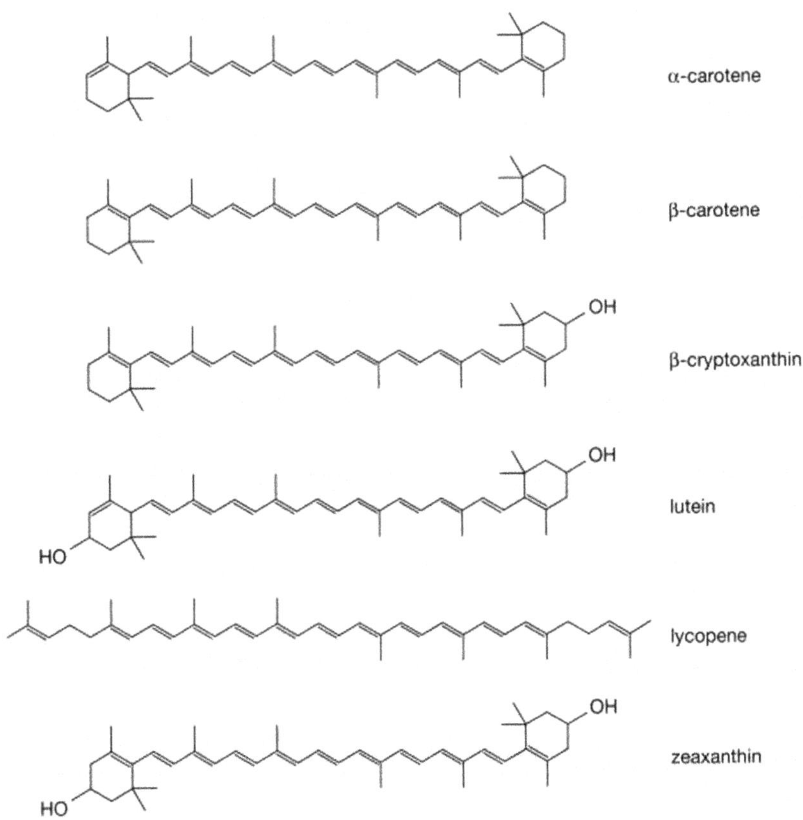

Figure 2 Chemical structure of common dietary carotenoids (Ellison, 2016).

dietary antioxidants. The intestinal cells convert carotenoids to retinoids (Bohn et al., 2019). It is then passed along with fat through the lymphatic system into the blood stream.

Carotenoids are typically indispensable within mammals, which makes it essential that they must ingest these molecules within their diet. As they are lipid-soluble, they are best absorbed through a source of fat. Lutein and zeaxanthin are relatively polar carotenoid pigments. They are found at high levels in parsley, spinach, and kale (Abdel-Aal et al., 2013). Other common sources of dietary carotenoids are found within fruits and vegetables which have bright yellow, orange, and red pigmentations such as carrots, sweet potatoes, and peppers; however, they are also prominent within dark leafy greens (Nabi et al., 2020).

2.3 Vitamins

There are three major antioxidant vitamins: vitamin A (beta-carotene), vitamin C, and vitamin E. Beta-carotene is known as the precursor of vitamin A. It is produced by the human body to promote normal function of the visual system (Johra et al., 2020), healthy skin, efficient mucosal membranes (Bohn et al., 2019), and the maintenance of cell function for growth. Beta-carotene is found in fruits and vegetables, and it is recommended to individuals ingest at least 5 servings per day, allowing the body to absorb at least 3–6 mg of beta-carotene daily (Müller, 1996). A deficiency of vitamin A can lead to a decrease in infection resistance. It has been discussed that dietary vitamin A is more efficient and protective than supplementation (Aserese et al., 2020). Vitamin A is rarely ever absorbed into the bloodstream. Alternatively, it is excreted from the body within 1–2 days.

Most dietary sources of vitamin A come from green leafy vegetables, such as kale and spinach, as well as from fruits and vegetables which contain both red and orange pigments, for example, peppers, carrots, and squash.

Vitamin C is a powerful antioxidant which has the ability to donate hydrogen atoms and form a relatively stable ascorbyl-free radical. It is a dibasic acid built with an enediol group on C2 and C3 of a heterocyclic lactone ring (Pehlivan, 2017). It is water-soluble and needs to be replaced within the body daily. It is essential for the creation of collagen, protein metabolism, and wound healing. It also plays a vital role in the absorption of non-heme iron, commonly found within vegetables. It is overall best known for its antioxidant properties, with studies finding that it has been responsible for the regeneration of other antioxidant vitamins such as vitamin E (Pehlivan, 2017). Vitamin C is a water-soluble substance and is actively absorbed within the small intestine. Reversible oxidation takes place here, where dehydroascorbic acid is produced (Dewhirst and Fry, 2018).

The most common sources of vitamin C within the human diet are citrus fruits, such as oranges, lemons, and limes, which are naturally very rich in ascorbate (Food and Nutrition Board, 2000). Other common sources include tomatoes and tomato juice. Fruits such as strawberries, mangoes, papaya, and watermelon also have variable amounts of vitamin C. It is recommended to consume five varied servings of fruits and vegetables a day, which can help to provide more than 200 mg of vitamin C (Olson and Hodges, 1987).

Vitamin E is a fat-soluble compound that can be stored within the body, meaning that it does not need to be restored on a daily basis. It is categorized as a lipophilic, naturally occurring compound whose molecular structure is comprised of a chromanol ring with a side chain located at the C2 position and includes four tocopherols and four tocotrienols (Niki and Abe, 2019). The chromanol ring is responsible for its antioxidative properties. Vitamin E prevents non-enzymatic oxidations of multiple cell components by molecular oxygen or free radicals such as hydrogen peroxide. It is essential for membrane structure and integrity, making it suitably known as a membrane antioxidant (DiPasquale et al., 2020).

It is typically absorbed with lipids within the upper small intestine, where it combines with micelles, also known as bile salts, to from mixed micelles. These are then taken up by mucosal cells, where they are incorporated into chylomicrons. The vitamin E is transported within the chylomicrons to the peripheral tissue or liver where it is stored (Pinto et al., 2020).

Common sources of vitamin E within the diet include dark leafy green vegetables, which contain a significant number of dietary antioxidants. Other sources include pumpkins, peppers, and vegetable oils. Vegetable oils would provide the richest source of vitamin E within the diet due to a-Tocopherol being the major contributor to the total vitamin E activity in some of these oils (Bramley et al., 2000).

2.4 Selenium

Selenium is an essential trace mineral and antioxidant which is commonly found in soil. It can form molecules of the ring structure, consisting of eight atoms and chain molecules that are characterized by a considerable length (Kieliszek et al., 2019). It is vital for good health and plays an active role in the formation of selenoproteins. This aids in the production of glutathione peroxidase, which are molecules that prevent cell damage. Glutathione peroxidase is responsible for the conversion of cellular toxins into harmless by-products for elimination. Selenium can be split into two forms: inorganic (selenate and selenite) and organic (selenomethionine and selenocysteine) (Avery and Hoffmann, 2018), with both being very good dietary sources. Selenium is mainly absorbed from the duodenum. It is then transported across the intestinal brush border, in the

Table 1 Recommended daily intake of antioxidants found in fruits and vegetables

Antioxidant	RDA - men (μg/day)	RDA - women (μg/day)	RDA - children (μg/day)
Vitamin C - ascorbic acid	90	75	40-65
Vitamin A - beta-carotene	900	700	300-600
Vitamin E	15	15	6-11
Polyphenols	650	650	500
Carotenoids	30-300	30-300	30-150
Selenium	50-60	50-60	20-40
Zinc	11	8	2-5

Sources: Gibson et al. (2016), Grosso et al. (2014), Lykkesfeldt et al. (2014), Risvi et al. (2014), Ross (2010), Stoffanellar and Morse (2015), Toti et al. (2018).

form of methionine analogue. Following absorption, it is bound to plasma proteins, for example, BETA-lipoproteins, and transported across the body.

Common dietary sources of selenium include beans and lentils, with lentils including 6 μg per serving. Another excellent source is dark leafy greens such as spinach, with up to 11 μg per serving.

2.5 Zinc

Zinc is an essential trace metal which is required for human health. Although it is a redox-inert metal, it functions as an antioxidant through the catalytic action of copper/zinc-superoxide dismutase, stabilization of membrane structure, protection of the protein sulfhydryl groups, and upregulation of the expression of metallothionein (Lee, 2018). In addition to this, zinc has been discovered to suppress inflammatory responses within the body, caused by oxidative stress. As an antioxidant, the consumption of foods containing zinc presents a number of benefits to human health. For example, it is used to treat and prevent diarrhoea in infants and children throughout the world (Prasad, 2014) and as an effective therapeutic agent for the treatment of Wilson's disease.

Sources of zinc within the human diet are pumpkin seeds, chickpeas, and peas. However, phytates that are found in legumes have been discovered to bind zinc and inhibit its absorption. Due to this, animal products are considered better sources of zinc (Olza et al., 2017) (Table 1).

3 Mechanism of action of fruit and vegetable antioxidants

Natural antioxidants present in fruit and vegetables possess several mechanisms in which they inhibit oxidative damage. There are numerous ways of classifying these antioxidants, with classification typically occurring according to these mechanisms of action and their origin. Antioxidants can be classified as either enzymatic or non-enzymatic and endogenous, originating in the body, or exogenous, obtained from outside the body typically through diet. However, antioxidants obtained from fruit and vegetables are non-enzymatic and exogenous compounds. Antioxidants can then be further classified as primary and secondary antioxidants based on their principal mechanisms. Primary refer to chain-breaking antioxidants, and secondary refer to peroxide-scavenging antioxidants. Secondary antioxidants also encompass chelating agents and oxygen-scavenging radicals (Aziz et al., 2019).

3.1 Primary antioxidants

The mode of action of primary antioxidants is a chain-breaking mechanism. Chain reactions in oxidation are divided into the initiation, propagation, and termination stages with primary antioxidants acting by impeding the initiation or propagation stages. Primary antioxidants react with peroxyl, alkyl, or hydroxy radicals and scavenge free radical species resulting in the formation of more stable radicals or non-radical species such as water and inert alcohols (Pisoschi et al., 2021). A recent study showed curcumin to have notable chain-breaking antioxidant activity in several free radical scavenging and reducing assays (Bisset et al., 2020).

3.2 Secondary antioxidants

Antioxidants derived from fruits and vegetables may exert their activity through a secondary mechanism. Secondary antioxidants encompass several mechanisms of action for protecting oxidative damage. One mechanism is singlet oxygen ($^1O_2$) quenching to prevent the degradation of biomolecules such as DNA and proteins. Dietary antioxidants, such as vitamin E, polyphenols, and carotenoids, may quench singlet oxygen via chemical or physical means. However, many antioxidants can quench singlet oxygen via both methods. Chemical quenching involves the oxidation of the quencher by singlet oxygen. On the other hand, physical quenching may occur via energy or charge transfer which deactivates the singlet oxygen (Petrou et al., 2018). Zeaxanthin extracted from *Lycium barbarum* was found to be highly effective in quenching $^1O_2$ in lipid membranes, offering high protection against oxidative damage (Aziz et al., 2019) (Fig. 3).

Lipid peroxidation induced by ROS generation is a form of oxidative stress that results in the production of undesirable lipid peroxide products such as malondialdehyde and damage to phospholipid membranes and other intracellular biomolecules. Lipid peroxidation and its products have several adverse effects not only in disease pathology but also in food systems (Su et al.,

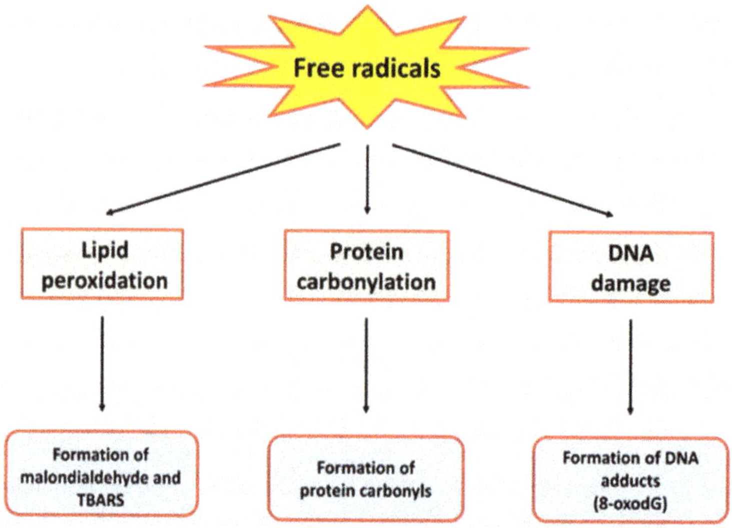

Figure 3 Deleterious effects of free radicals and oxidative stress on biomolecules. Source: Azat Aziz et al. (2019).

2019). The decomposition of lipid peroxides into stable products is an important secondary antioxidant mechanism. Lipid peroxidation products have been implicated in several diseases, particularly neurodegenerative diseases due to the high lipid content of the brain. Numerous dietary antioxidants, particularly carotenoids, have a role in inhibiting lipid peroxidation in disease. Several systematic reviews detail the benefits of dietary antioxidants against lipid peroxidation in neurodegenerative diseases (Bhatt and Patel, 2020; Petrovic et al., 2020; Ambrosone et al., 2020a). In food systems, lipid peroxidation is responsible for the production of a wide array of unwanted oxidation compounds in addition to accelerating food spoilage and quality deterioration. To prevent this, fruit and vegetable antioxidants are being incorporated into functional food products. A recent study reported gallic acid extracted from guava leaves to not only retard lipid oxidation in fresh pork sausages but also retain the colour over 14 days (Tran et al., 2020).

As certain metallic ions may have pro-oxidant activity, another mechanism of action of natural antioxidants is metal chelation. Antioxidants derived from fruit and vegetables such as flavonoids interact with metal ions to form complexes. These complexes protect against oxidative damage as they possess free radical scavenging properties. A study investigating the antioxidant capacity of bioactive compounds extracted from *Tetrapleura tetraptera* fruit observed the extracted phytochemicals, including flavonoids, phenols, and alkaloids, to possess significant metal chelating capacity compared to etylemediaminetetraacetic acid (EDTA) (Adusei et al., 2019). A recent study demonstrated wild and cultivated extracts of *Origanum vulgare* L. to exhibit significant metal scavenging capacity in both methanolic and ethanolic extracts, ranging from 50.2% to 140.9% in wild extracts while cultivated extracts possessed activities ranging from 11% to 90%. However, the activity of these extracts was substantially lower than the EDTA standard which possessed a metal chelating capacity of 345.5% (Jan et al., 2020).

4 Antioxidants in human health and disease

Oxidative stress is an imbalance between oxidants and the antioxidant defence system caused by excessive free radical generation and elevated reactive species, notably ROS. Normal physiological operations and signalling cascades are modulated by these ROS, reactive lipid species (RLS), and reactive nitrogen species (RNS) with disturbances to the redox system and subsequent oxidative stress resulting in oxidative damage to cellular components such as proteins, lipids, and DNA. Damage to these cellular components impacts critical cellular functioning and is the underlying pathophysiological cause of numerous diseases including cancer, inflammatory diseases, and neurodegenerative diseases. Antioxidants play a vital role in maintaining cellular functioning and consequently in the prevention and treatment of these diseases through their action in inhibiting oxidative damage.

In recent years, the antioxidants present in fruits and vegetables have emerged as major players for preventative strategies or therapeutics of several diseases. Scientific evidence has revealed these antioxidants to exert anti-inflammatory and anti-tumour effects, among others. This section will discuss the role of dietary antioxidants from fruits and vegetables in human diseases such as cancer, CVD, and obesity (Fig. 4).

4.1 Antioxidants and cancer

Oxidative stress induced by ROS production has been implicated in the aetiology of cancer, the second leading cause of death globally, primarily by action of cellular damage and DNA mutations. Research has identified genetic mutations to play a significant role in cancer pathogenesis, and the genetic changes induced by oxidative damage stimulate oncogene generation. Elevated ROS levels and oxidative damage further promotes cancer progression through their role in several signalling pathways, resulting in uncontrolled cell proliferation, apoptosis, and expression of pro-oxidant enzymes (Perillo et al., 2020).

Several sources of ROS and oxidative stress exist within cancer cells, with mitochondria presenting the most prominent source. Mitochondria generate ROS, which are critical for cell proliferation and homeostasis, as a result of the

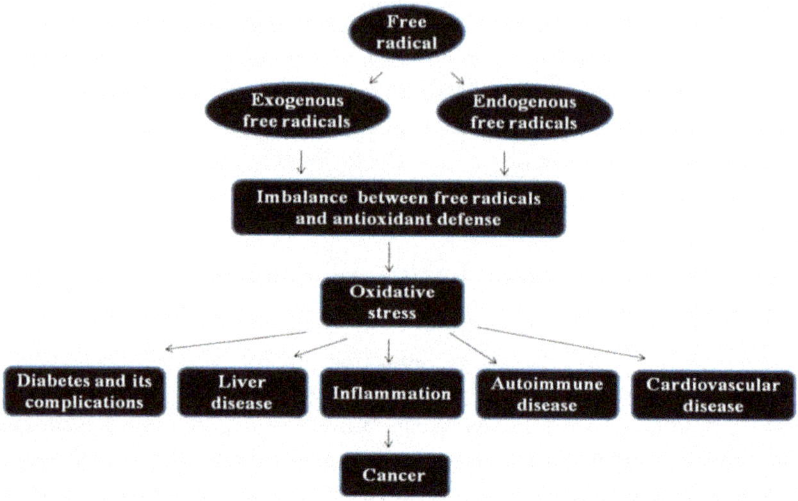

Figure 4 Impact of redox imbalance on health and disease outcomes. Source: Kumar and Pandey (2015).

electron transport chain (ETC) through oxidative phosphorylation. Caveolin 1 (cav-1) loss, which influences tumour recurrence, metastasis, and tumour growth, is regulated by mitochondria-generated ROS. Inactivation of the mitochondrial antioxidant system results in increased oxidative stress in cancer cells (Arfin et al., 2021).

Due to the involvement of ROS and oxidative stress in the varying cancer initiation, promotion, and progression stages, antioxidants have both a preventative and therapeutic role in cancer. Several classes of antioxidants such as vitamins, polyphenols, and alkaloids have been applied in clinical settings as chemopreventative agents. Research has shown various vitamin and non-vitamin antioxidants to have chemopreventative activity. The green tea catechin epigallocatechin gallate (EGCG) has several anticancer mechanisms including inhibition of cell proliferation, carcinogenic activity and tumourigenesis, and induction of cell apoptosis. A recent study observed a significant inhibition of cell proliferation, migration, and invasion of human lung cancer cells by nano emulsified EGCG (Chen et al., 2020).

However, recent studies have reported antioxidant compounds to contribute to cancer pathogenesis when present in high concentrations or to not have a chemopreventative effect at all. Although evidence predominantly indicates vitamin C to be a strong antioxidant compound, a Canadian case-control study reported no association between vitamin C intake and prostate cancer incidence or aggressiveness (Parent et al., 2018). Furthermore, antioxidant compounds interact with certain chemotherapies and reduce treatment efficacy. An observational study of a clinical trial reported antioxidant supplementation, prior to and during chemotherapy, to increase the risk of breast cancer recurrence by 41% (Ambrosone et al., 2020a,b). Evidence has demonstrated the relationship between antioxidants and ROS levels in cancer to be highly complex. Fruit and vegetable antioxidants have a beneficial effect on cancer but also contribute to the progression of cancer.

4.2 Antioxidants in neurodegenerative diseases

Neurodegenerative diseases, characterized by progressive deterioration of the central nervous or peripheral nervous systems, encompass several disorders including Alzheimer's disease (AD) and Parkinson's disease (PD). Heightened ROS production and elevated oxidative stress have been implicated in the pathogenesis of multiple neurodegenerative diseases, such as AD, making antioxidants invaluable for the prevention and treatment of these disorders.

AD has become the most prevalent neurodegenerative disorder, affecting a large proportion of the global population. This cognitive disease is clinically identified by memory loss and loss of executive functioning (Butterfield and Halliwell, 2019) and generally affects the elderly population. AD is a complex multifactorial disease; however, research has revealed oxidative stress to be a significant contributor to the pathogenesis of this disease, particularly early progression. A hallmark of AD is the deposition of amyloid-beta (Aβ) plaques in aggregated form in the brain whereas Aβ plaques are present in the soluble form in healthy brains. Although mechanisms have yet to be fully elucidated, elevated ROS levels have been implicated in membrane-associated oxidative stress resulting in lipid peroxidation and the production of a neurotoxic compound commonly as known as aldehyde 4-hydroxynonenal (HNE) (Chooi et al., 2019). This compound and other lipid peroxidation compounds such as malondialdehyde are detected in the early stages of AD and are connected to the progression of this disorder. Beyond the excessive generation of ROS, the endogenous antioxidant defence system is weakened in AD which contributes to the progression of this disease. Lycopene, a red carotenoid, is a naturally occurring antioxidant in several fruits and vegetables, such as tomatoes and watermelons, with potential application in the treatment of AD. A study conducted on M146L cells found treatment with lycopene to inhibit Aβ-induced oxidative stress through activation of the phosphatidyl inositol 3-kinase (PI3K)/protein kinase B (Akt)/nuclear factor erythroid 2-related factor 2 (Nrf2) signalling pathway (Fang et al., 2020) in AD. Ginkgolide and bilobalide terpenoid lactones extracted from Gingko biloba were shown to inhibit oxidative stress in cerebral ischaemia reperfusion through modulation of the Akt/Nrf2 signalling pathway and enhanced expression of antioxidant enzymes including superoxide dismutase (SOD) and hemeoxygenase-1 (HO-1) (Liu et al., 2019).

PD is another common neurodegenerative disease that may benefit from the neuroprotective effects of dietary antioxidants. This disease is a progressive neurological disease characterized by progressive degeneration of dopaminergic (DA) neurons. Impaired motor functions, tremors, bradykinesia, and rigidity are clinical features of this disorder in addition to conditions such as depression and anosmia (Duarte-Jurado et al., 2021). Extensive research has revealed ROS production and oxidative stress to contribute to DA neuron degradation and the pathogenesis of PD. The production of dopamine's metabolites, such as 3,4-dihydroxyphenlacetic acid (DOPAC) by monoamine oxidase (MAO), as a result of dopamine oxidation, deteriorates DA neurons. Neuroinflammation and the activation of microglial cells are a protective response by the central nervous system (CNS). Microglial cells release pro-inflammatory cytokines and produce ROS. However, excessive inflammatory responses have been linked to Parkinson's Disease (PD) (Kim et al., 2020b). Vitamin E, encompassing all the tocopherol and tocotrienol compounds present in fruit and vegetable sources, has been demonstrated to have antioxidant effects in neurodegenerative diseases (Park and Ellis, 2020). In their seminal study, Park and co-workers (2019) detailed the prevention of ΔN-Bcl-xL formation and subsequent oxidative stress by α-tocotrienol treatment of rat hippocampal neurons, a novel mechanism in which α-tocotrienol exerts its neuroprotective action. Vitamin C is highly recognized for its potent antioxidant properties. However, clinical studies supporting supplementation of vitamin C for reducing PD risk are limited.

4.3 Antioxidants and inflammatory diseases

Chronic inflammatory disorders, such as rheumatoid arthritis and inflammatory bowel disease, affect a significant number of the global population. Chronic inflammation is a prolonged state of inflammation, largely resulting from the immune system's inability to regulate acute inflammation (Mazarakis et al., 2020). Substantial evidence has revealed excessive ROS generation and oxidative stress to be critical factors in the pathogenesis and progression of several chronic inflammatory diseases. The interdependent relationship between oxidative stress and inflammation appears to be quite complex with oxidative damage being both a cause and effect of inflammation. Targeting oxidative stress is a promising method of preventing inflammatory damage.

Rheumatoid arthritis (RA), a complex systemic inflammatory disease, is characterized by swelling, pain, and tenderness of synovial joints. Studies have reported higher ROS production, oxidative stress, and lipid peroxidation in individuals with rheumatoid arthritis. Angiogenesis, a complex mechanism involving the development of new blood vessels from pre-existing vessels, plays a critical role in the pathogenesis of RA and other inflammatory diseases. In RA, angiogenesis in synovial tissues results in a perpetual state of inflammation and synovial hyperplasia. Like the relationships seen in other diseases, ROS production and oxidative stress have an intricate relationship with angiogenesis and present a double-edged sword. Low concentrations of ROS stimulate signalling pathways whereas exorbitant ROS levels result in pathological damage. Balogh and colleagues (2018) revealed oxidative stress to impact bioenergetic profiles by triggering anaerobic glycolysis from OXPHOS, thereby promoting angiogenesis and chronic inflammation. ROS also impacts RA and angiogenesis through signalling pathways including the vascular endothelial growth factor (VEGF) and MAPK pathways. The JNK and p38 MAPK pathways are critical for numerous regulatory processes such as cell development, growth, and cell death. These findings display the importance of targeting ROS in RA treatment due to their role in activating or deactivating signalling molecules (Phull et al., 2018).

Mateen and colleagues (2019) reported cinnamaldehyde and eugenol to display considerable antioxidant activity in peripheral blood mononuclear cells (PBMCs) obtained from RA patients. The researchers observed cinnamaldehyde and eugenol to abate ROS formation, lipid peroxidation, and nitric oxide, glutathione, and pro-inflammatory cytokine levels in the PBMCs. Additionally, treatment with these dietary antioxidants enhanced the activity of the antioxidant enzyme SOD, whereas they were ineffective in improving catalase and GPx activity. Recent research detailed the benefits of the natural antioxidant resveratrol in alleviating RA severity by downregulating MDA and SOD levels while also decreasing expression of pro-inflammatory cytokines such as interleukin-6 (IL-6) and tumour-necrosis factor α (TNF-α) in male rats with BIIC-induced rheumatoid arthritis (Yang et al., 2018a,b). Furthermore, these *in vitro* studies revealed that treatment with resveratrol suppressed IL-1β induced HIF-1α upregulation and activation of the JNK and p38 MAPK pathways in RSC-364 cell lines.

Similarly, oxidative stress has been implicated in the pathophysiology of inflammatory bowel disease (IBD). IBD is a chronic inflammatory disorder of the gastrointestinal tract and is a complex interplay of the immune system, the gut microbiome, genetics, and environmental factors. Overproduction of ROS and subsequent oxidative stress cause extensive damage to the mucosal lining, diminishing its functioning (Bourgonje et al., 2020). Polyphenols extracted from *Passiflora subpeltata* Ortega were shown to have protective effects in indomethacin-induced ulcerative colitis rats and RAW 264.7 cells. In addition to suppressing pro-inflammatory nitric oxide and TNF-α mediators, these extracts stimulated the activity of antioxidant enzymes including catalase and SOD (Shanmugam et al., 2020). Similarly, the dietary anthocyanin, pelargonidin

3-glucoside, was reported to increase the mucosal epithelium, villi length, crypt depth, and goblet cell production in DSS-induced IBD Wistar rats. These improvements were hypothesized to be a result of inhibition of free radicals and increased IL-10 concentration (Ghattamaneni et al., 2020). Research regarding the application of antioxidant compounds as therapeutics in IBD is required as current studies appear to be largely limited to animal models. Clinical trials are necessary to determine the efficacy of these compounds in human patients.

4.4 Antioxidants and cardiovascular diseases

CVDs, a cluster of diseases impacting the heart or blood vessels, are the leading cause of death worldwide. Elevated ROS and oxidative stress have been identified as key players in cardiac diseases including atherosclerosis, cardiac hypertrophy, coronary artery disease, and heart failure, among others. ROS imbalance and impaired antioxidant capacity result in cardiac dysfunction. In recent years, dietary components have emerged as promising interventions and multiple studies have established the role of fruit and vegetable antioxidants in reducing CVD risk through various mechanisms.

Similar to neurogenerative diseases, there are several origins of ROS within the heart which contribute to the progression of CVD. Recent research has revealed the importance of mitochondria in oxidative stress due to their role in molecular signalling and cell survival. Mitochondria synthesize ATP through oxidative phosphorylation and electron transfer in the electron transfer chain (ETC), which produces intracellular ROS as a by-product. Complexes I and III of the ETC are the primary sources of ROS. Additionally, other mitochondrial proteins such as $p66^{shc}$ have been implicated in ROS formation. This cytosolic protein has been found to trigger H_2O^2 production by oxidizing cytochrome c (D'Oria et al., 2020). An association between vascular endothelial dysfunction and several risk factors for CVD, such as hypertension and stroke, has been identified through research. Increased ROS generation and oxidative stress can increase CVD risk through inducing endothelial dysfunction. Oxidative stress results in endothelial nitric oxide synthase (eNOS) uncoupling, a critical hallmark in the pathogenesis of most CVDs. eNOS is an isoform of nitric oxide synthase, largely found in coronary arteries, cardiomyocytes, and endothelial cells, responsible for the release of nitric oxide from the microvascular endothelium. Uncoupled eNOS alters the nitrosoredox balance and induces oxidative stress through overproduction of ROS with less NO generation. This negatively impacts the cardiovascular system which results in cardiac dysfunction and degradation of the extracellular matrix (D'Oria et al., 2020). This has made targeting eNOS uncoupling a viable therapeutic target for ameliorating CVD risk. Lee and colleagues (2020) demonstrated anthocyanin-rich mulberry extracts to downregulate oxidative stress in the aortas of Sprague-Dawley rats and to alleviate endothelial senescence by increasing eNOS phosphorylation.

Other dietary antioxidants in fruits and vegetables, particularly flavonoids, have been implicated in inhibiting endothelial dysfunction. Quercetin, a dietary flavonoid present in fruits and vegetables such as berries and onions, was found to ameliorate oxidative stress and endothelial cell injury in HUVECs by inducing HMOX1 and the ERK/Nrf2 signalling pathway (Tian et al., 2019). Similarly, a study reported the polyphenol EGCG to protect HUVEC cells against H_2O_2-induced oxidative damage and apoptosis, making it a promising antioxidant in the treatment of oxidative stress-induced CVDs (Meng et al., 2020).

Recent research has highlighted the importance of microRNAs (miRNAs) in maintaining endothelial cell functioning and modulation of CVD risk. Novel research has demonstrated natural antioxidants in fruit and vegetables to impact CVD risk through their interaction with cardiovascular-related miRNAs. Daimiel and colleagues (2020) established an association between polyphenols and circulating miRNA, particularly miR-17-92, following consumption of polyphenol-rich extra virgin olive oil.

4.5 Antioxidants in obesity

Obesity, characterized by excessive accumulation of adipose tissue, is a complex multifactorial disease that has become a major public health issue today. Over 1.9 billion adults are classified as being overweight or obese, with individuals with a body mass index (BMI) of >25 kg/m^2 being classed as overweight whereas those with a BMI >30 kg/m^2 are considered obese (Chooi et al., 2019). This condition predisposes individuals to various non-communicable diseases such as type 2 diabetes, CVDs, and dyslipidaemia making it a major burden on both the health and economic systems. Research has revealed oxidative stress to have a significant role in the pathogenesis of obesity, and antioxidant pathways have become a promising target for tackling obesity prevalence (Kim et al., 2020a). Mitochondrial dysfunction, caused by oxidative stress, results in decreased antioxidant capacity and elevated ROS production in white adipose tissue consequently impacting the functioning of adipose cells. Evidence suggests that oxidative stress also contributes to obesity development through its role in the stimulation of adipogenesis including preadipocyte proliferation and differentiation. In pre-adipocytic adipogenesis, redox imbalance of glutathione has been shown to stimulate adipogenesis and lipid accumulation by increasing expression of the CCAAT/enhancer-binding protein-a (C/EBPA) LAP/LIP and peroxisome proliferator-activated receptor γ (PPARγ) (Tobore, 2020; Čolak and Pap, 2021).

Studies have reported on the relationship between oxidative stress and inflammation in obesity. Obesity progression is accompanied by both increased inflammation and ROS production. Elevated adipose tissue levels stimulate pro-inflammatory cytokine production such as TNF-α and IL-6. TNF-α increases lipogenesis, insulin sensitivity, and ROS generation (Masschelin et al., 2019).

Antioxidants derived from fruits and vegetables have gained considerable attention for their therapeutic applications in preventing obesity and mitigating associated obesity comorbidities such as type 2 diabetes mellitus (T2DM). A recent review analysing studies reported the various mechanisms in which gallic acid exerted its anti-obesity benefits. Although mechanisms are yet to be fully elucidated, gallic acid has been demonstrated to attenuate insulin-induced lipogenesis, improve insulin signalling, and mitigate oxidative stress and pro-inflammatory responses (Dludla et al., 2018). A study reported the anti-obesity effects of anthocyanin and non-anthocyanin polyphenols, obtained from lingonberry fruit, to be exerted through the inhibition of pro-oxidant enzymes, including NOX4 and iNOS, alongside elevated expression of the antioxidant defence enzyme, SOD2, in 3T3-L1 mouse cells (Kowalska et al., 2021).

The polyphenolic dietary antioxidant, piceatannol, was shown to have various antiadipogenic effects in human mesenchymal stem cells (hMSC). This compound effectively inhibited lipid accumulation and impaired glucose transport and lipogenesis in adipocytes (Carpéné et al., 2018). Curcumin, naturally present in turmeric roots, has gained considerable attention for its antioxidant capacity. A randomized placebo-controlled clinical trial established that curcumin supplementation improved both inflammation and oxidative stress markers such as malondialdehyde, therefore reducing obesity risk in postpubescent obese girls (Saraf-Bank et al., 2019). Similarly, the incorporation of açaí pulp, rich in anthocyanins and other polyphenols, to a hypo-energetic diet significantly reduced 8-isoprostanc lipid peroxidation markers in overweight adults over 90 days (Aranha et al., 2020).

4.6 Antioxidants and diabetes mellitus

Diabetes mellitus is a chronic metabolic disorder characterized by elevated blood glucose levels (hyperglycaemia) and disturbances to the metabolism of macronutrients due to dysregulated insulin action. Pancreatic cells either secrete insufficient insulin or cells respond inadequately to secreted insulin, resulting in systemic damage to various organs. Diabetes mellitus is associated with severe comorbidities such as diabetic nephropathy, neuropathy, retinopathy, obesity, and CVDs (Tran et al., 2020). Researchers have hypothesized several mechanisms in which oxidative stress impacts diabetes, most notably through hyperglycaemic dysfunction and insulin resistance. Hyperglycaemic dysfunction causes oxidative stress in pancreatic β cells, which possess poor antioxidant capacity, making them highly sensitive to oxidation. Cellular components such as lipids, proteins, and nuclei acids become damaged by elevated ROS levels resulting in cell deterioration, apoptosis, and necrosis which impair insulin production and functioning (Sun et al., 2021).

- β cell functioning has been shown to be highly susceptible to oxidative stress. Evidence has shown several critical pathways to be altered by oxidative stress including the AMP-activated protein kinase (AMPK) and c-Jun N-terminal kinase (JNK) pathways. Overexpression of pAMPK, stimulated by ROS, has been shown to increase β cell apoptosis, reduce insulin secretion, and impair β cell proliferation (Eguchi et al., 2021). Similarly, ROS-induced JNK activation impairs β cell functioning through various mechanisms. JNK activation has been shown to induce β cell apoptosis in human and animal models with several studies reporting inactivation of this pathway to increase β cell functioning and to inhibit apoptosis (Liu et al., 2020; Yu et al., 2019). Furthermore, oxidative stress may affect β cell functioning through its role in reducing the transcriptional activity of regulatory insulin genes. Recent evidence shows oxidative stress to enhance nuclear translocation of forkhead box protein O1 (FOXO1) which competes with transcriptional pancreas duodenal homeobox factor 1 (PDX-1) for DNA binding. Reduced PDX-1 expression inhibits β cell proliferation and growth and impairs insulin secretion (Zhang et al., 2020a,b).

Excessive free fatty acid (FFA) plasma levels in the blood have emerged as a cause of insulin resistance in diabetes. This heightened FFA plasma concentration results in inordinate ROS production, oxidative stress, and mitochondrial dysfunction (Park and Park, 2021). Excessive ROS production and FFA plasma levels stimulate pro-inflammatory signalling proteins and hinder insulin signalling. ROS-mediated activation of the serine-threonine kinase pathways, p38 MAPK, KNKI AND IKKβ, results in insulin resistance caused by suppressed insulin signalling pathways due to degradation of insulin receptor substrate (IRS) (Zhang et al., 2020a,b).

Although there is no current cure for diabetes mellitus, dietary antioxidants have emerged as a valuable preventative and therapeutic tool for this disease. Research has established dietary antioxidants to exert their effect through the protection of beta cell functioning. Phytochemical extracts of the plant *Ocimum canum* exhibited potent antioxidant activity in Wistar rats. The antioxidant-rich extracts demonstrated considerable antioxidant capacity both *in vivo* and *in vitro* in several assays including the DPPH assay, phosphomolybdate assay, and superoxide ion assay (Ononamadu et al., 2019). Similarly, EGCG inhibited cell apoptosis and oxidative stress in αTC1-6 β cells treated with H_2O_2. Furthermore, this study showed EGCG prevented β cell dysfunction by inactivation of the P38 and JNK MAPK pathways (Cao et al., 2018). A cohort study of 5796 individuals from the prospective Rotterdam Study reported an

Table 2 Nutraceutical properties of fruit and vegetable antioxidants

Antioxidant	Source	Role in health and disease	Reference
Polyphenols	Berries, plums, cherries, apples, apricots, peaches Onions, carrots, beans, tomatoes, spinach, broccoli, asparagus, potatoes, artichokes Turmeric, celery, cocoa powder, rosemary, oregano, thyme, cumin, basil, cinnamon Soybeans, black beans, hazelnuts, almonds, walnuts, flaxseed	• Protect against diverse diseases induced by oxidative stress including cancer, diabetes, neurodegenerative and cardiovascular diseases • Immunomodulatory and anti-inflammatory action • Improve cardiometabolic risk factors such as ventricular health and vascular endothelial health • Possess anti-obesogenic properties such as preventing lipogenesis and inflammation • Can modulate composition of the human gut microbiome and can stimulate growth of beneficial bacteria • Possess antimicrobial properties	Cory et al. (2018), Fraga et al. (2019), Mithul Aravind et al. (2021), D'Angelo (2020), Yan et al. (2020)
Carotenoids	Mangoes, watermelon, tomatoes, oranges, cantaloupe, kiwi, grapes, honeydew melon Spinach, kale, bell peppers, carrots, pumpkin, sweet potatoes, squash, corn, Brussels sprouts Egg yolks	• Potent free radical scavengers and inhibit lipid peroxidation • Pro-vitamin A precursors • Protective effect in age-related eye diseases • Cancer prevention • Prevents against neurodegeneration • Reduce lipid accumulation, lipogenesis, and insulin resistance • Hepatoprotective effects against development of liver diseases • Photoprotective effect against several skin disorders such as erythema and photo carcinogenesis	Bungau et al. (2019), Elvira-Torales et al. (2019), Mohd Hassan et al. (2019)
Vitamin A	Mango, apricots, papaya Spinach, carrots, red peppers, sweet potato, pumpkin Parsley, cayenne, coriander, basil, thyme, paprika, chilli powder	• Therapeutic effect in infectious diseases • Essential for normal immune function • Essential for cell proliferation, differentiation, and growth • Critical role in maintaining healthy vision • Modulation of signalling pathways	Huang et al. (2018), Meléndez-Martínez (2019), Fairulnizal et al. (2019)

Vitamin C	Citrus fruits, kiwi, strawberries, pineapple, cantaloupe, mango, watermelon, berries Spinach, cauliflower, broccoli, cabbage, bell peppers, Brussels sprouts, winter squash, turnips	• Free radical scavenging, protects intracellular components from oxidative damage • Prevents cellular lipid peroxidation • Exhibits powerful anti-tumour activity • Potent anticarcinogenic activity • Epigenetic regulation of DNA and histone methylation-demethylation and potential role in developmental programming • Immunomodulatory role • Improves gut-liver functioning and consequent vitamin E adequacy • Exhibits antiviral activity	Lee Chong et al. (2019), Colunga Biancatelli et al. (2020), Ngo et al. (2019), Traber et al. (2019)
Vitamin E	Spinach, avocados, almonds, broccoli. vegetable oil, sunflower seeds	• Protects cells from free radical damage • Prevents lipid peroxidation • Reduces expression of pro-inflammatory cytokines • Reduces risk of neuroinflammation and onset of neurodegenerative diseases	Chin and Ima-Nirwana (2018), Asbaghi et al. (2020), Browne et al. (2019)
Selenium (exogenous)	Tuna, sardines, oysters, crab, and salmon Brazil nuts, cashews Beef, pork, turkey, chicken Lentils, sunflower seeds, spinach	• Decreases CVD risk factors • Mitigates oxidative damage to DNA • Maintain endogenous antioxidant activity • Regulates thyroid metabolism • Selenium deficiency contributes to increased inflammation	Jenkins et al. (2020); Hu et al. (2021); Pisoschi et al. (2021)
Zinc	Oysters, crab, lobster Spinach, beans, mushrooms, asparagus, broccoli, kale Almonds, cashews, Poultry, beef, pork Yogurt, milk, cheese	• Maintains endogenous antioxidant activity • Reduces lipid peroxidation • Inhibits pathological angiogenesis • Anti-inflammatory benefits • Deficiency impairs B cell and T cell development, impairs immune function	Jenkins et al. (2020), Mousavi et al. (2018), Wessels et al. (2021)

Source: Adapted from Pisoschi et al. (2021).

inverse relationship between dietary antioxidant capacity and diabetes risk, but not pre-diabetes risk, over 15 years. In both normoglycaemic and pre-diabetic individuals, those with a high dietary antioxidant capacity were at a lower risk of developing type 2 diabetes (van der Schaft et al., 2019). A systematic review of seven randomized controlled trials examining the effect of blueberry and cranberry consumption demonstrated blueberry supplementation (9.1-9.8 mg of anthocyanins) for 8–12 weeks and daily cranberry juice intake (240 mL) for 12 weeks to be beneficial for managing glucose parameters in type 2 diabetic individuals (Rocha et al., 2018) (Table 2).

5 Applications of fruit and vegetable antioxidants

Bioactive compounds from fruits and vegetables modulate physiological and metabolic processes in the body while exerting health benefits. Recent studies showed that antioxidants from natural sources such as fruits and vegetables are safer and more desirable than synthetic antioxidants and have numerous applications in the food industry as food additives. This section will discuss the applications of dietary antioxidants from fruits and vegetables as functional foods and their applications in the industry.

5.1 Application as natural pigments

In the food industry, antioxidants can be added to prevent colour changes due to oxidation. Carotenoids are pigments that can be extracted from fruit and vegetables. Carotenoids are added to the foods to enhance the existing colour, to add colour, to replace colour loss, and to reduce colour variation. Carotenoids are added to foods such as pasta, margarine, cheese, sausages, and fruit juice (Mezzomo and Ferreira, 2016). In fresh juice, the addition of antioxidants can improve the stability of the carotenoid pigments.

Carotenoids are also used to restore or standardize the colour of foods and indirectly intensify the colour of foods through their application in animal feed. The addition of beta-carotene in animal feed aids in achieving the characteristic yellow-orange colour of egg yolks (Moreno et al., 2020) and chicken skin (Díaz-Gómez et al., 2017). This also intensifies the colour of fish (Jiang et al., 2019) and milk, with the carotenoid concentration in milk depending on carotenoid forage intake. The application of carotenoids in animal feed is a commonly used alternative to colour additives in the food industry.

Mandarin epicarp is a source of carotenoids that can be used as a natural colourant in baked products such as cake and bread. The use of natural colouring additives reduces the use of synthetic dyes while increasing the carotenoid pigments of the final products (Ordóñez-Santos et al., 2021).

Anthocyanins are polyphenols that can be extracted from fruits and vegetables. They are a promising alternative to synthetic food dyes due to its high stability (Giusti and Wrolstad, 2003). Anthocyanin pigments appear red in acidic conditions and purple in alkaline conditions. This makes anthocyanins suitable as a food colourant in beverages and as a purple food additive in jam, confectionaries, and beverages (Khoo et al., 2017). It can be incorporated into beverages as a natural pigment. Selected potato varieties with high anthocyanin content are incorporated into soft drinks as natural red and purple colourants. This acts as a substitute for existing synthetic colouring agents with suitable sensory profiles and high stability (Sampaio et al., 2021).

The anthocyanin from purple carrots can be used as a red food colourant in hard candy and jelly. The natural pigment has no significant differences in colour, taste, and odour when compared to carmine as synthetic red food dye (Assous et al., 2014). Passion fruit pericarp is a source of anthocyanins which can be used as a natural colouring agent as it is stable against light, heat, and storage. The pigment extract can be used as a pink/red colourant in processed foods such as jelly (Kawasoe et al., 2021). A study carried out by Tereucan et al. (2021) found that potatoes anthocyanins can be added to dairy products as a natural pigment (Tereucan et al., 2021). They are a stable natural alternative to synthetic dyes for use in the food industry. An anthocyanin called betacyanin also has applications as a natural food pigment. Betacyanins from red pitahaya are incorporated into mildly acidic foods such as yogurt with a high stability compared to commercially used food colourants (Gengatharan et al., 2017). Anthocyanin-rich extracts obtained from red radish are incorporated into soybean-based yogurt products to produce a pink colour (Dias et al., 2020).

5.2 Application as natural preservatives

Lipid oxidation from oxygen or sunlight exposure causes deterioration and leads to rancidity in many foods. This leads to changes in food characteristics such as odour and taste. The addition of antioxidants to foods acts as preservatives to retard lipid oxidation which can help maintain the quality and appearance while prolonging the shelf life of foods containing oxidized lipids such as vegetable oils and processed meats.

Antioxidants can also be used in ready-to-eat food products for the preservation and quality maintenance. Grapefruit seed extract can be used on vegetables as a natural preservative. Grapefruit seed extract is rich in flavonoids which are effective against pathogens and can inhibit the growth of bacteria such as *Salmonella* spp. and *Listeria monocytogenes*. This can extend the shelf life of ready-to-eat lettuce and cucumber (Xu et al., 2007).

Natural antioxidants from fruit and vegetables can be used in meat and meat products. This is because meat fat is highly susceptible to oxidation leading to rancidity. Myoglobin is responsible for pigmentation. The addition of antioxidants such as phenols and carotenoids can minimize metmyoglobin formation and can inhibit oxidation. Myoglobin is responsible for pigmentation. This prevents undesirable changes in quality and flavour (Karre et al., 2013).

Pomegranate extract is a natural preservative used in ready-to-eat meats. This is an alternative approach to control the growth of microorganisms while working as both an antimicrobial and an antioxidant (Hayrapetyan et al., 2012). Grape seed extract can be used on ready-to-eat meats such as chicken and beef as a natural preservative. This acts as a natural antimicrobial by inhibiting the growth and recontamination of *L. monocytogenes* activity on turkey frankfurters (Sivarooban et al., 2007).

Antioxidants can indirectly extend the shelf life of meat products through their applications in animal feed. The addition of dietary citrus pulp and grape pomace in animal feed acts as a natural preservative by increasing antioxidant activity and decreasing lipid oxidation (Tayengwa et al., 2020). The addition of pomegranate peel powder into the feed for broiler birds led to an increased antioxidant capacity and improved the quality of the chicken breast (Akuru et al., 2020). The application of antioxidants in animal feed is a natural alternative to retard oxidation in the food industry.

Phenolic acid is a polyphenol that can be extracted from fruit and vegetables. In the food industry, phenolic acid controls rancidity, maintains nutritional quality, extends the shelf life, and slows the formation of toxic oxidation. Phenolic acid is used as an antimicrobial agent in food as it provides a protective effect against deterioration which preserves the food and extends the shelf life (Martillanes et al., 2017). Phenolic acid acts as a bio-preservative by inhibiting oxidation and the growth of microorganisms. The phenolic compounds in grape pomace enabled the inhibition of lipid oxidation in frozen fish muscle (Pazos et al., 2005).

Flavonoids are a group of phytonutrients in fruit and vegetables which are useful as a novel ingredient in many foods to fight against food spoilage and extend the shelf life of perishable foods. Multigrain bread enriched with onionskin powder has increased total phenols, flavonoids, and antioxidant activity. The onionskin powder acts as a natural preservative and improves the shelf life of the baked product by 11 days in ambient conditions and 13 days in refrigerated conditions (Sagar and Pareek, 2021).

Passion fruit can be used as a preservative in the production of Coalho cheese. Ground passion fruit inhibits microbial growth and can reduce *Listeria* spp, *Staphylococcus aureus*, and lactic acid bacteria in the cheese. Therefore, passion fruit is a natural bio-preservative that can be used on dairy products to extend the shelf life by controlling pathogenic bacteria growth (Costa et al., 2020).

5.3 Application as edible films and coatings

Antioxidants can be used as protective active barriers when applied to the surface to food products as edible coatings. Antioxidants such as phenolic acid are used in packaging and edible films. This is important to reduce the negative impact of oxygen on food. They are effective against enzymatic browning and oxidative rancidity which can prolong the shelf life of food products due to their antioxidant and antimicrobial potential. The application of bioactive packaging will prolong the shelf life of food due to its antioxidant and antimicrobial potential (Arcan and Yemenicioglu, 2011).

Red pitaya peel extract which is rich in betalains has been used to develop an edible film. The pitaya peel extract was added to a starch/polyvinyl alcohol matrix. The film enhances the water vapour barrier and UV-visible light barrier while icing the antimicrobial and antioxidant activity. The film is used to extend the shelf life of shrimp but can also be used on animal products (Qin et al., 2020).

Date fruit syrup has been used to develop an edible gelatin film. The date extract and gelatin film blend are used on oil to extend its shelf life. The film increases phenolic compounds and antioxidant activity which enhances its preservation properties (Rangaraj et al., 2021). Corn-starch gelatin film enriched with mango peel and pulp and pineapple pomace has also been used to make an edible film. The enrichment of the film improves properties such as moisture content and thickness. Antioxidants, antimicrobial, and phenolic content also increased. This biodegradable edible film can be used on different food products in order to extend the shelf life (Susmitha et al., 2021).

Fish gelatin films enriched with rowanberry, blue-berried honeysuckle, and chokeberry pomace have been used to make an edible film. The enrichment of the gelatin film with all extracts increased antimicrobial activity against *Escherichia coli*, *Pseudomonas fluorescens*, *S. aureus*, and *Listeria innocua* and the enrichment with blue-berried honeysuckle had the highest increase in antioxidant activity due to its high anthocyanin content. This edible film can be used on different food products to extend their shelf life (Staroszczyk et al., 2020). *Gelidium corneum* gelatin blend films enriched with grapefruit seed extract or green tea extract can be used as an edible film. This decreases the population of bacteria on pork and improves the quality during storage (Hong et al., 2009). Gelatin/polyethylene bilayer films incorporated with fruit peels such as pomegranate, papaya, and jackfruit peel powders can be used as an edible film. The enrichment of the film improved properties such as moisture content and film thickness. Pomegranate peel powder had the highest antioxidant and antimicrobial activity. This edible film can be used on different food products to extend the shelf life (Nur Hanani et al., 2018).

Chitosan coating containing pomegranate peel can be used as an edible coating. The natural plant extract can be used to maintain the quality of apricot fruit during refrigerated storage. This reduces fruit decay while increasing antioxidant and carotenoid activity which extends the shelf life of the product (Gull et al., 2021). A chitosan-pullulan coating enriched with pomegranate peel has also been used on green bell pepper. This retained phenolic and antioxidant properties while maintaining quality for 18 days. The natural edible coating extended the shelf life and maintained quality (Kumar et al., 2021). A grape seed extract has been incorporated into chitosan-gelatin coating to extend the shelf life of pork. The edible coating reduced oxidation and microbial spoilage during cold storage for 20 days. Antioxidant activity was also enhanced which further preserves the meat product (Xiong et al., 2020).

5.4 Application as natural emulsifiers and stabilizers

Vitamin C is used in the food industry to inhibit the deterioration of highly unsaturated fatty acids due to oxidation. Vitamin C is also added to products to improve flavour, colour, and stability of products. It can also extend the shelf life of products such as vegetable oils (Xiao and Li, 2020). Natural antioxidants from fruits and vegetables can be used in edible oils to enhance stability. This is by improving its hydrolytic stability leading to the inhibition of thermal deterioration (Mohdaly et al., 2010).

Quince seed gum can be used as a stabilizer in an oil-in-water emulsion. The stability of the emulsion increases as quince seed gum concentration increases. This shows that quince seed gum is a promising natural emulsifier and stabilizer for the food industry (Yao et al., 2021). Quince seed mucilage powder can also be used as a stabilizer when added to yogurt. The quince seed mucilage powder improved the quality and consistency of the yogurt product as well as odour and flavour (Gurbuz et al., 2021).

Apple pomace can be used as a natural stabilizer when added to stirred yogurt. The apple pomace stabilizes the yogurt while altering the structure. The apple pomace also increases the yogurt product's fibre and antioxidant levels (Wang et al., 2020). Apples rich in polysaccharides can be used as an emulsifier. The polysaccharides from 'Jinshihi' apples compared to 'Qinyang' or 'Pinklady' have the highest emulsifying capacity. This showed that 'Jinshihi' polysaccharides have the most potential as a natural emulsifying agent for use in the food industry (Hou et al., 2019).

Eggplant flesh pulp can be used as a natural emulsifier when added to meatballs. The eggplant flesh pulp can stabilize oil-in-water emulsions due to the presence of polysaccharides and the formation of a surface film adhering to oil droplets. The concentration of the emulsifier is low compared to other polysaccharide emulsions. Eggplant flesh pulp is a natural plant-based emulsifier that can be used in the food industry in emulsion-based foods and beverages (Zhu et al., 2020). Black tomato pomace rich in polysaccharides can be used as an emulsifier. The pectin can stabilize emulsions containing a 50% oil phase. This is due to its ability to lower surface tension and increase viscosity. This showed that black tomato pomace polysaccharides have the most potential as a natural emulsifying agent for use in the food industry (Zhang et al., 2020a,b).

Pectin extracted from persimmon peel can be used as an emulsifier. Persimmon peel pectin had improved emulsifying capacity compared to commercial citrus pectin. Its increased stability is due to its acetyl group's hydrophobic interactions. Its antioxidant activity can also stabilize oil phase lipid oxidation which can prevent lipid peroxidation. Therefore, persimmon peel pectin is a natural polysaccharide emulsifier that can be used as an alternative to commercial emulsifiers in the food industry (Jiang et al., 2020). Pectin extracted from watermelon rind, citrus, or apples can be used as an emulsifier. Watermelon rind pectin had improved emulsifying capacity compared to citrus or apple pectin. Its structure and composition such as its high protein content led to increased stability of oil droplets (Mendez et al., 2021). Therefore, watermelon rind pectin is a natural emulsifier with potential applications in the food industry. Pomegranate peel pectin with a high amount of ester groups can be used as an emulsifier. The pectin can stabilize emulsions containing a 50% oil phase. This is due to its hydrophobic interactions, viscosity-enhancing capacity, and the presence of protein fractions. Therefore, pomegranate peel pectin is a natural emulsifier that can be used in the food industry (Yang et al., 2018a,b).

5.5 Application as nutraceuticals

Nutraceuticals are substances that provide physiological benefits and are used to improve health. An imbalance between free radicals and antioxidants in the body leads to oxidative stress and inflammation. These damaging free radicals contribute to the development of many chronic diseases such as CVD, cancer, and inflammatory diseases (Pizzino et al., 2017).

Antioxidants found in many fruits and vegetables include flavonoids, phenolic acids, and tannins (refer to section 2) which have many biological effects (refer to section 4). Antioxidants protect against potentially damaging free radicals by donating hydrogen to the lipid or peroxide free radical. This inhibits cellular damage through preventing the formation or by its scavenging activity (Lobo et al., 2010). Antioxidants play an important role in the prevention of cardiovascular and neurological diseases caused by lipid peroxidation and free radicals. Antioxidants can also reduce plasma levels of inflammatory markers such as IL-6 and TNF-α (Kurutas, 2016). This occurs through a decrease in lipid peroxidation and upregulation of antioxidant enzymes such as total antioxidant capacity while also decreasing oxidative stress and serum malondialdehyde levels.

Flavonoids reduce oxidative stress and low-density lipoprotein oxidation due to their antioxidant properties. This reduces tissue damage that leads to diseases such as cancer, CVD, hypertension, and metabolic syndromes. This includes anthocyanins, genistein, quercetin, and catechins. Similarly, anthocyanins protect against damaging free radicals by preventing the formation of prostaglandins linked to inflammation. Prostaglandins act like messenger molecules in inflammation through the inhibition of the enzyme cyclooxygenase (Levers et al., 2016). They can also reduce plasma levels of inflammatory markers which occur through decreasing lipid peroxidation and upregulating antioxidant enzymes while decreasing oxidative stress (Khoo et al., 2017). This can help protect against many diseases such as CVD and diabetes (Rechner and Kroner, 2005). Anthocyanins also have anti-carcinogen effects as they induce apoptosis and inhibit tumour growth and metastasis through the disruption of mitochondrial pathways and increased activation of caspase 3 (Mantena et al., 2006). Their low toxicity makes them attractive for use as a pharmacotherapeutic.

Phenolic acid has been linked to a reduction in cardiovascular and neurodegenerative disease. Its phytochemical neuroprotective effects have been found to prevent the neurotoxicity of cells by protecting neuron cells from oxidative stress-induced neurotoxicity. This helps reduce neurodegenerative disorders such as Alzheimer's disease (Kumar et al., 2012).

Carotenoids are powerful antioxidants that have many health benefits including anticancer, anti-inflammatory, and antibacterial. They are known for their free radical scavenging activity and for protecting the body from the action of ROS by inhibiting oxidative stress. This lowers the risk of CVD, cancer, and diabetes.

Carotenoid's antioxidant effects also lower the risk of osteoporosis. This is by counteracting oxidative stress which stimulates osteoblastic bone formation and inhibits osteoclastic bone resorption (Yamaguchi, 2012). Beta-carotene is a type of carotenoid that protects the body from damaging free radicals which lead to disease. This is through the inhibition of oxidation. Beta-carotene aids the immune system and lowers the risk of chronic illnesses such as heart disease and cancer. Oxidative stress is also linked with cognitive decline. The antioxidants in beta-carotene can help prevent cognitive deterioration.

Vitamins are organic compounds required by the body for growth, obtained through the diet. Vitamin C is a powerful antioxidant that protects tissues, cell membranes, and DNA from oxidative stress. This is through its free radical scavenging activity. It can reduce UVB-induced oxidative damage and UVA-induced lipid peroxidation through decreasing oxidative stress and serum malondialdehyde levels. Therefore, it can be used to treat hypertension and diabetes. Similarly, vitamin E is an antioxidant that can reduce the formation of radical oxygen species, reduce free radicals, and reduce apoptotic cells which protect the skin from UVB damage.

5.6 Other applications

Highly consumed foods can also be transformed into healthy foods by the addition of natural antioxidants. This includes foods such as bread, juices, and dairy products. Antioxidants are added to bakery products to control the autoxidation of fats present in baked goods to extend the shelf life (Reddy et al., 2005).

6 Conclusion

Fruits and vegetables are a rich source of numerous bioactive compounds such as polyphenols, carotenoids, vitamins such as vitamin A, C, and E, selenium, and zinc. The consumption of fruits and vegetables can reduce the risk of oxidative damage and contribute towards health promotion. Substantial research has shown the critical role oxidative stress and redox imbalance play in the pathogenesis of several non-communicable diseases. Dietary antioxidants from fruits and vegetables have emerged as promising preventatives and treatments. Although the evidence regarding the benefits of dietary antioxidants in disease in animal models and cell lines is convincing, clinical evidence regarding their nutraceutical properties is lacking. Despite fruit and vegetable antioxidants being natural products, these compounds are still pharmacological treatments. The discrepancies seen between studies of the clinical benefits of dietary antioxidants may be due to the administration, form, and dosage of these substances. Further research should be conducted to establish an effective dose.

7 Where to look for further information

7.1 Further reading

- Jaiswal, A. K. (Ed.). (2020). *Nutritional Composition and Antioxidant Properties of Fruits and Vegetables*. Academic Press.
- Nayik, G. A. and Gull, A. (Eds.). (2020). *Antioxidants in Fruits: Properties and Health Benefits*. Springer.
- Nayik, G. A. and Gull, A. (Eds.). (2020). *Antioxidants in Vegetables and Nuts-Properties and Health Benefits*. Springer.
- Laura, A., Alvarez-Parrilla, E. and González-Aguilar, G. A. (Eds.). (2009). *Fruit and Vegetable Phytochemicals: Chemistry, Nutritional Value and Stability*. John Wiley & Sons.

7.2 Key journals

- *Food Chemistry.*
- *Food Research International.*
- *Antioxidants.*
- *Trends in Food Science and Technology.*
- *Journal of Food Science and Technology.*
- *Critical Reviews in Food Science and Nutrition.*

8 References

Abdel-Aal, El-S. M., Akhtar, H., Zaheer, K. and Ali, R. 2013. Dietary sources of lutein and zeaxanthin carotenoids and their role in eye health. *Nutrients* 5(4), 1169–1185. DOI: 10.3390/nu5041169.

Adusei, S., Otchere, J. K., Oteng, P., Mensah, R. Q. and Tei-Mensah, E. 2019. Phytochemical analysis, antioxidant and metal chelating capacity of *Tetrapleura tetraptera*. *Heliyon* 5(11), e02762. DOI: 10.1016/j.heliyon.2019.e02762.

Akuru, E. A., Oyeagu, C. E., Mpendulo, T. C., Rautenbach, F. and Oguntibeju, O. O. 2020. Effect of pomegranate (*Punica granatum L*) peel powder meal dietary supplementation on antioxidant status and quality of breast meat in broilers. *Heliyon*. Cell Press 6(12), e05709. DOI: 10.1016/j.heliyon.2020.e05709.

Ambrosone, C. B., Zirpoli, G. R., Hutson, A. D., McCann, W. E., McCann, S. E., Barlow, W. E., Angelova, P., Esteras, N. and Abramov, A. 2020a. Mitochondria and lipid peroxidation in the mechanism of neurodegeneration: finding ways for prevention. *Medicinal Research Reviews* 41(2), 770–784. DOI: 10.1002/med.21712.

Ambrosone, C. B., Zirpoli, G. R., Hutson, A. D., McCann, W. E., McCann, S. E., Barlow, W. E., Kelly, K. M., Cannioto, R., Sucheston-Campbell, L. E., Hershman, D. L., Unger, J. M., Moore, H. C. F., Stewart, J. A., Isaacs, C., Hobday, T. J., Salim, M., Hortobagyi, G. N., Gralow, J. R., Budd, G. T. and Albain, K. S. 2020b. Dietary supplement use during chemotherapy and survival outcomes of patients With breast cancer enrolled in a cooperative group clinical trial (SWOG S0221). *Journal of Clinical Oncology* 38(8), 804–814. DOI: 10.1200/JCO.19.01203.

Aranha, L. N., Silva, M. G., Uehara, S. K., Luiz, R. R., Nogueira Neto, J. F., Rosa, G. and Moraes de Oliveira, G. M. 2020. Effects of a hypoenergetic diet associated with açaí (Euterpe oleracea Mart.) pulp consumption on antioxidant status, oxidative stress and inflammatory biomarkers in overweight, dyslipidemic individuals. *Clinical Nutrition* 39(5), 1464–1469. DOI: 10.1016/j.clnu.2019.06.008.

Arcan, I. and Yemenicioglu, A. 2011. Incorporating phenolic compounds opens a new perspective to use zein films as flexible bioactive packaging materials. *Food Research International* 44(2), 550–556. DOI: 10.1016/j.foodres.2010.11.034.

Arfin, S., Jha, N. K., Jha, S. K., Kesari, K. K., Ruokolainen, J., Roychoudhury, S., Rathi, B. and Kumar, D. 2021. Oxidative stress in cancer cell metabolism. *Antioxidants* 10(5), 642. DOI: 10.3390/antiox10050642.

Asbaghi, O., Sadeghian, M., Nazarian, B., Sarreshtedari, M., Mozaffari-Khosravi, H., Maleki, V., Alizadeh, M., Shokri, A. and Sadeghi, O. 2020. The effect of vitamin E supplementation on selected inflammatory biomarkers in adults: a systematic review and meta-analysis of randomized clinical trials. *Scientific Reports* 10(1), 17234. DOI: 10.1038/s41598-020-73741-6.

Aserese, A. D., Atenafu, A., Sisay, M., Sorrie, M. B., Yirdaw, B. W. and Zegeye, M. K. 2020. Adequate vitamin A rich food consumption and associated factors among lactating mothers visiting child immunization and post-natal clinic at health institutions in Gondar Town, Northwest Ethiopia. *PLoS ONE* 15(9), e0239308. DOI: 10.1371/journal.pone.0239308.

Assous, M. T. M., Abdel-Hady, M. M. and Medany, G. M. 2014. Evaluation of red pigment extracted from purple carrots and its utilization as antioxidant and natural food colorants. *Annals of Agricultural Sciences* 59(1), 1-7. DOI: 10.1016/j.aoas.2014.06.001.

Avery, J. C. and Hoffmann, P. R. 2018. Selenium, selenoproteins, and immunity. *Nutrients* 10(9). DOI: 10.3390/nu10091203.

Aziz, M. A., Diab, A. S. and Mohammed, A. A. 2019. *Antioxidant Categories and Mode of Action*. Antioxidants. IntechOpen. https://doi.org/10.5772/intechopen.83544.

Balogh, E., Veale, D. J., McGarry, T., Orr, C., Szekanecz, Z., Ng, C. T., Fearon, U. and Biniecka, M. 2018. Oxidative stress impairs energy metabolism in primary cells and synovial tissue of patients with rheumatoid arthritis. *Arthritis Research and Therapy* 20(1), 95. DOI: 10.1186/s13075-018-1592-1.

Behl, T., Bungau, S., Kumar, K., Zengin, G., Khan, F., Kumar, A., Kaur, R., Venkatachalam, T., Tit, D. M., Vesa, C. M., Barsan, G. and Mosteanu, D. E. 2020. Pleotropic effects of polyphenols in cardiovascular system. *Biomedicine and Pharmacotherapy* 130, 110714. DOI: 10.1016/j.biopha.2020.110714.

Bhatt, T. and Patel, K. 2020. Carotenoids: potent to prevent diseases review. *Natural Products and Bioprospecting* 10(3), 109-117. DOI: 10.1007/s13659-020-00244-2.

Bisset, S., Sobhi, W., Bensouici, C. and Khenchouche, A. 2020. Chain-breaking/preventive antioxidant, urate-lowering, and anti-inflammatory effects of pure curcumin. *Current Nutrition and Food Science* 17(1), 66-74. DOI: 10.2174/1573401316999200421095134.

Boeing, H., Bechthold, A., Bub, A., Ellinger, S., Haller, D., Kroke, A., Leschik-Bonnet, E., Müller, M. J., Oberritter, H., Schulze, M., Stehle, P. and Watzl, B. 2012. Critical review: vegetables and fruit in the prevention of chronic diseases. *European Journal of Nutrition* 51(6), 637-663. DOI: 10.1007/s00394-012-0380-y.

Bohn, T., Desmarchelier, C., El, S. N., Keijer, J., van Schothorst, E., Rühl, R. and Borel, P. 2019. β-carotene in the human body: metabolic bioactivation pathways - from digestion to tissue distribution and excretion. *Proceedings of the Nutrition Society* 78(1), 68-87. DOI: 10.1017/S0029665118002641.

Bourgonje, A. R., Feelisch, M., Faber, K. N., Pasch, A., Dijkstra, G. and van Goor, H. 2020. Oxidative stress and redox-modulating therapeutics in inflammatory bowel disease. *Trends in Molecular Medicine* 26(11), 1034-1046. DOI: 10.1016/j.molmed.2020.06.006.

Bramley, P. M., Elmadfa, I., Kafatos, A., Kelly, F. J., Manios, Y., Roxborough, H. E., Schuch, W., Sheehy, P. J. A. and Wagner, K.-H. 2000. Vitamin E. *Journal of the Science of Food and Agriculture* 80(7), 913-938. DOI: 10.1002/(SICI)1097-0010(20000515)80:7<913::AID-JSFA600>3.0.CO;2-3.

Browne, D., McGuinness, B., Woodside, J. V. and McKay, G. J. 2019. Vitamin E and Alzheimer's disease: what do we know so far? *Clinical Interventions in Aging* 14, 1303-1317.

Bungau, S., Abdel-Daim, M. M., Tit, D. M., Ghanem, E., Sato, S., Maruyama-Inoue, M., Yamane, S. and Kadonosono, K. 2019. Health benefits of polyphenols and carotenoids in age-related eye diseases. *Oxidative Medicine and Cellular Longevity* 2019, 9783429.

Businesswire. 2020. $722+ billion nutraceutical market size and share breakdown by product and region. Available at: https://www.businesswire.com/news/home/20200520005477/en/722-Billion-Nutraceutical-Market-Size-and-Share-Breakdown-by-Product-and-Region---ResearchAndMarkets.com.

Butterfield, D. A. and Halliwell, B. 2019. Oxidative stress, dysfunctional glucose metabolism and Alzheimer disease. *Nature Reviews. Neuroscience* 20(3), 148–160. DOI: 10.1038/s41583-019-0132-6.

Calado, A., Neves, P. M., Santos, T. and Ravasco, P. 2018. The effect of flaxseed in breast cancer: a literature review. *Frontiers in Nutrition* 5, 4. DOI: 10.3389/fnut.2018.00004.

Cao, T., Zhang, X., Yang, D., Wang, Y. Q., Qiao, Z. D., Huang, J. M. and Zhang, P. 2018. Antioxidant effects of epigallocatechin-3-gallate on the aTC1-6 pancreatic alpha cell line. *Biochemical and Biophysical Research Communications* 495(1), 693–699.

Carpéné, C., Pejenaute, H., del Moral, R., Boulet, N., Hijona, E., Andrade, F., Villanueva-Millán, M. J., Aguirre, L. and Arbones-Mainar, J. M. 2018. The dietary antioxidant piceatannol inhibits adipogenesis of human adipose mesenchymal stem cells and limits glucose transport and lipogenic activities in adipocytes. *International Journal of Molecular Sciences* 19(7), 2081.

Chen, B. H., Hsieh, C. H., Tsai, S. Y., Wang, C. Y. and Wang, C. C. 2020. Anticancer effects of epigallocatechin-3-gallate nanoemulsion on lung cancer cells through the activation of AMP-activated protein kinase signaling pathway. *Scientific Reports* 10(1), 5163. DOI: 10.1038/s41598-020-62136-2.

Chin, K. Y. and Ima-Nirwana, S. 2018. The role of vitamin E in preventing and treating osteoarthritis – a review of the current evidence. *Frontiers in Pharmacology* 9, 946.

Chooi, Y. C., Ding, C. and Magkos, F. 2019. The epidemiology of obesity. *Metabolism: Clinical and Experimental* 92, 6–10.

Čolak, E. and Pap, D. 2021. The role of oxidative stress in the development of obesity and obesity-related metabolic disorders. *Journal of Medical Biochemistry* 40(1), 1–9.

Colunga Biancatelli, R. M. L., Berrill, M. and Marik, P. E. 2020. The antiviral properties of vitamin C. *Expert Review of Anti-Infective Therapy* 18(2), 99–101.

Cory, H., Passarelli, S., Szeto, J., Tamez, M. and Mattei, J. 2018. The role of polyphenols in human health and food systems: a mini-review. *Frontiers in Nutrition* 5, 87. DOI: 10.3389/fnut.2018.00087.

Costa, C. F., Fusieger, A., Andretta, M., Camargo, A. C., Carvalho, A. F., Menezes, D. R. and Nero, L. A. 2020. Short communication: potential use of passion fruit (Passiflora cincinnata) as a biopreservative in the production of coalho cheese, a traditional Brazilian cheese. *Journal of Dairy Science* 103(4), 3082–3087. DOI: 10.3168/jds.2019-17791.

D'Oria, R., Schipani, R., Leonardini, A., Natalicchio, A., Perrini, S., Cignarelli, A., Laviola, L. and Giorgino, F. 2020. The role of oxidative stress in cardiac disease: from physiological response to injury factor. *Oxidative Medicine and Cellular Longevity* 2020, 5732956. DOI: 10.1155/2020/5732956.

Daimiel, L., Micó, V., Valls, R. M., Pedret, A., Motilva, M. J., Rubió, L., Fitó, M., Farrás, M., Covas, M. I., Solá, R. and Ordovás, J. M. 2020. Impact of phenol-enriched virgin olive oils on the postprandial levels of circulating microRNAs related to cardiovascular disease. *Molecular Nutrition and Food Research* 64(15), e2000049. DOI: 10.1002/mnfr.202000049.

D'Angelo, S. 2020. Current evidence on the effect of dietary polyphenols intake on brain health. *Current Nutrition and Food Science* 16(8), 1170–1182.

Dewhirst, R. A. and Fry, S. C. 2018. The oxidation of dehydroascorbic acid and 2,3-diketogulonate by distinct reactive oxygen species. *Biochemical Journal* 475(21), 3451–3470. DOI: 10.1042/BCJ20180688.

Dewhirst, R. A., Murray, L., Mackay, C. L., Sadler, I. H. and Fry, S. C. 2020. Characterisation of the non-oxidative degradation pathway of dehydroascorbic acid in slightly acidic aqueous solution. *Archives of Biochemistry and Biophysics* 681, 108240. DOI: 10.1016/j.abb.2019.108240.

Dias, S., Castanheira, E. M. S., Fortes, A. G., Pereira, D. M. and Gonçalves, M. S. T. 2020. Natural pigments of anthocyanin and betalain for coloring soy-based yogurt alternative. *Foods* 9(6), 717. DOI: 10.3390/foods9060771.

Díaz-Gómez, J., Moreno, J. A., Angulo, E., Sandmann, G., Zhu, C., Ramos, A. J., Capell, T., Christou, P. and Nogareda, C. 2017. High-carotenoid biofortified maize is an alternative to color additives in poultry feed. *Animal Feed Science and Technology* 231, 38–46. DOI: 10.1016/j.anifeedsci.2017.06.007.

DiPasquale, M., Nguyen, M. H. L., Rickeard, B. W., Cesca, N., Tannous, C., Castillo, S. R., Katsaras, J., Kelley, E. G., Herberle, F. A. and Marquardt, D. 2020. The antioxidant vitamin E as a membrane raft modulator: tocopherols do not abolish lipid domains. *Biochimica et Biophysica Acta. Biomembranes* 1862(8), 183189. DOI: 10.1016/j.bbamem.2020.183189.

Dludla, P. V., Nkambule, B. B., Jack, B., Mkandla, Z., Mutize, T., Silvestri, S., Orlando, P., Tiano, L., Louw, J. and Mazibuko-Mbeje, S. E. 2018. Inflammation and oxidative stress in an obese state and the protective effects of gallic acid. *Nutrients* 11(1), 23.

Duarte-Jurado, A. P., Gopar-Cuevas, Y., Saucedo-Cardenas, O., Loera-Arias, M. J., Montes-de-Oca-Luna, R., Garcia-Garcia, A. and Rodriguez-Rocha, H. 2021. Antioxidant therapeutics in Parkinson's disease: current challenges and opportunities. *Antioxidants* 10(3), 453.

Eguchi, N., Vaziri, N. D., Dafoe, D. C. and Ichii, H. 2021. The role of oxidative stress in pancreatic β cell dysfunction in diabetes. *International Journal of Molecular Sciences* 22(4), 1509. DOI: 10.3390/ijms22041509.

Ellison, S. 2016. Carotenoids: physiology. *Encyclopedia of Food and Health*, 670-675. DOI: 10.1016/B978-0-12-384947-2.00120-3.

Elvira-Torales, L. I., García-Alonso, J. and Periago-Castón, M. J. 2019. Nutritional importance of carotenoids and their effect on liver health: a review. *Antioxidants* 8(7), 229.

Fairulnizal, Md Noh, Devi, M., Nair Gunasegavan, R. and Mustar, S. 2019. Vitamin A in health and disease. *Vitamin A*.

Fang, Y., Ou, S., Wu, T., Zhou, L., Tang, H., Jiang, M., Xu, J. and Guo, K. 2020. Lycopene alleviates oxidative stress via the PI3K/Akt/Nrf2pathway in a cell model of Alzheimer's disease. *PeerJ* 8, e9308. DOI: 10.7717/peerj.9308.

Food & Nutrition Board. 2000. *Dietary Reference Intakes for Vitamin C, Vitamin E, Selenium, and Carotenoids, Institute of Medicine*. DOI: 10.17226/9810.

Fraga, C. G., Croft, K. D., Kennedy, D. O. and Tomás-Barberán, F. A. 2019. The effects of polyphenols and other bioactives on human health. *Food and Function* 10(2), 514–528.

Gao, L., Zhu, Y. and Xu, L. 2021. Mechanisms of protective effects of astaxanthin in nonalcoholic fatty liver disease. *Hepatoma Research*. DOI: 10.20517/2394-5079.2020.150.

Gengatharan, A., Dykes, G. A. and Choo, W. S. 2017. The effect of pH treatment and refrigerated storage on natural colourant preparations (betacyanins) from red pitahaya and their potential application in yoghurt. *LWT* 80, 437–445. DOI: 10.1016/j.lwt.2017.03.014.

Ghattamaneni, N. K., Sharma, A., Panchal, S. K. and Brown, L. 2020. Pelargonidin 3-glucoside-enriched strawberry attenuates symptoms of DSS-induced inflammatory bowel disease and diet-induced metabolic syndrome in rats. *European Journal of Nutrition* 59(7), 2905–2918. DOI: 10.1007/s00394-019-02130-1.

Gibson, R. S., King, J. C. and Lowe, N. 2016. A review of dietary zinc recommendations. *Food and Nutrition Bulletin* 37(4), 443–460. DOI: 10.1177/0379572116652252.

Giusti, M. M. and Wrolstad, R. E. 2003. Acylated anthocyanins from edible sources and their applications in food systems. *Biochemical Engineering Journal* 14(3), 217–225. DOI: 10.1016/S1369-703X(02)00221-8.

Grosso, G., Stepaniak, U., Topor-Madry, R., Szafraniec, K. and Pajak, A. 2014. Estimated dietary intake and major food sources of polyphenols in the Polish arm of the HAPIEE study. *Nutrition* 30(11–12), 1398–1403. PMID: 25280419.

Gull, A., Bhat, N., Wani, S. M., Masood, F. A., Amin, T. and Ganai, S. A. 2021. Shelf life extension of apricot fruit by application of nanochitosan emulsion coatings containing pomegranate peel extract. *Food Chemistry* 349, 129149. DOI: 10.1016/j.foodchem.2021.129149.

Gurbuz, Z., Erkaya-Kotan, T. and Şengül, M. 2021. Evaluation of physicochemical, microbiological, texture and microstructure characteristics of set-style yoghurt supplemented with quince seed mucilage powder as a novel natural stabiliser. *International Dairy Journal* 114, 104938. DOI: 10.1016/j.idairyj.2020.104938.

Hayrapetyan, H., Hazeleger, W. C. and Beumer, R. R. 2012. Inhibition of *Listeria monocytogenes* by pomegranate (*Punica granatum*) peel extract in meat paté at different temperatures. *Food Control* 23(1), 66–72. DOI: 10.1016/j.foodcont.2011.06.012.

Hernández-Rodríguez, P., Baquaro, L. and Larrota, H. 2019. Flavonoids. *Bioactive Compounds, Health Benefits and Potential Applications*, 265–288. DOI: 10.1016/B978-0-12-814774-0.00014-1.

Hong, Y. H., Lim, G. O. and Song, K. B. 2009. Physical properties of Gelidium corneum-gelatin blend films containing grapefruit seed extract or green tea extract and its application in the packaging of pork loins. *Journal of Food Science* 74(1), C6–C10. DOI: 10.1111/j.1750-3841.2008.00987.x.

Hou, Y., Gong, T., Zhang, J., Yang, X. and Guo, Y. 2019. Structural characterization and emulsifying properties of thinned-young apples polysaccharides. *Biochemical and Biophysical Research Communications* 516(4), 1175–1182. DOI: 10.1016/j.bbrc.2019.07.019.

Hu, W., Zhao, C., Hu, H. and Yin, S. 2021. Food sources of selenium and its relationship with chronic diseases. *Nutrients* 13(5), 1739.

Huang, Z., Liu, Y., Qi, G., Brand, D. and Zheng, S. G. 2018. Role of vitamin A in the immune system. *Journal of Clinical Medicine* 7(9), 258.

Jan, S., Rashid, M., Abd Allah, E. F. and Ahmad, P. 2020. Biological efficacy of essential oils and plant extracts of cultivated and wild ecotypes of origanum vulgare L. *BioMed Research International* 2020, 8751718. DOI: 10.1155/2020/8751718.

Janabi, A., Kamboh, A., Saeed, M., Xiaoyu, L., BiBi, J., Majeed, F., Naveed, M., Mughal, M., Korejo, N., Kamboh, R., Alagawany, M. and Lv, H. 2019. Flavonoid-rich foods (FRF): a promising nutraceutical approach against lifespan-shortening diseases, *Iranian Journal of Basic Medical Sciences*, 23(2), 141–153. DOI: https://dx.doi.org/10.22038/ijbms.2019.35125.8353.

Jenkins, D. J. A., Kitts, D., Giovannucci, E. L., Sahye-Pudaruth, S., Paquette, M., Blanco Mejia, S., Patel, D., Kavanagh, M., Tsirakis, T., Kendall, C. W. C., Pichika, S. C. and Sievenpiper, J. L. 2020. Selenium, antioxidants, cardiovascular disease, and all-cause mortality: a systematic review and meta-analysis of randomized controlled trials. *The American Journal of Clinical Nutrition* 112(6), 1642–1652.

Jiang, J., Nuez-Ortin, W., Angell, A., Zeng, C., de Nys, R. and Vucko, M. J. 2019. Enhancing the colouration of the marine ornamental fish Pseudochromis fridmani using natural and synthetic sources of astaxanthin. *Algal Research* 42, 101596. DOI: 10.1016/j.algal.2019.101596.

Jiang, Y., Xu, Y., Li, F., Lia, D. and Huang, Q. 2020. Pectin extracted from persimmon peel: a physicochemical characterization and emulsifying properties evaluation. *Food Hydrocolloids* 101, 105561. DOI: 10.1016/j.foodhyd.2019.105561.

Johra, F. T., Bepari, A. K., Bristy, A. T. and Reza, H. M. 2020. A mechanistic review of β-carotene, lutein, and zeaxanthin in eye health and disease. *Antioxidants* 9(11). DOI: 10.3390/antiox9111046.

Karre, L., Lopez, K. and Getty, K. J. K. 2013. Natural antioxidants in meat and poultry products. *Meat Science* 94(2), 220–227. DOI: 10.1016/j.meatsci.2013.01.007.

Kaur, C. and Kapoor, H. C. 2008. Antioxidants in fruits and vegetables - the millennium's health. *International Journal of Food Science and Technology* 36(7), 703–725. DOI: 10.1111/j.1365-2621.2001.00513.x.

Kaurinovic, B. and Vastag, D. 2019. Flavonoids and phenolic acids as potential natural antioxidants. *Antioxidants*. DOI: 10.5772/intechopen.83731.

Kawabata, K., Yoshioka, Y. and Terao, J. 2019. Role of intestinal microbiota in the bioavailability and physiological functions of dietary polyphenols. *Molecules* 24(2). DOI: 10.3390/molecules24020370.

Kawasoe, H., Wakamatsu, M., Hamada, S., Arata, Y., Nagayoshi, K., Uchida, R., Yamashita, R., Kishita, T., Yamanochi, H., Minami, Y. and Kajiya, K. 2021. Analysis of natural colourant extracted from the pericarp of passion fruit. *LWT* 136(2), 110412. DOI: 10.1016/j.lwt.2020.110412.

Kaźmierczak-Barańska, J., Boguszewska, K., Adamus-Grabicka, A. and Karwowski, B. T. 2020. Two faces of Vitamin C – antioxidative and pro-oxidative agent. *Nutrients* 12(5), 1501.

Khoo, H. E., Azlan, A., Tang, S. T. and Lim, S. M. 2017. Anthocyanidins and anthocyanins: colored pigments as food, pharmaceutical ingredients, and the potential health benefits. *Food and Nutrition Research* 61(1), 1361779. DOI: 10.1080/16546628.2017.1361779.

Kieliszek, M. and Blazejak, S. 2016. Current knowledge on the importance of selenium in food for living organisms: a review. *Molecules* 21(5), 605. DOI: 10.3390/molecules21050609.

Kieliszek, M., Blazejak, S., Bzducha-Wróbel, A. and Kot, A. M. 2019. Effect of selenium on growth and antioxidative system of yeast cells. *Molecular Biology Reports* 46(2), 1797–1808. DOI: 10.1007/s11033-019-04630-z.

Kim, S. A., Kim, J., Jun, S., Wie, G. A., Shin, S. and Joung, H. 2020a. Association between dietary flavonoid intake and obesity among adults in Korea. *Applied Physiology, Nutrition, and Metabolism* 45(2), 203-212. DOI: 10.1139/apnm-2019-0211.

Kim, T. Y., Leem, E., Lee, J. M. and Kim, S. R. 2020b. Control of reactive oxygen species for the prevention of Parkinson's disease: the possible application of flavonoids. *Antioxidants* 9(7), 583.

Koch, W. 2019. Dietary polyphenols–important non-nutrients in the prevention of chronic noncommunicable diseases: a systematic review. *Nutrients* 11(5). DOI: 10.3390/nu11051039.

Kowalska, K., Dembczyński, R., Gołąbek, A., Olkowicz, M. and Olejnik, A. 2021. ROS modulating effects of lingonberry (*Vaccinium vitis-idaea* L.) polyphenols on obese adipocyte hypertrophy and vascular endothelial dysfunction. *Nutrients* 13(3). DOI: 10.3390/nu13030885.

Kumar, K. P. S., Bhowmik, D., Duraivel, S. and Umadevi, M. 2012. Traditional and medicinal uses of banana. *Journal of Pharmacognosy and Phytochemistry* 1(3). ISSN: 2278- 4136.

Kumar, N. and Goel, N. 2019. Phenolic acids: natural versatile molecules with promising therapeutic applications. *Biotechnology Reports* 24, e00370. DOI: 10.1016/j.btre.2019.e00370.

Kumar, N., Pratibha, N., Ojha, A., Upadhyay, A., Singh, R. and Kumar, S. 2021. Effect of active chitosan-pullulan composite edible coating enrich with pomegranate peel extract on the storage quality of green bell pepper. *LWT* 138, 110435. DOI: 10.1016/j.lwt.2020.110435.

Kumar, S. and Pandey, A. 2015. Free radicals: health implications and their mitigation by herbals. *British Journal of Medicine and Medical Research* 7(6), 438–457.

Kurutas, E. B. 2016. The importance of antioxidants in cellular response against oxidative/nitrosative stress: current state. *Nutrition Journal* 15(1), 71.

Laxa, M., Liebthal, M., Telman, W., Chibani, K. and Dietz, K. J. 2019. The role of the plant antioxidant system in drought tolerance. *Antioxidants* 8(4). DOI: 10.3390/antiox8040094.

Lee Chong, T., Ahearn, E. L. and Cimmino, L. 2019. Reprogramming the epigenome with vitamin C. *Frontiers in Cell and Developmental Biology* 7, 128.

Lee, D., Kim, Y., Jo, H., Go, C., Jeong, Y., Jang, Y., Kang, D., Park, K., Kim, Y. S. and Kang, J. S. 2021. The anti-inflammatory effect of Aptamin C on house dust mite extract-induced inflammation in keratinocytes via regulation of IL-22 and GDNF production. *Antioxidants* 10(6). DOI: 10.3390/antiox10060945.

Lee, G. H., Hoang, T. H., Jung, E. S., Jung, S. J., Han, S. K., Chung, M. J., Chae, S. W. and Chae, H. J. 2020. Anthocyanins attenuate endothelial dysfunction through regulation of uncoupling of nitric oxide synthase in aged rats. *Aging Cell* 19(12), e13279.

Lee, S. R. 2018. Critical role of zinc as either an antioxidant or a prooxidant in cellular systems. *Oxidative Medicine and Cellular Longevity* 2018, 9156285. DOI: 10.1155/2018/9156285.

Levers, K., Dalton, R., Galvan, E., O'Connor, A., Goodenough, C., Simbo, S., Mertens-Talcott, S. U., Rasmussen, C., Greenwood, M., Riechman, S., Crouse, S. and Kreider, R. B. 2016. Effects of powdered Montmorency tart cherry supplementation on acute endurance exercise performance in aerobically trained individuals. *Journal of the International Society of Sports Nutrition* 13(22), 22. DOI: 10.1186/s12970-016-0133-z.

Liu, Q., Jin, Z., Xu, Z., Yang, H., Li, L., Li, G., Li, F., Gu, S., Zong, S., Zhou, J., Cao, L., Wang, Z. and Xiao, W. 2019. Antioxidant effects of ginkgolides and bilobalide against cerebral ischemia injury by activating the Akt/Nrf2 pathway in vitro and in vivo. *Cell Stress and Chaperones* 24(2), 441–452. DOI: 10.1007/s12192-019-00977-1.

Liu, X., Li, Q., Cheng, X., Liu, Z., Zhao, X., Zhang, S., Yu, G., Zhao, X. and Hao, J. 2020. Oligomannuronate prevents mitochondrial dysfunction induced by IAPP in RINm5F islet cells by inhibition of JNK activation and cell apoptosis. *Chinese Medicine* 15(1), 27.

Lobo, V. et al. 2010. Free radicals, antioxidants and functional foods: impact on human health. *Pharmacognosy Reviews* 4(8), 118–126.

Lykkesfeldt, J., Michels, A. J. and Frei, B. 2014. Vitamin C. *Advances in Nutrition* 5(1), 16–18. DOI: 10.3945/an.113.005157.

Lykkesfeldt, J. and Tveden-Nybord, P. 2019. The pharmacokinetics of vitamin C. *Nutrients* 11(10). DOI: 10.3390/nu11102412.

Mantena, S. K., Baliga, M. S. and Katiyar, S. K. 2006. Grape seed proanthocyanidins induce apoptosis and inhibit metastasis of highly metastatic breast carcinoma cells. *Carcinogenesis* 27(8), 1682–1691. DOI: 10.1093/carcin/bgl030.

Martillanes, S., Pimienta, J. R., Cabrera-Banegil, M., Martin-Vertedor, D. and Delgado, J. 2017. Application of phenolic compounds for food preservation: food additive and active packaging. *Phenolic Compounds-Biological Activity*. DOI: 10.5772/66885

Masschelin, P. M., Cox, A. R., Chernis, N. and Hartig, S. M. 2019. The impact of oxidative stress on adipose tissue energy balance. *Frontiers in Physiology* 10, 1638.

Mateen, S., Rehman, M. T., Shahzad, S., Naeem, S. S., Faizy, A. F., Khan, A. Q., Khan, M. S., Husain, F. M. and Moin, S. 2019. Anti-oxidant and anti-inflammatory effects of cinnamaldehyde and eugenol on mononuclear cells of rheumatoid arthritis patients. *European Journal of Pharmacology* 852, 14–24. DOI: 10.1016/j.ejphar.2019.02.031.

Mazarakis, N., Snibson, K., Licciardi, P. V. and Karagiannis, T. C. 2020. The potential use of l-sulforaphane for the treatment of chronic inflammatory diseases: a review of the clinical evidence. *Clinical Nutrition* 39(3), 664–675. DOI: 10.1016/j.clnu.2019.03.022.

McCall, B., McPartland, C. K., Moore, R., Frank-Kamenetskii, A. and Booth, B. W. 2018. Effects of astaxanthin on the proliferation and migration of breast cancer cells in vitro. *Antioxidants* 7(10). DOI: 10.3390/antiox7100135.

Meléndez-Martínez, A. J. 2019. An overview of carotenoids, apocarotenoids, and vitamin A in Agro-Food, nutrition, health, and disease. *Molecular Nutrition and Food Research* 63(15), e1801045.

Mendez, D. A., Fabra, M. J., Martínez-Aba, A., Martínez-Sanz, M, Gorria, M. and López-Rubio, A. 2021. Understanding the different emulsification mechanisms of pectin: comparison between watermelon rind and two commercial pectin sources. *Food Hydrocolloids* 120, 106957. DOI: 10.1016/j.foodhyd.2021.106957.

Meng, J., Chen, Y., Wang, J., Qiu, J., Chang, C., Bi, F., Wu, X. and Liu, W. 2020. EGCG protects vascular endothelial cells from oxidative stress-induced damage by targeting the autophagy-dependent PI3K-AKT-mTOR pathway. *Annals of Translational Medicine* 8(5), 200–200. DOI: 10.21037/atm.2020.01.92.

Mezzomo, N. and Ferreira, S. R. S. 2016. Carotenoids functionality, sources, and processing by supercritical technology: a review. *Journal of Chemistry*. RS 2016, 1-16. DOI: 10.1155/2016/3164312.

Mithul Aravind, S., Wichienchot, S., Tsao, R., Ramakrishnan, S. and Chakkaravarthi, S. 2021. Role of dietary polyphenols on gut microbiota, their metabolites and health benefits. *Food Research International* 142, 110189.

Mohd Hassan, N., Yusof, N. A., Yahaya, A. F., Mohd Rozali, N. N. and Othman, R. 2019. Carotenoids of capsicum fruits: pigment profile and health-promoting functional attributes. *Antioxidants* 8(10), 469.

Mohdaly, A. A. A., Sarhan, M. A., Mahmoud, A., Ramadan, M. F. and Smetanska, I. 2010. Antioxidant efficacy of potato peels and sugar beet pulp extracts in vegetable oils protection. *Food Chemistry* 123(4), 1019–1026. DOI: 10.1016/j.foodchem.2010.05.054.

Moreno, J. and Peinado, R. 2012. Polyphenols. *Enological Chemistry (1)*. ISBN: 9780123884398.

Moreno, J. A., Díaz-Gómez, J., Fuentes-Font, L., Angulo, E., Gosálvez, L. F., Sandmann, G., Portero-Otin, M., Capell, T., Zhu, C., Christou, P. and Nogareda, C. 2020. Poultry diets containing (keto)carotenoid-enriched maize improve egg yolk color and maintain quality. *Animal Feed Science and Technology* 260, 114334. DOI: 10.1016/j.anifeedsci.2019.114334.

Mousavi, S. N., Faghihi, A., Motaghinejad, M., Shiasi, M., Imanparast, F., Amiri, H. L. and Shidfar, F. 2018. Zinc and selenium co-supplementation reduces some lipid peroxidation and angiogenesis markers in a rat model of NAFLD-fed high fat diet. *Biological Trace Element Research* 181(2), 288–295.

Müller, H. 1996. Daily intake of carotenoids (carotenes and xanthophylls) from total diet and the carotenoid content of selected vegetables and fruit. *Zeitschrift Fur Ernahrungswissenschaft* 35(1), 45–50. DOI: 10.1007/BF01612027.

Nabi, F., Arain, M. A., Rajput, N., Alagawany, M., Soomro, J., Umer, M., Soomro, F., Wang, Z., Ye, R. and Lie, J. 2020. Health benefits of carotenoids and potential application in poultry industry: a review. *Journal of Animal Physiology and Animal Nutrition* 104(6), 1809–1818. DOI: 10.1111/jpn.13375.

Ngo, B., Van Riper, J. M., Cantley, L. C. and Yun, J. 2019. Targeting cancer vulnerabilities with high-dose vitamin C. *Nature Reviews. Cancer* 19(5), 271–282.

Niki, E. and Abe, K. 2019. Vitamin E: structure, properties and functions, *vitamin E: chemistry and nutritional benefits, food chemistry function and analysis. Food Chemistry, Function and Analysis*, 1–11. DOI: 10.1039/9781788016216-00001.

Nur Hanani, Z. A., Aelma Husna, A. B., Nurul Syahida, S., Nor Khaizura, M. A. B. and Jamilah, B. 2018. Effect of different fruit peels on the functional properties of gelatin/polyethylene bilayer films for active packaging. *Food Packaging and Shelf Life* 18, 201–211. DOI: 10.1016/j.fpsl.2018.11.004.

Olson, J. A. and Hodges, R. E. 1987. Recommended dietary intakes (RDI) of vitamin C in humans. *The American Journal of Clinical Nutrition* 45(4), 693–703. DOI: 10.1093/ajcn/45.4.693.

Olza, J., Aranceta-Bartrina, J., González-Gross, M., Ortega, R. M., Serra-Majem, L., Varela-Moreiras, G. and Gil, Á. 2017. Reported dietary intake and food sources of zinc, selenium, and vitamins A, E and C in the Spanish population: findings from the ANIBES study. *Nutrients* 9(7). DOI: 10.3390/nu9070697.

Ononamadu, C. J. et al. 2019. In vitro and in vivo anti-diabetic and antioxidant activities of methanolic leaf extracts. *Caspian Journal of Internal Medicine* 10(2), 162–175.

Oomah, B. and Mazza, G. 2000. Functional foods. In Francis, F. (ed.), *Encyclopedia of Food Science and Technology Vol 2: 1176-1182*. John Wiley and Sons Ltd, NY.

Ordóñez-Santos, L. E., Esparza-Estrada, J. and Vanegas-Mahecha, P. 2021. Ultrasound-assisted extraction of total carotenoids from mandarin epicarp and application as natural colorant in bakery products. *LWT* 139, 110598. DOI: 10.1016/j.lwt.2020.110598.

Pandey, K. B. and Rizvi, S. I. 2009. Plant polyphenols as dietary antioxidants in human health and disease. *Oxidative Medicine and Cellular Longevity* 2(5), 270-278. DOI: 10.4161/oxim.2.5.9498.

Parent, M. E., Richard, H., Rousseau, M. C. and Trudeau, K. 2018. Vitamin C intake and risk of prostate cancer: the Montreal PROtEuS study. *Frontiers in Physiology* 9, 1218. DOI: 10.3389/fphys.2018.01218.

Park, H. A. and Ellis, A. C. 2020. Dietary antioxidants and Parkinson's disease. *Antioxidants* 9(7), 570. DOI: 10.3390/antiox9070570.

Park, H. A., Mnatsakanyan, N., Broman, K., Davis, A. U., May, J., Licznerski, P., Crowe-White, K. M., Lackey, K. H. and Jonas, E. A. 2019. Alpha-tocotrienol prevents oxidative stress-mediated post-translational cleavage of Bcl-xL in primary hippocampal neurons. *International Journal of Molecular Sciences* 21(1), 220. DOI: 10.3390/ijms21010220.

Park, S. and Park, S. Y. 2021. Can antioxidants be effective therapeutics for type 2 diabetes? *Yeungnam University Journal of Medicine* 38(2), 83-94. DOI: 10.12701/yujm.2020.00563.

Pazos, M., Gallardo, J. M., Torres, J. L. and Medina, I. 2005. Activity of grape polyphenols as inhibitors of the oxidation of fish lipids and frozen fish muscle. *Food Chemistry* 92(3), 547-557. DOI: 10.1016/j.foodchem.2004.07.036.

Pehlivan, F. 2017. Vitamin C: an antioxidant agent. *Vitamin C*. DOI: 10.5772/intechopen.69660.

Peoples, J. N., Saraf, A., Ghazal, N., Pham, T. T. and Kwong, J. Q. 2019. Mitochondrial dysfunction and oxidative stress in heart disease. *Experimental and Molecular Medicine* 51(12), 1-13.

Perillo, B., Di Donato, M., Pezone, A., Di Zazzo, E., Giovannelli, P., Galasso, G., Castoria, G. and Migliaccio, A. 2020. ROS in cancer therapy: the bright side of the moon. *Experimental and Molecular Medicine* 52(2), 192-203. DOI: 10.1038/s12276-020-0384-2.

Petrou, A. L., Petrou, P. L., Ntanos, T. and Liapis, A. 2018. A possible role for singlet oxygen in the degradation of various antioxidants: a meta-analysis and review of literature data. *Antioxidants* 7(3), 35. DOI: 10.3390/antiox7030035.

Petrovic, S., Arsic, A., Ristic-Medic, D., Cvetkovic, Z. and Vucic, V. 2020. Lipid peroxidation and antioxidant supplementation in neurodegenerative diseases: a review of human studies. *Antioxidants* 9(11), 1128. DOI: 10.3390/antiox9111128.

Phull, A. R., Nasir, B., Haq, I. U. and Kim, S. J. 2018. Oxidative stress, consequences and ROS mediated cellular signaling in rheumatoid arthritis. *Chemico-Biological Interactions* 281, 121-136.

Pinto, M., Benfeito, S., Fernandes, C. and Borges, F. 2020. Antioxidant therapy, oxidative stress, and blood-brain barrier: the road of dietary antioxidants. *Oxidative Stress and Dietary Antioxidants in Neurological Diseases*. DOI: 10.1016/B978-0-12-817780-8.00009-8.

Pisoschi, A. M., Pop, A., Iordache, F., Stanca, L., Predoi, G. and Serban, A. I. 2021. Oxidative stressmitigation by antioxidants - an overview on their chemistry and influences on health status. *European Journal of Medicinal Chemistry* 209, 112891. DOI: 10.1016/j.ejmech.2020.112891.

Pizzino, G. et al. 2017. Oxidative stress: harms and benefits for human health. *Oxidative Medicine and Cellular Longevity* 2017, 8416763.

Prasad, A. S. 2014. Zinc is an antioxidant and anti-inflammatory agent: its role in human Health, Frontiers. *Frontiers in Nutrition* 1, 14. DOI: 10.3389/fnut.2014.00014.

Qin, Y., Liu, Y., Zhang, X. and Liu, J. 2020. Development of active and intelligent packaging by incorporating betalains from red pitaya (Hylocereus polyrhizus) peel into starch/polyvinyl alcohol films. *Food Hydrocolloids* 100, 105410. DOI: 10.1016/j.foodhyd.2019.105410.

Rangaraj, V. M., Rambabu, K., Banat, F. and Mittal, V. 2021. Effect of date fruit waste extract as an antioxidant additive on the properties of active gelatin films. *Food Chemistry* 355, 129631. DOI: 10.1016/j.foodchem.2021.129631.

Reboul, E. 2017. Vitamin E bioavailability: mechanisms of intestinal absorption in the spotlight. *Antioxidants* 6(4). DOI: 10.3390/antiox6040095.

Rechner, A. R. and Kroner, C. 2005. Anthocyanins and colonic metabolites of dietary polyphenols inhibit platelet function. *Thrombosis Research* 116(4), 327-334. DOI: 10.1016/j.thromres.2005.01.002.

Reddy, V., Urooj, A. and Kumar, A. 2005. Evaluation of antioxidant activity of some plant extracts and their applications in biscuits. *Food Chemistry* 90(1-2), 317-321. DOI: 10.1016/j.foodchem.2004.05.038.

Risvi, S., Raza, S. T., Ahmed, F., Ahmad, A., Abbas, S. and Mahdi, F. 2014. The role of vitamin E in human health and some diseases. *Sultan Qaboos University Medical Journal* 14(2), e157-e165. PMID: 24790736.

Rocha, D., Caldas, A., da Silva, B., Hermsdorff, H. and Alfenas, R. 2018. Effects of blueberry and cranberry consumption on type 2 diabetes glycemic control: a systematic review. *Critical Reviews in Food Science and Nutrition* 59(11), 1816-1828. DOI: 10.1080/10408398.2018.1430019.

Ross, A. C. 2010. Diet in vitamin A research. *Methods in Molecular Biology* 652, 295-313. DOI: 10.1007/978-1-60327-325-1_17.

Sagar, N. A. and Pareek, S. 2021. Fortification of multigrain flour with onion skin powder as a natural preservative: effect on quality and shelf life of the bread. *Food Bioscience* 41, 100992. DOI: 10.1016/j.fbio.2021.100992.

Sakaki, J., Melough, M., Lee, S., Pounis, G. and Chun, O. 2019. Polyphenol-rich diets in cardiovascular disease prevention, analysis of nutrition research. *Principles of Statistical Methodology and Interpretation of the Results*, 259-298. DOI: 10.1016/B978-0-12-814556-2.00010-5.

Sampaio, S. L., Lonchamp, J., Dias, M. I., Liddle, C., Petropoulos, S. A., Glamočlija, J., Alexopoulos, A., Santos-Buelga, C., Ferreira, I. C. F. R. and Barros, L. 2021. Anthocyanin-rich extracts from purple and red potatoes as natural colourants: bioactive properties, application in a soft drink formulation and sensory analysis. *Food Chemistry* 342, 128526. DOI: 10.1016/j.foodchem.2020.128526.

Saraf-Bank, S., Ahmadi, A., Paknahad, Z., Maracy, M. and Nourian, M. 2019. Effects ofcurcumin supplementation on markers of inflammation and oxidative stress among healthy overweight and obese girl adolescents: a randomized placebo-controlled clinical trial. *Phytotherapy Research* 33(8), 2015-2022. DOI: 10.1002/ptr.6370.

Scalbert, A. and Williamson, G. 2000. Dietary intake and bioavailability of polyphenols. *The Journal of Nutrition* 130(8S Suppl), 2073S-2085S. DOI: 10.1093/jn/130.8.2073S.

Scott, M. and Stuart, L. 2020. Preconception and pregnancy, *Textbook of Natural Medicine* 5. DOI: 10.1016/B978-0-323-43044-9.00211-9.

Shanmugam, S., Thangaraj, P., dos Santos Lima, B., Trindade, G. G. G., Narain, N., Mara de Oliveirae Silva, A., Santin, J. R., Broering, M. F., Serafini, M. R., Quintans-Júnior, L. J. and Antunes de Souza Araújo, A. 2020. Protective effects of flavonoid composition rich *P. subpeltata* Ortega. on indomethacin induced experimental ulcerative colitis in rat models of inflammatory bowel diseases. *Journal of Ethnopharmacology* 248, 112350. DOI: 10.1016/j.jep.2019.112350.

Simpson, D. and Amos, S. 2017. Chapter 12. *Other Plant Metabolites, Pharmacognosy*, 267–280. DOI: 10.1016/B978-0-12-802104-0.00012-3.

Singh, A. et al. 2019. Oxidative stress: a key modulator in neurodegenerative diseases. *Molecules* 24, 1583.

Singla, R. K., Dubey, A. K., Garg, A., Sharma, R. K., Fiorino, M., Ameen, S. M., Haddad, M. A. and Al-Hiary, M. 2019 Natural polyphenols: chemical classification, definition of classes, subcategories, and structures. *Journal of AOAC International* 102(5), 1397–1400. DOI: 10.1093/jaoac/102.5.1397.

Sivarooban, T., Hettiarachch, N. S. and Johnson, M. G. 2007 Inhibition of Listeria monocytogenes using nisin with grape seed extract on turkey frankfurters stored at 4 and 10 degrees C. *Journal of Food Protection* 70(4), 1017–1020. DOI: 10.4315/0362-028X-70.4.1017.

Spencer, J. P., Abd El Mohsen, M. M., Minihane, A. M. and Mathers, J. C. 2008. Biomarkers of the intake of dietary polyphenols: strengths, limitations and application in nutrition research. *British Journal of Nutrition* 99(1), 12–22. DOI: 10.1017/S0007114507798938.

Staroszczyk, H., Kusznierewicz, B., Malinowska-Pańczyk, E., Sinkiewicz, I., Gottfried, K. and Kołodziejska, I. 2020. Fish gelatin films containing aqueous extracts from phenolic-rich fruit pomace. *LWT* 117, 108613. DOI: 10.1016/j.lwt.2019.108613.

Stoffanellar, R. and Morse, N. L. 2015. A review of dietary selenium intake and selenium status in Europe and the Middle East. *Nutrients* 7(3), 1494–1537. DOI: 10.3390/nu7031494.

Su, L. J., Zhang, J. H., Gomez, H., Murugan, R., Hong, X., Xu, D., Jiang, F. and Peng, Z. Y. 2019. Reactive oxygen species-induced lipid peroxidation in apoptosis, autophagy, and ferroptosis. *Oxidative Medicine and Cellular Longevity* 2019, 5080843. DOI: 10.1155/2019/5080843.

Sun, C., Liu, Y., Zhan, L., Rayat, G. R., Xiao, J., Jiang, H., Li, X. and Chen, K. 2021. Anti-diabetic effects of natural antioxidants from fruits. *Trends in Food Science and Technology* 117, 3–14. DOI: 10.1016/j.tifs.2020.07.024.

Susmitha, A., Sasikumar, K., Rajan, D., Padmakumar, A. and Nampoothiri, K. M. 2021. Development and characterization of corn starch-gelatin based edible films incorporated with mango and pineapple for active packaging. *Food Bioscience* 41, 100977. DOI: 10.1016/j.fbio.2021.100977.

Tayengwa, T., Chikwanha, O. C., Gouws, P., Dugan, M. E. R., Mutsvangwa, T. and Mapiye, C. 2020. Dietary citrus pulp and grape pomace as potential natural preservatives for extending beef shelf life. *Meat Science* 162, 108029. DOI: 10.1016/j.meatsci.2019.108029.

Tereucan, G., Ercoli, S., Cornejo, P., Winterhalter, P., Contreras, B. and Ruiz, A. 2021. Stability of antioxidant compounds and activities of a natural dye from coloured-flesh potatoes in dairy foods. *LWT* 144, 111252. DOI: 10.1016/j.lwt.2021.111252.

Tian, R., Yang, Z., Lu, N. and Peng, Y. Y. 2019. Quercetin, but not rutin, attenuated hydrogen peroxide-induced cell damage via heme oxygenase-1 induction in endothelial cells. *Archives of Biochemistry and Biophysics* 676, 108157. DOI: 10.1016/j.abb.2019.108157.

Tobore, T. O. 2020. Towards a comprehensive theory of obesity and a healthy diet: the causal role of oxidative stress in food addiction and obesity. *Behavioural Brain Research* 384, 112560.

Toti, E., Chen, C. O., Palmery, M., Valencia, D. and Peluso, I. 2018. Non-provitamin A and provitamin A carotenoids as immunomodulators: recommended dietary allowance, therapeutic index, or personalized nutrition? (Special Issue: Antioxidant, Anti-Inflammatory, and Microbial-Modulating Activities of Nutraceuticals and Functional Foods 2018). *Oxidative Medicine and Cellular Longevity* 2018, 4637861. DOI: 10.1155/2018/4637861.

Traber, M. G., Buettner, G. R. and Bruno, R. S. 2019. The relationship between vitamin C status, the gut-liver axis, and metabolic syndrome. *Redox Biology* 21, 101091.

Tran, T. T. T., Ton, N. M. N., Nguyen, T. T., Le, V. V. M., Sajeev, D., Schilling, M. W. and Dinh, T. T. N. 2020. Application of natural antioxidant extract from guava leaves (*Psidium guajava* L.) in fresh pork sausage. *Meat Science* 165, 108106. DOI: 10.1016/j.meatsci.2020.108106.

van der Schaft, N., Schoufour, J. D., Nano, J., Kiefte-de Jong, J. C., Muka, T., Sijbrands, E. J. G., Ikram, M. A., Franco, O. H. and Voortman, T. 2019. Dietary antioxidant capacity and risk of type 2 diabetes mellitus, prediabetes and insulin resistance: the Rotterdam Study. *European Journal of Epidemiology* 34(9), 853–861.

Wang, X., Kristo, E. and LaPointe, G. 2020. Adding apple pomace as a functional ingredient in stirred-type yogurt and yogurt drinks. *Food Hydrocolloids* 100, 105453. DOI: 10.1016/j.foodhyd.2019.105453.

Wcislo, G. and Szarlej-Wcsilo, K. 2014. Colorectal cancer prevention by wheat consumption. *Wheat and Rice in Disease Prevention and Health*, 91–111. DOI: 10.1016/B978-0-12-401716-0.00008-8.

Wessels, I., Fischer, H. J. and Rink, L. 2021. Dietary and physiological effects of zinc on the immune system. *Annual Review of Nutrition* 41(1), 133–175.

Widomska, J., Welc, R. and Gruszecki, W. I. 2019. The effect of carotenoids on the concentration of singlet oxygen in lipid membranes. *Biochimica et Biophysica Acta. Biomembranes* 1861(4), 845–851. DOI: 10.1016/j.bbamem.2019.01.012.

Xiao, S. and Li, J. 2020. Study on functional components of functional food based on food vitamins. *Journal of Physics: Conference Series* 1549(3), 032002. DOI: 10.1088/1742-6596/1549/3/032002.

Xiong, Y., Chen, M., Warner, R. D. and Fang, Z. 2020. Incorporating nisin and grape seed extract in chitosan-gelatine edible coating and its effect on cold storage of fresh pork. *Food Control* 110, 107018. DOI: 10.1016/j.foodcont.2019.107018.

Xu, W., Qu, W., Huang, K., Guo, F., Yang, J., Zhao, H. and Luo, Y. 2007. Antibacterial effect of Grapefruit Seed Extract on food-borne pathogens and its application in the preservation of minimally processed vegetables. *Postharvest Biology and Technology* 45(1), 126–133. DOI: 10.1016/j.postharvbio.2006.11.019.

Yamaguchi, M. 2012. Role of carotenoid β-cryptoxanthin in bone homeostasis. *Journal of Biomedical Science* 19(1), 36. DOI: 10.1186/1423-0127-19-36.

Yan, Z., Zhong, Y., Duan, Y., Chen, Q. and Li, F. 2020. Antioxidant mechanism of tea polyphenols and its impact on health benefits. *Animal Nutrition* 6(2), 115–123.

Yang, G., Chang, C. C., Yang, Y., Yuan, L., Xu, L., Ho, C. T. and Li, S. 2018a. Resveratrol alleviates rheumatoid arthritis via reducing ROS and inflammation, inhibiting MAPK signaling pathways, and suppressing angiogenesis. *Journal of Agricultural and Food Chemistry* 66(49), 12953–12960.

Yang, X., Nisar, T., Hou, Y., Gou, X., Sun, L. and Guo, Y. 2018b. Pomegranate peel pectin can be used as an effective emulsifier. *Food Hydrocolloids* 85, 30–38. DOI: 10.1016/j.foodhyd.2018.06.042.

Yao, Y. T., Wang, W. Y., Liu, H. M., Hou, L. X. and Wang, X. D. 2021. Emulsifying properties of Chinese quince seed gum in oil-in-water emulsions. *LWT* 147, 111560. DOI: 10.1016/j.lwt.2021.111560.

Yoon, J. H. and Baek, S. J. 2005. Molecular targets of dietary polyphenols with anti-inflammatory properties. *Yonsei Medical Journal* 46(5), 585–596. DOI: 10.3349/ymj.2005.46.5.585.

Young, A. J. and Lowe, G. L. 2018. Carotenoids - antioxidant. *Antioxidants* 7(2). DOI: 10.3390/antiox7020028.

Yu, C. Y., Yang, C. Y. and Rui, Z. L. 2019. MicroRNA-125b-5p improves pancreatic β-cell function through inhibiting JNK signaling pathway by targeting DACT1 in mice with type 2 diabetes mellitus. *Life Sciences* 224, 67–75.

Yuzbasiyan-Gurkan, V., Grider, A., Nostrant, T., Cousins, R. J. and Brewer, G. J. 1992. Treatment of Wilson's disease with zinc: X. Intestinal metallothionein induction. *The Journal of Laboratory and Clinical Medicine* 120(3), 380-386. PMID: 1517684.

Zhang, P., Li, T., Wu, X., Nice, E. C., Huang, C. and Zhang, Y. 2020a. Oxidative stress and diabetes: antioxidative strategies. *Frontiers of Medicine* 14(5), 583-600. DOI: 10.1007/s11684-019-0729-1.

Zhang, W., Fan, X., Gu, X., Gong, S., Wu, J., Wang, Z., Wang, Q. and Wang, S. 2020b. Emulsifying properties of pectic polysaccharides obtained by sequential extraction from black tomato pomace. *Food Hydrocolloids* 100, 105454. DOI: 10.1016/j.foodhyd.2019.105454.

Zhu, Y., Ren, X., Bao, Y., Li, S., Peng, Z., Zhang, Y. and Zhou, G. 2020. Emulsification of oil-in-water emulsions with eggplant (Solanum melongena L.). *Journal of Colloid and Interface Science* 563, 17–26. DOI: 10.1016/j.jcis.2019.12.055.

Chapter 2

Advances in understanding phenolic compounds in fruits and vegetables

Cristine Vanz Borges, São Paulo State University (UNESP), Brazil; Fabio Vianello, University of Padua (UNIPD), Italy; Ricardo Alfredo Kluge, University of São Paulo (USP), Brazil; and Giuseppina Pace Pereira Lima, São Paulo State University (UNESP), Brazil

1 Introduction

The demand for health-promoting 'functional' food compounds has been increasing both from consumers in search of improved health and from industry for phytochemicals as food additives and food supplements (Galanakis, 2012). Many companies now include information on compounds with antioxidant potential in food labelling as an indicator of the health-promoting properties of their food products. The addition of bioactive compounds can even make processed food richer in antioxidants than some fresh, unprocessed foods.

Nutraceuticals are compounds found naturally in foods that demonstrate preventive and therapeutic effects against chronic diseases in particular (Brito et al., 2021). In 2021, the nutraceutical market was worth US$382.51 billion and is expected to reach US$722.49 billion by 2026, with an estimated growth of 8.3% over the next 5 years (https://www.marketdataforecast.com/market-reports/global-nutraceuticals-market). A range of nutraceutical supplements is now available in the form of pills, capsules or liquids, some including phenolic compounds (Shahidi, 2009). These compounds have been widely used in the

http://dx.doi.org/10.19103/AS.2022.0101.01

Table 1 Phenolic compounds content in fruit, vegetables and by-products

Food	Range of phenolic compounds			References
	Total phenols (mg/100 g)	Phenolic acids (mg/100 g)	Flavonoids (mg/100 g)	
		Fruits		
Eureka lemons' pulps	-	4-hydroxy-3-methoxycinnamic acid, 13.62–54.45	-	Dong et al. (2019)
Physalis angulata	-	Ferulic acid, 17.0 (green fruit) to 13.0 (ripe fruit)	-	Oliveira et al. (2020)
Fruits of the Indian Himalayan Region (fw)	0.92–76.67	Gallic acid, 5.88–281.98 Chlorogenic acid, 11.24–196.64 Caffeic acid, 4.13–137.28 Coumaric acid, 0.38–57.75	Catechin, 6.36–76.54	Bhatt et al. (2017)
Banana fruis pulp (dw)	74.6–506.12	Gallic acid, 67.8–254.4 Hydroxybenzoic acid, 10.7–72.7	Epigallocatechin, 11.2–170.3 Catechin, 24.0–641.6 Quercetin, 6.9–153.0	Borges et al. (2020)
Banana fruits pulp, Ney Poovan During ripening stage	-	Phenolic acids (mean) 42.3 (stage 2), 170.2 (stage 5) and 126.9 (stage 7)	Flavonoids (mean) 130.3 (stage 2), 149.0 (stage 5) and 142.6 (stage 7)	Borges et al. (2020)
Banana fruits pulp, Tiparot During ripening stage		Phenolic acids (mean) 190.3 (stage 2), 153.7 (stage 5) and 111.1 (stage 7)	Flavonoids (mean) 240.7 (stage 2), 314.8 (stage 5) and 89.9 (stage 7)	
Persimmon (*Diospyros kaki* Thunb.) (fw)	-	Ferulic acid, 10.29 Gallic acid, 21.9 Protocatechuic acid, 6.5 Vanillic acid, 0.71 p-Coumaric acids, 58.2	-	Park et al. (2008)
Yerba mate (Ilex paraguariensis) (fw)	-	3-Caffeoylquinic acid, 2700 4-Caffeoylquinic acid, 680.6 5-Caffeoylquinic acid, 1207.0 Total chlorogenic acids, 9188.6	-	Meinhart et al. (2017)

Eggplant (fw)		Chlorogenic acids, 10.0–190.0 (fresh) Chlorogenic acids, 216.0 (grilled)		Šilarová et al. (2019)
Mango Mature green	19.91–44.47	Benzoic acid, 7.83–11.25 p-Coumaric acid, 3.8–5.2 Caffeic acid, 0.1–0.6 Ferulic acid, 1–1.8 Syringic acid, 0.9–2.2	Vanillin, 12.9–18.8 Apigenin, 0.8–2.4 Mangiferin,0.8–9.7	Gentile et al. (2019)
Mature ripe	17.99–49.13	Benzoic acid, 7.3–14 p-Coumaric acid, 3.8–4.9 Caffeic acid, 0.3–0.6 Ferulic acid, 1.1–1.9 Syringic acid, 0.9–1.8	Vanillin, 13.6–18.8 Apigenin, 0.7–1.6 Mangiferin, 0.8–12.8	
Mango fruit (*Mangifera indica* L., cv. Ataulfo) During ripening stage	<150 (RS4) to 174 (RS2,3)	-	-	Palafox-Carlos et al. (2012)
Carrot, unpeeled (storage for 0 and 48 h at 15°C before blanching), dw	301.9-513.4	Chlorogenic acid, 56.2-124.2	-	Santana-Gálvez et al. (2019)
Cenoura, peeled (storage for 48 h at 15°C before blanching), dw	160.8–320.7	Chlorogenic acid, 73-661	-	
Banana cooking (Pelipita)	-	Gallic acid, 75.5 (raw) to 160.6 (microwave with peel) Hydroxybenzoic acid, 43.9 (raw) to 97.8 (microwave with peel)	Epigallocatechin, 27.8 (raw) to 55.9 Catechin, 79.8 (raw) to 83.5 (microwave with peel) Quercetin, 38.9 (raw) to 60.0 (microwave with peel)	Borges et al. (2020)

(*Continued*)

Table 1 (*Continued*)

	Range of phenolic compounds			
Food	Total phenols (mg/100 g)	Phenolic acids (mg/100 g)	Flavonoids (mg/100 g)	References
		By-products and wastes (flours, seed, peel)		
Eureka lemons peels	-	4-Hydroxy-3-methoxycinnamic acid, 69.28–148.73	-	Dong et al. (2019)
Cantaloupe melon seeds	-	4-Hydroxy-3-methoxycinnamic acid, 151	-	Vella et al. (2019)
Banana fruis peels (dw)	374.2–703.0	-	-	Borges et al. (2020)
Mango peel During ripening stage (0, 3 and 6 days)	-	Gallic acid, 0.01 (0d) to 0.42 (3d)	Mangiferin, 4.3 (6 days) to 5526.1 (0 day) Catechin, 0.4 (6 days) to 252.8 (0 day) Quercetin-3-D-galactosid, 59.2 (6 days) to 239.9 (0 day) Quercetin-3-glucoside, 10.0 (6 days) to 175.9 (0 day) Epicathechin, 0.04 (0 day) to 0.33 (6 days)	Monribot-Villanueva et al. (2019)
Residual grape pomaces flour (dw)*	2.3–4.9	3-Hydroxytyrosol, 0.06–083 p-Coumaric,0.13–0.4 Caffeic, 0.13–1.23 trans-Ferulic, 0.08–0.36	Rutin, 0.37–1.2 Kampferol, 0.01–0.29 Catechin, 3.2–4.8 Cyanidin 3,5-diglucoside, 2.8–10.6	Monteiro et al. (2021)
Eggplant flour	123.18 (control, wheat flour) to 322.40 (eggplant flour)	-	-	Uthumporn et al. (2015)

* fw, fresh weight; dw, dry weight; nd, not identified; grape pomaces flour [seeds (39%), skins (5%) and pulp (56%)].

pharmaceutical and health food industries as well as by the food industry, for example, as natural food preservatives, dyes and additives (Kumar and Goel, 2019). In addition to their potential nutraceutical benefits, phenolics are used as functional additives to enhance food quality (Galanakis, 2013), for example, as microbial growth retardants and to inhibit lipid oxidation so as to extend the shelf life of some processed foods (Galanakis, 2018). An area of growing interest is cosmetics. As an example, grapes and their derivatives, including grape-skin flour (Monteiro et al., 2021), have been identified as a source of bioactive compounds with photoprotective potential against UV radiation and potential use as a herbal cosmetic (Hu et al., 2017).

A number of studies have focused on obtaining bioactive compounds from fruit and vegetable by-products (Galanakis, 2012). These include materials that are usually discarded during domestic or industrial processing, for example, peel, seeds and stems, as well as products falling below commercial standards, either because they are of the incorrect size/shape or have cosmetic defects. These discarded parts/products of plants may be important sources of bioactive compounds such as biogenic amines (Lima et al., 2008; Monteiro et al., 2021), carotenoids and phenolic compounds (Monaco et al., 2016). By-products from processing fruits such as grapes, for example, can be a valuable source of phytochemicals (Soto et al., 2015). Grape pomace (the pulpy residue left after the juice has been extracted) can, for example, be turned into flour containing varying levels of phenolic compounds depending on the genotype used. Monteiro et al. (2021) found catechin (flavan-3-ol) and cyanidin 3,5-diglucoside (anthocyanin) as the main compounds in grape pomace flour (Table 1). However, even though food wastes are a source of valuable components, there is still a need for technologies that can separate target compounds effectively and turn them into stable and functional additives (Galanakis, 2012).

2 Characteristics of phenolic compounds: flavonoids and non-flavonoids

Fruits and vegetables are sources of bioactive compounds such as provitamin A (carotenoids), bioactive amines (melatonin and serotonin, among others) and vitamin C. The antioxidant activity of foods depends particularly on the type and content of phenolic compounds found in plant cells (Demiray et al., 2009; Anyasi et al., 2018). Regular consumption of phenolic compounds has been associated with a decrease in cardiovascular diseases, some types of cancer, metabolic syndromes and diabetes, among other chronic diseases (Yeon et al., 2015; Čanadanović-Brunet et al., 2017; Caleja et al., 2017).

Phenolic compounds show a diversity of structures, from simple molecules to polyphenols. Phenolic compounds consist of one or more hydroxyl groups bonded directly to an aromatic hydrocarbon ring (Swallah et al., 2020). Some phenolic compounds have simple molecular structures, such as gallic acid, or they can be complex molecules (polyphenols), such as flavonoids, stilbenes, anthocyanins or condensed tannins of high molecular weight (Balasundram et al., 2006).

The antioxidant properties of phenolic compounds are attributed to the presence of hydroxyl (OH) groups, which have the ability to bind to free radicals present in the body, preventing damage to, or oxidation of, cellular components. Free radical scavenging properties are generally influenced by chemical structure, the position and number of the OH group, glycosylation or other forms of substitution (Cai et al., 2006). The greater the number of dissociable OH groups in the structure of the polyphenolic compound, the greater its activity as a H^+ donor agent (Riihinen et al., 2008). This mechanism of action reduces lipid oxidation in tissues, reducing the risk of developing pathologies such as atherosclerosis and cancer (Ramarathnam et al., 1995). Phenolic compounds protect cells against the harmful effects of reactive oxygen species (ROS) (Haminiuk et al., 2012) due to their high redox potential and their ability to chelate metals (Ignat et al., 2011).

Phenolic compounds can be divided into two main subgroups:

1 flavonoids; and
2 non-flavonoids.

These are discussed in more detail below.

2.1 Flavonoids

Flavonoids can be divided into several classes according to the degree of oxidation of the oxygen heterocycle:

- flavones;
- flavonols;
- isoflavones;
- anthocyanins;
- proanthocyanidins; and
- flavanones.

As an example, quercetin, the main flavonol in our diet, is present in many fruits and vegetables and is associated with the induction of apoptosis in cancer cells (Niedzwiecki et al., 2016). It has also been found to have antioxidant and anti-inflammatory properties in preventing diabetes and some neurodegenerative diseases (Zizkova et al., 2017).

A sub-class of flavonoids that has attracted attention as nutraceuticals is prenylated flavonoids. These molecules contain a flavonoid skeleton with a lipophilic prenyl side chain (prenyl, geranyl) (Yang et al., 2015). They have been found to have antioxidant, antibacterial, estrogenic, anti-inflammatory and anti-allergic properties (Gomez-Gomez et al., 2018). The largest sources of prenylated flavonoids are leaves, roots and seeds of species of the Moraceae, Leguminosae and Asteraceae families (Yang et al., 2015). Among prenylated flavonoids, 8-prenylnaringenin stands out as a nutraceutical compound found in hops and beer (Yang et al., 2015).

Some flavonoids, such as flavan-3-ols and their derivatives, have been found to have neuroprotective properties (Grassi et al., 2015). These flavonoids include catechin, epicatechin, epigallocatechin and epigallocatechin gallate (Makkar et al., 2020). Catechin, epicatechin and proanthocyanidins are found in white and red grapes, mainly in skin and seeds, and in products such as juices and wines. In grapes (*Vitis vinífera*) grown in the Tuscan region (Italy), levels of catechin in seeds range from 60.3 (ISV RC1) to 205.7 mg/100 g (Montepulciano), while in skin the content of resveratrol varies between 0.7 (AP SG 1) and 25.5 mg/100 g in the 'Cabernet Sauvignon' grape (Iacopini et al., 2008).

Resveratrol is a stilbene able to scavenge free radicals (ROS and reactive nitrogen species). In the Brazilian grape variety *Vitis labrusca*, levels of trans-resveratrol were found to vary, depending on altitude as well as on annual cultivation cycles (Gomes et al., 2021). de Oliveira et al. (2019) detected values from 4.11 to 8.17 mg/kg in the Syrah grape cultivated in Brazilian tropical regions (Bahia and Pernambuco). Gomez-Gomez et al. (2018) found levels of 0.5 mg/L of *t*-resveratrol in juice and wine produced from *V. labrusca* in Brazil.

2.2 Non-flavonoids

Along with flavonoids, phenolic compounds include simple phenols, phenolic acids, coumarins, stilbenes, hydrolyzable and condensed tannins, lignans and lignins. These abundant secondary metabolites are produced mainly from L-phenylalanine and L-tyrosine via a shikimate pathway (Kumar and Goel, 2019) (see Fig. 1).

Phenolic acids are divided into two main subgroups:

1 hydroxybenzoic acid; and
2 hydroxycinnamic acid.

These groups are distinguished by distinct carbon structures as well as the positioning and number of OH groups in the aromatic ring. Hydroxybenzoic acids are derived from benzoic acid with seven carbon atoms in the C6-C1 structure. Gallic, p-hydroxybenzoic, salicylic, ellagic, gentisic, protocatechuic, synergic and vanillic acids are the main hydroxybenzoic acids and differ from each other in the modification of the aromatic ring (Rashmi and Negi, 2020).

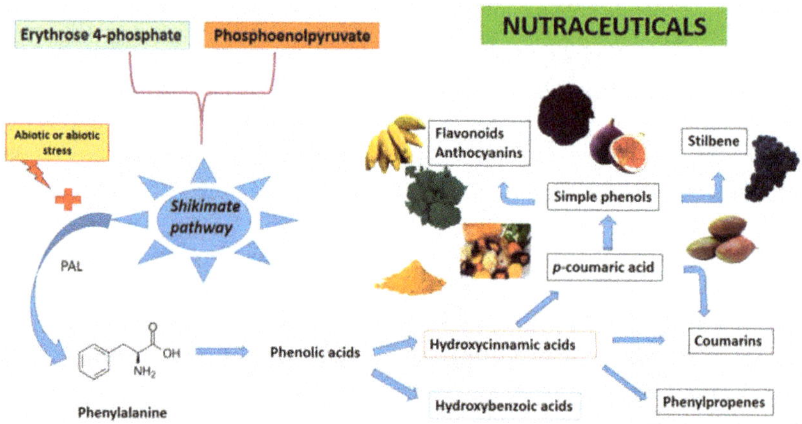

Figure 1 The biosynthetic pathway of the phenolic compounds in plant-based foods.

Hydroxycinnamic acids are more abundant than hydroxybenzoic acids in nature and occur generally in various conjugated forms. Hydroxycinnamic acids include caffeic, p-coumaric, ferulic and sinapic acids. Vegetables tend to have lower levels of hydroxycinnamic acids compared to fruits. Hydroxybenzoic acids may be found in free form or in conjugates. Caffeic acid is commonly found in fruits and vegetables. Caffeic and quinic acids may combine to form chlorogenic acid which is present in many fruits and some vegetables (Clifford et al., 2020).

Ferulic acid is associated with dietary fibre and may be linked by ester bonds to hemicelluloses. One of the main dietary sources of ferulic acid is wheat bran (5 mg/g) (Kroon et al., 1997). However, ferulic acid (4-hydroxy-3-methoxycinnamic acid) can also be found in a variety of fruits and vegetables, such as in Cantaloupe melon seeds (Vella et al., 2019a) and in Eureka lemon pulp and peel (Table 1) (Dong et al., 2019). Like other phenolic compounds, levels of ferulic acid are affected by the degree of ripening in fruit, making the timing of harvesting a key factor in optimizing content (Table 1) (de Oliveira et al., 2020; Gomes et al. (2021). Ferulic acid has been described as being protective against ROS, thereby preventing disorders such as Alzheimer's disease, diabetes (Chowdhury et al., 2019) and atherosclerosis (Gu et al., 2021), making this a safe, effective and easily produced nutraceutical compound (Zhao and Moghadasian, 2008).

Many fruits are rich in gallic acid, for example, Himalayan wild edible fruit (Bhatt et al., 2017) (Table 1). In the ten fruits they analysed, Bhatt et al. (2017) found that gallic acid was the main phenolic compound present, with levels ranging from 5.88 (*Ficus palmata*) to 281.98 mg/100 g (*Terminalia chebula*). In persimmon, for example, Park et al. (2008) found high levels of ferulic acid and gallic acid which were associated with a lower risk of atherosclerosis in animal studies. Persimmon also contains high levels of p-coumaric acid (Table 1) which is known to reduce the risk of stomach cancer by reducing the formation of carcinogenic nitrosamines (Bhatt et al., 2017). It is important to note that, as with other phenolic compounds, both ferulic acid and gallic acid may induce toxic effects in *in vitro* cell models at high concentrations (Truzzi et al., 2020).

Chlorogenic acid is considered a potent antioxidant with high bioavailability, anti-inflammatory, antidiabetic and antihypertensive properties as well as assisting in weight loss (Thom, 2007; Ong et al., 2013). A study of 100 species by Meinhart et al. (2017) found that yerba mate had the highest levels of chlorogenic acid (Table 1), making it a good potential source of phenolic compounds with antioxidant and nutraceutical potential. Chlorogenic acid occurs at higher levels in the Solanaceae family of plants such as potatoes, tomatoes and eggplant. Šilarová et al. (2019) demonstrated that the levels of chlorogenic acid in some eggplant cultivars increased on thermal processing compared to the raw samples (Table 1). Uthumporn et al. (2015) demonstrated that the use of 10–15% of eggplant flour in the formulation of cookies increased phenolic content by up to 2.6-fold as well as increasing shelf-life (Table 1).

3 Effects of cultivation and post-harvest operations on phenolic compounds

The phenolic content of fruits and vegetables depends on factors such as stage of ripening and on storage conditions. Palafox-Carlos et al. (2012) found that the contents of phenolic compounds in 'Ataulfo' mango pulp vary significantly during the ripening process, with greater antioxidant activity in the pulp during the second and third stages of ripening (Table 1). Similar results were found for 'Ubá' mango pulp, where the accumulation of phenolic compounds was higher in the third ripening stage (Oliveira et al., 2016). However, some authors have reported that the content of phenolic compounds may decrease during post-harvest ripening of mangoes (Table 1). According to Gentile et al. (2019), this effect may be related to metabolic processes including hydrolysis of tannins and the increased activity of polyphenol oxidase, typical of ripe fruit. Mango peel, which is usually discarded, is a natural source of phenolic compounds (such as flavonol *O*- and xanthone *C*-glycosides) which persist during post-harvest storage (Berardini et al., 2005).

The functionality and stability of phytochemicals in the human body depend not only on the quantity but also on the binding and/or on the interaction of these compounds with other molecules, on their location in the food matrix and on the presence of other bioactive compounds in fruits and vegetables. Phytochemicals may be bound to cell membranes or be in free form. Food processing operations such as heating or freezing can disrupt the cell membrane leading to the release of these membrane-bound compounds and increased bioavailability.

The amount of phytochemicals retained in fruits and vegetables after processing also depends on the stability of these compounds during post-harvest processing. This is mainly dependent on the sensitivity of the compounds to oxidation and/or isomerization (Leong and Oey, 2012). While many fruits are eaten when fresh, most vegetables are usually consumed after thermal processing to improve palatability, flavour and texture. Thermal processes include boiling, frying, cooking and baking (using a traditional oven or microwave) (Palermo et al., 2014). The advantages of cooking are inactivation of microorganisms, promotion of the bioavailability of some nutrients and reduction of anti-nutritional factors. However, there may be losses of some bioactive compounds such as polyphenols depending on the type of cooking (Mazzeo et al., 2011) in addition to the formation of undesirable compounds (e.g. acrylamide) (van Boekel et al., 2010).

Recent studies demonstrate that thermal processing using high temperatures can in some cases increase the nutritional value by releasing bioactive compounds, resulting in an increase in antioxidant activity (Borges et al., 2020). High temperatures disrupt cell membranes leading to the release of phytochemicals and an increase in bioavailability (Lemmens et al., 2009; Borges et al., 2020). In bananas (plantains and cooking bananas), for example, boiling increases the content of secondary metabolites (Table 1) (Borges et al., 2020).

However, changes in polyphenol content after thermal processing are complex since these compounds have different structures and solubilities influencing bioavailability. In carrots, for example, boiling results in significant losses of total phenolic compounds (49%), mainly of phenolic acids, except for p-coumaric acid (Mazzeo et al., 2011). Decreases in phenolic compounds may be due to isomerization processes caused by high temperature, as described in green tea by Wang and Ho (2009). These authors found a decrease in epicatechin due to isomerization (alteration of carbon 2) and an increase in catechin after thermal processing.

4 Phenolic compounds and COVID-19

In December 2019, the world became aware of a new coronavirus responsible for acute respiratory syndrome coronavirus 2 (SARS-CoV-2), leading to a global pandemic being declared by the World Health Organization, resulting in over 160 million cases and over 3 million deaths by May 2021 (Cucinotta and Vanelli, 2020; WHO, 2021). While vaccines are considered the main measure to combat COVID-19, a potential secondary approach is to promote intake of phytochemicals such as polyphenols which have been shown to increase general immunity to viral infections, particularly as they have fewer side effects than many synthetic drugs (Fig. 2) (Chojnacka et al., 2020). There is evidence of the ability of some polyphenols to boost immune responses due to the antiviral, antimicrobial, anti-inflammatory and cytotoxic effects of flavonoids (González-Gallego et al., 2010). According to Dhar and Bhattacharjee (2021), curcumin, a naturally occurring polyphenol in *Curcuma longa*, may help to prevent COVID-19 infections by blocking autophagosomes–lysosome fusion by stabilizing pathogen-containing vacuoles and restricting genome replication.

Resveratrol, a triphenolic stilbene occurring in grapes and their by-products (Gomez-Gomez et al., 2019), or pterostilbene (resveratrol analogue), can be used as an adjuvant therapy when combined with zinc in preventing progression to moderate-severe disease due to COVID-19 (Kelleni, 2021). Other flavonoids such as quercetin and kaempferol have been tested by research groups around the world in the treatment of SARS-CoV-2 by inhibiting the activity of the SARS 3-chymotrypsin protease (3CLpro), a vital enzyme related to the replication of COVID-19 (Yang et al., 2020). Brito et al. (2021) have shown that quercetin inhibits the entry of SARS-CoV into host cells, while Colunga Biancatelli et al. (2020) report evidence that co-administration of vitamin C and quercetin exerts a synergistic antiviral action.

Other polyphenols have been described as possible adjuvants in the treatment of COVID-19, such as luteolin and catechins [(-)-epigallocatechin-3-gallate (EGCG), (-)-epicatechin-3-gallate, (-)-epigallocatechin and (-)-epicatechin] (Levy et al., 2020; Yang et al., 2020; Ghosh et al., 2020). These alternatives may perhaps be safer as adjuvants in the treatment of COVID-19 compared to some recommended drugs such as chloroquine and

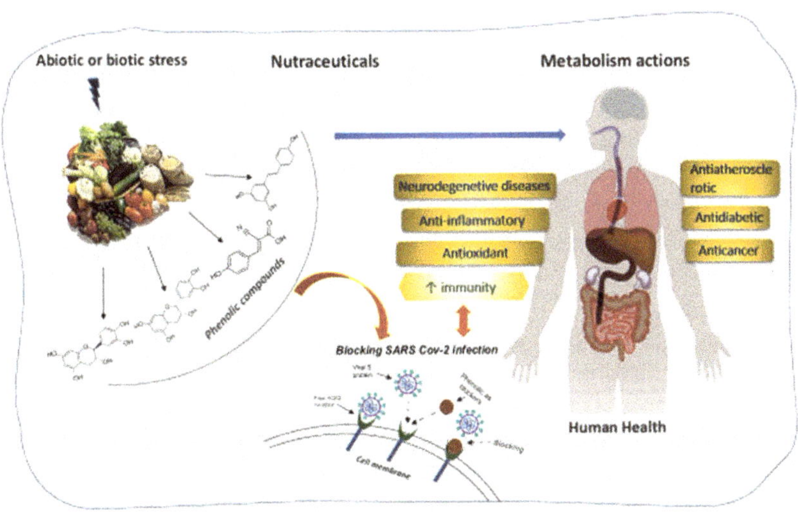

Figure 2 Effect of phenolic compounds in human health.

hydroxychloroquine which have been shown to have significant side effects (Caponi et al., 2021; Melo et al., 2021).

5 Improving phenolic compounds in fruits and vegetables

The availability of foods rich in bioactive compounds is essential in developing countries since many families lack the financial resources to ensure a sufficient and nutritionally balanced diet. Even in more affluent countries, dietary choices are often poor, increasing rates of obesity and vulnerability to chronic diseases such as diabetes, cardiovascular disease and cancer. This makes the supply of antioxidant-rich foods an important public health issue.

Several studies, including those from our laboratory, have both been investigating plant genotypes with a high content of bioactive compounds, as well as the best ways to process these plant-based foods in a way that optimizes the bioavailability of these compounds. One potential approach is the controlled application of abiotic stress during post-harvest processing. These stresses may be caused by various post-harvest processes used to prepare food such as mechanical processing (e.g. cutting and peeling) or the use of temperature, modified atmospheres or ultraviolet light for decontamination and to extend shelf life (Santana-Gálvez et al., 2016; Santana-Gálvez et al., 2019). It has been found, for example, that mechanical processing (e.g. cutting and shredding) can increase phenolic compounds in carrots by activating the phenylpropanoid pathway (Jacobo-Velázquez et al., 2015) The authors observed that the phenolic content increases with the intensity of mechanical processing (whole carrots < slices < shreds). This increased bioactivity after injury has been used to produce a carrot powder rich in nutraceuticals which can later be incorporated into products such as corn tortillas (Santana-Gálvez et al., 2016) and carrot sausages (Alvarado-Ramírez et al., 2018) increasing the nutritional and nutraceutical quality of these foods (Santana-Gálvez et al., 2019). An important factor is which parts of a plant have the greatest concentration of phenolic compounds. Many fruits and vegetables, for example, contain high levels of phenolic compounds in the peel. It has been found, for example, that juices prepared with whole carrots (i.e. with intact skins) had higher levels of chlorogenic acid (174%), compared to those made with peeled carrots. Another important finding was that stored and bleached carrot juices showed 662% higher chlorogenic acid content (Table 1) (Santana-Gálvez et al., 2019).

Another focus of recent research has been bananas and plantains. In a study of phenolic acids present in 22 genotypes of dessert and cooking bananas and plantains, Borges et al. (2020) found the main compound to be gallic acid with catechin being the main flavonoid present (Table 1). Catechin and quercetin contributed most to the antioxidant activity of *Musa* spp. germplasm (Ferric

Reducing Antioxidant Power: r = 0.92 and 0.85; 1,1-diphenyl-2-picryl-hydrazil: r = 0.85 and 0.80; 2,2′-azinobis (3-ethylbenzothiazoline 6-sulfonate: r = 0.78 and 0.77, $P < 0.05$, respectively). The dessert banana 'Ney Poovan' and the cooking bananas 'Tiparot' and 'Pelipita' were found to have particularly high concentrations of phenolic compounds (Borges et al., 2020). *Musa* spp. fruits with light-coloured pulp were also found to have higher levels of flavonoids in comparison with other antioxidant compounds: carotenoids and pro-vitamins (Borges et al., 2019b). In addition, the peel was found to contain several phenolic antioxidants in higher amounts than those found in the pulp (Table 1) (Borges et al., 2019a/b, 2020).

These results were maintained even after thermal processing. The highest concentrations of phenolic compounds were found in ripe fruits (stage 5 – yellow with green) (Table 1). Differences in fruit firmness in *Musa* spp., even within the same species, may also influence the content of phenolic compounds, particularly after cooking (Borges et al., 2020). For example, cooking the less firm 'Pelipita' (cooking banana) increased the availability of phenolic acids and flavonoids (epigallocatechin and quercetin) compared to the firmer 'D'Angola' (plantain) (Borges et al., 2019b).

Cultivation practices can have a significant effect on phenolic content. Gomes et al. (2021), for example, found that the use of 1 and 2 mmol L^{-1} salicylic acid (SA), a low-cost plant growth regulator, in the cultivation of 'Niagara Rosada' table grape promoted an increase in the level of phenolic compounds, especially chlorogenic and gallic acids, rutin, cyanidin-3,5-diglucoside and 3-O-glycoside delphinidin. The application of exogenous SA during pre-harvest also promoted an increase in shelf-life, decreasing berry drop and decay. Simões et al. (2020) also found, for example, that delaying harvesting of orange-fleshed sweet potatoes by 30–60 days improves visual quality, phenolic compound content and antioxidant activity. On the other hand, cream-fleshed sweet potatoes showed a decrease in total phenolic compounds if harvest was delayed. Camu-camu fruit [*Myrciaria dubia* H. B. K. (McVough)] harvested from 88 to –116 days after anthesis (DAA) showed a linear decline in total phenolic content over a 6–10-day storage period (Neves et al., 2017). The authors recommended harvesting at 88 DAA and refrigeration for up to 5 days to preserve antioxidant activity.

Post-harvest storage conditions and length also affect levels of antioxidant compounds in fruits and vegetables. Most of the phenolic compounds in mango fruit studied by Monribot-Villanueva et al. (2019) showed a reduction in content during storage, except for gallic acid and (–)epicatechin in 'Ataulfo' mango peel (Table 1). A study by Galani et al. (2017) of 19 fruits and vegetables found decreases in phenolic content during storage, though increases were observed in the phenolic content of table grapes during storage (Gomes et al., 2021).

6 Summary and future trends

Safety regulations governing the use of synthetic antioxidants make natural antioxidants from food sources such as fruits and vegetables an attractive alternative. Optimizing their use requires a number of steps. These include identification of sources of bioactive substances in particular types of fruit and vegetable as well as identification of genotypes and/or cultivars in germplasm banks with the highest levels of phenolic compounds for potential use in genetic improvement programs. It also requires identifying the stages during growth, cultivation or post-harvest processing at which concentrations may be at their highest or at which appropriate interventions can enhance phenolic content. Examples include the use of growth regulators during cultivation or controlled application of stress during post-harvest processing to optimize bioactive content and release.

This understanding will make it possible to design horticultural systems focused on optimizing the nutraceutical properties of selected fruit and vegetable cultivars and producing standardized raw material from fresh products or their by-products from which nutraceutical compounds can be extracted for use as food ingredients or in the development of nutraceutical supplements. In this view, the full exploitation of the entire production chain during food elaboration, such as the use of usually discarded by-products (e.g. peels), will involve rural producers and industries. However, further research is needed in areas such as the bioavailability of bioactive compounds, and more detailed dose-response analyses are needed to validate the large-scale use of nutraceuticals. Research has identified antiviral activity by phenolic compounds which suggests a more immediate potential role in dealing with the ongoing COVID pandemic as well as helping to protect the population from future viruses.

7 Where to look for further information

Soto-Hernández, M., García-Mateos, R. and Palma-Tenango, M. (2019). Plant Physiological Aspects of Phenolic Compounds. IntechOpen, London, UK, https://www.intechopen.com/books/7688.

The Food and Agriculture Organization of the United Nations (FAO) and World Health Organization (WHO) offers several guidelines related to food nutraceuticals and biofortification.

Biofortification of staple crops. WHO. https://www.who.int/elena/titles/biofortification/en/.

FAO. https://agris.fao.org/agris-search/search.do?recordID=US201700045261.

The HarvestPlus programme (biofortification) https://www.harvestplus.org https://www.harvestplus.org/biofortification-nutrition-revolution-now.

8 Acknowledgements

The authors gratefully acknowledge the financial support from National Council for Scientific and Technological Development, Brazil (CNPq) (grant number 307571/2019-0) and São Paulo Research Foundation (FAPESP) (grant number 2016/22665-2 and 2019/27227-1).

9 References

Alvarado-Ramírez, M., Santana-Gálvez, J., Santacruz, A., Carranza-Montealvo, L. D., Ortega-Hernández, E., Tirado-Escobosa, J., Cisneros-Zevallos, L. and Jacobo-Velázquez, D. A. (2018). Using a functional carrot powder ingredient to produce sausages with high levels of nutraceuticals, *Journal of Food Science* 83(9), 2351-2361. doi: 10.1111/1750-3841.14319.

Anyasi, T. A., Jideani, A. I. O. and Mchau, G. R. A. (2018). Phenolics and essential mineral profile of organic acid pretreated unripe banana flour, *Food Research International* 104, 100-109. doi: 10.1016/j.foodres.2017.09.063.

Balasundram, N., Sundram, K. and Samman, S. (2006). Phenolic compounds in plants and agri-industrial by-products: antioxidant activity, occurrence, and potential uses, *Food Chemistry* 99(1), 191-203. doi: 10.1016/j.foodchem.2005.07.042.

Berardini, N., Knödler, M., Schieber, A. and Carle, R. (2005). Utilization of mango peels as a source of pectin and Polyphenolics, *Innovative Food Science and Emerging Technologies* 6(4), 442-452.

Bhatt, I. D., Rawat, S., Badhani, A. and Rawal, R. S. (2017). Nutraceutical potential of selected wild edible fruits of the Indian Himalayan region, *Food Chemistry* 215, 84-91. doi: 10.1016/j.foodchem.2016.07.143.

Borges, C. V., Belin, M. A. F., Amorim, E. P., Minatel, I. O., Monteiro, G. C., Gomez Gomez, H. A., Monar, G. R. S. and Lima, G. P. P. (2019a). Bioactive amines changes during the ripening and thermal processes of bananas and plantains, *Food Chemistry* 298, 125020. doi: 10.1016/j.foodchem.2019.125020.

Borges, C. V., Minatel, I. O., Amorim, E. P., Belin, M. A. F., Gomez-Gomez, H. A., Correa, C. R. and Lima, G. P. P. (2019b). Ripening and cooking processes influence the carotenoid content in bananas and plantains (Musa spp.), *Food Research International* 124, 129-136. doi: 10.1016/j.foodres.2018.08.022.

Borges, C. V., Maraschin, M., Coelho, D. S., Leonel, M., Gomez, H. A. G., Belin, M. A. F., Diamante, M. S., Amorim, E. P., Gianeti, T., Castro, G. R. and Lima, G. P. P. (2020). Nutritional value and antioxidant compounds during the ripening and after domestic cooking of bananas and plantains, *Food Research International* 132, 109061. doi: 10.1016/j.foodres.2020.109061.

Brito, J. C. M., Lima, W. G. and Nizer, W. S. da C. (2021). Quercetin as a potential nutraceutic against coronavirus disease 2019 (COVID-19) TT - la Quercetina como un potencial nutracéutico contra la enfermedad por coronavirus 2019 (COVID-19), *Ars Pharmaceutica* 62(1), 89-95. doi: 10.30827/ars.v62i1.15684.

Cai, Y. Z., Sun, M., Xing, J., Luo, Q. and Corke, H. (2006). Structure-radical scavenging activity relationships of phenolic compounds from traditional Chinese medicinal plants, *Life Sciences* 78(25), 2872-2888. doi: 10.1016/j.lfs.2005.11.004.

Caleja, C., Ribeiro, A., Barreiro, M. F. and Ferreira, I. C. F. R. (2017). Phenolic compounds as nutraceuticals or functional food ingredients, *Current Pharmaceutical Design* 23(19), 2787–2806. doi: 10.2174/1381612822666161227153906.

Čanadanović-Brunet, J., et al. (2017). Phenolic profile, antiradical and antitumour evaluation of raspberries pomace extract from Serbia, *Iranian Journal of Pharmaceutical Research: I.J.P.R.* 16(Suppl), 142–152. doi: 10.22037/ijpr.2017.1999.

Caponi, S., Brzozowski, F. S., Hellmann, F. and Bittencourt, S. C. (2021). O uso político da cloroquina: COVID-19, negacionismo e neoliberalismo / The political use of chloroquine: COVID-19, denialism and neoliberalism, *Revista Brasileira de Sociologia - R.B.S.* 9(21), 78–102. doi: 10.20336/rbs.774.

Chojnacka, K., Witek-Krowiak, A., Skrzypczak, D., Mikula, K. and Młynarz, P. (2020). Phytochemicals containing biologically active polyphenols as an effective agent against Covid-19-inducing coronavirus, *Journal of Functional Foods* 73, 104146. doi: 10.1016/j.jff.2020.104146.

Chowdhury, S., Ghosh, S., Das, A. K. and Sil, P. C. (2019). Ferulic acid protects hyperglycemia-induced kidney damage by regulating oxidative insult, inflammation and autophagy, *Frontiers in Pharmacology* 10, 27. doi: 10.3389/fphar.2019.00027.

Clifford, M. N., Kerimi, A. and Williamson, G. (2020). Bioavailability and metabolism of chlorogenic acids (acyl-quinic acids) in humans, *Comprehensive Reviews in Food Science and Food Safety* 19(4), 1299–1352. doi: 10.1111/1541-4337.12518.

Colunga Biancatelli, R. M. L., Berrill, M., Catravas, J. D. and Marik, P. E. (2020). Quercetin and vitamin C: an experimental, synergistic therapy for the prevention and treatment of SARS-CoV-2 related disease (COVID-19), *Frontiers in Immunology* 11(June), 1451. doi: 10.3389/fimmu.2020.01451.

Cucinotta, D. and Vanelli, M. (2020). Of WHO declares COVID-19 a pandemic, *Acta Bio-Medica: Atenei Parmensis* 91(1), 157–160. doi: 10.23750/abm.v91i1.9397.

Demiray, S., Pintado, M. E. and Castro, P. M. L. (2009). Evaluation of phenolic profiles and antioxidant activities of Turkish medicinal plants: Tilia argentea, crataegi folium leaves and Polygonum bistorta roots 3(6), 74–79.

Dhar, S. and Bhattacharjee, P. (2021). Promising role of curcumin against viral diseases emphasizing COVID-19 management: a review on the mechanistic insights with reference to host-pathogen interaction and immunomodulation, *Journal of Functional Foods* 82, 104503. doi: 10.1016/j.jff.2021.104503.

de Oliveira, A. M., Malunga, L. N., Perussello, C. A., Beta, T. and Ribani, R. H. (2020). Phenolic acids from fruits of Physalis angulata L. in two stages of maturation, *South African Journal of Botany* 131, 448–453. doi: 10.1016/j.sajb.2020.02.029.

de Oliveira, J. B., Egipto, R., Laureano, O., de Castro, R., Pereira, G. E. and Ricardo-da-Silva, J. M. (2019). Climate effects on physicochemical composition of Syrah grapes at low and high altitude sites from tropical grown regions of Brazil, *Food Research International* 121, 870–879. doi: 10.1016/j.foodres.2019.01.011.

Dong, X., Hu, Y., Li, Y. and Zhou, Z. (2019). The maturity degree, phenolic compounds and antioxidant activity ofEureka lemon, *Scientia Horticulturae* 243, 281–289.

Galanakis, C. M. (2012). Recovery of high added-value components from food wastes: conventional, emerging technologies and commercialized applications, *Trends in Food Science and Technology* 26(2), 68–87. doi: 10.1016/j.tifs.2012.03.003.

Galanakis, C. M. (2013). Emerging technologies for the production of nutraceuticals from agricultural by-products: a viewpoint of opportunities and challenges, *Food and Bioproducts Processing* 91(4), 575–579. doi: 10.1016/j.fbp.2013.01.004.

Galanakis, C. M. (2018). Phenols recovered from olive mill wastewater as additives in meat products, *Trends in Food Science and Technology* 79, 98–105. doi: 10.1016/j.tifs.2018.07.010.

Galani, J. H. Y., Patel, J. S., Patel, N. J. and Talati, J. G. (2017). Storage of fruits and vegetables in refrigerator increases their phenolic acids but decreases the total phenolics, anthocyanins and vitamin C with subsequent loss of their antioxidant capacity, *Antioxidants* 6(3). doi: 10.3390/antiox6030059.

Gentile, C., Di Gregorio, E., Di Stefano, V., Mannino, G., Perrone, A., Avellone, G., Sortino, G., Inglese, P. and Farina, V. (2019). Food quality and nutraceutical value of nine cultivars of mango (Mangifera indica L.) fruits grown in Mediterranean subtropical environment, *Food Chemistry* 277, 471–479. doi: 10.1016/j.foodchem.2018.10.109.

Ghosh, R., Chakraborty, A., Biswas, A. and Chowdhuri, S. (2020). Evaluation of green tea polyphenols as novel corona virus (SARS CoV-2) main protease (Mpro) inhibitors-an in silico docking and molecular dynamics simulation study, *Journal of Biomolecular Structure and Dynamics* 39(12), 4362–4374. doi: 10.1080/07391102.2020.1779818.

Gomes, E. P., Vanz Borges, C., Monteiro, G. C., Filiol Belin, M. A., Minatel, I. O., Pimentel Junior, A., Tecchio, M. A. and Lima, G. P. P. (2021). Preharvest salicylic acid treatments improve phenolic compounds and biogenic amines in "Niagara Rosada" table grape, *Postharvest Biology and Technology* 176. doi: 10.1016/j.postharvbio.2021.111505.

Gomez-Gomez, H. A., Borges, C. V., Minatel, I. O., Luvizon, A. C. and Lima, G. P. P. (2018) Heath benefits of dietary phenolic compounds and biogenic amines. In: (1st edn.) Mérillon, J.-M., Ramawat, K. G. (Eds) *Bioactive Molecules in Food*, 1-25. doi: 10.1007/978-3-319-54528-8_27-1.

Gomez-Gomez, H. A., Borges, C. V., Minatel, I. O., Luvizon, A. C. and Lima, G. P. P. (2019). Health benefits of dietary phenolic compounds and biogenic amines, *Reference Series in Phytochemistry*, 3–27. doi: 10.1007/978-3-319-78030-6_27.

González-Gallego, J., García-Mediavilla, M. V., Sánchez-Campos, S. and Tuñón, M. J. (2010). Fruit polyphenols, immunity and inflammation, *British Journal of Nutrition* 104 (Suppl. 3), S15–S27. Enhanced Reader.pdf.

Grassi, D., Ferri, C. and Desideri, G. (2015). Brain protection and cognitive function: cocoa flavonoids as nutraceuticals, *Current Pharmaceutical Design* 22(2), 145–151. doi: 10.2174/1381612822666151112145730.

Gu, Y., Zhang, Y., Li, M., Huang, Z., Jiang, J., Chen, Y., Chen, J., Jia, Y., Zhang, L. and Zhou, F. (2021). Ferulic acid ameliorates atherosclerotic injury by modulating gut microbiota and lipid metabolism, *Frontiers in Pharmacology* 12(March), 621339. doi: 10.3389/fphar.2021.621339.

Haminiuk, C. W. I., Maciel, G. M., Plata-Oviedo, M. S. V. and Peralta, R. M. (2012). Phenolic compounds in fruits - an overview, *International Journal of Food Science and Technology* 47(10), 2023–2044. doi: 10.1111/j.1365-2621.2012.03067.x.

Hu, S., Zhang, X., Chen, F. and Wang, M. (2017). Dietary polyphenols as photoprotective agents against UV radiation, *Journal of Functional Foods* 30, 108–118. doi: 10.1016/j.jff.2017.01.009.

Iacopini, P., Baldi, M., Storchi, P. and Sebastiani, L. (2008). Catechin, epicatechin, quercetin, rutin and resveratrol in red grape: content,in vitroantioxidant activity and interactions, *Journal of Food Composition and Analysis* 21(8), 589–598.

Ignat, I., Volf, I. and Popa, V. I. (2011). A critical review of methods for characterisation of polyphenolic compounds in fruits and vegetables, *Food Chemistry* 126(4), 1821–1835. doi: 10.1016/j.foodchem.2010.12.026.

Jacobo-Velázquez, D. A., González-Agüero, M. and Cisneros-Zevallos, L. (2015). Cross-talk between signaling pathways: the link between plant secondary metabolite production and wounding stress response, *Scientific Reports* 5, 8608. doi: 10.1038/srep08608.

Kelleni, M. T. (2021). Resveratrol-zinc nanoparticles or pterostilbene-zinc: potential COVID-19 mono and adjuvant therapy, *Biomedicine and Pharmacotherapy = Biomedecine and Pharmacotherapie* 139, 111626. doi: 10.1016/j.biopha.2021.111626.

Kroon, P. A., Faulds, C. B., Ryden, P., Robertson, J. A. and Williamson, G. (1997). Release of covalently bound ferulic acid from fiber in the human colon, *Journal of Agricultural and Food Chemistry* 45(3), 661–667. doi: 10.1021/jf9604403.

Kumar, N. and Goel, N. (2019). Phenolic acids: natural versatile molecules with promising therapeutic applications, *Biotechnology Reports* 24, e00370. doi: 10.1016/j.btre.2019.e00370.

Lemmens, L., Tibäck, E., Svelander, C., Smout, C., Ahrné, L., Langton, M., Alminger, M., Van Loey, A. and Hendrickx, M. (2009). Thermal pretreatments of carrot pieces using different heating techniques: effect on quality related aspects, *Innovative Food Science and Emerging Technologies*. Elsevier Ltd 10(4), 522–529. doi: 10.1016/j.ifset.2009.05.004.

Leong, S. Y. and Oey, I. (2012). Effects of processing on anthocyanins, carotenoids and vitamin C in summer fruits and vegetables, *Food Chemistry*. Elsevier Ltd 133(4), 1577–1587. doi: 10.1016/j.foodchem.2012.02.052.

Levy, E., Delvin, E., Marcil, V. and Spahis, S. (2020). Can phytotherapy with polyphenols serve as a powerful approach for the prevention and therapy tool of novel coronavirus disease 2019 (COVID-19)?, *American Journal of Physiology. Endocrinology and Metabolism* 319(4), E689–E708. doi: 10.1152/ajpendo.00298.2020.

Lima, G. P. P., da Rocha, S. A., Takaki, M., Ramos, P. R. R. and Ono, E. O. (2008). Comparison of polyamine, phenol and flavonoid contents in plants grown under conventional and organic methods, *International Journal of Food Science and Technology* 43(10), 1838–1843. doi: 10.1111/j.1365-2621.2008.01725.x.

Makkar, R., Behl, T., Bungau, S., Zengin, G., Mehta, V., Kumar, A., Uddin, M. S., Ashraf, G. M., Abdel-Daim, M. M., Arora, S. and Oancea, R. (2020). Nutraceuticals in neurological disorders, *International Journal of Molecular Sciences* 21(12), 1–19. doi: 10.3390/ijms21124424.

Mazzeo, T., N'Dri, D., Chiavaro, E., Visconti, A., Fogliano, V. and Pellegrini, N. (2011). Effect of two cooking procedures on phytochemical compounds, total antioxidant capacity and colour of selected frozen vegetables, *Food Chemistry*. Elsevier Ltd 128(3), 627–633. doi: 10.1016/j.foodchem.2011.03.070.

Meinhart, A. D., Damin, F. M., Caldeirão, L., da Silveira, T. F. F., Filho, J. T. and Godoy, H. T. (2017). Chlorogenic acid isomer contents in 100 plants commercialized in Brazil, *Food Research International* 99(1), 522–530. doi: 10.1016/j.foodres.2017.06.017.

Melo, J. R. R., Duarte, E. C., Moraes, M. V., Fleck, K., Silva, A. S. D. N. E. and Arrais, P. S. D. (2021). Adverse drug reactions in patients with COVID-19 in Brazil: analysis of spontaneous notifications of the Brazilian pharmacovigilance system, *Cadernos de Saúde Pública* 37(1), e00245820.

Monaco, K., Costa, S. M., Minatel, I. O., Correa, C. R., Calero, F. A., Vianello, F. and Lima, G. P. P. (2016). Influence of ozonated water sanitation on postharvest quality of conventionally and organically cultivated mangoes after postharvest storage, *Postharvest Biology and Technology*. Elsevier B.V. 120, 69–75. doi: 10.1016/j.postharvbio.2016.05.003.

Monribot-Villanueva, J. L., Elizalde-Contreras, J. M., Aluja, M., Segura-Cabrera, A., Birke, A., Guerrero-Analco, J. A. and Ruiz-May, E. (2019). Endorsing and extending the repertory of nutraceutical and antioxidant sources in mangoes during postharvest shelf life, *Food Chemistry* 285, 119–129. doi: 10.1016/j.foodchem.2019.01.136.

Monteiro, G. C., Minatel, I. O., Junior, A. P., Gomez-Gomez, H. A., de Camargo, J. P. C., Diamante, M. S., Pereira Basílio, L. S., Tecchio, M. A. and Pereira Lima, G. P. (2021). Bioactive compounds and antioxidant capacity of grape pomace flours, *LWT* 135. doi: 10.1016/j.lwt.2020.110053.

Neves, L. C., de Campos, A. J., Cisneros-Zevallos, L., Colombo, R. C. and Roberto, S. R. (2017). Postharvest behavior of camu-camu fruits based on harvesting time and nutraceutical properties, *Scientia Horticulturae* 217, 276–284. doi: 10.1016/j.scienta.2017.01.030.

Niedzwiecki, A., Roomi, M. W., Kalinovsky, T. and Rath, M. (2016). Anticancer efficacy of polyphenols and their combinations, *Nutrients* 8(9). doi: 10.3390/nu8090552.

Oliveira, B. G., Costa, H. B., Ventura, J. A., Kondratyuk, T. P., Barroso, M. E. S., Correia, R. M., Pimentel, E. F., Pinto, F. E., Endringer, D. C. and Romão, W. (2016). Chemical profile of mango (Mangifera indica L.) using electrospray ionisation mass spectrometry (ESI-MS), *Food Chemistry* 204, 37–45. doi: 10.1016/j.foodchem.2016.02.117.

Ong, K. W., Hsu, A. and Tan, B. K. H. (2013). Anti-diabetic and anti-lipidemic effects of chlorogenic acid are mediated by ampk activation, *Biochemical Pharmacology* 85(9), 1341–1351. doi: 10.1016/j.bcp.2013.02.008.

Palafox-Carlos, H., Yahia, E. M. and González-Aguilar, G. A. (2012). Identification and quantification of major phenolic compounds from mango (Mangifera indica, cv. Ataulfo) fruit by HPLC-DAD-MS/MS-ESI and their individual contribution to the antioxidant activity during ripening, *Food Chemistry*. Elsevier Ltd 135(1), 105–111. doi: 10.1016/j.foodchem.2012.04.103.

Palermo, M., Pellegrini, N. and Fogliano, V. (2014). The effect of cooking on the phytochemical content of vegetables, *Journal of the Science of Food and Agriculture* 94(6), 1057–1070. doi: 10.1002/jsfa.6478.

Park, Y. S., Leontowicz, H., Leontowicz, M., Leontowicz, Namieśnik, J., Jesion, I. and Gorinstein, S. (2008). Nutraceutical value of persimmon (Diospyros kaki Thunb.) and its influence on some indices of atherosclerosis in an experiment on rats fed cholesterol-containing diet, *Advances in Horticultural Science* 22(4), 250–254. doi: 10.1400/100650.

Ramarathnam, N., Osawa, T., Ochi, H. and Kawakishi, S. (1995). The contribution of plant food antioxidants to human health, *Trends in Food Science and Technology* 6(3), 75–82. doi: 10.1016/S0924-2244(00)88967-0.

Rashmi, H. B. and Negi, P. S. (2020). Phenolic acids from vegetables: a review on processing stability and health benefits, *Food Research International* 136(April), 109298. doi: 10.1016/j.foodres.2020.109298.

Riihinen, K., Jaakola, L., Kärenlampi, S. and Hohtola, A. (2008). Organ-specific distribution of phenolic compounds in bilberry (Vaccinium myrtillus) and "northblue" blueberry (Vaccinium corymbosum x V. angustifolium), *Food Chemistry* 110(1), 156-160. doi: 10.1016/j.foodchem.2008.01.057.

Santana-Gálvez, J., Pérez-Carrillo, E., Velázquez-Reyes, H. H., Cisneros-Zevallos, L. and Jacobo-Velázquez, D. A. (2016). Application of wounding stress to produce a nutraceutical-rich carrot powder ingredient and its incorporation to nixtamalized corn flour tortillas, *Journal of Functional Foods* 27, 655–666. doi: 10.1016/j.jff.2016.10.020.

Santana-Gálvez, J., Cisneros-Zevallos, L. and Jacobo-Velázquez, D. A. (2019). A practical guide for designing effective nutraceutical combinations in the form of foods, beverages, and dietary supplements against chronic degenerative diseases, *Trends in Food Science and Technology* 88, 179–193. doi: 10.1016/j.tifs.2019.03.026.

Shahidi, F. (2009). Nutraceuticals and functional foods: whole versus processed foods, *Trends in Food Science and Technology* 20(9), 376–387. doi: 10.1016/j.tifs.2008.08.004.

Šilarová, P., Boulekbache-Makhlouf, L., Pellati, F. and Česlová, L. (2019). Monitoring of chlorogenic acid and antioxidant capacity of Solanum melongena L. (eggplant) under different heat and storage treatments, *Antioxidants* 8(7), 1–11. doi: 10.3390/antiox8070234.

Simões, A. do N., Almeida, S. L. d., Van Borges, C., Fonseca, K. S., Barros Júnior, A. P., Albuquerque, J. R. T. d., Corrêa, C. R., Minatel, I. O., Morais, M. A. d. S., Diamante, M. S. and Lima, G. P. P. (2020). Delaying the harvest induces bioactive compounds and maintains the quality of sweet potatoes, *Journal of Food Biochemistry* 44(8), 1–13. doi: 10.1111/jfbc.13322.

Soto, M. L., Falqué, E. and Domínguez, H. (2015). Relevance of natural phenolics from grape and derivative products in the formulation of cosmetics, *Cosmetics* 2(3), 259–276. doi: 10.3390/cosmetics2030259.

Swallah, M. S., Sun, H., Affoh, R., Fu, H. and Yu, H. (2020). Antioxidant potential overviews of secondary metabolites (Polyphenols) in fruits, *International Journal of Food Science* 2020, 9081686.

Thom, E. (2007). The effect of chlorogenic acid enriched coffee on glucose absorption in healthy volunteers and its effect on body mass when used long-term in overweight and obese people, *Journal of International Medical Research* 35(6), 900–908. doi: 10.1177/147323000703500620.

Truzzi, F., Valerii, M. C., Tibaldi, C., Zhang, Y., Abduazizova, V., Spisni, E. and Dinelli, G. (2020). Are supplements safe? effects of Gallic and ferulic acids on in vitro cell models, *Nutrients* 12(6), 1–13. doi: 10.3390/nu12061591.

Uthumporn, U., Woo, W. L., Tajul, A. Y. and Fazilah, A. (2015). Physico-chemical and nutritional evaluation of cookies with different levels of eggplant flour substitution, *CyTA – Journal of Food* 13(2), 220–226. doi: 10.1080/19476337.2014.942700.

van Boekel, M., Fogliano, V., Pellegrini, N., Stanton, C., Scholz, G., Lalljie, S., Somoza, V., Knorr, D., Jasti, P. R. and Eisenbrand, G. (2010). A review on the beneficial aspects of food processing, *Molecular Nutrition and Food Research* 54(9), es215–1247. doi: 10.1002/mnfr.200900608.

Vella, F. M., Cautela, D. and Laratta, B. (2019). Characterization of polyphenolic compounds in cantaloupe melon by-products, *Foods* 8(6). doi: 10.3390/foods8060196.

Wang, Y. and Ho, C. T. (2009). Polyphenols chemistry of tea and coffee: a century of progress, *Journal of Agricultural and Food Chemistry* 57(18), 8109–8114. doi: 10.1021/jf804025c.

Yang, X., Jiang, Y., Yang, J., He, J., Sun, J., Chen, F., Zhang, M. and Yang, B. (2015). Prenylated flavonoids, promising nutraceuticals with impressive biological activities, *Trends in Food Science and Technology* 44(1), 93–104. doi: 10.1016/j.tifs.2015.03.007.

Yang, Y., Islam, M. S., Wang, J., Li, Y. and Chen, X. (2020). Traditional Chinese medicine in the treatment of patients infected with 2019-new coronavirus (SARS-CoV-2): a review and perspective, *International Journal of Biological Sciences* 16(10), 1708–1717. doi: 10.7150/ijbs.45538.

Yeon, J. Y., Bae, Y. J., Kim, E. Y. and Lee, E. J. (2015). Association between flavonoid intake and diabetes risk among the Koreans, *Clinica Chimica Acta; International Journal of Clinical Chemistry* 439, 225–230. doi: 10.1016/j.cca.2014.10.042.

Zhao, Z. and Moghadasian, M. H. (2008). Chemistry, natural sources, dietary intake and pharmacokinetic properties or ferulic acid: a review, *Food Chemistry* 109(4), 691–702.

Zizkova, P., Stefek, M., Rackova, L., Prnova, M. and Horakova, L. (2017). Novel quercetin derivatives: from redox properties to promising treatment of oxidative stress related diseases, *Chemico-Biological Interactions* 265, 36–46. doi: 10.1016/j.cbi.2017.01.019.

Chapter 3

Understanding the nutraceutical properties of flavonoids in fruits and vegetables: chemical structure and groups

A. D. Diwan, S. N. Harke and A. N. Panche, MGM Institute of Biosciences & Technology, Mahatma Gandhi Mission University, India

1 Introduction

Medicinal plants and herbal medicines have traditionally been used for the treatment of various diseases and disorders such as neurological ailments and cancers, including as part of the centuries-old 'Ayurveda' system of medicine in India (Umashanker and Shruti, 2011). Medicinal herbs have also been used effectively as veterinary drugs for treatment of conditions such as mastitis and foot and mouth diseases (Rahal et al., 2014). The World Health Organization (WHO) has recorded 21 000 species of medicinal plants from around the world. Among these, vegetables and fruits have been used as both a food source and a natural source of medicinal compounds for thousands of years. Fruits and vegetables are also a common source of carbohydrates, proteins, fiber, vitamins, minerals and essential amino acids in providing a healthy diet (Murphy et al., 2012; Bumgarner et al., 2012).

Recent advances in medical and nutrition sciences have highlighted the role of phytonutrients from foods such as fruits and vegetables in boosting immune

http://dx.doi.org/10.19103/AS.2022.0101.04

function and preventing chronic diseases without the side effects associated with synthetic drugs (Berger and Shenkin, 2006; Bagchi, 2006; Ramaa et al., 2006). Nutraceuticals, a term first coined by Stephen DeFelice in 1989, are products derived from food sources with health-promoting properties in addition to meeting basic nutritional requirements (Chauhan et al., 2013; Ronis et al., 2018). DeFelice defined a nutraceutical as 'a food or it can be part of a food that provides medical or health benefits, as well as the prevention and/or treatment of a disease' (Brower, 1998). An analysis by the Business Communication Company suggests that the global nutraceutical market is expanding at a compound annual growth rate of 7.5%, and from an estimated value of US$198.7 billion in 2016 could reach a value of US$285.0 billion by the end of 2021 (Golla, 2018).

Fruits and vegetables have been shown to contain varying amounts of different nutraceutical compounds (Weingartner et al., 2009; Goff and Klee, 2006). Fruits and vegetables are sources of groups of phytochemicals such as antioxidants, flavonoids and carotenoids (Goff and Klee, 2006). A range of nutraceuticals have, for example, been found in fruits like berries, bananas (*Musa spp.*), grapes (*Vitis vinifera*), watermelon (*Citrullus lanatus*), citrus fruits like orange (*Citrus sinensis*) and lemon (*Citrus limon*) and vegetables like tomato (*Solanum lycopersicum*), carrot (*Daucus carota*), bael (*Aegle marmelos*), pomegranate (*Punica granatum*) and ginger (*Zingiber officinale*) (Yao et al., 2004; Tikunov et al., 2010). Nutraceutical compounds in vegetable crops include lycopene in tomatoes, curcumin in turmeric, gingerol in ginger, organo-sulphur compounds in allium species, and omega-3 fatty acids in cucurbitaceous vegetables. Harnly et al. (2006) analyzed 20 flavonoids from more than 60 fruits and vegetable species collected from different regions of the United States. They found limited seasonal variation in flavonoid content with some exceptions, such as blueberry. Each fruit or vegetable contains a unique combination of nutraceuticals and a diversity of fruits and vegetables should be eaten to ensure the range of nutraceuticals needed to optimize health benefits (Hayat et al., 2017; Sakthinathan and Nandhini, 2017).

This chapter provides an overview of flavonoids present in fruits and vegetables. It reviews flavonoid chemical structure and groups. Groups of flavonoids discussed are:

- flavones;
- flavonols;
- flavanones;
- flavanonols;
- isoflavones;
- neoflavonoids;
- flavanols;
- anthocyanidins; and
- chalcones.

The chapter concludes with a summary of the range of flavonoids in fruits and vegetables and their role as nutraceuticals. An accompanying chapter looks in more detail at modes of action in delivering health benefits (Box 1).

Box 1: Why we need flavonoids

A diet rich in plant foods benefits the body in many ways. Phytonutrients like flavonoids have beneficial anti-inflammatory effects and they protect your cells from oxidative damage that can lead to disease. These dietary antioxidants can prevent the development of cardiovascular disease, diabetes, cancer and cognitive diseases like Alzheimer's and dementia.

2 Flavonoids

Flavonoids constitute one of the most significant groups of plant phenolics, with about 9000 varieties of flavonoids identified from different plant sources (Ahmad et al., 2015). Nobel laureate Albert Szent-Gyorgyi, who discovered vitamin C, isolated flavonoids (proanthocyanidins) in the 1930s, which were initially recognized as vitamin P. Flavonoids attracted particular attention from researchers with the discovery of the French Paradox: the lower incidence of cardio-vascular disease observed in the Mediterranean population despite higher red wine consumption and a greater amount of saturated fat in the average diet compared to other countries. This discrepancy was increasingly linked to a higher consumption of fruits and vegetables containing phytochemicals such as flavonoids (Tapas et al. 2008).

Flavonoids are a class of low molecular weight phenolic compounds that are widely distributed in the plant kingdom. Flavonoid compounds are found in parts of plants such as leaves and fruits and support plant growth (Pedro et al., 2016; Tsuchiya, 2010; Ahmad et al., 2015; Panche et al., 2016). Many flavonoids are easily recognized as flower pigments in most angiosperm families. However, their occurrence is not restricted to flowers and can occur in many other parts of plants (Dewick, 2009a,b).

Flavonoids have long been known to be synthesized in particular sites in plants. In their work on pomegranates, for example, Kolar et al. (2021) found the maximum concentration of flavonoids in the leaves and arils (which cover seeds) with differing degrees of antioxidant activity associated with different parts of plants. As well as being responsible for colour and aroma in flowers and fruits to attract pollinators, flavonoids also play a role in spore germination, growth and development of seedlings (Jorgensen, 1995; Griesbach, 2005; Dias et al., 2021). Flavonoids also protect plants from different biotic and abiotic stresses, including as antimicrobial defensive compounds, and act as a unique UV-filter (Takahashi and Ohnishi, 2004). Flavonoids have important roles in frost hardiness, drought resistance and may play a functional role in plant heat acclimation and freezing tolerance (Samanta et al., 2011).

Different vegetables, fruits and berries are good sources of anthocyanidins, while parsley, celery and herbs are excellent sources of flavones (Crozier et al., 2009). Genistein and daidzein are major isoflavones found in *Genista tinctoria*, a Chinese medicinal herb, and other leguminous plants (Veitch, 2013). Flavonols such as quercetin, kaempferol, isorhamnetin and fisetin can be commonly found throughout the plant kingdom (Babu et al., 2013). Hesperidin and naringin are examples of flavanones mostly present in citrus fruits. Naringin gives a bitter taste to grapefruit (Jung et al., 2006). Sources of anthocyanidins and proanthocyanidins include cabbage, currants, barley, banana, berries (strawberries, rasberries, cranberries, blueberries, blackberries), chocolate, tea (black and green), wine, beer, spices, onions, plums, peas, grapes, peaches, nuts (walnuts, peanuts, cashews, pistachio, almonds), mangoes and lentils (Kruger et al., 2014).

As noted, given their range of function in plants and potential health-promoting properties, flavonoids have become a growing subject of study as shown by reviews by, for example, Dixon and Pasinetti (2010) and Panche et al. (2016). This includes research on biological and health-promoting properties (Kaleem and Ahmad, 2018; Dias et al., 2021).

3 Flavonoid chemical structure and groups

Flavonoids are the largest class of polyphenols. Chemically, they may be defined as a group of polyphenolic compounds consisting of substances that have two substituted benzene rings connected by a chain of three carbon atoms and an oxygen bridge, as shown in Fig. 1. Their basic structure is a skeleton of diphenylpropane, namely two benzene rings A and B (Fig. 1) linked by a three-carbon chain that forms a closed pyran ring. The B ring is mainly attached to position 2 of the C ring, but it can also bind in position 3 or 4.

Antioxidant and other properties depend both on structural characteristics and the pattern of glycosylation (Babu et al., 2013; D'Amelia et al., 2018; Dias et al., 2021). The structural features of the B ring and the patterns of glycosylation and hydroxylation of the three rings make flavonoids one of the larger and more diversified groups of phytochemicals (Figure 2). The position of the catechol B-ring on the pyran C-ring, and the number and position of hydroxy groups on the catechol group of the B-ring, influence flavonoid antioxidant capacity (D'Amelia et al., 2018). The functional hydroxy groups in flavonoids can donate electrons through resonance to stabilize free radicals and mediate antioxidant protection (Šamec et al., 2021).

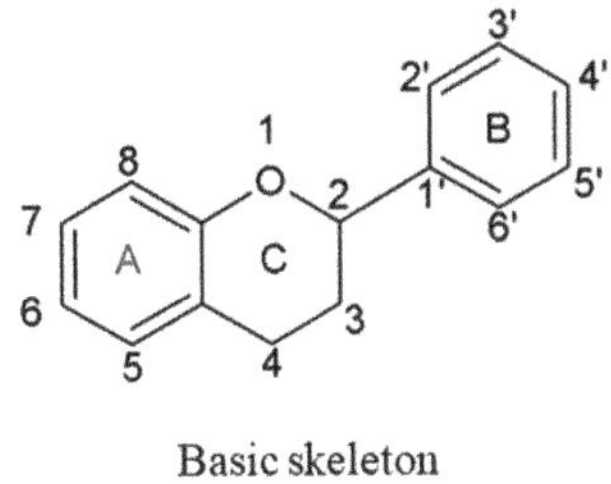

Figure 1 Basic structure of flavonoids.

Flavonoids are grouped according to the presence of different substituents on the rings and the degree of ring saturation. They are frequently attached with a sugar moiety to increase their water solubility (Stump and Conn, 1981). Flavonoids have several subgroups, which include chalcones, flavones, flavonols and isoflavones. Flavonoids can be subdivided into different subgroups depending on the carbon of the C ring on which the B ring is attached, and the degree of unsaturation and oxidation of the C ring. Flavonoids can be classified into six major subgroups, based on their molecular structure. Flavonoids in which the B ring is linked in position 3 of the ring C are called isoflavones and those in which the B ring is linked in position 4 are called neoflavonoids. The flavonoids in which the B ring is linked in position 2 can be further subdivided

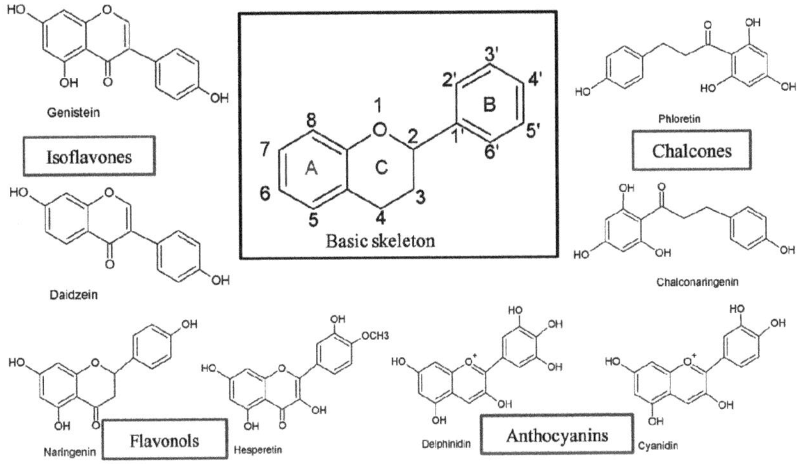

Figure 2 Structures of the major classes of flavonoids.

into several subgroups on the basis of the structural features of the C ring. These subgroups are:

- flavones;
- flavonols;
- flavanones;
- flavanonols;
- flavanols or catechins; and
- anthocyanins.

Finally, flavonoids with an open C ring are called chalcones (Andersen and Markham, 2005; Panche et al., 2016; Kaleem and Ahmad, 2018; Šamec et al., 2021) (Fig. 2). Flavonoids within these subclasses play a beneficial and sometimes key role in a number of physiological processes (Box 2).

Box 2

Flavonoids are plant compounds with a variety of health benefits. There are six primary types of flavonoids, each with health-promoting effects. These are:

- Flavonols
- Flavones
- Flavan-3-ols
- Flavanones
- Anthocyanidins
- Isoflavones

The best way to obtain all six types of flavonoids is to consume a variety of fruits and vegetables. Many plant-based foods and beverages like tea and wine contain flavonoids. Numerous studies have shown the many benefits of these phytonutrients. Researchers have found that eating a diet rich in flavonoids reduces the risks of diabetes, cardiovascular disease and some cancers.

4 Flavones and flavonols

4.1 Flavones

Flavones are one of the important subgroups of flavonoids and are widely present in leaves, flowers and fruits as glucosides. Celery, parsley, red peppers, chamomile, mint and *Ginkgo biloba* are among the major sources of flavones. Luteolin, apigenin and tangeritin belong to this sub-class of flavonoids. The

Figure 3 Structure and sources of flavones.

peel of citrus fruits is rich in the polymethoxylated flavones tageretin, nobiletin and sinensetin (Manach et al., 2004).

Flavones have a double bond between positions 2 and 3 and a ketone in position 4 of the C ring (Fig. 3). Most flavones of vegetables and fruits have a hydroxyl group in position 5 of the A ring. Hydroxylation in other positions, for the most part in position 7 of the A ring or 3 and 4 of the B ring, may vary according to the taxonomic classification of the particular vegetable or fruit. Glycosylation occurs primarily on positions 5 and 7, while methylation and acylation occur on the hydroxyl groups of the B ring. Some flavones, such as nobiletin and tangeretin, are polymethoxylated. Iwashina (2013) has reviewed the flavonoid properties of five families newly incorporated into the order Caryophyllales.

4.2 Flavonols

Flavonols are flavonoids containing a ketone group. They are building blocks of proanthocyanins. These compounds occur widely in a variety of fruits and vegetables. The most studied flavonols are kaempferol, quercitin, myricetin and fisetin. Onions, kale, lettuce, tomato, apple, grape and berries are good sources (Fig. 4). Apart from fruits and vegetables, beverages such as tea and red wine are also important sources of flavonols. Intake of flavonols is found to be associated with a wide range of health benefits related to antioxidant potential, including reduced risk of vascular disease. Compared to flavones, they have a hydroxyl group in position 3 of the C ring, which may also be glycosylated. Like

Figure 4 Structure and sources of flavonols.

flavones, flavonols are very diverse in methylation and hydroxylation patterns. Given the different glycosylation patterns, they are perhaps the most common and largest subgroup of flavonoids in fruits and vegetables. Quercitin, for example, is present in many plant foods (Iwashina, 2013).

5 Flavanones and flavanonols

5.1 Flavanones

Flavanones, also called dihydroflavonols, are another important class of flavonoids generally present in citrus fruits, such as oranges, lemons and grapes. Hesperitin, naringenin and eriodictyol are examples of this class of flavonoids. These compounds are responsible for the bitter taste of the juice and peel of citrus fruits (Fig. 5). Flavonones are associated with a number of health benefits because of their free-radical scavenging properties. Citrus flavonoids have antioxidant, anti-inflammatory, blood lipid- and cholesterol-lowering properties. Flavanones have the saturated C ring (Fig. 5). Unlike flavones, the double bond between positions 2 and 3 is saturated, and this is the only structural difference between the two subgroups of flavonoids.

Flavanones can be multi-hydroxylated, and several hydroxyl groups can be glycosylated or methylated or both. Some flavanones have unique patterns of substitution, like furanoflavanones, prenylated flavanones, pyranoflavanones or benzylated flavanones, giving a great number of substituted derivatives (Iwashina, 2013).

5.2 Flavanonols

Flavanonols, also called dihydroflavonols, are the 3-hydroxy derivatives of flavanones. They are a highly diversified and multisubstituted subgroup (Fig. 6). Flavanonols have been reported to be responsible for increasing Bifidobacterium levels in patients, potentially correcting the dysbiosis associated with systemic lupus erythematosus (Cuervo et al., 2015).

Figure 5 Structure and sources of flavanones.

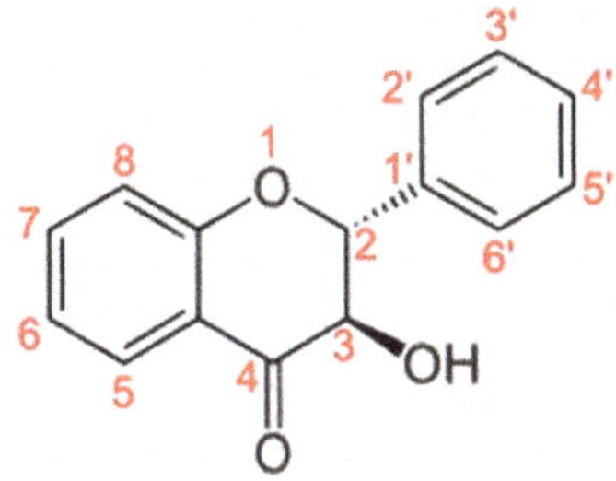

Figure 6 Structure and sources of flavanonols.

6 Isoflavones and neoflavonoids

6.1 Isoflavones

Isoflavones, also known as isoflavonoids, are a large and very distinctive subgroup of flavonoids. They are a subgroup of flavonoids in which the B ring is attached to position 3 of the C ring. Isoflavonoids have only a limited distribution in the plant kingdom and are predominantly found in soya beans and other leguminous plants as well as non-plant sources (Matthies et al., 2008) (Fig. 7). They play an important role as precursors for the development of phytoalexins during plant-microbe interactions (Aoki et al., 2000; Dixon and Ferreira, 2002).

Isoflavonoids have a chemical structure similar to the plant hormone oestrogen and are therefore known as phytoestrogens. Isoflavones such as genistein and daidzein are commonly regarded to be phytoestrogens because of their oestrogenic activity in certain animal models, which has been shown to influence various disease pathways (Szkudelska and Nogowski, 2007).

6.2 Neoflavonoids

While flavonoids have the 2-phenylchromen-4-one backbone, neoflavonoids have the 4-phenylcoumarin backbone ($C_{15}H_{12}O_2$) with no hydroxyl group substitution at position 2 (Fig. 8). The first neoflavone isolated from natural

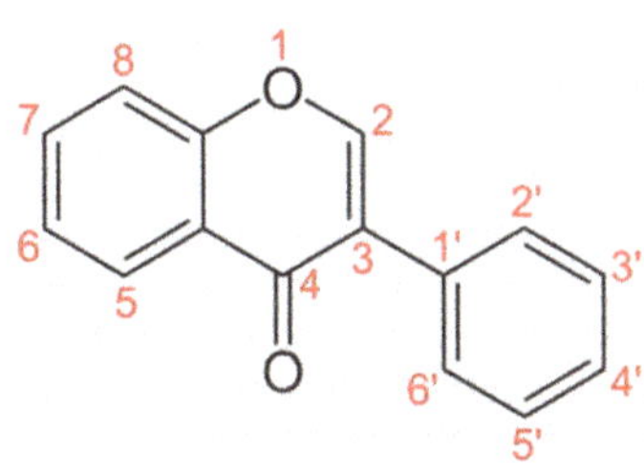

Figure 7 Structure and sources of isoflavones.

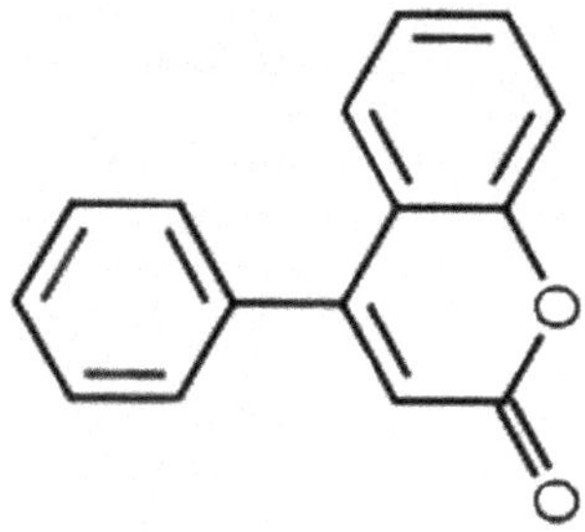

Figure 8 General structure of neoflavonoids.

sources was calophyllolide, isolated in 1951 from *Calophyllum inophyllum* seeds (Linuma et al., 1987; Nishimura et al., 2000; Garazd et al., 2003).

There is evidence that neoflavonoids and tetrahydroquinolones exert therapeutic effects such as the inhibition of activation of nuclear factor kappa-light-chain-enhancer of activated B cells (NFκB), inhibition of aromatase activity and inducing of QR1 expression, which are key factors in carcinogenesis (Luqman et al., 2012). Neoflavonoids have also been reported to have anti-Tat activity which inhibits HIV replication (Olmedo et al., 2017).

7 Flavanols, anthocyanidins and chalcones

7.1 Flavanols

Flavanols are also known to possess flavan-3-ols as the hydroxyl group is almost always bound to position 3 of the C ring. They are also called catechins. Unlike many flavonoids, there is no double bond between positions 2 and 3. Another distinctive feature of flavonols is the lack of a carbonyl group, which is a keto group, in position 4 which is present in flavanonols at position 3. This particular chemical structure allows flavanols to have two chiral centres in the molecule, on positions 2 and 3, and four possible diastereoisomers. Epicatechin is the isomer with the *cis* configuration and catechin is the isomer with the trans configuration. Each of these configurations has two stereoisomers:

1 (+)-epicatechin and (−)-epicatechin
2 (+)-catechin and (−)-catechin

(+)-Catechin and (−)-epicatechin are the two isomers most often present in edible plants.

Another important feature of flavanols, particularly of catechin and epicatechin, is the ability to form polymers called proanthocyanidins or

Figure 9 Structure and sources of flavanols.

condensed tannins. The name 'proanthocyanidins' is because an acid-catalyzed cleavage produces anthocyanidins. Proanthocyanidins typically contain 2–60 monomers of flavanols. Monomeric and oligomeric flavanols (containing two to seven monomers) are strong antioxidants (Fig. 9).

7.2 Anthocyanins

Anthocyanins are pigments responsible for colours in flowers and fruits. Cyanidin, delphinidin, malvidin, pelargonidin and peonidin are the most commonly studied anthocyanins. They occur predominantly in the outer cell layers of various fruits such as cranberry, black currant, red grape, raspberry, strawberry, blueberry, bilberry and blackberry (Fig. 10).

Anthocyanins are glycosides of anthocyanidins. Sugar units are bound mostly to position 3 of the C ring and they are often conjugated with phenolic acids such as ferulic acid (Fig. 10). Chemically, anthocyanidins are flavylium cations and are generally present as chloride salts. They are the only group of flavonoids that give plants colour (all other flavonoids are colourless). Anthocyanidins show coloured pigments in highly oxidized form.

Their stability coupled with health benefits means they are used in the food industry in a variety of applications (Vaccaro et al., 2017). Anthocyanins display a wide range of biological activities including antioxidant, anti-inflammatory, anti-microbial and anti-carcinogenic activities, as well as reducing the risk of coronary heart disease (Khoo et al., 2017; Ciumărnean et al., 2020; Cassidy, 2018; Rolnik et al., 2020).

Anthocyanins include cyanidin, responsible for red to magenta colours; delphinidin, responsible for purple to blue colours; and pelargonidin, responsible for orange to pink colours. This colour differentiation is associated with attracting different insects and other species for pollination. The presence of a sugar moiety promotes changes in colour brightness. The most common sugar is glucose with a β-linkage, but galactose, rhamnose and xylose are also found. These sugar moieties can have acyl substituents, highlighting cinnamic

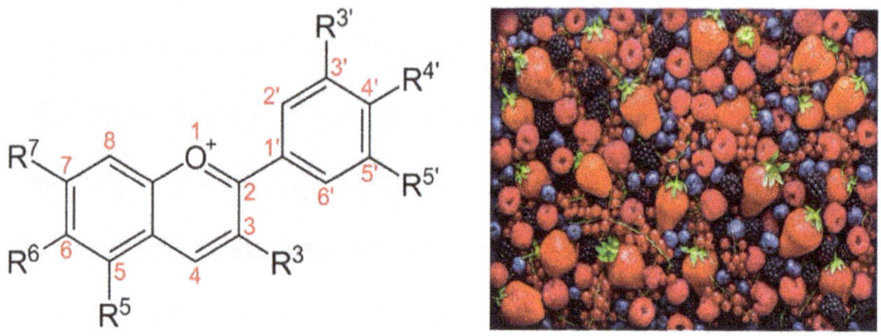

Figure 10 Structure and sources of anthocyanins.

acyl derivatives, such as caffeic, ferulic and p-coumaric acids (Dias et al., 2021). The colour of the anthocyanin depends on the pH and also on methylation or acylation at the hydroxyl groups on the A and B rings (Iwashina, 2013).

7.3 Chalcones

Chalcones are characterized by the absence of 'ring C' of the basic flavonoid skeleton structure, as shown in Fig. 11. They are also referred to as open-chain flavonoids. Major examples of chalcones include phloridzin, arbutin, phloretin and chalconaringenin. Chalcones occur in significant amounts in tomatoes, pears, strawberries, bearberries and certain wheat products (Fig. 11). Chalcones and their derivatives have attracted considerable attention because of their numerous nutritional and biological benefits.

8 Conclusion

Figure 12 summarizes the main groups of flavonoids and their main fruit and vegetable sources. Flavonoids found in different vegetables are shown in Table 1. Flavonoids found in fruits are summarized in Table 2.

Figure 11 Structure and sources of chalcones.

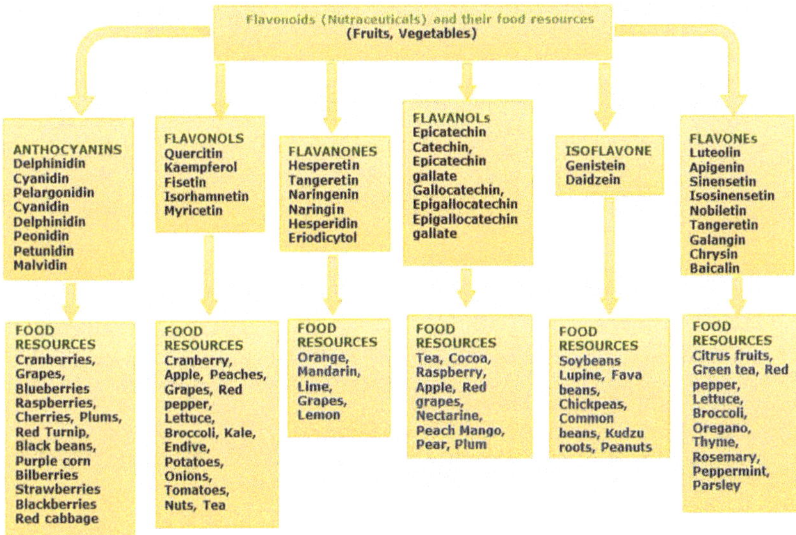

Figure 12 Flavonoids and their food sources.

The protective value of antioxidants is expressed in terms of oxygen radical absorption capacity (ORAC) units. The highest protection is achieved when our daily diet provides 5000 ORACs. The intensity of colour is generally a good indicator in determining the ORAC value, because the deeper the colour of a fruit or vegetable, the higher its ORAC score (Bernaert et al., 2012). Hence, it is advisable to include a range of highly coloured fruits and vegetables in the diet to ensure a protective effect.

It is well known that the colour of fruits and vegetables is due to the presence of the different types of pigments such as chlorophyll, anthocyanin and carotenoids. In general terms, the darker the colour, the higher the content of these compounds. A variety of vegetables such as asparagus, broccoli, Brussels sprouts, Chinese cabbage, cucumber, green beans, green cabbage, green onion, green peppers, lettuce, okra, peas and spinach can, for example, be identified as a source of nutraceuticals because of their dark green colour (Fig. 12).

Yellow and orange vegetables are rich in pigments such as lutein, zeaxanthin, carotenoids, flavonoids, lycopene, potassium and vitamin C. Zeaxanthin is one of the most common carotenoids in nature. It is synthesized in plants and gives the characteristic colour to vegetables and fruits such as paprika, corn and saffron. A red colour suggests the presence of anthocyanins and lycopene, including red-coloured vegetables such as watermelon, tomato, red spinach, red cabbage and carrot.

Table 1 Flavonoids sources in vegetables

Sr.no.	Flavonoids	Class	Dietary sources	Reference
1	Abyssinones	Flavanone	French Bean Seeds	Rathmell and Bendall, 1971; Cruickshank et al., 1974
2	Glcosides of apigenin and luteolin, apigenin-7-O-glycoside	Flavone	Celery, parsley, lettuce, broccoli, rosemary	Panche et al., 2016
3	Quercetin	Flavonols	Vegetables	Hertog et al., 1993; Justesen and Knuthsen, 2001; Stewart et al., 2000; Zheng and Wang, 2001
4	Hesperetin 7-O-rutinoside	Flavanone	Oranges, mandarins, limes, lemons	Terahara, 2015; Agrawal, 2011
5	Luteolin, apigenin	Flavone	Mentha	Dukic et al., 1996
6	Eriocitrin, eriodictyol, eriodictyol 7-O-rutinoside	Flavanone	Lemons	Terahara, 2015; Agrawal, 2011
7	Luteolin	Flavone	Celery, broccoli, green pepper, parsley, thyme, dandelion, perilla, chamomile, carrots, peppermint, rosemary, navel oranges, oregano	Shimoi et al., 1998; Lopez-Lazaro, 2009
8	Kaempferol 3-O-glycosides	Flavonol	Broccoli, kale, endive, potatoes, onions, tomatoes	Panche et al., 2016
9	Eriodictyol	Flavanone	Lemons, rose hips	Hvattum, 2002
10	Myricetin	Flavonol	Vegetables	Ross and Kasum, 2002; Basli et al., 2012
11	Rutin, spinacetin glycosides and patuletin glycosides	Flavonol	Spinach	Panche et al., 2016
12	Cyanidin 3-O-glucoside	Anthocynin	Black beans, purple corn	Terahara, 2015
13	Biochanin	Isoflavone	Soy, alfalfa sprouts, peanuts, chickpea	Medjakovic and Jungbauer, 2008
14	Acylated anthocyanins	Anthocynin	Red cabbages, red turnips, purple sweet potatoes	Dias et al., 2021
17	Kaempferol	Flavonol	Tomatoes, potatoes, onions, broccoli, brussels, sprouts, squash, cucumbers, lettuce, green beans, spinach	Calderón-Montaño et al., 2011; Liu, 2013; Kim and Choi, 2013
18	Fisetin	Flavonol	Onions, cucumbers	Sahu et al., 2014

Table 2 Flavonoids sources in fruits

Sr.no.	Flavonoids	Class	Dietary sources	Reference
1	Quercetin	Flavonols	Fruits, beverages, fruit juices	Hertog et al., 1993; Justesen and Knuthsen, 2001; Stewart et al., 2000; Zheng and Wang, 2001
2	Rutin	Flavonol	Citrus fruits, apple, berries, peach	Cruickshank et al. 1974; Chang et al. 2000
3	Luteolin	Flavone	Olive, navel oranges	Shimoi et al., 1998; Lopez-Lazaro, 2009
4	Naringenin	Flavanone	Grapes	Felgines et al., 2000
5	Glycosides of luteolin and apigenin	Flavone	Citrus fruits	Panche et al., 2016
6	Myricetin	Flavonol	Fruits, nuts, berries	Ross and Kasum, 2002; Basli et al., 2012
7	Taxifolin	Flavanonol	Citrus fruits	Grayer and Veitch, 2006; Kawaii et al., 1999
8	Rutin	Flavonol	Grape seeds, apple, citrus fruits, berries, peaches	Atanassova and Bagdassarian, 2009; Chang et al., 2000; Malagutti et al., 2006
9	Macluraxanthone	Xanthone	*Maclura tinctoria* (dyer's mulberry), *Maclura pomifera* (hedge apple)	Khan et al., 2009
10	Peonidin	Anthocyanidin	Cranberries, blueberries, plums, grapes, and cherries,	Truong et al., 2010
11	Kaempferol	Flavonol	Apples, grapes, peaches, blackberries, raspberries	Calderón-Montaño et al., 2011; Liu, 2013; Kim and Choi, 2013
12	Fisetin	Flavonol	Strawberries, apples, persimmons	Sahu et al., 2014
13	Hesperidin	Flavanone	Bitter orange,orange, orange juice, lemon, lime	Khan et al., 2009
14	(+)-Catechin, (−)-epicatechin, and (−)-epigallocatechin	Flavonol	Apples, red grapes, peaches, mangoes, pears, plums, nectarins, raspberries	Dias et al., 2021
15	Kaempferol 3-O-glycosides	Flavonol	Grapes, tomatoes	Panche et al., 2016

White-coloured vegetables contain flavonoids like quercetin and epicatechin. Examples include cauliflower, garlic, mushrooms, onions, potato, shallots, turnips and radish. Blue/purple vegetables are rich in lutein, zeaxanthin, resveratrol, vitamin C, fibre, flavonoids, ellagic acid and quercetin. Examples include eggplant, purple asparagus, purple cabbage, purple onion, purple broccoli, purple kohlrabi and purple broad bean. It is important to identify traits associated with the presence and concentration of nutraceutical compounds and use these in breeding cultivars with improved nutritional attributes through conventional and molecular breeding approaches (Boxes 3 and 4) (Rai et al., 2009).

Box 3: Fruits and vegetables high in flavonoids nutraceuticals

Berries

All berries contain flavonoids, but certain varieties are more potent than others. Blackberries are particularly powerful and include all six types of flavonoids. Blueberries, cherries and raspberries also contain all flavonoids. Strawberries have moderate amounts of anthocyanidins.

Red cabbage

Another great dietary source of anthocyanidins is red cabbage. Aiithocyanidins, in particular, have been studied for their protective effects against cancer, cardiovascular disease, diabetes and age related cognitive disorders.

Onions

Onions form the basis for a multitude of cuisines, and it's no wonder why. This humble vegetable is a powerhouse of nutrients and adds flavor to any dish. Onions are a great source of flavonols, which can reduce the risk of prostate cancer.

Kale

Another great source of flavonols is kale. Kale leaves make an excellent base for salads and can be added to soups and stews to boost their nutritional value. If you don't care for the taste, add kale in smoothies and protein shakes to hide the taste.

Parsley

Parsley provides more flavonols in the American diet than any other food. Parsley contains over 130 milligrams of flavonols per gram. Add it to soups and sauces, or sprinkle over dishes before serving.

Box 4: Fruits and vegetables high in flavonoids nutraceuticals

Tea

The easiest way to add flavonoids to your diet is to drink tea. Green, oolong and black teas all contain high levels of flavanols, which have been studied for their benefits to cardiovascular and cognitive health.

Red wine

Another great source of flavanols is red wine. Red wine in moderation has multiple health benefits, especially with lowering risks of cardiovascular disease.

Dark chocolate

Chocolate and cocoa are both high in flavanols. Cocoa, in particular, has been studied for its cognitive-boosting properties and its protective effect on the cardiovascular system.

Citrus fruits

Citrus fruits like oranges, grapefruit, tangerines, lemons and limes contain flavanones. Juicing these fruits results in even more concentrated availability of these healthy plant compounds. You can also squeeze fresh lemon or lime juice into ice water to add nutritional value.

Soya beans

Soya beans come in a variety of different forms and are the best source of isoflavones. Eating edamame, tofu, tempeh and soya sauce are great ways to increase isoflavones in your diet. Isoflavones have been studied for their protective effects against reproductive cancers like breast, ovarian, prostate and testicular cancer.

Flavonols such as quercetin, kaempferol and myricetin are widely prevalent in vegetables. Sources with abundant flavonols include onions, hot peppers, kale, broccoli, rutabagas and spinach. Onions, lettuce, tomatoes, celery, hot peppers, spring onions and broccoli are also major contributors of flavonol compounds to the diet. Legumes are the only vegetables that contain flavan-3-ol compounds, catechins and epicatechin. Many herbs, as well as edible leaves, also contain high levels of flavonols and flavones. Parsley contains very high amounts of apigenin, a flavone, while celery hearts and rutabagas are the other vegetable sources of this type of flavonoid. The major contributors of apigenin to the diet are parsley, celery and lettuce. Thyme is very high in luteolin, which is also present in beets, Brussels sprouts, cabbage and cauliflower. Major contributors of luteolin to the diet include celery, chilli peppers, sweet pepper, lettuce and spinach. Red potatoes and red onions are the only vegetables containing anthocyanidins. Vegetables are not particular sources of flavanones, except eriodictyol which is found in high amounts in peppermint (Haytowitz et al., 2002).

A close relationship has been established between diet and chronic diseases such as cardiovascular disease, diabetes, inflammatory and neurodegenerative diseases as well as various types of cancer (Sandoval et al., 2020). As a result of this link, compounds such as flavanones are increasingly viewed as nutraceutical compounds for the prevention and treatment of chronic diseases (Tapas et al., 2008; Panche et al., 2016; Golla, 2018). As has been noted, due to their antioxidant and other properties, flavonoids have been linked to a range of health benefits (Middleton, 1996; Panche et al., 2016). Flavonoids have been reported to exhibit immuno-modulatory (Catoni et al., 2008), anti-carcinogenic (Knekt et al., 2002), anti-allergenic (Liang et al., 2017) and anti-viral (Saravanan et al., 2015) properties, particularly in chronic disease prevention (AlDrak et al., 2018). Mechanisms of action include the production of lymphocyte, macrophages and natural killer cells and are also found to increase phagocytosis and interferon synthesis (Aghsaghali Mirzaei, 2012). These mechanisms of action are discussed in more depth in an accompanying chapter.

9 References

Aghsaghali Mirzaei, A. Importance of medical herbs in animal feeding: a review. *Ann. Biol. Res.* 2012 3(2):918-923.

Agrawal, A. D. Pharmacological activity of flavonoids: a review. *Int. J. Pahrmaceut Sci. Nano* 2011 4:1394-1398.

Ahmad, A., Kaleem, M., Ahmed, Z. and Shafiq, H. Therapeutic potential of flavonoids and their mechanism of action against microbial and viral infections - a review. *Food Res. Int.* 2015 77:221-235.

AlDrak, N., Abudawood, M., Hamed, S. S. and Ansar, S. Effect of rutin on proinflammatory cytokines and oxidative stress in toxin-mediated hepatotoxicity. *Toxin Rev.* 2018 37(3):223-230.

Andersen, O. M. and Markham, K. R. *Flavonoids: Chemistry, Biochemistry and Applications* (1st edn.). CRC Press 2005.

Aoki, T., Akashi, T. and Ayabe, S. Flavonoids of leguminous plants: structure, biological activity, and biosynthesis. *J. Plant Res.* 2000 113(4):475-488.

Atanassova, M. and Bagdassarian, V. Rutin content in plant products. *J. Univ. Chem. Technol. Metall.* 2009 44(2):201-203.

Babu, P. V. A., Liu, D. and Gilbert, E. R. Recent advances in understanding the anti-diabetic actions of dietary flavonoids. *J. Nutr. Biochem.* 2013 24(11):1777-1789.

Bagchi, D. Nutraceuticals and functional foods regulations in the United States and around the world. *Toxicology* 2006 221(1):1-3.

Basli, A., Soulet, S., Chaher, N., Mérillon, J. M., Chibane, M., Monti, J. P. and Richard, T. Wine polyphenols: potential agents in neuroprotection. *Oxid. Med. Cell. Longev.* 2012 2012:805762.

Berger, M. M. and Shenkin, A. Vitamins and trace elements: practical aspects of supplementation. *Nutr. (Burbank, Los Angel Cty, Calif)* 2006 22(9):952-955.

Bernaert, N., De Paepe, D., Bouten, C., De Clercq, H., Stewart, D., Van Bockstaele, E., De Loose, M. and Van Droogenbroeck, B. Antioxidant capacity, total phenolic and ascorbate content as a function of the genetic diversity of leek (Allium ampeloprasum var. porrum). *Food Chem.* 2012 134(2):669-677.

Brower, V. Nutraceuticals: poised for a healthy slice of the healthcare market? *Nat. Biotechnol.* 1998 16(8):728-731.

Bumgarner, N., Scheerens, J. and Kleinhenz, M. Nutritional yield: a proposed index for Fresh Food improvement illustrated with leafy vegetable data. *Plant Foods Hum. Nutr.* 2012 67:215-222.

Calderón-Montaño, J. M., Burgos-Morón, E., Pérez-Guerrero, C. and López-Lázaro, M. A review on the dietary flavonoid kaempferol. *Mini Rev. Med. Chem.* 2011 11(4):298-344.

Cassidy, A. Berry anthocyanin intake and cardiovascular health. *Mol. Asp. Med.* 2018 61:76-82.

Catoni, C., Schaefer, H. M. and Peters, A. Fruit for health: the effect of flavonoids on humoral immune response and food selection in a frugivorous bird. *Funct. Ecol.* 2008 22(4):649-654.

Chang, S., Tan, C., Frankel, E. N. and Barrett, D. M. Low-density lipoprotein antioxidant activity of phenolic compounds and polyphenol oxidase activity in selected clingstone peach cultivars. *J. Agric. Food Chem.* 2000 48(2):147-151.

Chauhan, B., Kumar, G., Kalam, N. and Ansari, S. H. Current concepts and prospects of herbal nutraceutical: a review. *J. Adv. Pharm. Technol. Res.* 2013 4(1):4-8.

Ciumărnean, L., Milaciu, M. V., Runcan, O., Vesa, Ş. C., Răchis, A. L., Negrean, V., Perné, M. G., Donca, V. I., Alexescu, T. G. M., Para, I. and Dogaru, G. The effects of flavonoids in cardiovascular diseases. *Molecules* 2020 25(18):4320.

Crozier, A., Jaganath, I. B. and Clifford, M. N. Dietary phenolics: chemistry, bioavailability and effects on health. *Nat. Prod. Rep.* 2009 26(8):1001-1043.

Cruickshank, I. A. M., Biggs, D. R., Perrin, D. R. and Whittle, C. P. Phaseollin and phaseollidin relationships in infection-droplets on endocarp of Phaseolus vulgaris. *Physiol. Plant Pathol.* 1974 4(2):261-276.

Cuervo, A., Hevia, A., López, P., Suárez, A., Sánchez, B., Margolles, A. and González, S. Association of polyphenols from oranges and apples with specific intestinal microorganisms in systemic lupus erythematosus patients. *Nutrients* 2015 7(2):1301-1317.

D'Amelia, V., Aversano, R., Chiaiese, P. and Carputo, D. The antioxidant properties of plant flavonoids: their exploitation by molecular plant breeding. *Phytochem. Rev.* 2018 17(3):611-625.

Dewick, P. The shikimate pathway: aromatic amino acids and phenylpropanoids. In: *Medicinal Natural Products: A Biosynthetic Approach* (3rd edn.). Wiley Online Library 2009a;pp.137-184.

Dewick, P. M. *Medicinal Natural Products: A Biosynthetic Approach* (3rd edn.). Wiley 2009b.

Dias, M. C., Pinto, D. C. G. A. and Silva, A. M. S. Plant flavonoids: chemical characteristics and biological activity. *Molecules* 2021 26(17):5377.

Dixon, R. A. and Ferreira, D. Genistein. *Phytochemistry* 2002 60(3):205-211.

Dixon, R. A. and Pasinetti, G. M. Flavonoids and isoflavonoids: from plant biology to agriculture and neuroscience. *Plant Physiol.* 2010 154(2):453-457.

Dukic, M. N., Jakovljevic, V., Popovic, M., Gasic, O., Szabo, A. Pharmacological study of *Mentha longifolia* phenolic extracts. *Int J Pharmacognosy* 1996; 34: 359-64.

Felgines, C., Texier, O., Morand, C., Manach, C., Scalbert, A., Régerat, F. and Rémésy, C. Bioavailability of the flavanone naringenin and its glycosides in rats. *Am J Physiol Gastrointest Liver Physiol* 2000; 279, G1148-G1154.

Garazd, M. M., Garazd, Y. L. and Khilya, V. P. Natural distribution and spectral and biological properties. *Chem. Nat. Compd* 2003 39(1):54-121.

Goff, S. A. and Klee, H. J. Plant volatile compounds: sensory cues for health and nutritional value? *Science* 2006 311(5762):815-819.

Golla, U. Emergence of nutraceuticals as the alternative medications for pharmaceuticals. *Int. J. Compl. Alt Med.* 2018 11(3):155-158.

Grayer, R. J. and Veitch, N. C. Flavanones and dihydroflavonols. In: Anderson, O. M. and Markham, K. R. (Eds), *Flavonoids: Chemistry, Biochemistry and Applications*. Boca Raton, FL: CRC Press/Taylor & Francis Group: 2006:918-1002.

Griesbach, R. Biochemistry and genetics of flower color. *Plant Breed. Rev.* 2005 25:89-114.

Harnly, J. M., Robert, F., Gary, D., Joanne, R. B., David, M. H., Seema, B. H. and Susan, G. Flavonoid content of U. S. fruits, vegetables, nut. *J. Agric. Food Chem.* 2006 54:9966-9977.

Hayat, M., Abbas, M., Munir, F., Qasim Hayat, M., Keyania, R. and Amir, R. Potential of plant flavonoids in pharmaceutics and nutraceutics. *J. Biomol. Biochem.* 2017 1(1):12-17.

Haytowitz, D. B., Pehrsson, P. R. and Holden, J. M. The identification of key foods for food composition research. *J. Food Compos. Anal.* 2002 15(2):183-194.

Hertog, M. G. L., Hollman, P. C. H. and van de Putte, B. Content of potentially anticarcinogenic flavonoids of tea infusions, wines, and fruit juices. *J. Agric. Food Chem.* 1993 41(8):1242-1246.

Hvattum, E. Determination of phenolic compounds in rose hip (Rosa canina) using liquid chromatography coupled to electrospray ionisation tandem mass spectrometry and diode-array detection. *Rapid Commun. Mass Spectrom. RCM* 2002 16(7):655-662.

Iwashina, T. Flavonoid properties of five families newly incorporated into the order caryophyllales. *Bull. Natl Mus. Nat Sci.* 2013 39:25-51.

Jorgensen, R. A. Cosuppression, flower color patterns, and metastable gene expression States. *Science* 1995 268(5211):686-691.

Jung, U. J., Lee, M. K., Park, Y. B., Kang, M. A. and Choi, M. S. Effect of citrus flavonoids on lipid metabolism and glucose-regulating enzyme mRNA levels in type-2 diabetic mice. *Int. J. Biochem. Cell Biol.* 2006 38(7):1134-1145.

Justesen, U. and Knuthsen, P. Composition of flavonoids in fresh herbs and calculation of flavonoid intake by use of herbs in traditional Danish dishes. *Food Chem.* 2001 73(2):245-250.

Kaleem, M. and Ahmad, A. Flavonoids as nutraceuticals. In: *Therapeutic, Probiotic, and Unconventional Foods*. Elsevier 2018; pp.137-154.

Kawaii, S., Tomono, Y., Katase, E., Ogawa, K. and Yano, M. Quantitation of flavonoid constituents in citrus fruits. *J. Agric. Food Chem.* 1999 47(9):3565-3571.

Khan, M. T. H., Orhan, I., Senol, F. S., Kartal, M., Sener, B., Dvorská, M., Smejkal, K. and Slapetova, T. Cholinesterase inhibitory activities of some flavonoid derivatives and chosen xanthone and their molecular docking studies. *Chem. Biol. Interact.* 2009 181(3):383–389.

Khoo, H. E., Azlan, A., Tang, S. T. and Lim, S. M. Anthocyanidins and anthocyanins: colored pigments as food, pharmaceutical ingredients, and the potential health benefits. *Food Nutr. Res.* 2017 61(1):1361779.

Kim, S. H. and Choi, K. C. Anti-cancer effect and underlying mechanism(s) of kaempferol, a phytoestrogen, on the regulation of apoptosis in diverse cancer cell models. *Toxicol. Res.* 2013 29(4):229–234.

Knekt, P., Kumpulainen, J., Järvinen, R., Rissanen, H., Heliövaara, M., Reunanen, A., Hakulinen, T. and Arpo, A. Flavonoid intake and risk of chronic diseases. *Am. J. Clin. Nutr.* 2002 76(3):560–568.

Kolar, F. R., Lingasur, S. M., Kumathalli, T. M. and Gurikar, S. A. Comparative analysis of Phytochemical Constituents and Antioxidant activity of two pomegranate (*Punica granatum* L.) cultivars. *Isr. J. Plant Sci.* 2021 69(1-2):100–109 https://doi.org/10.1163/22238980-bja10042.

Kruger, M. J., Davies, N., Myburgh, K. H. and Lecour, S. Proanthocyanidins, anthocyanins and cardiovascular diseases. *Food Res. Int.* 2014 59(59):41–52.

Liang, Q., Chen, H., Zhou, X., Deng, Q., Hu, E., Zhao, C. and Gong, X. Optimized microwave-assistant extraction combined ultrasonic pretreatment of flavonoids from Periploca forrestii Schltr. and evaluation of its anti-allergic activity. *Electrophoresis* 2017 38(8):1113–1121.

Linuma, M., Tanaka, T., Hamada, K., Mizuno, M., Asai, F., Reher, G.and Kraus, L. Revised structure of neoflavone in *Coutarea hexandra*. *Phytochemistry* 1987 26(11):3096–3097.

Liu, R. H. Health-promoting components of fruits and vegetables in the diet. *Adv. Nut.r Bethesda Md.* 2013 4(3):384S–92S.

Lopez-Lazaro, M. Distribution and biological activities of the flavonoid luteolin. *Mini Rev. Med. Chem.* 2009 9(1):31–59.

Luqman, S., Meena, A., Singh, P., Kondratyuk, T. P., Marler, L. E., Pezzuto, J. M. and Negi, A. S. Neoflavonoids and tetrahydroquinolones as possible cancer chemopreventive agents. *Chem. Biol. Drug Des.* 2012 80(4):616–624.

Malagutti, A. R., Zuin, V. G., Cavalheiro, É. T. G. and Mazo, L. H. Determination of Rutin in green tea infusions using square-wave voltammetry with a rigid carbon-polyurethane composite electrode. *Electroanalysis* 2006 18(10):1028–1034.

Manach, C., Scalbert, A., Morand, C., Rémésy, C. and Jiménez, L. Polyphenols: food sources and bioavailability. *Am. J. Clin. Nutr.* 2004 79(5):727–747.

Matthies, A., Clavel, T., Gütschow, M., Engst, W., Haller, D., Blaut, M. and Braune, A. Conversion of daidzein and genistein by an anaerobic bacterium newly isolated from the mouse intestine. *Appl. Environ. Microbiol.* 2008 74(15):4847–4852.

Medjakovic, S. and Jungbauer, A. Red clover isoflavones biochanin A and formononetin are potent ligands of the human aryl hydrocarbon receptor. *J. Steroid Biochem. Mol. Biol.* 2008 108(1-2):171–177.

Middleton, E. Biological properties of plant flavonoids: an overview. *Int. J. Pharmacogn.* 1996 34(5):344–348.

Murphy, M. M., Barraj, L. M., Herman, D., Bi, X., Cheatham, R. and Randolph, R. K. Phytonutrient intake by adults in the United States in relation to fruit and vegetable consumption. *J. Acad. Nutr. Diet.* 2012 112(2):222–229.

Nishimura, S., Taki, M., Takaishi, S., Iijima, Y. and Akiyama, T. Structures of 4-aryl-coumarin (neoflavone) dimers isolated from Pistacia chinensis BUNGE and their estrogen-like activity. *Chem. Pharm. Bull. (Tokyo)* 2000 48(4):505–508.

Olmedo, D. A., López-Pérez, J. L., Del Olmo, E., Bedoya, L. M., Sancho, R., Alcamí, J., Muñoz, E., Feliciano, A. S. and Gupta, M. P. Neoflavonoids as inhibitors of HIV-1 replication by targeting the tat and NF-κB pathways. *Molecules* 2017 22(2):321. https://doi.org/10.3390/molecules22020321. PMID: 28218730; PMCID: PMC6155902.

Panche, A. N., Diwan, A. D. and Chandra, S. R. Flavonoids: an overview. *J. Nutr. Sci.* 2016 5:e47.

Pedro, A. C., Granato, D. and Rosso, N. D. Extraction of anthocyanins and polyphenols from black rice (Oryza sativa L.) by modeling and assessing their reversibility and stability. *Food Chem.* 2016 191:12–20.

Rahal, A., Mahima, Verma, A. K. and Kumar, A. Phytonutrients and nutraceuticals in vegetables and their multi-dimensional medicinal and health benefits for humans and their companion animals: a review. *J. Biol. Sci.* 2014 14(1):1–19.

Rai, P. K., Jaiswal, D., Mehta, S. and Watal, G. Anti-hyperglycemic potential of Psidium guajava raw fruit peel. *Indian J. Med. Res.* 2009 129(5):561–565.

Ramaa, C. S., Shirode, A. R., Mundada, A. S. and Kadam, V. J. Nutraceuticals-an emerging era in the treatment and prevention of cardiovascular diseases. *Curr. Pharm. Biotechnol.* 2006 7(1):15–23.

Rathmell, W. G. and Bendall, D. S. Phenolic compounds in relation to phytoalexin biosynthesis in hypocotyls of Phaseolus vulgaris. *Physiol. Plant Pathol.* 1971 1(3):351–362.

Rolnik, A., Zuchowski, J., Stochmal, A. O. and Olas, B. Quercetin and kaempferol derivatives isolated from aerial parts of Lens culinaris Medik as modulators of blood platelet functions. *Ind. Crops Prod.* 2020 152, 112536.

Ronis, M. J. J., Pedersen, K. B. and Watt, J. Adverse effects of nutraceuticals and dietary supplements. *Annu. Rev. Pharmacol. Toxicol.* 2018 58:583–601.

Ross, J. A. and Kasum, C. M. Dietary flavonoids: bioavailability, metabolic effects, and safety. *Annu. Rev. Nutr.* 2002 22:19–34.

Sahu, B. D., Kalvala, A. K., Koneru, M., Kumar, J. M., Kuncha, M., Rachamalla, S. S. and Sistla, R. Ameliorative effect of fisetin on cisplatin-induced nephrotoxicity in rats via modulation of NF-κB activation and antioxidant defence. *PLoS ONE* 2014 9(9):e105070.

Sakthinathan, B. and Nandhini, D. U. Phytochemicals-A nutraceutical source of vegetables. *Chem. Sci. Rev. Lett.* 2017 6(24):2133–2137.

Samanta, A., Das, G. and Das, S. Roles of flavonoids in plants. *Int. J. Pharm Sci. Technol.* 2011 6:12–35.

Šamec, D., Karalija, E., Šola, I., Bok, V. V. and Salopek-Sondi, B. The role of polyphenols in abiotic stress response: the influence of molecular structure. *Plants (Basel)* 2021 10(1):118.

Sandoval, V., Sanz-Lamora, H., Arias, G., Marrero, P. F., Haro, D. and Relat, J. Metabolic impact of flavonoids consumption in obesity: from central to peripheral. *Nutrients* 2020 12(8):2393.

Saravanan, D., Thirumalai, D. and Asharani, I. V. Anti-HIV flavonoids from natural products: a systematic review. *Int. J. Res. Pharm. Sci.* 2015 6(3):248–255.

Shimoi, K., Okada, H., Furugori, M., Goda, T., Takase, S., Suzuki, M., Hara, Y., Yamamoto, H. and Kinae, N. Intestinal absorption of luteolin and luteolin 7-O-[beta]-glucoside in rats and humans. *FEBS Lett.* 1998 438(3):220–224.

Stewart, A. J., Bozonnet, S., Mullen, W., Jenkins, G. I., Lean, M. E. and Crozier, A. Occurrence of flavonols in tomatoes and tomato-based products. *J. Agric. Food Chem.* 2000 48(7):2663–2669.

Stump, P. K. and Conn, E. E. *Secondary Plant Products, the Biochemistry of Plants*. New York: Academic Press: 1981:7.

Szkudelska, K. and Nogowski, L. Genistein-a dietary compound inducing hormonal and metabolic changes. *J. Steroid Biochem. Mol. Biol.* 2007 105(1–5):37–45.

Takahashi, A. and Ohnishi, T. The significance of the study about the biological effects of solar ultraviolet radiation using the Exposed Facility on the International Space Station. *Uchu Seibutsu Kagaku* 2004 18(4):255–260.

Tapas, A., Sakarkar, D. and Kakde, R. Flavonoids as nutraceuticals: a review. *Trop. J. Pharm. Res.* 2008 7(3):1089–1099.

Terahara, N. Flavonoids in foods: a review. *Nat. Prod. Commun.* 2015 10(3):521–528.

Tikunov, Y. M., de Vos, R. C. H., González Paramás, A. M., Hall, R. D. and Bovy, A. G. A role for differential Glycoconjugation in the emission of phenylpropanoid volatiles from tomato fruit discovered using a metabolic data fusion approach. *Plant Physiol.* 2010 152(1):55–70.

Truong, V. D., Deighton, N., Thompson, R. T., McFeeters, R. F., Dean, L. O., Pecota, K. V. and Yencho, G. C. Characterization of anthocyanins and anthocyanidins in purple-fleshed sweet potatoes by HPLC-DAD/ESI-MS/MS. *J. Agric. Food Chem.* 2010 58(1):404–410.

Tsuchiya, H. Structure-dependent membrane interaction of flavonoids associated with their bioactivity. *Food Chem.* 2010 120(4):1089–1096.

Umashanker, M. and Shruti, S. Traditional Indian herbal medicine used as antipyretic, antiulcer, anti-diabetic and anticancer: a review. *Int. J. Res. Pharm. Chem.* 2011 1(4):1152–1159.

Vaccaro, M. C., Alfieri, M., Malafronte, N., De Tommasi, N. and Leone, A. Increasing the synthesis of bioactive abietane diterpenes in Salvia sclarea hairy roots by elicited transcriptional reprogramming. *Plant Cell Rep.* 2017 36(2):375–386.

Veitch, N. C. Isoflavonoids of the Leguminosae. *Nat. Prod. Rep.* 2013 30(7):988–1027.

Weingartner, O., Böhm, M. and Laufs, U. Controversial role of plant sterol esters in the management of hypercholesterolaemia. *Eur. Heart J.* 2009 30(4):404–409.

Yao, L. H., Jiang, Y. M., Shi, J., Tomas-Barberan, F. A., Datta, N., Singanusong, R. and Chen, S. S. Flavonoids in food and their health benefits. *Plant Foods Hum. Nutr.* 2004 59(3):113–122.

Zheng, W. and Wang, S. Y. Antioxidant activity and phenolic compounds in selected herbs. *J. Agric. Food Chem.* 2001 49(11):5165–5170.

Chapter 4

Understanding the nutraceutical properties of flavonoids in fruits and vegetables: mechanisms of action

A. D. Diwan, S. N. Harke and A. N. Panche, MGM Institute of Biosciences & Technology, Mahatma Gandhi Mission University, India

1 Introduction

Flavonoids constitute one of the most significant groups of plant phenolics, with about 9000 varieties of flavonoids identified from different plant sources (Ahmad et al. 2015). Figure 1 summarizes the main groups of flavonoids and common fruit and vegetable sources. An accompanying chapter reviews the chemical structure of flavonoids. This chapter focuses on the mechanisms of action of flavonoids in delivering health benefits in preventing or treating a range of diseases.

The chapter first looks at the antioxidant properties of flavonoids and their role in preventing auto-immune diseases. It then looks at the antimicrobial, antifungal, and antiviral activity of flavonoids and their role in the treatment of diabetes. The chapter goes on to discuss the anticancer properties of flavonoids and their anti-neoplastic activity in tumour suppression. It also reviews their role in preventing cardiovascular disease and anti-thrombogenic

http://dx.doi.org/10.19103/AS.2022.0101.21

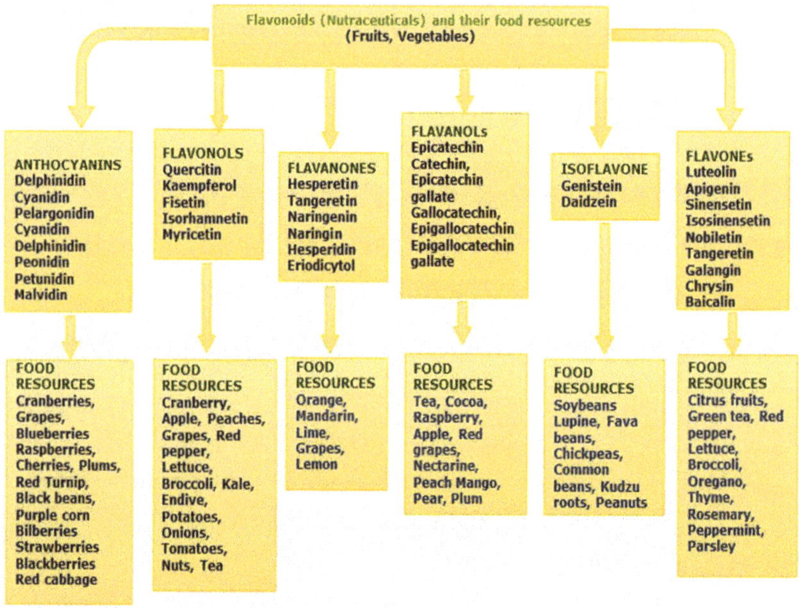

Figure 1 Flavonoids and their food sources.

activity of flavonoids, as well as their neuro-protective, anti-ulcerogenic, anti-inflammatory, and hepato-protective activity. Finally, the chapter discusses biotechnological approaches for the enhanced production of nutraceuticals in fruits and vegetables.

2 Antioxidant properties and role in preventing auto-immune diseases

Plant flavonoids are present in many fruits and vegetables as, for example, flavones and catechins, which are important sources of antioxidants (Dias et al. 2021). The adverse effects of oxidative processes on organic molecules like carbohydrates, lipids, DNA, and proteins in biological systems are reduced by a wide range of nutraceuticals. It is well known that the presence of reactive oxygen species (ROS) in the body has the capacity to damage DNA molecules in cells. They are almost always present in the human body since they are produced by cellular metabolism in response to toxic factors. Superoxide anion radicals (O_2^-), hydrogen peroxide (H_2O_2), hydroxyl radicals (OH), and singlet oxygen (O_2) are just some of the chemical products belonging to the ROS family (Casciaro et al. 2017). Body cells and tissues are continuously threatened by the damage caused by free radicals and ROS which are produced during normal oxygen metabolism (De Groot 1994; Grace 1994). Free radicals and

ROS have been implicated in a large number of human diseases (Wegener and Fintelmann 1999; Ares and Outt 1998).

Antioxidant activity displays a double action by scavenging ROS and by inhibiting oxidases (Dias et al. 2021). One important antioxidant is quercetin, which scavenges highly reactive species such as peroxynitrite and hydroxyl radicals. Genistein, an isoflavanone, is also reported as an important antioxidant (Unnikrishnan et al. 2014). Antioxidants scavenge free radicals by conjugating with glucuronic acid and glutathione with increased uridine diphosphate glucuronosyl transferase (UDPGT) activity and glutathione S-transferase activity, respectively (Ajiboye et al. 2011). The iron chelation activity of quercetin works to reduce oxidative injury induced in the erythrocyte membranes (Procházková et al. 2011). This injury is induced by a number of oxidizing agents such as phenyl hydrazine and acrolein. Some nutraceuticals also demonstrate chelation with metal ions like copper and iron to reduce free radical development.

Different diseases can also be prevented by the intake of antioxidants in the form of flavonoids and other nutraceuticals in fruits and vegetables. Consumption of flavonoids can help in the reduction of the risk of chronic diseases like cancer, heart disease, liver disorders, diabetes, and neurodegenerative diseases. (Cassidy 2018; Rodríguez-García et al. 2019; Ciumărnean et al. 2020). Interaction of biomolecules with antioxidants in food enhances the activation of certain enzymes as well as inhibition of other enzymes (Ajiboye et al. 2010). It has been reported that flavones and catechins are the most powerful nutraceuticals for protecting the body against ROS.

Antioxidants like quercetin, kaempferol, apigenin, morin, myricetin, and rutin have been shown to demonstrate anti-inflammatory, antiallergic, antiviral as well as anticancer activities (Shukla et al. 2019; Ginwala et al. 2019; Tavsan and Kayali 2019; Zakaryan et al. 2017; Lalani and Poh 2020). They have also been suggested to play a protective role in liver diseases, cataracts, and cardiovascular diseases. Quercetin and silybin, acting as free radical scavengers, were shown to exert a protective effect on liver reperfusion ischaemic tissue damage (Fraga et al. 1987).

In terms of dosage, the toxicity of flavonoids is very low in animals. For rats, the LD50 is 2–10 g per animal for most nutraceutical compounds. As a precaution, doses less than 1 mg per adult per day have been recommended for humans (Stavric 1984). Dunnick and Hailey reported that high doses of quercetin over several years might result in the formation of tumours in mice (Dunnick and Hailey 1992). However, in other long-term studies, no carcinogenicity was found (Plakas et al. 1985). Moreover, as described elsewhere in this chapter, quercetin has been reported to be anti-mutagenic in in vivo studies.

2.1 Antioxidants in preventing auto-immune diseases

Antioxidant properties can play a crucial role in controlling auto-immune-related abnormalities. The antioxidant mechanisms are linked to enzymes which play an important role in neutralizing oxidative stress. Most of these enzymes are common in fruits and vegetables. Oxidative stress is associated with the pathogenesis of several auto-immune diseases. The precise details of how the two pathways integrate with each other are not yet completely understood (Wang et al. 2019; Ruggeri et al. 2016). Antioxidant mechanisms involve enzymes which counterbalance ROS effects. The most studied of these enzymes are superoxide dismutase (SOD), catalase (CAT), glutathione peroxidase (GPx), and myeloperoxidase (MPO). These enzymes are involved in the transformation of radicals into less harmful molecules.

In addition to enzymes, there are also some non-enzymatic and exogenous molecules which reduce oxidative stress (Vaccaro et al. 2017; Cannavo et al. 2019). These include vitamins E and C, carotenoids, and flavonoids (Carr and Maggini 2017). Most of these substances are responsible for the nutraceutical properties of fruits and vegetables in reducing chronic illnesses and increasing longevity (Martel et al. 2019). The efficacy of these exogenous antioxidants has not completely been demonstrated in relation to the immune system. Mannucci et al. (2021) observed that exogenous dietary supplementation supports the treatment of diverse auto-immune disorders such as rheumatoid arthritis, lupus, diabetes, and multiple sclerosis. Increasing consumption of fruits and vegetables resulted in a significant reduction of oxidative stress parameters.

3 Antimicrobial, antifungal and antiviral activity

The world is now facing a serious problem of antibiotic resistance which makes several infections harder to treat. Research suggests that natural plant products have antifungal, antiviral, and antibacterial properties (Atoui et al. 2005). Plant extracts have been used to improve the human immune system to combat disease (Vaquero et al. 2007). Plants from different species rich in nutraceuticals have been found to exhibit enhanced antibacterial activity (Cushnie and Lamb 2005; Lalani and Poh 2020; Al Aboody and Mickymaray 2020; Redondo-Blanco et al. 2020).

Numerous nutraceuticals such as apigenin, galangin, glycosides, flavones, isoflavones, chalcones, flavanones, and flavonol have shown effective antibacterial activity (Mishra et al. 2009; Xu and Lee 2001). Esters of phenolic acids have also been found to possess antibacterial, antifungal, and antiviral activities.

3.1 Antimicrobial activity

Research has highlighted the antimicrobial properties of different nutraceuticals such as quercetin, rutin, vanillic acid, gallic acid, and caffeic acid in suppressing pathogenic micro-organisms (Ma et al. 2011). The antimicrobial activity of naringin and quercetin has also been reported (El-Moez and Abdelmonem 2013). Quercetin has, for example, been reported to completely inhibit the growth of the bacterium *Staphylococcus aureus*. Most flavanones having no sugar moiety show antimicrobial activities whereas none of the flavonols and flavonolignans tested have shown inhibitory activity against micro-organisms (Havsteen 1983). Flavonoids show unique molecular activity in forming non-specific bonding with proteins using, for example, hydrophobic, hydrogen, and covalent bonding. There is also evidence that lipophilic flavonoids may also disrupt microbial membranes (Xu and Lee 2001). Antimicrobial properties may include minimizing microbial adhesion and intracellular transport proteins.

Nutraceuticals may also be effective against antibiotic-resistant bacteria. The flavonoid myricetin has been shown to inhibit the growth of the multi-drug resistant (MDR) bacteria *Burkholderia cepacia* (Xu and Lee 2001) due to the inhibition of protein synthesis by *B. cepacia* (Kaul et al. 1985). Similar results were observed in studying the effect of flavonoids such as quercetin on the replication and infectivity of human viruses such as herpes simplex virus type 1 (HSV-1) and respiratory syncytial virus (RSV) (van Dam et al. 2013).

3.2 Antifungal activity

A number of nutraceuticals isolated from the peelings of tangerine orange have been found to exhibit antifungal activity (Ben-Aziz 1967). The nutraceuticals nobiletin and langeritin exhibited strong and weak antifungal activities, respectively, while hesperidin stimulated fungal growth slightly. Chlorflavonin was the first chlorine-containing flavonoid-type antifungal antibiotic produced by strains of *Aspergillus candidus* (Tencate et al. 1973). Antifungal activities have been reported in flavonoids such as neohesperidin, hesperidin, naringin, and derivatives isolated from the citrus plants. These have been modified enzymatically and are being studied on four types of fungal species that usually contaminate food: *Fusarium semitectum*, *Aspergillus parasiticus*, *Penicillium expansum*, and *Aspergillus flavus*. Mycelia growth of *Penicillium expansum* is inhibited by the flavonoid hesperetin. Growth of *Fusarium semitectum*, *Aspergillus parasiticus*, and *Aspergillus flavus* can also be inhibited by another flavonoid called 'prunin decanoate' (Lackeman et al. 1986).

Huesken et al. (1995) reported that the growth of *Aspergillusflavus* is affected by flavones extracted from *Artemisia giraldii*. Similarly, *Eysenhardtia texana* has prenylated flavanones that work against *Candida albicans* (Bast et al.

2007). Antifungal activity has also been reported in flavonoids extracted from citrus fruits while bergamot peel is used to inhibit *Saccharomyces cerevisiae* (Kontruck et al. 1986). A flavanol commonly present in propolis, a chemical produced by bees, has the potential for use against moulds (Izzo et al. 1994).

Grapes are a rich source of nutraceuticals and their pomaces are used to stop the growth of *Zygosaccharomyces bailii* and *Zygosaccharomyces rouxii* (Murakami et al. 1992), Chilean grape pomace extract is reported to have antifungal activity against *Botrytis cinerea* (Kim et al. 1993). *Candida albicans* can be inhibited by extracts from Brazilian grapes which are widely known for their flavonoid content (Gill et al. 1994). Flavonoids also enhanced the antifungal action of fluconazole (an antifungal drug) (Hirano et al. 1994). Modes of action for antifungal activity include induction of apoptosis, ROS accumulation, DNA fragmentation, and mitochondrial damage.

3.3 Antiviral activity

Quercetin, morin, rutin, dihydroquercetin (taxifolin), apigenin, catechin, and hesperidin have been reported to possess antiviral activity against several types of viruses (Selway 1986; Zakaryan et al. 2017; Lalani and Poh 2020). The antiviral activity appears to be associated with nonglycosidic compounds, with hydroxylation at the 3-position apparently a prerequisite for antiviral activity. It has been found that flavonols are more active than flavones against HSV-1and the order of importance was galangin > kaempferol > quercetin (Thomas et al. 1988). A plant flavonoid polymer of molecular weight 2100 daltons was found to have antiviral activity against two strains of type 1 and type 2 herpes simplex viruses (Loewenstein 1979).

Because of the worldwide spread of HIV since the 1980s, the investigation of the antiviral activity of flavonoids has mainly focused on HIV (Ng et al. 1997). Out of twenty-eight flavonoids tested, flavones were found to be generally more effective than flavanones in selective inhibition of HIV-1, HIV-2 and similar immunodeficiency virus infections (Lalani and Poh 2020). Flavonoids can block the binding and penetration of viruses into cells, interfere with viral replication or translation, and prevent the release of the virus. Apigenin, for example, was demonstrated to be active against several DNA and RNA viruses, herpes simplex virus types 1 and 2, hepatitis C and B viruses, and the African swine fever virus by suppressing viral protein synthesis (Yanez et al. 2013).

Kaempferol can also inhibit HIV replication in target cells (Behbahani et al. 2014) and block herpes simplex virus types 1 and 2 by attaching and entering the host cell (Zakaryan et al. 2017). The antiviral activity of quercetin, kaempferol, and epigallocatechin gallate against several influenza virus strains was demonstrated by Wu et al. (2015). It has been reported that nutraceuticals like hesperetin and hesperidin from citrus fruits work against Flavivirus proteases. A number of viruses like dengue virus, yellow fever virus, and West Nile virus belong to the genus Flavivirus. Eberle et al. (2021) found that hesperetin and hesperidin were potentially effective in controlling these viruses. Numerous laboratory studies have shown that certain flavonoids prevent cell replication of H1N1 flu, HIV, SARS, and RSV viruses. Further research is needed to determine how flavonoids work in the body against viruses, and whether they could be an effective preventive measure.

4 Role in treatment of diabetes

Diabetes mellitus (DM) is a chronic metabolic disorder characterized by defective insulin secretion or function, or both. This disorder is prevalent worldwide and has been reported in all age groups. It has been estimated that its prevalence in all age groups may reach upto 4.4% globally by 2030. There are two types of diabetes:

1. Type 1 diabetes is caused by the degradation of pancreatic β- cells, which results in lack of insulin in the body.
2. Type 2 diabetes occurs when the body becomes resistant to insulin or the pancreas loses its ability to produce enough insulin.

It has been suggested that this disease can be treated by nutraceuticals found in plants such as shamimin, daidzein, epicatechin, myricetin, epigallocatechin, hesperidin, naringenin, hesperidin, chrysin, apigenin, genistein, kaempferol, luteolin, and quercetin (Kaul and Ramarao 2001). It has been shown that flavonoids inhibit the activity of the catalyst aldose reductase which converts sugars to sugar alcohols (Tadera et al. 2006). Another study has also shown the role of flavonoids in the inhibition of the enzymes α-glucosidase and α-amylase, which are the key elements in carbohydrate digestion (Kobayashi et al. 2000). The inhibition leads to suppression of carbohydrate digestion which consequently affects glucose absorption with hypoglycaemic effects.

Quercetin is reported to have the ability to regenerate pancreatic islets and also increases insulin release in streptozotocin-induced diabetic rats (Vessal et al. 2003). It has also been reported that quercetin stimulates insulin release and also enhances Ca_2+ uptake from isolated islets cells (Hii and Howell 1984). Nutraceuticals are also known to reduce hyperglycaemia by the interruption of glucose absorption from the intestine. Catechin has been known to inhibit the sodium-dependent glucose transporter SGLT1 (Cherukupalli et al. 2017). Nutraceuticals may also retard the development of cataracts in individuals with inborn errors in sugar metabolism such as diabetes by blocking aldose reductase (van Dam et al. 2013).

5 Anticancer properties and anti-neoplastic activity in tumour suppression

5.1 Anticancer properties

Cancer is a multifactorial heterogeneous chronic disease, which is the main cause of death in some countries together with cardiovascular diseases and is expected (according to some estimates) to increase by about 70% in the next 20 years. The disease has been related to dietary factors, including reduced consumption of vegetables, limited physical activity and consumption of alcohol and tobacco (Ruiz-Cruz et al. 2017). According to some studies, flavonoids have pharmacological properties that inhibit cell damage (Benavente-Garcia and Castillo 2008; Tavsan and Kayali 2019; Rodríguez-García et al. 2019; Peter et al. 2021). Herrera et al. (2009) has reviewed the potential of antioxidant supplementation in inhibiting cancer risk in healthy subjects. Other research suggests that antioxidants may have a role in treating cancer (Hollman and Katan 1999).

Some animal studies have shown good results in the prevention and treatment of various types of cancer using nutraceutical compounds, including in inhibiting initiation, promotion, and progression of the disease. Flavonoids can inactivate the carcinogen, inhibit cell proliferation, repair DNA processes, and reduce oxidative stress at the initiation stage of cancer; they may also exhibit antioxidant activity, induce apoptosis, and develop cytotoxic action against cancer cells at the progression stage (Kozlowska and Szostal-Wegierek 2014; Sokolov et al. 2013; Dias et al. 2021). As inflammation is closely related to tumour promotion, bioactive compounds like flavonoids are expected to exert chemo preventive effects against carcinogenesis, especially in the promotion and progression stages (Kang et al. 2011; Rodríguez-García et al. 2019).

Consumption of flavonoid-rich foods has been reported to lower the risk of cancer. Suggested mechanisms of action include cell cycle arrest, inhibiting proliferation, antioxidation, and induction of apoptosis. HeLa cells treated with flavonoid extracts from *Andrographis glandulosa* exhibited loss of mitochondrial membrane potential (MMP) and apoptosis. This cytotoxic activity of flavonoid extracts makes them promising candidates for the production of anticancer drugs (Majewska and Czeczot 2012). Flavonoids have the potential to treat cancer at each stage of the disease (Sengupta et al. 2004).

It is necessary to identify and separate anticancer constituents from plants for potential treatments (Matthies et al. 2008a,b). *Scutellaria baicalensis* infusions have been reported to cause growth inhibition in several cancer cell lines such as breast cancer, colon cancer, hepatocellular carcinoma, and squamous cell carcinoma (Syed et al. 2011). Treatment of human melanoma cells with the flavonol fisetin reduced cell viability by moderating G1 phase arrest (Funakoshi-Tago et al. 2011). Al-Ishaq et al. (2021) reported that flavonoids from fruits and vegetables undergo enzymatic metabolism with the help of microbiota available in the gut intestine which inhibits gastrointestinal cancer. Sindhu et al. (2021) reported that nutraceuticals like flavan-3-Ol, procyanidin, isoflavone, flavone, flavanol, flavanones, lignans, anthocyanidins, and catechin all have anticancer properties and can be used in prevention and treatment of breast cancer. The effect of these nutraceuticals as anticancer drugs differs with the type of cancer, doses, and the cell lines.

5.2 Anti-neoplastic activity in tumour suppression

A number of flavonoids have exhibited anti-neoplastic activity. A number of studies have shown that quercetin inhibits cell growth and colony formation in tumours (Kontruck et al. 1986; Izzo et al. 1994; Murakami et al. 1992) Flavonoids such as kaempferol, catechin, taxifolin, and fisetin can suppress the cell growth (Kim et al. 1993; Gill et al. 1994). The screening studies carried out for the anti-leukaemic efficacy of 28 natural and synthetic flavonoids on human promyelocytic leukaemic HL-60 cells revealed that the isoflavone genistein exhibited a particularly strong effect (Hirano et al. 1994; Wei et al. 1990).

6 Role in preventing cardiovascular disease and anti-thrombogenic activity

Heart disease is a major cause of mortality worldwide and its prevalence has been linked to patterns of fruit and vegetable consumption (Ruiz-Cruz et al. 2017). Studies have reported that flavonoids help reduce blood cholesterol and glucose levels in humans. Sufficient intake of flavonoids has been reported to reduce the effects of coronary heart disease (Bohn et al. 2012). Flavonoids have been shown to reduce LDL cholesterol and regulate anti-inflammatory and antioxidant activities (Novotny et al. 2015; Shukla et al. 2019; Dias et al. 2021). The antioxidant and chelating properties of flavonoids inhibit or inactivate ROS, which play an important role in the cardiovascular system (Benavente-García et al. 1997; Sokolov et al. 2013; Ciumărnean et al. 2020). Mechanisms of action of flavonoids in preventing cardiovascular disease include improving coronary vasodilatation, decreasing the ability of platelets in the blood to clot, preventing LDLs from oxidizing, inhibiting inflammation propagation, anti-apoptotic, anti-necrotic, free radical scavenging abilities, and effects on capillary blood vessel (Benavente-García et al. 1997; Arct and Pytkowska 2008; Soobrattee et al. 2005). Flavonoids block the angiotensin-converting enzyme (ACE) that is responsible for raising blood pressure. Platelet stickiness and aggregation are prevented by blocking the cyclo-oxygenase enzyme that breaks down prostaglandins (van Dam et al. 2013).

Quercetin from cranberries can help lower blood pressure (Nichols and Morimoto 2000). Flavonoids present in *Camellia sinensis* (tea plant) are known to prevent cardiovascular diseases. Tea also contains flavonoids which reduce levels of cholesterol in the blood, damage caused by oxidative stress, lower blood pressure, and reduce inflammation. Many studies have suggested that flavonoids in tea also elevate endothelial function (McCullough et al. 2012). Specific hydroxyl (-OH) groups in isoflavones have been shown to be critical in inhibiting phosphodiesterase isoenzymes (Mohan et al. 2014). This may also explain the therapeutic effects of flavonoids on platelet aggregability and blood pressure.

A study by Rice-Evans et al. (1996) on the link between intake of flavonoids and risk of coronary heart disease showed that intake of natural flavonoids can reduce the risk of cardiovascular diseases by a factor of 2.4. The study suggested that antioxidant and anti-thrombotic properties contribute to this improved protection. Ahmed (2021) reported that intake of nutraceuticals available in fruits, vegetables, tea, coffee, and wine reduces risk factors associated with heart disorders. Besides cardiovascular health benefits, they also emphasized other health benefits like vasodilatory effects, prevention of endothelial dysfunction, inhibition of platelets aggregation, and smooth muscle cell proliferation along with antioxidant, anti-inflammatory, anti-obesity, anti-diabetic, and anti-atherosclerotic effects.

6.1 Anti-thrombogenic activity

Platelet aggregation plays a pivotal role in the physiology of thrombotic diseases. Activated platelets adhering to vascular endothelium generate lipid peroxides and oxygen-free radicals which inhibit the endothelial formation of prostacyclin and nitrous oxide. Flavonoids such as quercetin, kaempferol, and myricetin have been proved to be effective against platelet aggregation (Osman et al. 1998). Flavonols possess this anti-thrombotic property because they directly scavenge free radicals and are thus able to maintain a proper concentration of endothelial prostacyclin and nitric oxide (Gryglewski et al. 1987). Tea pigments, for example, have been shown to reduce blood coagulability, increase fibrinolysis, and prevent platelet adhesion and aggregation (Lou et al. 1989).

7 Neuro-protective activity

Flavonoids compounds have a broad range of biological activities including antidepressant (Paulke et al. 2008), cytotoxic, anti-tumour (Murakami et al. 2004), antioxidant (Ali 2011), and anti-inflammatory (Araujo and Leon 2001; De Felice 1995; Ginwala et al. 2019; Shukla et al. 2019) activities. The flavonoid folecitin has been found to have neuro-protective effects (Farooq et al. 2021).

Camomile (*Matricaria recutita*) flowers have been used for their relaxing properties due to the presence of flavone apigenin (Jager and Saaby 2011). *Tanacetum parthenium* commonly known as everfew has also been utilized for the treatment of headaches (Aguirre-Hernández et al. 2010). Tilia sp. such as Linden blooms have been employed used as a tranquilizer, involving flavonols such as kaempferol and quercetin (Saaby et al. 2009). Heather (*Calluna vulgaris*) containing quercetin had been utilized as a nerve soothing agent (Fisher et al. 2006).

Cocoa flavonoid nutraceuticals have been identified as having neuro-protective properties, including increased cerebral blood flow (CBF). Human trials reported that a 900 mg/day treatment with cocoa for a week increased CBF in grey matter (Buitrago-Lopez et al. 2011). A meta-analysis of three individual studies with over 114 000 participants reported a reduction in risk of stroke by 29% in consumers who consumed higher levels of chocolate (Salas et al. 2011).

The consumption of flavonoids has been linked to neuro-protective effects in preventing diseases related to the nervous system such as Alzheimer's disease, Parkinson's disease, and dementia (Grassi et al. 2016). It has been proposed that the neurobiological actions of flavonoids may occur in two major ways (Sokolov et al. 2013):

- The first is regulation of neuronal signal cellular cascades, which are caused by neurotoxic substances and may damage neurogenesis, neuronal function, and brain connectivity.
- Second, flavonoids seem to improve blood flow towards the brain and exert beneficial effects on the peripheral and central nervous system.

There is evidence that antioxidants from diets rich in flavonoids (at low concentration) help to maintain human cognitive functions like memory, protect vulnerable neurons, stimulate neuronal regeneration, and prevent oxidative neuronal damage (Huntley 2009). Studies carried out in rats, for example, showed that natural extracts rich in polyphenolic compounds improve cognitive and other functions (Pandey and Rizvi 2009).

8 Anti-ulcerogenic, anti-inflammatory and hepato-protective activity

8.1 Anti-ulcerogenic activity

Some studies suggest that flavonoids possess anti-ulcerogenic activity. Flavonoids of *Ocimum basilicum* have been shown to decrease ulcer index as well as inhibit gastric acid and pepsin secretions in rats (Alarcon et al. 1994).

Quercetin, rutin, and kaempferol are reported to inhibit gastric damage (Izzo et al. 1994).

8.2 Anti-inflammatory activity

Flavonoids also possess anti-inflammatory properties (Tanaka and Takahashi 2013 Maleki et al. 2019; Shukla et al. 2019). Fisetin, luteolin, and apigenin are reported to have good anti-inflammatory properties. The anti-inflammatory properties of fisetin have been shown to the diminish effects of asthma, a disease caused by airway inflammation (Wang et al. 2011). In China and Japan, most of the herbs used in medicines and infusions are obtained from the roots of *Scutellaria baicalensis*. These roots have several groups of flavonoids such as wogonin, baicalein, and baicalin. *Scutellaria baicalensis* infusions have been used for hyperlipidemia, inflammatory diseases, allergies and arteriosclerosis (Houghton et al. 2006). Of the 350 different species of desmodium, 28 are found only in China (Dzoyem et al. 2013). These have been used to isolate nearly 200 compounds, notably alkaloids and flavonoids (Manthey 2000). Desmodium extracts with these compounds have been found to have anti-inflammatory and other protective properties (Ma et al. 2011).

The anti-inflammatory properties of nutraceuticals involve the biosynthesis of protein cytokines that moderate the attachment of circulating leukocytes to the location of the injury. Some nutraceuticals are potent inhibitors of synthesis of dominant pro-inflammatory molecules called 'prostaglandins' (Beretz and Cazenave 1988). Several nutraceuticals are involved in platelet adhesion, aggregation, and secretion (Kumar and Pandey 2013). The effects of nutraceuticals on platelets have been related to the carbon monoxide prohibition of arachidonic acid metabolism (Sarris et al. 2013).

Flavone/flavonol glycosides and flavonoid aglycons have been reported to exert significant anti-inflammatory activity in animal models (Lee et al. 1993; Hang et al. 2002). Hesperidin, a citrus flavonoid, possesses significant anti-inflammatory and analgesic effects (Shahid et al. 1998). Nutraceuticals like apigenin, luteolin, and quercetin have been reported to exhibit anti-inflammatory activity (Farmica and Regelson 1995; Shukla et al. 2019; Ginwala et al. 2019). A number of reports have been published which demonstrate that flavonoids can modulate arachidonic acid metabolism via the inhibition of cyclooxygenase (COX) and lipooxygenase activity (LO). It has been suggested that the anti-inflammatory and antiallergic properties of flavonoids are the consequence of their inhibitory actions on arachidonic acid metabolism (Ferrandiz and Alcaraz 1991). Kaempferol, quercetin, myricetin, and fisetin were reported to possess LO and COX-inhibitory activities (Kim et al. 1998; Jachak 2001).

8.3 Hepato-protective activity

The liver is subject to acute and potentially lethal injury by several substances including phalloidin (the toxic constituent of the mushroom, *Amanita phalloides*), CCl_4, galactosamine, ethanol, and other compounds. Flavonoids have also been found to possess hepato-protective activity. In a study carried out to investigate the flavonoid derivatives silymarin, apigenin, quercetin, and naringenin, silymarin was found to be most effective against microcrystin LR-induced hepatotoxicity (Di Carlo et al. 1993). The flavonoids rutin and venoruton showed regenerative and hepato-protective effects in experimental cirrhosis (Lorenz et al. 1994).

9 Biotechnological approaches for enhanced production of nutraceuticals in fruits and vegetables

Various ways are being tried to enhance the production of plant secondary metabolites such as flavonoids. The use of plant flavonoids is challenging in part because they are generally found in low and variable concentrations (Leonard et al. 2007; Chandra 2012).

Biotechnological approaches such as metabolic engineering and plant cell and tissue culture techniques provide the potential for improved production of secondary metabolites in plants for drug production (Verpoorte and Memelink 2002). Challenges in the use of biotechnological techniques include screening, identification, and selection of high-yielding secondary metabolites, plant cell immobilization, culture media composition, optimizing parameters to ensure a high yield of secondary metabolites, elicitation techniques, large-scale cultivation, and production using bioreactor systems (Halder et al. 2019).

Many approaches can be applied such as over-expression of regulatory genes, reducing competitive pathways, overcoming rate-limiting steps, and so on. Metabolic engineering is an emerging technique for enhanced production of a specific secondary metabolite. Several methods are being used for altering genes, enzymes, and proteins involved in the synthesis of a metabolite. Competitive pathways can be blocked by anti-sense genes which result in higher production of desired secondary metabolites (Verpoorte et al. 2000).

Utilizing micro-organisms for over-expression of a plant gene is one method, involving bioconversion of precursors into desired chemicals (Howat et al. 2014). Cell and tissue culture is another approach for the extraction of target chemical compounds for drug production (Amer 2018). However, there are problems associated with plant cell culture which include genetic instabilities, slow and variable culture growth as well as increased susceptibility to stress and aggregation (Lee et al. 2010). The problem has been addressed by the initiation of culture using undifferentiated cambial meristematic cells

(CMC), which are multipotent with stem-cell-like properties rather than being dedifferentiated cells (DDC) (Zhang et al. 2006). The utilization of CMCs can provide a strong foundation for future metabolic engineering strategies in flavonoid biosynthesis.

Another novel approach is biochemical synthesis using micro-organisms (Chandra 2012). Both yeast and bacteria have been used as model organisms for bioreactor-based flavonoid production (Chemler et al. 2006). The gut bacteria *Escherichia coli* and the bread and wine yeast *Saccharomyces cerevisiae* (Beekwilder et al. 2006) have been utilized for the biosynthesis of different flavonoids. This has been made possible through the simultaneous co-expression of several downstream enzymatic activities in a flavanone biosynthetic pathway (Sut et al. 2016). Fowler and Koffas (2009) have reported that flavonoid nutraceuticals can be derived from micro-organisms, with a focus on heterologous protein expression. Such processes appear as attractive production alternatives for the commercial synthesis of these high-value compounds.

Olaiya et al. (2015) have highlighted the use of plant hormones. They also reported that manipulation in soilless culture solutions can promote the antioxidant content of tomatoes, including vitamin C, flavonoids, lycopene, and β-carotene. In addition, the spraying of nutrients such as potassium in field conditions has a strong stimulatory effect on the lycopene content of tomatoes.

Transgenic technology offers a rapid way to develop desirable traits. An important trend is the shift from enhancing single nutritional compounds towards enhancing multiple nutrients and phytochemicals in order to harness their synergistic interactions. This could be achieved by the use of strategies having pleiotropic effects such as bio-regulators and multigene engineering. However, the full potential of these technologies has yet to be realized.

Biotechnological approaches using hairy root culture have greatly enhanced the production of flavonoid compounds without the loss of concentration frequently observed in cell suspension cultures. Because hairy roots originate from a single plant cell infection by *Agrobacterium rhizogenes*, they are usually considered genetically stable. Nanoparticles also offer a new strategy in enhancing secondary metabolite production. However, further research is required to elucidate the effects of nanoparticles in the mechanisms of secondary metabolite synthesis in medicinal plants.

Elicitation is also one of the widely used methods for accelerating the process of biosynthesis and yield of secondary metabolites (Ramakrishna and Ravishankar 2011; Wang and Wu 2013). In this method, different types of elicitors are used. These molecules possess the ability to induce or enhance the biosynthesis of specific secondary metabolites (Namdeo 2007; Ramakrishna and Ravishankar 2011; Wang and Wu 2013). The use of precise parameters impacts the elicitation process and consequently determines the effectiveness of the technique (Kaur and Pati 2018; Dhiman et al. 2018; Naik and Al-Khayri 2016). These parameters include elicitor type, concentrations, duration of exposure, treatment schedule, culture type, cell line, medium composition, presence or absence of growth regulation, and age or stage of the culture at the time of elicitor treatment. These elicitors act as signalling molecules. They initiate the elicitation process through interaction with elicitor-specific receptors present on the plant cell membrane which ultimately initiate the signal transduction. This in turn changes the expression levels of the regulatory transcription factors/genes of the specific secondary metabolic pathway, resulting in increased synthesis of secondary metabolites (Wang and Wu 2013; Vasconsuelo and Boland 2007; Mishra et al. 2012; Zhai et al. 2016).

10 Conclusion and future trends in research

As populations become more affluent and live longer, there is a growing burden of chronic diseases including cancer, cardiovascular disease, diabetes, hypertension, arthritis, obesity, and allergy. Nutraceutical compounds such as flavonoids have been found to have bioactive properties such as anti-inflammatory, anticancer, anti-ageing, cardio-protective, neuro-protective, immune-modulatory, anti-diabetic, anti-parasitic, and antiviral properties (Saini et al. 2017; Jucá et al. 2020; Fraga et al. 2019; Shukla et al. 2019; Sandoval et al. 2020) (Fig. 2) (Box 1). From physiological to psychological health, nutraceuticals have the potential to not only treat a wide array of illnesses but also boost energy, relieve anxiety, improve overall health, reduce the effects of ageing and enhance sleep quality. Nutraceuticals can be used as an

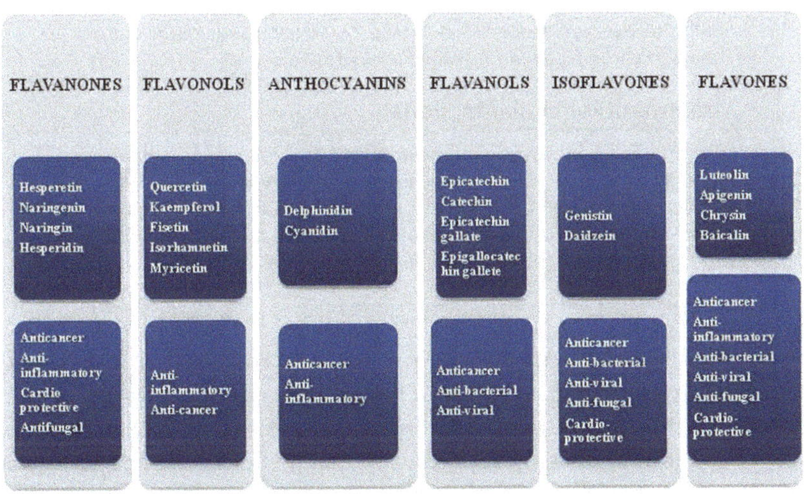

Figure 2 Flavonoids and their possible therapeutic applications.

additional treatment alongside more conventional therapies. Consumers are attracted to the use of flavonoids and other nutraceuticals as an alternative or complementary treatment to conventional synthetic pharmaceuticals in the prevention and treatment of these chronic diseases. As a result, an Assocham report in 2018 predicted the Indian nutraceuticals market alone could grow from US$4 billion in 2017 to US$18 billion in 2025. The use of nutraceuticals as a complementary therapy may, for example, be beneficial in the prevention and management of COVID-19. The main nutraceuticals under evaluation for their potential preventive and treatment uses are zinc, vitamin C, vitamin D, selenium, glutathione, curcumin, and omega-3 fatty acid.

Box 1: Other health benefits flavonoids/ nutraceuticals

Cancer prevention

A published review of all flavonoid studies over eleven years concluded that a diet rich in flavonoids leads to a reduced risk of several different cancers. These studies indicate the antioxidant activity of flavonoids protect against breast, prostate, and colorectal cancers.

It's important to note that these studies suggest that different flavonoids have a protective effect against specific cancer types. For example, anthocyanidins decrease lung cancer risk, while flavonols reduce the risk of prostate cancer. Therefore, it's best to consume various plant food sources to obtain different flavonoid subtypes.

Another medical review evaluated the anti-inflammatory and pain-relieving properties of flavonoids, as demonstrated in several studies. Studies have shown that flavonoids reduce the cellular response to pain. Researchers believe flavonoids could be used medicinally to manage chronic pain and treat inflammatory diseases.

Treatment for viral infections

Flavonoids have proven antibacterial and antiviral effects. Numerous laboratory studies have shown that certain flavonoids prevent cell replication of H1M1 flu, HIV, SARS, and RSV viruses. Further research is needed to determine how flavonoids work in the body against viruses, and whether they could be an effective preventative measure.

It has been suggested that a diet rich in bright fruits and vegetables like grapes, carrots, and sweet potatoes could cut the risk of many neurological disorders including Alzheimer's disease. A 20-year study tracking 77 000 older people found that those who ate most yellow and orange fruits and vegetables were 38% less likely to suffer mental decline (Baroni et al. 2021; Krikorian et al.

2010). Resveratrol in grapes is known to promote anti-ageing effects and protect against β-amyloid peptide formation in the brain and it can increase the expression of nuclear factor-related genes which has neuro-protective effects (Lee et al. 2017; Buendia et al. 2016). Consumption of grapes can also create positive effects on cognitive health including the modulation of CBF, the modulation of glucose metabolism, and the inhibition of monoamine oxidase (MAO) activity (Haskell-Ramsay et al. 2017).

Berries contain large amounts of anthocyanins, flavonols (such as quercetin, myricetin, and kaempferol), proanthocyanidins, ellagitannins, and phenolic acids (Rajaram et al. 2019). Blackberries are a particularly rich source of anthocyanins and flavonols. Research has shown that consumption of blackberries helps in the restoration of memory (Haskell-Ramsay et al. 2017; Bell et al. 2015). Their beneficial effects have been attributed not only to their antioxidant action but also to a decrease in blood pressure, mitigation of neuroinflammation, protection against cardiovascular risk, interaction with gut microbiota, increased neurogenesis, and modulated glucose regulation (Rajaram et al. 2019; Haskell-Ramsay et al. 2017). High intake of blueberries, blackberries, and cherries has been associated with a 24% reduced risk of memory loss while apples or a handful of strawberries cut the risk by 20% (Kent et al. 2017). Consumption of cherry juices has been shown to improve memory, learning, visual attention, and verbal fluency in older adults (Kent et al. 2017). Cherries have been shown to scavenge free radicals, to have anti-inflammatory effects, and to improve antioxidant defences in older adults (Kent et al. 2017).

Pomegranates are rich in anthocyanins and hydrolyzable tannins, especially glycosides of ellagic acid (Rajaram et al. 2019). Oranges are a rich source of flavanones (hesperidin) and flavonols (rutin and quercetin). Rutin has been shown to decrease and reverse amyloid β-protein fragment fibril formation and in vitro aggregation, prevent mitochondrial damage, reduce OxS marker generation, and enhance antioxidant enzymes. Orange juice appears to benefit cognitive functions and psychomotor performance across age groups (Bell et al. 2015). Apples provide quercetin, glycosides, and epicatechin (Bondonno et al. 2014). Supplementation with apple juice could help maintain cognitive performance (Bondonno et al. 2014).

Onion is an excellent source of flavonoids, particularly quercetin. It has neuro-protective and anti-dysglycemic effects. It has been suggested that the ingestion of quercetin-rich onions improves cognitive function and reduces cognitive decline in elderly people (Balakrishnan et al. 2021).

Looking to the future, although some information is available on the number of flavonoids in different fruits and vegetables, we still need more data on concentrations in different plants. More clinical trials are necessary to generate reliable data to validate therapeutic applications of flavonoids. There needs to be mechanism and physiological functions, side effects and shelf life. Challenges include the need to accurately test the efficacy and safety of such chemical compounds, identify their exact mode of action, evaluate their bioavailability, and study possible interactions with various body organs and systems.

11 References

Aguirre-Hernández, E., González-Trujano, M. E., Martínez, A. L., Moreno, J., Kite, G., Terrazas, T. and Soto-Hernández, M. HPLC/MS analysis and anxiolytic-like effect of quercetin and kaempferol flavonoids from Tilia americana var. mexicana. *J. Ethnopharmacol.* 2010 127(1):91-97.

Ahmad, A., Kaleem, M., Ahmed, Z. and Shafiq, H. Therapeutic potential of flavonoids and their mechanism of action against microbial and viral infections - a review. *Food Res. Int.* 2015 77:221-235.

Ahmed, A. Flavonoids and cardiovascular risk factors - a review. *Pharm. Adv.* 2021 3(3):521-547.

Ajiboye, T. O., Salau, A. K., Yakubu, M. T., Oladiji, A. T., Akanji, M. A. and Okogun, J. I. Acetaminophen perturbed redox homeostasis in Wistar rat liver: protective role of aqueous *Pterocarpus osun* leaf extract. *Drug Chem. Toxicol.* 2010 33(1):77-87.

Ajiboye, T. O., Salawu, N. A., Yakubu, M. T., Oladiji, A. T., Akanji, M. A. and Okogun, J. I. Antioxidant and drug detoxification potentials of Hibiscus sabdariffa anthocyanin extract. *Drug Chem. Toxicol.* 2011 34(2):109-115.

Al Aboody, M. S. and Mickymaray, S. Anti-fungal efficacy and mechanisms of flavonoids. *Antibiotics* 2020 9(2):45.

Alarcon, D. L., Martin, M. J., Locasa, C. and Motilva, V. Antiulcerogenic activity of flavonoids and gastric protection. *Ethnopharmacol* 1994 42:161-170.

Al-Ishaq, R. K., Liskova, A., Kubatka, P. and Büsselberg, D. Enzymatic metabolism of flavonoids by gut microbiota and its impact on gastrointestinal cancer. *Cancers* 2021 13(16):3934. https://doi.org/10.3390/cancers13163934.

Ali, G. Flavonoids and phenolic acids: role and biochemical activity in plants and human. *J. Med. Plants Res.* 2011 5(31): 1-5.

Amer, A. Biotechnology approaches for in vitro production of flavonoids. *J. Microbiol. Biotechnol. Food Sci.* 2018 7(5):457-468.

Araujo, C. C. and Leon, L. L. Biological activities of *Curcuma longa* L. *Mem. Inst. Oswaldo Cruz* 2001 96(5):723-728.

Arct, J. and Pytkowska, K. Flavonoids as components of biologically active cosmeceuticals. *Clin. Dermatol.* 2008 26(4):347-357.

Ares, J. J. and Outt, P. E. Gastroprotective agents for the prevention of NSAID-induced gastropathy. *Curr. Pharm. Des.* 1998 4(1):17-36.

Atoui, A. K., Mansouri, A., Boskou, G. and Kefalas, P. Tea and herbal infusions: their antioxidant activity and phenolic profile. *Food Chem.* 2005 89(1):27-36.

Balakrishnan, R., Duk-Yeon, S. A., Su-Kim, C. I. and Dong-Kug, C. and Choi, D. K. Natural phytochemicals as novel therapeutic strategies to prevent and treat parkinson's disease: current knowledge and future perspectives. *Oxid. Med. Cell. Longev.* 2021:32. https://doi.org/10.1155/2021/6680935.

Baroni, L., Sarni, A. R. and Zuliani, C. Plant foods rich in antioxidants and human cognition: a systematic review. *Antioxidants (Basel)* 2021 10(5):714. https://doi.org/10.3390/antiox10050714.

Bast, A., Kaiserová, H., den Hartog, G. J., Haenen, G. R. and van der Vijgh, W. J. Protectors against doxorubicin induced cardiotoxicity: flavonoids. *Cell Biol. Toxicol.* 2007 23(1):39–47.

Beekwilder, J., Wolswinkel, R., Jonker, H., Hall, R., Ric de vos, C. H. and Bovy, A. Production of resveratrol in recombinant microorganisms. *Appl. Environ. Microbiol.* 2006 72(8):5670–5672.

Behbahani, M., Sayedipour, S., Pourazar, A. and Shanehsazzadeh, M. In vitro anti-HIV-1 activities of kaempferol and kaempferol-7-O-glucosides isolated from Securigera Securidaca. *Res. Pharm. Sci.* 2014 9(6):463–469.

Bell, L., Lamport, D. J., Butler, L. T. and Williams, C. M. A review of the cognitive effects observed in humans following acute supplementation with flavonoids, and their associated mechanisms of action. *Nutrients* 2015 7(12):10290–10306.

Benavente-Garcia, O. and Castillo, J. Update on uses and properties of citrus flavonoids: new findings in anti-cancer, cardiovascular and anti-inflammatory activities. *J. Agric. Food Chem.* 2008 56(15): 6185–6205.

Benavente-García, O., Castillo, J., Marin, F. R., Ortuño, A. and Del Rio, J. A. Uses and properties of citrus flavonoids. *J. Agric. Food Chem.* 1997 45(12):4505–4515.

Ben-Aziz, A. Nobiletin is main fungistat in tangerines resistant to Mal Secco. *Science* 1967 155(3765):1026–1027.

Beretz, A. and Cazenave, J. P. The effect of flavonoids on blood-vessel wall interactions. *Prog. Clin. Biol. Res.* 1988 280:187–200.

Bohn, S. K., Ward, N. C., Hodgson, J. M. and Croft, K. D. Effects of tea and coffee on cardiovascular disease risk. *Food Funct.* 2012 3(6):575–591.

Bondonno, C. P., Downey, L. A., Croft, K. D., Scholey, A., Stough, C., Yang, X., Considine, M. J., Ward, N. C., Puddey, I. B., Swinny, E., Mubarak, A. and Hodgson, J. M. The acute effect of flavonoid-rich apples and nitrate-rich spinach on cognitive performance and mood in healthy men and women. *Food Funct.* 2014 5(5):849–858.

Buendia, I., Michalska, P., Navarro, E., Gameiro, I., Egea, J. and León, R. Nrf2–ARE pathway: an emerging target against oxidative stress and neuroinflammation in neurodegenerative diseases. *Pharmacol. Ther.* 2016 157:84–104.

Buitrago-Lopez, A., Sanderson, J., Johnson, L., Warnakula, S., Wood, A., Angelantonio, E. D. and Franco, O. H. Chocolate consumption and cardiometabolic disorders: systematic review and meta-analysis. *BMJ* 2011 343:d3488.

Cannavo, S. P., Riso, G., Casciaro, M., Di Salvo, E. and Gangemi, S. Oxidative stress involvement in psoriasis: a systematic review. *Free Radic. Res.* 2019 53(8):829–840.

Carr, A. C. and Maggini, S. Vitamin C and immune function. *Nutrients* 2017 9(11):E1211.

Casciaro, M., Di Salvo, E., Pace, E., Ventura-Spagnolo, E., Navarra, M. and Gangemi, S. Chlorinative stress in age-related diseases: a literature review. *Immun. Ageing* 2017 14:21.

Cassidy, A. Berry anthocyanin intake and cardiovascular health. *Mol. Asp. Med.* 2018 61:76–82.

Chandra, S. Natural plant genetic engineer agrobacterium rhizogenes: role of T-DNA in plant secondary metabolism. *Biotechnol. Lett.* 2012 34(3):407–415.

Chemler, J. A., Yan, Y. and Koffas, M. A. Biosynthesis of isoprenoids, polyunsaturated fatty acids and flavonoids in Saccharomyces cerevisiae. *Microb. Cell Factories* 2006 5(1):20.

Cherukupalli, N., Bhumireddy, S., Akella, S., Sataniya, A., Sripadi, P., Khareedu, V. and Vudem, D. Phytochemical profiling and in vitro anticancer activity of purified flavonoids of Andrographis glandulosa. *Planta Med. Int. Open* 2017 4(1):e24-34.

Ciumărnean, L., Milaciu, M. V., Runcan, O., Vesa, ŞC., Răchis, A. L., Negrean, V., Perné, M. G., Donca, V. I., Alexescu, T. G. M., Para, I. and Dogaru, G. The effects of flavonoids in cardiovascular diseases. *Molecules* 2020 25(18):4320.

Cushnie, T. P. T. and Lamb, A. J. Antimicrobial activity of flavonoids. *Int. J. Antimicrob. Agents* 2005 26(5):343-356.

De Felice, S. L. The nutraceutical revolution: its impact on food industry R&D. *Trends Food Sci. Technol.* 1995 6(2):59-61.

De Groot, H. Reactive oxygen species in tissue injury. *Hepatogastroenterology* 1994 41(4):328-332.

Dhiman, N., Patial, V. and Bhattacharya, A. The current status and future applications of hairy root cultures. In: Kumar, N. (Ed.), *Biotechnological Approaches for Medicinal and Aromatic Plants*. Springer, Singapore 2018;pp.87-155.

Di Carlo, G., Autore, G., Izzo, A. A., Maiolino, P., Mascolo, N., Viola, P., Diurno, M. V. and Capasso, F. Inhibition of intestinal motility and secretion by flavonoids in mice and rats: structure-activity relationships. *J. Pharm. Pharmacol.* 1993 45(12):1054-1059.

Dias, M. C., Pinto, D. C. G. A. and Silva, A. M. S. Plant flavonoids: chemical characteristics and biological activity. *Molecules* 2021 26(17):5377. https://doi.org/10.3390/molecules26175377.

Dunnick, J. K. and Hailey, J. R. Toxicity and carcinogenicity studies of quercetin, a natural component of foods. *Fundam. Appl. Toxicol.* 1992 19(3):423-431.

Dzoyem, J. P., Hamamoto, H., Ngameni, B., Ngadjui, B. T. and Sekimizu, K. Antimicrobial action mechanism of flavonoids from Dorstenia species. *Drug Discov. Ther.* 2013 7(2):66-72.

Eberle, R. J., Olivier, D. S., Amaral, M. S., Willbold, D., Arni, R. K. and Coronado, M. A. Promising natural compounds against Flavivirus proteases: citrus flavonoids hesperetin and hesperidin. *Plants* 2021 10:2183. https://doi.org/10.3390/plants10102183.

El-Moez, S. and Abdelmonem, M. Effect of sterilization on microbial contaminants in medicinal herbs and their antimicrobial activities against animal foodborne pathogens. *Anim. Foodborne Pathog.* 2013:208-228.

Farmica, J. V. and Regelson, W. Review of the biology of quercetin and related bioflavonoids. *Fd Chem. Toxicol.* 1995 33(12):1061-1080.

Farooq, U., Khan, T., Shah, S. A., Hossain, M. S., Ali, Y., Ullah, R., Raziq, N., Shahid, M. and Capasso, R. Isolation, characterization and neuroprotective activity of Folecitin: an in vivo study. *Life (Basel)* 2021 11(8):825. https://doi.org/10.3390/life11080825.

Ferrandiz, M. L. and Alcaraz, M. J. Anti-inflammatory activity and inhibition of arachidonic acid metabolism by flavonoids. *Agents Actions* 1991 32(3-4):283-288.

Fisher, N. D. L., Sorond, F. A. and Hollenberg, N. K. Cocoa flavanols and brain perfusion. *J. Cardiovasc. Pharmacol.* 2006 47(Suppl. 2):S210-S214.

Fowler, Z. L. and Koffas, M. A. G. Biosynthesis and biotechnological production of flavanones: current state and perspectives. *Appl. Microbiol. Biotechnol.* 2009 83(5):799-808.

Fraga, C. G., Martino, V. S., Ferraro, G. E., Coussio, J. D. and Boveris, A. Flavonoids as antioxidants evaluated by in vitro and in situ chemiluminescence. *Biochem. Pharmacol.* 1987 36(5):717-720.

Fraga, C. G., Croft, K. D., Kennedy, D. O. and Tomás-Barberán, F. A. The effects of polyphenols and other bioactives on human health. *Food Funct.* 2019 10(2):514–528.
Funakoshi-Tago, M., Nakamura, K., Tago, K., Mashino, T. and Kasahara, T. Anti-inflammatory activity of structurally related flavonoids, Apigenin, luteolin and fisetin. *Int. Immunopharmacol.* 2011 11(9):1150–1159.
Gill, B., Sanz, M. J., Terencio, M. C., Ferrandiz, M. L., Bustos, G. and Paya, M. The flavonoids. *Life Sci.* 1994 54:333–339.
Ginwala, R., Bhavsar, R., Chigbu, D. I., Jain, P. and Khan, Z. K. Potential role of flavonoids in treating chronic inflammatory diseases with a special focus on the anti-inflammatory activity of apigenin. *Antioxidants (Basel)* 2019 8(2):35.
Grace, P. A. Ischaemia-reperfusion injury. *Br. J. Surg.* 1994 81(5):637–647.
Grassi, D., Draijer, R., Schalkwijk, C., Desideri, G., D'Angeli, A., Francavilla, S., Mulder, T. and Ferri, C. Black tea increases circulating endothelial progenitor cells and improves flow mediated dilatation counteracting deleterious effects from a fat load in hypertensive patients: a randomized controlled study. *Nutrients* 2016 8(11).
Gryglewski, R. J., Korbut, R., Robak, J. and Swies, J. On the mechanism of antithrombotic action of flavonoids. *Biochem. Pharmacol.* 1987 36(3):317–322.
Halder, M., Sarkar, S. and Jha, S. Elicitation: a biotechnological tool for enhanced production of secondary metabolites in hairy root cultures. *Eng. Life Sci.* 2019 19(12):880–895.
Hang, T., Jin, J. B., Cho, S. and Cyang, J. C. Evaluation of the anti-inflammatory effects of baicalein on dextran sulfate sodium-induced colitis in mice. *Planta Med.* 2002 68(2):268–271.
Haskell-Ramsay, C. F., Stuart, R. C., Okello, E. J. and Watson, A. W. Cognitive and mood improvements following acute supplementation with purple grape juice in healthy Young adults. *Eur. J. Nutr.* 2017 56(8):2621–2631.
Havsteen, B. Flavonoids, a class of natural products of high pharmacological potency. *Biochem. Pharmacol.* 1983 32(7):1141–1148.
Herrera, C. M., de Vega, C., Canto, A. and Pozo, M. I. Yeasts in floral nectar: a quantitative survey. *Ann. Bot.* 2009 103(9):1415–1423.
Hii, C. S. and Howell, S. L. Effects of epicatechin on rat islets of Langerhans. *Diabetes* 1984 33(3):291–296.
Hirano, T., Gotoh, M. and Oka, K. Natural flavonoids and lignans are potent cytostatic agents against human leukemic HL-60 cells. *Life Sci.* 1994 55(13):1061–1069.
Hollman, P. C. and Katan, M. B. Dietary flavonoids: intake, health effects and bioavailability. *Food Chem. Toxicol.* 1999 37(9–10):937–942.
Houghton, P. J., Ren, Y. and Howes, M. J. Acetylcholinesterase inhibitors from plants and fungi. *Nat. Prod. Rep.* 2006 23(2):181–199.
Howat, S., Park, B., Oh, I. S., Jin, Y. W., Lee, E. K. and Loake, G. J. Paclitaxel: biosynthesis, production and future prospects. *New Biotechnol.* 2014 31(3):242–245.
Huesken, B. C. P., Dejong, J., Beekman, B., Onderwater, R. C. A., van der Vijgh, W. J. F. and Bast, A. Modulation of the in vitro cardiotoxicity of doxorubicin by flavonoids. *Cancer Chemother. Pharmacol.* 1995 37(1–2):55–62.
Huntley, A. L. The health benefits of berry flavonoids for menopausal women: cardiovascular disease, cancer and cognition. *Maturitas* 2009 63(4):297–301.
Izzo, A., Di Carlo, G., Mascolo, N., Capasso, F. and Autore, G. Antiulcer effect of flavonoids. *Role Endogenous PAF Pytotherapy Res.* 1994 8:179–181.
Jachak, S. M. Natural products: potential source of COX inhibitors. *CRIPS* 2001 2(1):12–15.

Jager, A. K. and Saaby, L. Flavonoids and the CNS. *Molecules* 2011 16(2):1471-1485.

Jucá, M. M., Filho, F. M. S. C., de Almeida, J. C., Mesquita, D. S., Barriga, J. R. M., Dias, K. C. F., Barbosa, T. M., Vasconcelos, L. C., Leal, L. K. A. M., Ribeiro, J. R. and Vasconcelos, S. M. M. Flavonoids: biological activities and therapeutic potential. *Nat. Prod. Res.* 2020 5:692-705.

Kang, N. J., Shin, S. H., Lee, H. J. and Lee, K. W. Polyphenols as small molecular inhibitors of signaling cascades in carcinogenesis. *Pharmacol. Ther.* 2011 130(3):310-324.

Kaul, C. L. and Ramarao, P. The role of aldose reductase inhibitors in diabetic complications: recent trends. *Methods Find. Exp. Clin. Pharmacol.* 2001 23(8):465-475.

Kaul, T. N., Middleton, E. and Ogra, P. L. Antiviral effect of flavonoids on human viruses. *J. Med. Virol.* 1985 15(1):71-79.

Kaur, K. and Pati, P. K. Stress-induced metabolite production utilizing plant hairy roots. In: Srivastava, V., Mehrotra, S. and Mishra, S. (Eds), *Hairy Roots- an Effective Tool of Plant Biotechnology*. Springer, Singapore 2018;pp.123-145.

Kent, K., Charlton, K., Roodenrys, S., Batterham, M., Potter, J., Traynor, V., Gilbert, H., Morgan, O. and Richards, R. Consumption of anthocyanin-rich cherry juice for 12 weeks improves memory and cognition in older adults with mild-to-moderate dementia. *Eur. J. Nutr.* 2017 56(1):333-341.

Kim, H. K., Namgoong, S. Y. and Kim, H. P. Biological actions of flavonoids-I. *Arch. Pharm. Res.* 1993 16(1):18-24.

Kim, H. P., Mani, I., Iversen, L. and Ziboh, V. A. Effects of naturally-occurring flavonoids and biflavonoids on epidermal cyclooxygenase and lipoxygenase from guinea-pigs. *Prostaglandins Leukot. Essent. Fatty Acids* 1998 58(1):17-24.

Kobayashi, Y., Suzuki, M., Satsu, H., Arai, S., Hara, Y., Suzuki, K., Miyamoto, Y. and Shimizu, M. Green tea polyphenols inhibit the sodium-dependent glucose transporter of intestinal epithelial cells by a competitive mechanism. *J. Agric. Food Chem.* 2000 48(11):5618-5623.

Kontruck, S. J., Radecki, T., Brozozowski, T., Drozdowicz, D., Piastucki, I., Muramatsu, M., Tanaka, M. and Aihara, H. Anti-ulcer and gastroprotective effects of solon, a synthetic flavonoid derivative of sophorandin. Role of endogenous prostaglandins. *Bur. J. Pharmac.* 1986 125:185-192.

Kozlowska, A. and Szostak-Wegierek, D. Flavonoids--food sources and health benefits. *Rocz. Panstw. Zakl. Hig.* 2014 65(2):79-85.

Krikorian, R., Shidler, M. D., Nash, T. A., Kalt, W., Vinqvist-Tymchuk, M. R., Shukitt-Hale, B. J. and Joseph, J. A. Blueberry supplementation improves memory in older adults. *J. Agric. Food Chem.* 2010 58(7):3996-4000.

Kumar, S. and Pandey, A. K. Chemistry and biological activities of flavonoids: an overview. *ScientificWorldJournal* 2013 2013:162750.

Lackeman, G. M., Claeys, M., Rwanagabo, P. C., Herman, A. G. and Vlietinck, A. Chronotropic effect of quercetin on guinea pig right atrium. *Planta Med.* 1986 52(6):433-437.

Lalani, S. and Poh, C. L. Flavonoids as antiviral agents for Enterovirus A71 (EV-A71). *Viruses* 2020 12(2):184.

Lee, E. K., Jin, Y. W., Park, J. H., Yoo, Y. M., Hong, S. M., Amir, R., Yan, Z., Kwon, E., Elfick, A., Tomlinson, S., Halbritter, F., Waibel, T., Yun, B. W. and Loake, G. J. Cultured cambial meristematic cells as a source of plant natural products. *Nat. Biotechnol.* 2010 28(11):1213-1217.

Lee, J., Torosyan, N. and Silverman, D. H. Examining the impact of grape consumption on brain metabolism and cognitive function in patients with mild decline in cognition: a double-blinded placebo controlled pilot study. *Exp. Gerontol.* 2017 87(A):121-128.

Lee, S. J., Son, K. H., Chang, H. W., Do, J. C., Jung, K. Y., Kang, S. S. and Kim, H. P. Anti-inflammatory activity of naturally occurring flavone and flavonolglycosides. *Arch. Pharm. Res.* 1993 16(1):25-28.

Leonard, E., Lim, K. H., Saw, P. N. and Koffas, M. A. G. Engineering central metabolic pathways for high-level flavonoid production in Escherichia coli. *Appl. Environ. Microbiol.* 2007 73(12):3877-3886.

Loewenstein, W. R. Junctional intercellular communication and the control of growth. *Biochim. Biophys. Acta* 1979 560(1):1-65.

Lorenz, W., Kusche, J., Barth, H. and Mathias, C. H. Action of several flavonoids on enzyme of histidine metabolism in vivo. In: Maslinski, Cz. (Ed.), *Histamine. Pennsylvania.* Hutchinson and Ross 1994;pp.265-269.

Lou, F. Q., Zhang, M. F., Zhang, X. G., Liu, J. M. and Yuan, W. L. A study on tea-pigment in prevention of atherosclerosis. *Chin. Med. J. (Engl.)* 1989 102(8):579-583.

Ma, X., Zheng, C., Hu, C., Rahman, K. and Qin, L. The genus Desmodium (Fabaceae)-traditional uses in Chinese medicine, phytochemistry and pharmacology. *J. Ethnopharmacol.* 2011 138(2):314-332.

Majewska, W. M. and Czeczot, H. Anticancer activity of flavonoids. *Pol. Merkuriusz Lek. Organ Pol. Tow. Lek.* 2012 33:364-369.

Maleki, S. J., Crespo, J. F. and Cabanillas, B. Anti-inflammatory effects of flavonoids. *Food Chem.* 2019 299:125124.

Mannucci, C., Casciaro, M., Sorbara, E. E., Calapai, F., Di Salvo, E., Pioggia, G., Navarra, M., Calapai, G. and Gangemi, S. Nutraceuticals against oxidative stress in autoimmune disorders. *Antioxidants (Basel)* 2021 10(2):261. https://doi.org/10.3390/antiox10020261.

Manthey, J. A. Biological properties of flavonoids pertaining to inflammation. *Microcirculation* 2000 7(6 Pt 2):S29-34.

Martel, J., Ojcius, D. M., Ko, Y. F., Ke, P. Y., Wu, C. Y., Peng, H. H. and Young, J. D. Hormetic effects of phytochemicals on health and longevity. *Trends Endocrinol. Metab.* 2019 30(6):335-346.

Matthies, A., Banbury, L., Bone, K. M., Leach, D. N. and Lehmann, R. P. Echinacea alkylamides modulate induced immune responses in T-cells. *Fitoterapia* 2008a 79(1):53-58.

Matthies, A., Clavel, T., Gütschow, M., Engst, W., Haller, D., Blaut, M. and Braune, A. Conversion of daidzein and genistein by an anaerobic bacterium newly isolated from the mouse intestine. *Appl. Environ. Microbiol.* 2008b 74(15):4847-4852.

McCullough, M. L., Peterson, J. J., Patel, R., Jacques, P. F., Shah, R. and Dwyer, J. T. Flavonoid intake and cardiovascular disease mortality in a prospective cohort of US adults. *Am. J. Clin. Nutr.* 2012 95(2):454-464.

Mishra, A., Sharma, K. and Mishra, R. Elicitor recognition, signal transduction and induced resistance in plants. *Plant-Microb Interact.* 2012 7(2):95-120.

Mishra, A. K., Mishra, A., Kehri, H. K., Sharma, B. and Pandey, A. K. Inhibitory activity of Indian spice plant Cinnamomum zeylanicum extracts against Alternaria solani and Curvularia lunata, the pathogenic dematiaceous moulds. *Ann. Clin. Microbiol. Antimicrob.* 2009 8:9.

Mohan, V., Pradeepa, R. G. and Anjana, R. Epidemiology and its application to clinical care in diabetes. In: *Improving Diabetes Care in the Clinic. Jaypee.* Digital Press 2014.

Murakami, S., Muramatsu, M. and Otomo, S. Gastric H+ /K+ ATPase inhibition by catechins. *J. Pharm. Pharmacol.* 1992 44(11):926–928.

Murakami, T., Ajima, K., Miyawaki, J., Yudasaka, M., Iijima, S. and Shiba, K. Drug-loaded carbon nanohorns: adsorption and release of dexamethasone *in vitro*. *Mol. Pharm.* 2004 1(6):399–405.

Naik, P. and Al-Khayri, J. Abiotic and biotic elicitors-role in secondary metabolites production through in vitro culture of medicinal plants. In: *Abiotic and Biotic Stress in Plants - Recent Advances and Future Perspectives*. IntechOpen 2016.

Namdeo, A. G. Plant cell elicitation for production of secondary metabolites: a review. *Pharmacogn. Rev.* 2007 1:69–79.

Ng, T. B., Huang, B., Fong, W. P. and Yeung, H. W. Anti-human immunodeficiency virus (anti-HIV) natural products with special emphasis on HIV transcriptase inhibitors. *Life Sci.* 1997 61(10):933–949.

Novotny, J. A., Baer, D. J., Khoo, C., Gebauer, S. K. and Charron, C. S. Cranberry juice consumption lowers markers of cardiometabolic risk, including blood pressure and circulating C-reactive protein, triglyceride, and glucose concentrations in adults. *J. Nutr.* 2015 145(6):1185–1193.

Olaiya, C. O., Ogunyem, O. and Karigidi, K. Biotechnological strategies for enhancing the nutritive and nutraceutical values of tomato (Solanum Lycopersicum). *Annals Uni Dunarea de Jos of Galati Fascicle - Food Tech.* 2015 39(2):9–19.

Osman, H. E., Maalej, N., Shanmuganayagam, D. and Folts, J. D. Grape juice but not orange or grapefruit juice inhibits platelet activity in dogs and monkeys. *J. Nutr.* 1998 128(12):2307–2312.

Pandey, K. B. and Rizvi, S. I. Plant polyphenols as dietary antioxidants in human health and disease. *Oxid. Med. Cell. Longev.* 2009 2(5):270–278.

Paulke, A., Nöldner, M., Schubert-Zsilavecz, M. and Wurglics, M. St. John's wort flavonoids and their metabolites show antidepressant activity and accumulate in brain after multiple oral doses. *Pharmazie* 2008 63(4):296–302.

Peter, I., Theophine, A., Samson, A. and Peter, A. Flavonoids and the mechanisms of their anticancer effects. *Trends Nat. Prod. Res.* 2021 2(1) https://doi.org/10.48245/tnpr-2734391.2021.2.101.

Plakas, S. M., Lee, T. C. and Wolke, R. E. Absence of overt toxicity from feeding the flavonol, quercetin, to rainbow trout (Salmo gairdneri). *Food Chem. Toxicol.* 1985 23(12):1077–1080.

Procházková, D., Boušová, I. and Wilhelmová, N. Antioxidant and prooxidant properties of flavonoids. *Fitoterapia* 2011 82(4):513–523.

Rajaram, S., Jones, J. and Lee, G. J. Plant-based dietary patterns, plant foods, and age-related cognitive decline. *Adv. Nutr.* 2019 10(Suppl_4):S422–S436.

Ramakrishna, A. and Ravishankar, G. A. Influence of abiotic stress signals on secondary metabolites in plants. *Plant Signal. Behav.* 2011 6(11):1720–1731.

Redondo-Blanco, S., Fernández, J., López-Ibáñez, S., Miguélez, E. M., Villar, C. J. and Lombó, F. Plant phytochemicals in food preservation: antifungal bioactivity: a review. *J. Food Prot.* 2020 83(1):163–171.

Rice-Evans, C. A., Miller, N. J. and Paganga, G. Structure-antioxidant activity relationships of flavonoids and phenolic acids. *Free Radic. Biol. Med.* 1996 20(7):933–956.

Rodríguez-García, C., Sánchez-Quesada, C. and Gaforio, J. J. Dietary flavonoids as cancer chemo preventive agents: an updated review of human studies. *Antioxidants* 2019 8:13.

Ruggeri, R. M., Vicchio, T. M., Cristani, M., Certo, R., Caccamo, D., Alibrandi, A., Giovinazzo, S., Saija, A., Campennì, A., Trimarchi, F. and Gangemi, S. Oxidative stress and advanced glycation end products in Hashimoto's thyroiditis. *Thyroid* 2016 26(4):504-511.

Ruiz-Cruz, S., Chaparroâ, S., Hernández-Ruiz, K. L., Cira-Chávez, L. A., Estrada-Alvarado, M. I., Gassos Ortega, L. E., Ornelas-Paz, J. J. and Lopez-Mata, M. A. Flavonoids: important biocompounds in food. In: *Flavonoids - From Biosynthesis to Human Health*. IntechOpen 2017.

Saaby, L., Rasmussen, H. B. and Jäger, A. K. MAO-A inhibitory activity of quercetin from *Calluna vulgaris* (L.) Hull. *J. Ethnopharmacol.* 2009 121(1):178-181.

Saini, N., Gahlawat, S. K. and Lather, V. Flavonoids: a nutraceutical and its role as anti-inflammatory and anticancer agent. In: Gahlawat, S., Salar, R., Siwach, P., Duhan, J., Kumar, S. and Kaur, P. (Eds), *Plant Biotechnology: Recent Advancements and Developments*. Springer, Singapore 2017.

Salas, M. P., Céliz, G., Geronazzo, H., Daz, M. and Resnik, S. L. Antifungal activity of natural and enzymatically-modified flavonoids isolated from citrus species. *Food Chem.* 2011 124(4):1411-1415.

Sandoval, V., Sanz-Lamora, H., Arias, G., Marrero, P. F., Haro, D. and Relat, J. Metabolic impact of flavonoids consumption in obesity: from central to peripheral. *Nutrients* 2020 12(8):2393.

Sarris, J., McIntyre, E. and Camfield, D. A. Plant-based medicines for anxiety disorders, part 2: a review of clinical studies with supporting preclinical evidence. *CNS Drugs* 2013 27(4):301-319.

Selway, J. W. T. Antiviral activity of flavones and flavans. In: Cody, V., Middleton, E. and Harborne, J. B. (Eds), *Plant Flavonoids in Biology and Medicine: Biochemical. Pharmacological and Structure Activity Relationships*. Alan R Liss Inc., New York 1986;pp.521-536.

Sengupta, S., Toh, S. A., Sellers, L. A., Skepper, J. N., Koolwijk, P., Leung, H. W., Yeung, H. W., Wong, R. N. S., Sasisekharan, R. and Fan, T. P. D. Modulating angiogenesis: the yin and the yang in ginseng. *Circulation* 2004 110(10):1219-1225.

Shahid, F., Yang, Z. and Saleemi, Z. O. Natural flavonoids as stabilizers. *J. Food Lipids* 1998 1:69-75.

Shukla, R., Pandey, V., Vadnere, G. P. and Lodhi, S. Role of flavonoids in management of inflammatory disorders. In: Watson, R. R. and Preedy, V. R. (Eds), *Bioactive Food as Dietary Interventions for Arthritis and Related Inflammatory Diseases*. Academic Press, Cambridge, MA 2019;pp. 293-322.

Sindhu, R. K., Verma, R., Salgotra, T., Rahman, M. H., Shah, M., Akter, R., Murad, W., Mubin, S., Bibi, P., Qusti, S., Alshammari, E. M., Batiha, G. E., Tomczyk, M. and Al-Kuraishy, H. M. Impacting the remedial potential of Nano delivery-based flavonoids for breast cancer treatment. *Molecules* 2021 26(17):5163. https://doi.org/10.3390/molecules26175163.

Sokolov, A. N., Pavlova, M. A., Klosterhalfen, S. and Enck, P. Chocolate and the brain: neurobiological impact of cocoa flavanols on cognition and behavior. *Neurosci. Biobehav. Rev.* 2013 37(10 Pt 2):2445-2453.

Soobrattee, M. A., Neergheen, V. S., Luximon-Ramma, A., Aruoma, O. I. and Bahorun, T. Phenolics as potential antioxidant therapeutic agents: mechanism and actions. *Mutat. Res.* 2005 579(1-2):200-213.

Stavric, B. Mutagenic food flavonoids. Symposium report. *Fed. Proc.* 1984 43(9):2454-2458.

Sut, S., Baldan, V., Faggian, M., Peron, G. and Dall Acqua, S. Nutraceuticals, A new challenge for medicinal chemistry. *Curr. Med. Chem.* 2016 23(28):3198-3223.

Syed, D. N., Afaq, F., Maddodi, N., Johnson, J. J., Sarfaraz, S., Ahmad, A., Setaluri, V. and Mukhtar, H. Inhibition of human melanoma cell growth by the dietary flavonoid fisetin is associated with disruption of Wnt/β-catenin signaling and decreased Mitf levels. *J. Invest. Dermatol.* 2011 131(6):1291-1299.

Tadera, K., Minami, Y., Takamatsu, K. and Matsuoka, T. Inhibition of alpha-glucosidase and alpha-amylase by flavonoids. *J. Nutr. Sci. Vitaminol. (Tokyo)* 2006 52(2):149-153.

Tanaka, T. and Takahashi, R. Flavonoids and asthma. *Nutrients* 2013 5(6):2128-2143.

Tavsan, Z. and Kayali, H. A. Flavonoids showed anticancer effects on the ovarian cancer cells: involvement of reactive oxygen species, apoptosis, cell cycle and invasion. *Biomed. Pharmacother.* 2019 116:109004.

Tencate, J. W., van Hoeringen, N. J., Gerritsen, J. and Glasius, E. Biological activity of a semisynthetic flavonoid O-(β-hydroxyethyl) rutosine: light scattering and metabolic studies of human red cells and platelets. *Clin. Chem.* 1973 19(1):31-35.

Thomas, P. R. S., Nash, G. B. and Dormandy, J. A. White cell accumulation in dependent legs of patients with venous hypertension: a possible mechanism for trophic changes in the skin. *Br. Med. J. (Clin Res Ed)* 1988 296(6638):1693-1695.

Unnikrishnan, M. K., Veerapur, V., Nayak, Y., Mudgal, P. P. and Mathew, G. Antidiabetic, antihyperlipidemic and antioxidant effects of the flavonoids. *Polyphenols Hum. Health Dis.* 2014;1:143-161.

Vaccaro, M. C., Alfieri, M., Malafronte, N., De Tommasi, N. and Leone, A. Increasing the synthesis of bioactive abietane diterpenes in Salvia sclarea hairy roots by elicited transcriptional reprogramming. *Plant Cell Rep.* 2017 36(2):375-386.

van Dam, R. M., Naidoo, N. and Landberg, R. Dietary flavonoids and the development of type 2 diabetes and cardiovascular diseases: review of recent findings. *Curr. Opin. Lipidol.* 2013 24(1):25-33.

Vaquero, M. J. R., Alberto, M. R. and de Nadra, M. C. M. Antibacterial effect of phenolic compounds from different wines. *Food Control* 2007 18(2):93-101.

Vasconsuelo, A. and Boland, R. Molecular aspects of the early stages of elicitation of secondary metabolites in plants. *Plant Sci. Int. J. Exp. Plant Biol.* 2007 172(5):861-875.

Verpoorte, R. and Memelink, J. Engineering secondary metabolite production in plants. *Curr. Opin. Biotechnol.* 2002 13(2):181-187.

Verpoorte, R., van der Heijden, R. and Memelink, J. Engineering the plant cell factory for secondary metabolite production. *Transgen. Res.* 2000 9(4-5):323-43; discussion 321.

Vessal, M., Hemmati, M. and Vasei, M. Antidiabetic effects of quercetin in streptozocin-induced diabetic rats. *Comp. Biochem. Physiol. Toxicol. Pharmacol.* 2003 135C(3):357-364.

Wang, C. Z., Mehendale, S. R., Calway, T. and Yuan, C. S. Botanical flavonoids on coronary heart disease. *Am. J. Chin. Med.* 2011 39(4):661-671.

Wang, J. W. and Wu, J. Y. Effective elicitors and process strategies for enhancement of secondary metabolite production in hairy root cultures. In: Doran, P. M. (Ed.), *Biotechnology of Hairy Root Systems. Advances in Biochemical Engineering/Biotechnology*. Springer, Berlin, Heidelberg 2013;pp.55-89.

Wang, Y., Li, S. and Li, C. Perspectives of new advances in the pathogenesis of vitiligo: from oxidative stress to autoimmunity. *Med. Sci. Monit. Int. Med. J. Exp. Clin. Res.* 2019 6(25):1017-1023.

Wegener, T. and Fintelmann, V. Flavonoids and bioactivity. *Wein Wochem Schr.* 1999 149:241-247.

Wei, H., Tye, L., Bresnick, E. and Birt, D. F. Inhibitory effect of apigenin, a plant flavonoid, on epidermal ornithine decarboxylase and skin tumor promotion in mice. *Cancer Res.* 1990 50(3):499-502.

Wu, W., Li, R., Li, X., He, J., Jiang, S., Liu, S. and Yang, J. Quercetin as an antiviral agent inhibits influenza A virus (IAV) entry. *Viruses* 2015 8(1):6.

Xu, H. X. and Lee, S. F. Activity of plant flavonoids against antibiotic-resistant bacteria. *Phytother. Res.* 2001 15(1):39-43.

Yanez, J. A. Connie, M., Remsberg J. K., Takemoto, K. R., Vega-Villa, P. K., Andrews, C. L., Sayre, S. E. M. and Davies, N. M. Polyphenols and flavonoids: an overview. In: Davies, N. M. and Yanez, J. A., (Eds), *Flavonoid Pharmacokinetics*. John Wiley and Son, New York 2013.

Zakaryan, H., Arabyan, E., Oo, A. and Zandi, K. Flavonoids: promising natural compounds against viral infections. *Arch. Virol.* 2017 162(9):2539-2551.

Zhai, X., Min, J., Chen, L., Zheng, C., Rahman, K., Han, T. and Qin, L. P. The regulatory mechanism of fungal elicitor-induced secondary metabolite biosynthesis in medical plants. *Crit. Rev. Microbiol.* 2016 43(2):238-261.

Zhang, Y., Li, S. Z., Li, J., Pan, X., Cahoon, R. E., Jaworski, J. G., Wang, X., Jez, J. M., Chen, F. and Yu, O. Using unnatural protein fusions to engineer resveratrol biosynthesis in yeast and Mammalian cells. *J. Am. Chem. Soc.* 2006 128(40):13030-13031.

Part 2

Phytochemicals in fruits and vegetables: glucosinolates and organosulfur compounds

Chapter 5

Health-promoting effects of glucosinolates and their breakdown products

Ella O'Grady, Greta Pileckaite, Almha Gilheany and Endrita Kucana, School of Food Science and Environmental Health, Technological University Dublin - City Campus, Ireland; and Swarna Jaiswal and Amit K. Jaiswal, School of Food Science and Environmental Health, Technological University Dublin - City Campus and Environmental Sustainability and Health Institute, Technological University Dublin - City Campus, Ireland

1 Introduction

Glucosinolates (GLs) belong to a wide range of secondary metabolites found in plants that have nutritional and physiologically active properties. They are mainly found in the *Brassicaceae* family, more common in edible plants such as broccoli, cabbage, cauliflower, rapeseed, etc. These sulphur-containing compounds are responsible for the pungent aromas and tastes of the *Brassica genera* and have been found to have anticancer, antimicrobial, anti-viral, anti-mutagenic, and anti-inflammatory effects along with decreasing the risk of cardiovascular disease (Putnik et al., 2019). Over 130 GL structures have been discovered and validated to date, and some common GLs include gluconapin, epiprogoitrin, and glucobrassicin (Blažević et al., 2020).

GLs are anions (β-thioglycoside N-hydroxysulphates) with a side chain (R) of alkyl, aralkyl, or indolyl and a sulphur linked β-d-glucopyranose moiety (Fig. 1) (Dini, 2018). Side-chain modification of the amino acid precursors prior

http://dx.doi.org/10.19103/AS.2022.0101.07

to the GLs formation and a wide range of secondary modifications, including oxidation, sulfation, hydroxylation, methoxylation, and glucosylation, as well as substitutions with acyl conjugation on the sugar moieties, are responsible for the high number of different GLs. GLs can be classified into three groups: aliphatics, aromatics, and indoles. Classification comes from the R chain of the compound which is derived from one of eight amino acids of the plant (aliphatic such as alanine, leucine, isoleucine, methionine, or valine; aromatic such as phenylalanine or tyrosine; or indole such as tryptophan) (Prieto et al., 2019).

GLs are hydrolyzed by the enzyme myrosinase (EC 3.2.1.147); when the plant tissue is injured, a protective response is induced by the plant which increases GLs content. While in the presence of water, myrosinase cleaves a glucose group from the GLs to form an isothiocyanate (ITC), a thiocyanate, or a nitrile (unstable aglucones) (Blažević et al., 2020). Myrosinases cleave the thioglucosidic bond as part of a two-step process. In the first step, at the anomeric carbon, a nucleophilic attack by a glutamic acid (Glu) residue in the enzyme liberates the aglucone. The glucose group is now covalently bound to the active site of the Glu residue of the enzyme. In the second step, the free enzyme is then restored by hydrolysis by the Glu-glucose bond aided by the activation of a water molecule (Blažević et al., 2020). These breakdown products have a protective effect on the plant but are also responsible for, along with their precursor GLs, the beneficial health effects associated with the *Brassica* plants (Lachance et al., 2020).

In this chapter, we have presented brief information on the natural source, recent advances in potential health effects of GLs, and its breakdown products together with antinutritional properties of GLs.

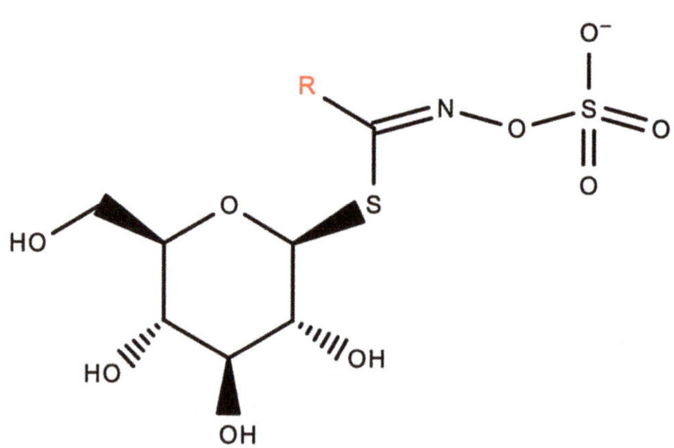

Figure 1 Generic structure diagram of a GL (the side group R varies).

2 Natural sources

GLs are secondary metabolites produced by the selected group of plants mostly occurring in the order Brassicales. Families of this order include Moringaceae, Cleomaceae, Limnathaceae, Caricaceae, Capparidaceae, and Brassicaceae (Blažević et al., 2015). Of these families, Brassicaceae is the most studied GL-producing group due to its economic importance. Well-known species of this family are broccoli, cabbage, kale, cauliflower, and turnip, vegetables that are grown and consumed across the globe. Outside the Brassica genus, GLs are produced by some plants of the Putranjivaceae, Violaceae, Euphorbiaceae, and Rubiaceae families (Blažević et al., 2015).

GLs are derivatives of cyanogenic glucosides. These glucosides are found in many plants; however, not all plants possess the biochemical pathway for the conversion of cyanogenic glucosides into GLs. Along with this, different species and genera produce different types of GLs depending on the modification of the side chain due to S-oxygenation, alkenylation, hydroxylation, and acyl substitution on the sugar moieties (Tacer-Caba, 2019).

The concentrations of the different GLs found in the plants are dependent on factors such as the variety of plants, genetics, plant nutrition, and the environment. GL levels within the plants differ between individual parts of the plant. Seeds, roots, leaves, and stems of a plant all contain different levels of GLs. The youngest tissues of a plant, especially seeds, have the highest GL levels (Tacer-Caba, 2019). Liu et al. (2020) found that in *Brassica napus* plants, GL levels were highest in the seed than in the leaves of the plant. The class of GL also varies between plant organs. For example, in both the leaves and seeds of *B. napus,* there are mostly aliphatic GLs; however, leaves contain more indole GLs than seeds (Liu et al., 2020). Nguyen et al. (2020) propose that the profile of GLs contained in an organ is directly related to the function of that organ. Abiotic stress causes a change in the GL profile of the plant's roots as it resists stressors such as drought and acidity (Nguyen et al., 2020). Roots have been found to contain a higher amount of GLs than leaves in stressful conditions, and Martínez-Ballesta et al. (2013) conclude that abiotic and biotic stressors reroute GLs to allow a plant to cope with stressful conditions.

A study on ten varieties of cabbage found that the plant contains ten GLs (glucoiberin, glucoiberverin, sinigrin, gluconapin, glucoraphanin, progoitrin, glucobrassicin, neoglucobrassicin, 4-methoxy glucobrassicin, and gluconasturtiin). The most abundant GL was glucobrassicin, contributing 22–53% of the total GLs content. Making up 15–20% of the total GLs, sinigrin and progoitrin were also present in cabbage in high amounts (Choi et al., 2014). Similarly, arugula, rucola, and roquette, collectively known as rocket, of *Eruca sativa* and *Diplotaxis tenuifolia*, contain high levels of vitamin C, GLs, flavonals, and phenolics. The three most abundant GLs of the species include

4-mercaptobutyl GLs (glucosativin), 4-methylthiobutyl GLs (glucoerucin), and 4-methylsulfinylbutyl GLs (glucoraphanin) (Bell and Wagstaff, 2014).

Mustard seed or *Brassica juncea* is another economically important Brassicaceae member which is used to produce the condiment mustard paste and is grown and consumed in most areas of Asia, Europe, America, and Africa. In the *B. juncea* plant, the main GL produced is sinigrin (Márquez-Lema et al., 2009). In a different study, carried out on *Sinapis* genus, the dominating GL is sinablin and, according to Sørensen et al. (2016), has a structure of interest and is absorbed and transformed to a sinalbin metabolite in the liver by glucuronidation by liver phase II enzymes with subsequent excretion to the urine.

Tropaeolum majus commonly known as nasturtium contains just one GLs, benzylglucosinolate or glucotropaeolin. It is used in traditional medicine to treat urinary tract infections with the basis being that the breakdown product of the GLs is benzyl ICT, which lends antimicrobial effects as it passes through the urinary tract following digestion of the plant (Kleinwächter et al., 2008). The plant has also been investigated to have a potential role in type two diabetes. The GL breakdown product, benzyl ICTs, can reduce hepatic glucose production in patients (Guzmán-Pérez et al., 2016).

Brassica Rapa or turnip has been consumed in Europe for approximately 4500 years as fresh vegetables or as fodder. The tuber of the vegetable alone is commonly eaten in Northern and Eastern Europe and Asia, whereas in Southern Europe, leaves, shoots, tuber, turnip top, and greens are all consumed. Unlike other vegetables, total GLs amounts in turnips are not affected by storage temperature, rather their composition changes. As aliphatic and indolic GLs respond differently to temperature, higher storage temperatures induced higher indolic GL levels and lowered aliphatic GL levels. The dominant GLs of *B. Rapa* include glucoraphanin, glucobrassicin, and gluconasturtiin (Tacer-Caba, 2019).

Moringa oleifera or moringa is a tree native to tropical and sub-tropical areas in the world. The plant is used as both human and animal feed and in traditional medicine. Leaves, roots, fruits, seeds, and bark of the plant are highly nutritious and have been used to treat inflammation and infectious diseases along with gastrointestinal, haematological, cardiovascular, and hepatorenal disorders. Seed extracts of moringa have been used for anti-bacterial and anticancer activity, fruit and bark extracts as anti-inflammatory and hepatoprotective agents, and leaf extracts have been shown to regulate cholesterol levels and stabilize thyroid status in rats. The Moringa genus is gaining interest due to its unusual sugar-substituted hydroxy-aromatic GLs. *M. oleifera* contains many peculiar GLs with uncommon properties as a result of the presence in their structure of a second saccharide residue in the aglycon side chain. The dominant GL in *M. oleifera* is 4-(α-L-rhamnopyranosiloxy)benzyl

GLs, known as glucomoringin. This particular GL with its atypical structure may exhibit different biological effects than seen in many other GLs (Maldini et al., 2014).

Armoracia rusticana belongs to the Brassicaceae family and its leaves and roots, with their spicy taste, have been consumed as food and as a medicinal herb. A study by Popović et al. (2020) found the most common GL gluconasturtiin (64.9%) present in the roots of wild horseradish, followed by sinigrin (33.91%) and glucobrassicin (1.19%). In leaves, sinigrin made up 99.9% of GLs, while three other GLs namely gluconapin, glucocochlearin, and glucobrassicanapin were found in trace amounts in all plant parts. The study also found that leaf and root extracts from *A. rusticana* had cytotoxic effects on human cancer cells and anti-bacterial properties inhibit the growth of methicillin-resistant *Staphylococcus aureus*, *Listeria monocytogenes*, clinical *Acinetobacter Baumannii*, and fungi (Popović et al., 2020).

3 Factors affecting the composition and levels of glucosinolates

The post-harvest processes of the vegetables have considerable effects on the composition and levels of GLs in the *Brassica* plants. Storage conditions and how the plants are eaten can have adverse effects on the GLs concentrations. The temperature and length of transport of the vegetables are important factors in the reduction of GLs levels. The loss of GLs can be mitigated by refrigeration during transport and reducing transport length (Esteve, 2020). A recent study by Cavaiuolo et al. (2017) found that exposure to cold temperature upregulated genes involved in the biosynthesis of indole GLs like indol-3-ylmethyl, while many other genes involved in the biosynthesis of aliphatic GLSs such as glucoraphanin, glucoiberin, and glucoalyssin were downregulated. Furthermore, tissue exposure to wounding, cold, and dark stress caused an upregulation of genes involved in GLs catabolism (Cavaiuolo et al., 2017).

Agricultural and environmental factors also affect the levels of GLs in *Brassica* plants. Increased levels of phytochemicals and GLs are observed in spring-grown plants which grow in times of longer sun hours, high light intensity, moderate temperatures, and lower rainfall. Autumn and winter crops have lower GLs and other phytochemical concentrations most likely due to growth at lower temperatures, low-intensity light, shorter sun hours, and higher rainfall levels (Biondi et al., 2021).

Attacks from pests such as aphids increase the levels of GLs produced as the plant mounts its defence system. Abiotic factors can also cause an increase in the concentration of GLs and other phytochemicals. A slight salinity in water can cause broccoli to increase the production of GLs by affecting the myrosinase–GL system, and recent research on the production of low potassium

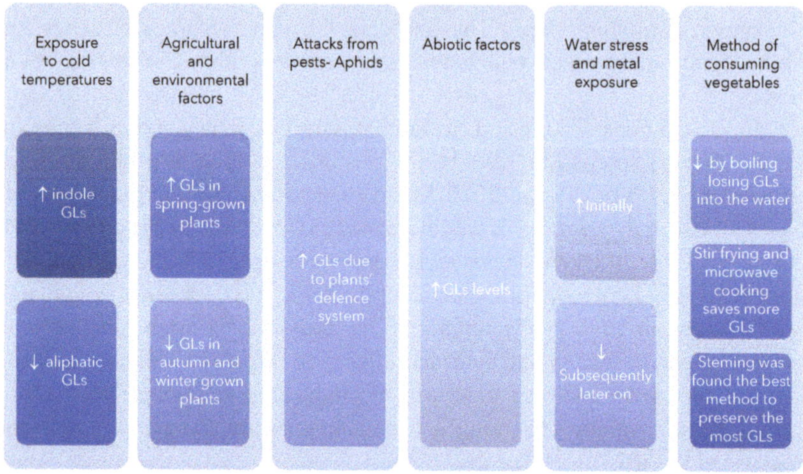

Figure 2 Summary of factors affecting levels of GLs.

kale for renal failure patients found that by decreasing potassium levels in the nutrient solution, there was a concurrent rise in GLs in the plant. The cultivation system, water, nutrient stress, and increased plant density all result in a general increase in GLs concentrations (Biondi et al., 2021).

Light-emitting diode technology can affect the levels of phytochemicals produced in plants. Aliphatic GL levels have been increased in broccoli tissues after exposure to blue light (Loi et al., 2020).

The method of consuming vegetables can also result in a loss of phytochemicals. Boiling of the vegetables leads to the significant loss of GLs due to leaching into the cooking water, followed by stir-frying and microwave cooking. Tabart et al. (2018) used broccoli and red cabbage to evaluate the influence of the cooking method on phytochemical content and total antioxidant capacity (TAC). It was determined that steaming and microwaving allowed to preserve most antioxidants and GLs. Steaming is determined to be the best way to cook the brassicas to reduce the loss of GLs before consumption (Esteve, 2020). Figure 2 presents a summary of factors affecting the levels of GLs in plants.

4 Potential health effects of glucosinolates

Some of the general health benefits of GLs include antimicrobial, anticarcinogenic, antidepressant, or anti-inflammatory activity of a compound, antidiuretic properties, and lower the risk of cardiovascular disease. Health effects often differ for each bioactive compound since they all have different mechanisms of action due to a variety of chemical structures.

4.1 Antimicrobial activity

Volatile compounds present in GLs have functions against fungi, bacteria, and other microorganisms. Currently, there is an increase in drug and pathogen resistance, as well as food spoilage around the world, all of which begin to pose a global issue (Maina et al., 2020). To counteract this, natural compounds like GLs are of particular interest, as they have natural antimicrobial potential and could help to counteract mentioned issues.

The antimicrobial activity of GLs is frequently associated with the *Brassicaceae* family, and more specifically with *Sinapis species* (*ssp*.), including *S. alba* and *S. nigra*. It was found that *Sinapis nigra* provides better antimicrobial activity when compared to *S. alba*, due to better inhibition against *S. aureus*, *L. monocytogenes*, *Salmonella typhimurium*, and *Escherichia coli*, with a minimum inhibitory concentration of 10 mg/mL of the *S. nigra* seed oil needed to inhibit the growth of *E. coli* (Boscaro et al., 2018). Current knowledge could be implemented in the production of ready-to-eat foods like freshly cut salad, as *Sinapis ssp*. seed oil spray could extend product shelf life by reducing the microbial activity of certain bacterial strains. Following this, hydroalcoholic seed extract of *S. nigra* and *S. alba* showed significant inhibition against *S. aureus* and *E. coli*, followed by lower anti-bacterial activity against *Bacillus cereus*, *Pseudomonas aeruginosa*, *Streptococcus pyogenes*, and *Candida albicans* (Boscaro et al., 2018). This provides promises for GLs incorporation in anti-bacterial drugs, with lower efficacy, in hopes of obtaining a synergistic effect between the drug and the bioactive compound and improving their working principle.

Popović et al. (2020) assessed volatile degradation products from the roots and leaves of horseradish (including ITCs, nitriles, and other GLs) for their antimicrobial and cytotoxic activity. 2-Phenylethyl isothiocyanate and 3-phenylpropanenitrile were used on a number of bacterial and fungi strains. Clinical antibiotic-resistant strains of bacteria (*Salmonella enterica, E. coli, Klebsiella pneumoniae, A. baumannii*) and strains isolated from food (*L. monocytogenes, S. aureus, Enterococcus faecalis, S. pyogenes,* and *B. cereus*) were used to assess minimal inhibitory concentration (MIC), which was ranging from 3.75 to 30 μg/mL depending on species and bioactive compound used. Meanwhile, for opportunistic pathogenic fungal strains (including *C. albicans*, *Penicillium citrinum*, and *Aspergillus niger*), MIC_{50} was found to be <0.12 and 0.47 μg/mL.

4.2 Antioxidant activity

Reactive oxygen species (ROS) and free radicals are often consumed through food and are naturally present in the human body. Presence of current compounds often leads to age-related illnesses. Consumed or elsehow implemented antioxidants in the diet bind and scavenge ROS and free radicals, remove them from the body, and eliminate damaging effects. The current strategy is often linked to GLs anticarcinogenic activity.

Vegetables from *Brassicaceae* family, for example, kale, cabbage, and broccoli contain a significant amount of GLs, mostly consisting of aromatic GLs (Chang et al., 2019). Glucoraphasatin is the main GLs responsible for the TAC of radish sprouts, and glucoerucin is a GL prevalent in the seeds and roots of rocket salad and contributes to TAC, as well as 4-methoxyglucobrassicin GLs in Chinese kale, which is also often linked to TAC (Chang et al., 2019). It can be seen that the prevalence, chemical composition, and activity of GLs differ between number of plants; however, most vegetables from *Brassicaceae* family contain at least one or more bioactive compounds responsible for antioxidant activity.

4.3 Anti-inflammatory activity

Inflammation occurs as a natural process in the body in order to try and fight against infection, injury, toxins, or cell damage. In the current scenario, the immune system gets triggered and releases chemicals to elute response against the detected damage and attempt to heal the body itself. However, when inflammation becomes chronic, cell mutation and proliferation can occur, leading to cancer. Anti-inflammatory substances, including different GLs, aim to reduce the inflammation and in turn reduce pain; however, their working principle differs between cells, as they work through different chemical pathways.

Isatis indigotica from *Brassicaceae* family was profiled and identified to exhibit anti-inflammatory activity; however, from 16 isolated GLs, only 3 showed significant *in vitro* anti-inflammatory effect. It was found that gluconapin presented IC_{50} of 1.90 mM, neoglucobrassicin with IC_{50} of 1.25 mM, (*R,S*)-goitrin was determined the most potent with IC_{50} of 7.01 mM, and lastly, total GLs of *Isatidis Radix* (dried root part of the *I. indigotica)* presented overall IC_{50} of 0.17 mg/mL (Guo et al., 2020). It can be seen that different GLs have different anti-inflammatory activity. Generally, isolated or pure GLs provide more potent effect, when compared to combination of total GLs present in the plant.

4.4 Anticarcinogenic activity

Different types of cancer have been linked to reduced intake of cruciferous vegetables including broccoli, cabbage, or Brussel sprouts due to their powerful antioxidant and anti-inflammatory activities from the presence of GLs. Colon cancer, prostate cancer, melanoma, and breast cancer are some examples that are linked to reduced or elsehow altered intake of cruciferous vegetables. GLs also enhance detoxification of the body, which mainly happens through the liver. This process reduces the presence of toxins and in turn reduces the probability of developing cancer. Overall process is influenced by phase I and phase II enzymes including glutathione-*S*-transferase and quinone reductase (NQO1) (Miękus et al., 2020; Maina et al., 2020).

ITC is the main GLs breakdown by-product responsible for anticarcinogenic activity in the body. It is linked to cell cycle arrest, induced antioxidant pathways, and cell apoptosis. Furthermore, sulforaphane (SFN) belongs to ITC by-products and works by inhibiting deacetylase activity and in turn increasing histone acetylation, which regulates cell response to stress like inflammation or cancer.

The intake of 4-methylsulphinylbutyl glucosinolate (further converted to sulforaphane) reduces the expression of genes linked to prostate cancer, which in turn limits cancer progression through compounds including 2-propenyl ITC and 3-butenyl ITC which inhibit cytochrome P450 1A enzyme activity and in turn inhibit DNA replication within cancer cells (Miękus et al., 2020; Maina et al., 2020).

Almuhayawi et al. (2020) investigated the use of elevated CO_2 for nutritional improvement in *Brassica oleracea L* during sprouting in order to stimulate anti-inflammatory and anti-cancer activity of GLs. It was found that treatment for 9 days with 620 ± 42 μmol CO_2/mol air yield higher levels of sulforaphane, which is linked to anticarcinogenic activity. In turn, detoxification enzymes were promoted which improved anti-cancer activity.

It was found that people with higher body mass index (BMI) < 26 kg/m^2 experience higher GLS effects, which were dose-dependent. It was confirmed by observing a decrease in pro-inflammatory cytokines and an increase in tumour-suppressor decorin in a dose-dependent fashion. Personal genotype also has an influence on the overall effect, where people with genotype GSTT1 were more susceptible to the anticancer treatment using 2-phenethyl ITC, followed by the downregulation of AMACR and ARLNC1 genes using broccoli sprout extract, which was linked to the implication of prostate cancer. Current knowledge could be useful in the development of targeted therapies by the use of GLs by-products including sulforaphane, 2-phenethyl ITC, or 4-methylsulphinylbutyl in combination with other treatments. Table 1 summarizes some of the clinical trials, where GLs were used as an active agent in order to elute an effect and obtain health benefits.

4.5 Cardiovascular protection

Broccoli sprouts are the most common nutritional source cited for the use of cardiovascular protection, where they work by decreasing oxidative stress

Table 1 Summary of the clinical trials with focus on the use of GLs

Nutrient	Dose	Subjects	Design	Duration	Control	Effect	Reference
Anticarcinogenic activity							
200 g of cooked broccoli and 20 g of raw daikon radish daily for 15 days; 100 g of broccoli and 10 g of daikon radish on day 16; and 200 g of broccoli and 20 g of daikon radish on day 17		18 healthy adults (age 37–65) with BMI 19–37 kg/m²	Randomized, single-blinded, crossover, intervention study	17 days	No broccoli for 16 days; on day 17, 200 g of broccoli and 20 g of daikon radish	For subjects with BMI > 26 kg/m², plasma AUC and urinary excretion rates were higher in control, than broccoli diet. For subjects with BMI < 26 kg/m², plasma AUC and urinary excretion rates were higher in broccoli, than control diet. Interaction with BMI but not with *GSTM1* genotype.	Charron et al. (2018)
Broccoli sprout extract containing sulforaphane	50 μg; 100 μg; 200 μg of sulforaphane	17 patients from 12 years of age, with atypical nevi and a prior history of melanoma. Abstained from GLs and ITC 3 days before study	Randomized, intervention study	28 days → with samples taken on days 1, 2, and 28	None	On day 28, skin SFN levels were 0, 3.1, and 34.1 ng/g for 50, 100, and 200 μmol, respectively. Plasma levels of pro-inflammatory cytokines were decreased and tumour-suppressor decorin was increased. Doses of 200 μmol/day were safe and produced dose-dependent effect.	Tahata et al. (2018)
2-Phenethyl isothiocyanate	10 mg in 1 mL of olive oil 4 times per day	79 patients (considered regular smokers)	Randomized, double-blind, crossover, placebo-controlled, interventional phase II clinical trial	1 week treatment period + 1 week washout period → total 5 weeks	Placebo (1 mL olive oil)	Subjects with null genotype GSTM1 and GSTT1 experienced biomarker (MHBMA) increase by 58.7 and 90%, respectively, but no effect on DHBMA. Treatment group experienced detoxification of 1,3-butadiene in cigarette smokers.	Boldry et al. (2019)

4-methylsulphinylbutyl glucosinolate (glucoraphanin) → in a regular 300 mL portion of soup made from a standard broccoli	Soup X: 72 μmol Soup Y: 214 μmol Soup Z: 492 μmol	49 men aged 18–80, with BMI 19.5–35 kg/m^2, with diagnosis of low or intermediate risk for prostate cancer and undergoing active surveillance	Randomized, double-blind, 3-arm parallel intervention study	12 months	300 mL portion of soup made from a standard broccoli	Gene expression in prostate and associated oncogenic pathways was reduced in men on the glucoraphanin-rich broccoli soup in a dose-dependent manner. Lead to reduced risk of cancer progression.	Traka et al. (2019)
Broccoli sprout extract	200 μmol daily (provided 100 μmol SFN per capsule)	98 men scheduled for prostate biopsy	Randomized, double-blind, placebo-controlled, intervention study	4–8 weeks, or when subjects prostate biopsy is completed	Placebo (gelatine capsule containing microcrystalline cellulose)	Downregulation of *AMACR* and *ARLNC1* genes implicated in prostate cancer. SFN was higher in treatment group, but no significant difference in blood histone deacetylase.	Zhang et al. (2020)
Cardiovascular protection							
Broccoli sprouts powder (BSP) group STT + BSP group	6 g/day, containing at least 22.5 μmol SFN per each gram	77 type 2 diabetic patients with positive *Helicobacter pylori* stool antigen test, aged 25–60	Randomized, intervention study	28 days BSP 14 days STT	Standard triple therapy (STT) group including omeprazole 20 mg, clarithromycin 500 mg, amoxicillin 1000 mg, twice a day for 14 days	*Helicobacter pylori* eradication rates were 89.3% in STT group, 56.0% in BSP, and 91.7% in STT + BSP. Systolic and diastolic blood pressure was decreased in STT + BSP group ($P = 0.05$). TG/HDL-C ratio increased in STT group ($P < 0.05$).	Mirmiran et al. (2014)

(Continued)

Table 1 (Continued)

Nutrient	Dose	Subjects	Design	Duration	Control	Effect	Reference
High glucoraphanin broccoli – with 21.6 ± 1.6 μmol/g DW glucoraphanin (4-methylsulphinylbutyl glucosinolate) and 4.5 ± 0.34 μmol/g DW of glucoibering (3-methylsulphinylpropyl glucosinolate) 21.6 ± 1.60 μmol/g dry weight glu-coraphanin (4-methylsulphinylbutyl glucosinolate) and 4.5 ± 0.34 μmol/g dry weight glucoiberin (3-methylsulphinylpropyl glucosinolate), 21.6 ± 1.60 μmol/g dry weight glu-coraphanin (4-methylsulphinylbutyl glucosinolate) and 4.5 ± 0.34 μmol/g dry weight glucoiberin (3-methylsulphinylpropyl glucosinolate), 21.6 ± 1.60 μmol/g dry weight glu-coraphanin (4-methylsulphinylbutyl glucosinolate) and 4.5 ± 0.34 μmol/g dry weight glucoiberin (3-methylsulphinylpropyl glucosinolate)	400 g	37 volunteers aged 50–77 with 10 years of cardiovascular profile	Randomized, double-blind, parallel dietary intervention studies	12 weeks treatment	400 g standard broccoli – with 6.9 ± 0.44 μmol/g DW glucoraphanin and 0.7 ± 0.33 μmol/g DW of glucoibering	High glucoraphanin diet reduced LDL cholesterol by 7.1%, standard diet by 1.8%. (LDL cholesterol reduction was significantly higher from high glucoraphanin broccoli than standard (*P* = 0.031) between results of two studies.)	Armah et al. (2015) (Study 1)

High glucoraphanin broccoli- with 21.6 ± 1.6 μmol/g DW glucoraphanin (4-methylsulphinylbutyl glucosinolate) and 4.5 ± 0.34 μmol/g DW of glucoibering (3-methylsulphinylpropyl glucosinolate)	400 g	93 healthy volunteers age <50	Randomized, double-blind, parallel dietary intervention studies	12 week treatment	400 g standard broccoli- with 6.9 ± 0.44 μmol/g DW glucoraphanin and 0.7 ± 0.33 μmol/g DW of glucoibering	High glucoraphanin diet reduced LDL cholesterol by 5.1%, standard diet by 2.5%. (LDL cholesterol reduction was significantly higher from high glucoraphanin broccoli than standard ($P = 0.031$) between results of two studies.)	Armah et al. (2015) (Study 2)
Semi-quantitative food frequency questionnaires		210 574 participants (men and women) from Nurses' Health Study	Observational semi-quantitative food frequency questionnaires	2–4 years during follow-up	None	Participants who consumed one or more servings per week of Brussels sprouts ($P < 0.001$) and cabbage ($P = 0.009$) had a significantly higher CHD risk than those who consumed these cruciferous vegetables less than once per month.	Ma et al. (2018a,b)
Central nervous system protection							
Sulforaphane derived from broccoli sprout extracts	-	29 young men aged 13–27 with moderate to severe autism spectrum disorder (ASD)	Randomized, double-blind, placebo-controlled, intervention study	18 weeks	Placebo	34% decrease in aberrant behaviour ($P < 0.001$) and 17% improvement in social responsiveness ($P = 0.017$). Improvement in social interaction, abnormal behaviour, and verbal communication was also observed.	Singh et al. (2014)
Sulforaphane	30 mg - three tablets	7 outpatients with schizophrenia, aged 20–65	Intervention	8 weeks	None	Potential to improve cognitive deficits in patients with schizophrenia. SFN attenuates oxidative stress caused by antipsychotics.	Shiina et al. (2015)

(Continued)

Table 1 (Continued)

Nutrient	Dose	Subjects	Design	Duration	Control	Effect	Reference
Sulforaphane	~2.5 μmol glucoraphanin (GR)/l b	15 children and young adults enrolled in a school for children with ASD and related neurodevelopmental disorders, aged 5-22	Open-label, intervention study	12 weeks	None	Social responsiveness was significantly improved by 9.7 points, followed by aberrant behaviour improvement by 7.1 points, which was not significant. Main pathways responsible for improvement were of oxidative stress, amino acid/gut microbiome, neurotransmitters, hormones, and sphingomyelin metabolism.	Bent et al. (2018)
Risperidone + sulforaphane (1-isothio cyanato-4-methylsulfinylb utane)	Maximum dose 1 mg for kids<20 kg, 2.5 mg for kids 20–45kg, and 3.5 mg for kids>45 kg 50 μmol for kids<45kg and 100 μmol for kids 45-90kg	60 drug-free patients aged 4–12 with ASD	Randomized, double-blind, placebo-controlled, intervention study	5 weeks	Risperidone+ placebo	Sulforaphane group: greater improvements in irritability score ($P = 0.001$) and significant time × treatment effect for irritability ($P = 0.007$) and hyperactivity/ noncompliance ($P = 0.008$). No difference in lethargy/social interaction score, stereotypic behaviour score, inappropriate speech score, and frequency of adverse events.	Momtaz-manesh et al. (2020)

Other health benefits							
High glucosinolate broccoli group	100 g	40 osteoarthritis patients referred for total knee replacement	Randomized, intervention study	14 days prior to surgery	Low glucosinolate group: washout diet continued – no cruciferous vegetables	Increased broccoli intake results in ITC uptake into the joint, with concomitant changes in the joint. ITCs reached the synovial fluid at concentrations with biological impact on the articular joint tissues and alter the synovial fluid protein profile. Chondroprotective properties can penetrate the knee in osteoarthritis.	Davidson et al. (2017)
Cooked broccoli Raw daikon radish	200 g 20 g	18 healthy adults, aged 21–70	Randomized, crossover, controlled feeding study	18-day treatment periods separated by a 24-day washout period - between broccoli and radish	None	Current intervention influenced microbiota by decreasing the relative abundance of Firmicutes by 9% ($P = 0.05$), increasing the relative abundance of Bacteroidetes by 10% ($P = 0.03$), and increasing Bacteroides by 8% ($P = 0.02$). participants with BMI < 26 kg/m^2 experienced stronger effects.	Kaczmarek et al. (2019)

DW, dry weight; AUC, area under the curve; BMI, Body Mass Index; ITC, isothiocyanate; MHBMA, monohydroxy 3-butenyl mercapturic acid; DHBMA, 1, 2- dihydroxybutyl mercapturic acid; SFN, sulforaphane; BSP, broccoli sprout extract; STT, standard triple therapy; TG, triglycerides; HDL-C, high-density lipoprotein cholesterol; LDL, low-density lipoprotein; CHD, coronary heart disease.

and protein nitrosation. ITCs improve ventricle function, decrease myocardial size and cardiomyocyte apoptosis, and prevent chronic inflammation of cardiovascular tissue (Miękus, et al., 2020). The main GLs metabolite responsible for the health effect is glutathione followed by enzymes glutathione peroxidase and glutathione reductase, which enhance endothelial aorta relaxation and maintain normal blood pressure (Miękus, et al., 2020).

There are a number of promising results obtained through *in vitro* and *in vivo* studies. Ma et al. (2018a,b) investigated if SFN could be used to improve cardiac function in a rabbit model with chronic heart failure. It was found that 0.5 mg of SFN per kg body weight (BW) corrected heart weight and left ventricular to BW ratio, the left ventricular end-diastolic and systolic diameter, plasma brain natriuretic peptide, atrial natriuretic peptide levels, apoptotic index, expression levels of collagen I, collagen III, TNF-α, interleukin-6 and malondialdehyde in the myocardial tissue, and a decrease in cardiac superoxide dismutase activity, all of which was linked to inhibition of oxidative stress and inflammation. Mitochondrial dysfunction is often linked to number of diseases, including heart disease due to insufficient supply of oxygen, which affect mitochondrial fusion, mitophagy, and ATP generation (Lian et al., 2021).

Rhoden et al. (2021) used three-dimensional engineered heart tissue to observe if SFN impairs contractility of the heart and mitochondrial function. SFN pre-treatment of the heart increased lactate formation, enhanced ROS production by mitochondria, and resulted in decreased mitochondrial membrane potential, which is a favourable outcome in case of heart, diabetes, neurodegeneration, or similar health condition. However, cardiac function should be monitored upon administration of SFN to avoid cardiotoxic side effects, occurring from excessive ROS which may be harmful to other biomolecules.

It was found that ~22.5 μmol of SFN can help to reduce systolic and diastolic blood pressure, while ~21.6 μmol of glucoraphanin can decrease low-density lipoprotein (LDL) cholesterol by 7.1% in people with cardiovascular profile and by 5.1% in healthy population. However, there are some studies which indicate opposite effects. It was reported that consumption of one or more portions of Brussel sprouts per week could increase chances of developing coronary heart disease (CHD) (Ma et al., 2018a,b) However, last results were taken from a dietary questionnaire; therefore, they could be influenced by other food sources, which could have an opposite effect of GLs. As seen in Table 1, a number of clinical trials have been carried out using broccoli sprouts to confirm the beneficial effects of SFN and glucoraphanin.

4.6 Benefits for diabetic patients

Diabetes is a metabolic disease that causes high blood sugar. Type 1 diabetes is characterized by the failure to produce insulin by β-cells of the pancreas leading to insulin deficiency in the cells, meanwhile, type 2 diabetes (T2D) is characterized by insulin resistance, where the body cells fail to respond to produce insulin, leading to increased blood glucose levels. Due to antioxidative and anti-inflammatory effects of GLs metabolites like ITC, they were linked to the possibility to modulate T2D through the activation of nuclear factor E2-related factor (Nrf2) and is further responsible for phase II enzyme activation which plays an important role in reducing insulin resistance.

There are a number of animal studies confirming the beneficial effects of GLs on diabetic mice; however, there is a lack of evidence from clinical trials. Sahin et al. (2019) found that 100 mg/kg BW of allyl ITC throughout 12-week period reduced blood glucose levels, total cholesterol, triglycerides, and creatine levels in diabetic mice. TAC was also significantly increased ($P = 0.001$).

Sinigrin is an aliphatic GLs, which gets absorbed in the intestines as allyl ITC and is then conjugated with glutathione. It was found that 15 µmol sinigrin/kg BW of mice for 21 days reduced plasma glucose levels and significantly improved insulin resistance in mice with T2D (Truong and Koyama, 2020).

Currently, there is an ongoing phase 2 clinical trial investigating if broccoli sprout extract, more particularly sulforaphane, could be used as a dietary supplement for glucose tolerance and insulin sensitivity. The current study is a double-blind, parallel assignment, placebo-controlled, randomized trial, involving 108 patients with T2D aged 35–75 over a 12-week treatment period. The overall principle of the trial is based on the knowledge that 'Sulforaphane suppressed glucose production from hepatic cells by nuclear translocation of NRF2 and decreased expression of key enzymes in gluconeogenesis' (Axelsson et al., 2017). Results for the current clinical trial are still pending; however, it looks very promising.

4.7 Central nervous system protection

Cruciferous vegetables have been linked to central nervous system (CNS) protection while following similar pathway as for cardiovascular and carcinogenic protection through the reduction of inflammation and oxidative stress. The main CNS health benefit was linked to the activation of transcription factor Nrf2 and reduced action of ROS. Shiina et al. (2015) reported beneficial results in schizophrenic patients when consuming SFN for 8 weeks, as seen in Table 1. There are some ongoing clinical trials investigating the effectiveness of SFN against Alzheimer's disease or depression, both of which are CNS conditions.

Extremely promising results were also observed in the treatment of autism spectrum disorder (ASD). From 3 ASD clinical trials summarized in Table 1 (Singh, et al., 2014; Bent, et al., 2018; Momtazmanesh, et al., 2020), results seem to agree and indicate a beneficial effect. It was found that social responsiveness and aberrant behaviour were improved. SFN was deemed to work by reversing ASD abnormalities, including oxidative stress, lower antioxidant capacity, depressed glutathione synthesis, reduced mitochondrial function and oxidative phosphorylation, increased lipid peroxidation, and neuroinflammation (Singh, et al., 2014).

Zhang et al. (2021) investigated if nano-sulforaphane could lower 2-amino-1-methyl-6-phenylimidazo[4,5-b]pyrimidine (PhIP) induced early or abnormal embryonic neuro-development in chicken embryo model with abnormal embryonic nervous system defects. Nano-sulforaphane was made by using biodegradable methoxy polyethylene glycol 5000-b-polyglutamic acid 10 000 (mPEG5K-PGA10K) as the substrate, following which nano-sulforaphane system proved positive results, where peripheral nervous system defects were prevented, protection against PhIP-induced CNS and neural tube defects was observed. Current results provide very positive outlook, as it could increase the embryo survival rate during human pregnancy; however, further research and clinical trials are needed.

4.8 Other health benefits

Other health benefits have also been reported, including neuropathy protection where GLs work by reducing pathological changes in glomerulus. Cruciferous vegetables have also alleviated symptoms of skin lesions and induced expressions of keratins 16 and 6, which is helpful for certain skin conditions (Więkus et al., 2020).

Table 1 includes clinical trials, where additional benefits linked to osteoarthritis (Davidson et al., 2017) and microbiome (Kaczmarek, et al., 2019) are summarized. Most of the health benefits are constantly linked to antioxidant and anti-inflammatory effects of GLs; however, there is a lack of the use of its antimicrobial activity.

5 Antinutritional properties of glucosinolates

Although GLs have many beneficial health effects as discussed in the previous section, these sulphur-containing compounds and their by-products may also have potentially toxic effects on humans. For example, a study by di Gioia et al. (2019) showed that GL decomposition products such as epithionitriles may have toxic effects on the kidneys and liver in mammals while at the same time having potential anticancer and therapeutic properties.

Some breakdown products of GLs include oxazolidine-2-thiones (e.g. progoitrin and its myrosinase-induced degradation product, goitrin, and glucomoringin) and thiocyanate ions, all of which interfere with thyroxine production in the body, reducing iodine supply to the thyroid gland resulting in goitre and other related conditions. Progoitrin, a thiocyanate high in Brassica plants such as Brussel sprouts, causes hypothyroidism as it is a competitive inhibitor of the sodium/iodide symporter on the basolateral membrane of the thyroid follicular cell.

Consuming large amounts of GLs can result in symptoms such as hyperthyroidism, reduced feed intake and performance in animals, enlarged thyroid gland, and reduced levels of circulating thyroid hormones (Embaby, 2010).

GLs are antinutritional substances found in abundance in the seed meal fraction of oilseed Brassica species. They are present in various amounts in different genotypes. Those genotypes with less than 30 mol/g GL content are called low/zero GL types and are recommended for edible applications due to their low pungency (Mawlong et al., 2017).

Some examples of sources where antinutrient properties are found mostly include garlic mustard plants where a large amount of alliarinoside is found, and gamma-hydroxynitrile glucoside structurally related to cyanogenic glycosides which are a potential source of highly toxic hydrogen cyanide (di Gioia et al., 2019). In another study by Embaby et al. (2010) on canola meals and their anti-nutritive effects, the researchers found that the level of anti-nutritive effects could be largely linked to the levels of aliphatic GLs rather than indole GLs.

Rapeseed meal is another example known for its high protein level and has the potential to be used in human nutrition, but its high levels of GLs render it unsuitable for consumption due to the anti-nutritive effects mentioned above. When ingested in large numbers, these antinutrients tend to alter the bioavailability and metabolism of different important nutrient components, and some antinutrient substances may be harmful to health (Chongtham et al., 2021) . Therefore, the existence of antinutritional GLs lowers their food and feed value significantly, which makes their market value decrease as well (Sawicka, 2020).

6 Dietary intake, absorption and digestion of glucosinolates

In plant cells, GLs are relatively stable. When GL-containing plant tissue is damaged, as in food preparation (cutting, chopping, mixing) or chewing, a thioglucosidase called myrosinase is released. Depending on the plant species, the enzyme is normally stored separately from GLs in different cells or intracellular compartments (Halkier and Gershenzon, 2006). Myrosinase hydrolyzes GLs to produce a molecule of D-glucose and an unstable aglycone, thiohydroximate-O-sulfonate. The impulsive reorganization of this intermediate (chemical Lossen rearrangement) results in the release of sulphate ions and the creation of metabolites, the structures of which are determined by the nature of the GLs side chain (R) and the physicochemical conditions of the medium.

6.1 Dietary intake: the bioavailability of glucosinolates in the human diet

The amount of GLs consumed is affected by variety, agronomic factors, and the storage and processing of vegetables prior to consumption (Mithen et al., 2000). Even though mechanical damage causes rapid hydrolysis and degradation of GLs, cutting and pre-harvest stress have been shown to increase the concentrations of indole GLs in cabbage under certain conditions (Verkerk et al., 2009). The history of plant tissue throughout the food chain, from grower to consumer, has a significant impact on its ultimate biological role in human nutrition. Regardless of the final level of GLs in the prepared vegetable, the absorption, metabolism, and delivery of GL breakdown products to target tissues are heavily reliant on the residual level of myrosinase activity (Dekker et al., 2000). When vegetables such as broccoli or cauliflower are eaten raw, entire GLs and active myrosinase are consumed at the same time, allowing the GLs to be broken down within the digestive tract. For example, when rats were fed benzyl GLs in the presence of active myrosinase obtained from Brussels sprouts, a significant portion of the administered dose was excreted in the urine as ITC excretion products (Rouzard et al., 2000). Many countries currently lack dietary GLs intake estimation due to a lack of adequate dietary GL composition data (Wu et al., 2017). In one study, GLs consumption in the German population was estimated to be 14.2 ± 1.1 mg/day for men and 14.8 ± 1.3 mg/day for women (Steinbrecher and Linseisen, 2009). In another study, GLs intake in a Spanish adult population was estimated to be 6.5 mg/day, with indole GLs accounting for 35% of that (Agudo et al., 2008). In the United Kingdom, the national mean daily intake was calculated to be 46.1 mg in fresh material and 29.4 mg in cooked material (Sones et al., 1984). These rough estimates of dietary intake, however, were based on very limited data on dietary exposure to *Brassica* vegetables. Moreover, the data were simply calculated using mean values that did not account for variation. Overall, the dietary intake of GLs does not account for the wide range of GLs concentrations caused by genetic background, cultivation, and cooking or processing (Wu et al., 2021).

6.2 Digestion: bioavailability of glucosinolates and their breakdown product

Human digestion is a multilevel, complex process. It entails the mechanical and chemical breakdown of foods, allowing embedded nutrients to be released and absorbed into the body via intestinal mucosal cells (Mennah-Govela and Bornhorst, 2017). The mouth and stomach reduce the size of food, whereas the small intestines are the primary site of nutrient absorption. Gastric acids, bile salts, and digestive enzymes are present in the stomach and operate to homogenize and transform food (Kong and Singh, 2008). The gastric digest is further dissolved in the intestine, and nutrients are absorbed through the intestinal walls.

Bioaccessibility is the amount of a food product that is released into the digestive tract and becomes available for absorption (Heaney, 2001). This definition also includes the digestive transformations of the food material until assimilation, as well as its enterocytic metabolism. Bioavailability, on the other hand, is a subcategory of absorption that refers to the proportion of administered molecules that are absorbed and reach the circulation system (Wood, 2005). It has been demonstrated that the indole-3-carbinol molecule condenses in acid medium, such as the gastric content, to form polycyclic aromatics. In general, each GLs can provide multiple aglycone structures at the same time (Holst and Williamson, 2004). However, depending on the structure of the GLs side chain and the environmental conditions, one of them is formed more frequently. The aroma of cruciferous vegetables is caused by the breakdown products of GLs.

Cooking the plant material causes the myrosinase to denature. Denaturation intensity is especially important when the applied temperature is high and the cooking time is long, whether baking with water, steam, or microwave. Because of their hydrophilic nature (thioglucose and sulphate group), GLs transit to the colon and are metabolized by the intestinal microbiota when myrosinase is inactivated. In the stomach, intact GLs may be partially absorbed; the remaining GLs will transit through the gastrointestinal tract to the small intestine, where they may be hydrolyzed by plant myrosinase and the breakdown products absorbed. The outstanding non-hydrolyzed GLs will then transport to the colon, where they will be hydrolyzed by bacterial myrosinase and the resulting breakdown molecules will be absorbed or excreted. It has been demonstrated that incubating human faeces with pure GLs or cruciferous vegetable juices whereby myrosinase has been inactivated by heating results in the formation of ITCs (Krul et al., 2002). Other breakdown products of GLs by intestinal microbiota are very likely but are still poorly documented. After incubating human faeces with GLs, the creation of amines from the secondary degradation of ITCs was demonstrated (Combourieu et al., 2001). *Bifidobacterium* strains from the human intestinal microbiota can metabolize GLs to nitriles *in vitro*, and traces of nitriles have been found in the urine of rats fed a pure GL. Many microorganisms are also known to be capable of converting nitriles into ammonia and organic acids (Cheng et al., 2004) (Fig. 3).

6.3 Absorption and post-absorptive metabolism

It has been suggested by various studies that when a food product contains active myrosinase, GLs are rapidly hydrolyzed in the proximal gut (small

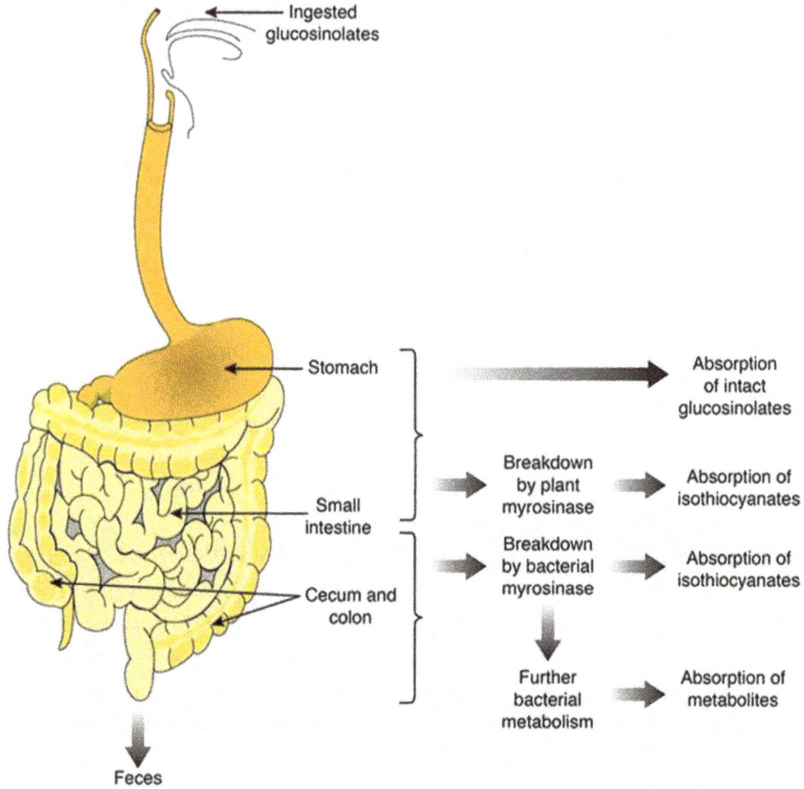

Figure 3 Summary of the role of GLs and their breakdown products in the human gut (adapted from Barba et al., 2016).

intestine). When myrosinase is inactivated (e.g. by cooking), intact GLs can reach the distal gut (colon) and can be metabolized by bacterial enzymes. Additionally, to myrosinase hydrolysis, some studies have shown that a small fraction of myrosinase can be absorbed in its native state by the small intestine lining (Bheemreddy and Jeffery, 2007). *In vivo*, this absorption results in the presence of native GLs in urine up to 5% of the ingested dose. *Ex vivo* studies using isolated rodent intestinal loops suggest a passive or facilitated transport, independent of the glucose uptake mechanism (Michaelsen et al. (1994).

The rapid absorption of ITCs from the upper gastrointestinal tract has been confirmed in pharmacokinetic studies using orally administered ITCs. After dosing rats with [^{14}C] phenethyl isothiocyanate or [^{14}C] allyl isothiocyanate (25–250 μmol kg^{-1}), the rate of appearance of ^{14}C in the blood is rapid, with a peak concentration of 10–100 nmol mL^{-1} occurring at ~3 h (Bollard et al., 1997). After absorption, the main route of ITC metabolism in humans is through the conversion of ITC to N-acetylcysteine derivatives (mercapturic acids or *N*-acetyl-*S*-(*N*-alkylthiocarbamoyl)-1-cysteine). This is accomplished through initial glutathione conjugation, which is aided by glutathione-*S*-transferases, followed by hydrolysis of the resulting conjugates to cysteine derivatives and final N-acetylation. Other than ITC, the post-absorptive fate of GLs derivatives has received relatively little attention. Glutathione-S-transferases can convert thiocyanates to cyanide and thiol derivatives, and epithionitriles can be excreted as mercapturic acids (Conaway et al., 1999).

7 Conclusion and future trends

Over 130 identified GLs structures were found in cruciferous plants, mainly from the *Brassicaceae* family which are hydrolyzed in the small intestine by myrosinase enzymes resulting in a majority of GLs by-products including aliphatic and indole GLs like indole-3-carbinol. Myrosinase get denatured by high temperatures during cooking or processing, when *Bifidobacterium* strains from the human intestinal microbiota takeover the GLs breakdown process. It is not yet known what are the best cruciferous vegetable growth conditions to increase the production of more nutritional compounds, i.e. ITC, and reduce the production of antinutritional compounds, i.e. oxazolidine-2-thiones, moreover, there is a number of known factors that can influence these changes, including the season, temperatures, water stress, and even the consumption method, i.e. boiling or steaming.

GLs and, in particular, SFN mechanism of action is linked to the upregulation of Nrf2 protein which has beneficial effects in the brain, pancreatin, and skin cancer, management of cardiovascular disease and diabetes, and neuroprotective effects in CNS. There is number of *in vitro* studies investigating the anticarcinogenic properties of GLs, which should induce further focus on GSL pharmacological pathways as there are no known short- and long-term human intervention studies. Known adverse effects of GLs are rare but can include mutagenicity, goitrogenicity, hepatotoxicity, and nephrotoxicity, and high levels of oxazolidine-2-thiones and thiocyanate ions were linked to reduced iodine supply to the thyroid gland resulting in goitre.

In the future, the phenomena of GLs metabolism should be further investigated to determine what percentage of bioactive compounds reach the active site and elute beneficial effect, followed by the consequences of denatured myrosinase and if it would be possible to create a nutritional supplements of GLs with myrosinase for better activity and more potent health benefit. Following this, studies using larger population sample size should be carried out to determine current intake, adverse effects from low intake, and amount of GLs needed for beneficial effect.

8 Where to look for further information

8.1 Further reading

- Galanakis, C. M. (Ed.). (2019). *Glucosinolates: Properties, recovery, and applications*. Academic Press.
- Bischoff, K. L. (2021). Glucosinolates. In *Nutraceuticals* (pp. 903-909). Academic Press.
- Wu, X., Huang, H., Childs, H., Wu, Y., Yu, L. and Pehrsson, P. R. (2021). Glucosinolates in Brassica vegetables: Characterization and factors that influence distribution, content, and intake. *Annual Review of Food Science and Technology*, *12*, 485-511.
- Akram, M., Jabeen, F., Riaz, M., Khan, F. S., Okushanova, E., Imran, M., Shariati, M. A., Egbuna, C. and Ezeofor, N. J. (2021). Health benefits of glucosinolate isolated from cruciferous and other vegetables. In *Preparation of Phytopharmaceuticals for the Management of Disorders* (pp. 361-371). Academic Press.
- Miękus, N., Marszałek, K., Podlacha, M., Iqbal, A., Puchalski, C. and Świergiel, A. H. (2020). Health benefits of plant-derived sulfur compounds, glucosinolates, and organosulfur compounds. *Molecules*, *25*(17), 3804.
- Barba, F. J., Nikmaram, N., Roohinejad, S., Khelfa, A., Zhu, Z. and Koubaa, M. (2016). Bioavailability of glucosinolates and their breakdown products: Impact of processing. *Frontiers in nutrition*, *3*, 24.

8.2 Key conferences

- International Conference on Glucosinolates and Glucosinolates In Brassica Crops, 20-21 September 2022 in Lisbon, Portugal.
- Australian Brassica Conference 'Growing the Future'. 7-8 September 2021. Hybrid.
- Brassica 2021 Conference, International crucifer genetics conference, Saskatoon, Canada.
- International Rapeseed Congress.
- Annual Brassica Workshops at Plant Animal Genome.

9 References

Agudo, A., Ibanez, R., Amiano, P., Ardanaz, E., Barricarte, A., Berenguer, A., Dolores Chirlaque, M., Dorronsoro, M., Jakszyn, P., Larrañaga, N., Martinez, C., Navarro, C., Pera, G., Quirós, J. R., Sanchéz, M. J., Tormo, M. J. and Gonzalez, C. A. (2008). Consumption of cruciferous vegetables and GLs in a Spanish adult population. *Eur. J. Clin. Nutr.* 62(3): 324–331.

Almuhayawi, M. S., AbdElgawad, H., Al Jaouni, S. K., Selim, S., Hassan, A. H. A. and Khamis, G. (2020). Elevated CO2 improves glucosinolate metabolism and stimulates anticancer and anti-inflammatory properties of broccoli sprouts. *Food Chem.* 328: 127102.

Armah, C. N., Derdemezis, C., Traka, M. H., Dainty, J. R., Doleman, J. F., Saha, S., Leung, W., Potter, J. F., Lovegrove, J. A. and Mithen, R. F. (2015). Diet rich in high glucoraphanin broccoli reduces plasma LDL cholesterol: evidence from randomised controlled trials. *Mol. Nutr. Food Res.* 59(5): 918-926.

Axelsson, A. S., Tubbs, E., Mecham, B., Chacko, S., Nenonen, H. A., Tang, Y., Fahey, J. W., Derry, J. M. J., Wollheim, C. B., Wierup, N., Haymond, M. W., Friend, S. H., Mulder, H. and Rosengren, A. H. (2017). Sulforaphane reduces hepatic glucose production and improves glucose control in patients with type 2 diabetes. *Sci. Transl. Med.* 9(394): 4477.

Barba, F. J., Nikmaram, N., Roohinejad, S., Khelfa, A., Zhu, Z. and Koubaa, M. (2016). Bioavailability of GLs and their breakdown products: impact of processing. *Front. Nutr.* 3(24).

Bell, L. and Wagstaff, C. (2014). GLs, myrosinase hydrolysis products, and flavonols found in rocket (Eruca sativa and Diplotaxis tenuifolia). *J. Agric. Food Chem.* 62(20): 4481-4492.

Bent, S., Lawton, B., Warren, T., Widjaja, F., Dang, K., Fahey, J. W., Cornblatt, B., Kinchen, J. M., Delucchi, K. and Hendren, R. L. (2018). Identification of urinary metabolites that correlate with clinical improvements in children with autism treated with sulforaphane from broccoli. *Mol. Autism* 9(1): 35.

Bheemreddy, R. M.and Jeffery, E. H. (2007). The metabolic fate of purified glucoraphanin in F344 rats. *J. Agric. Food Chem.* 55(8): 2861-2866.

Biondi, F., Balducci, F., Capocasa, F., Visciglio, M., Mei, E., Vagnoni, M., Mezzetti, B. and Mazzoni, L. (2021). Environmental conditions and agronomical factors influencing the levels of phytochemicals in Brassica vegetables responsible for nutritional and sensorial properties. *Appl. Sci.* 11(4).

Blažević, I., Montaut, S., Burčul, F. and Rollin, P. (2015). GLs: novel sources and biological potential. In: *Reference Series in Phytochemistry* Mérillon, J. M. and Ramawat, K. (Eds). Springer, 1-58.

Blažević, I., Montaut, S., Burčul, F., Olsen, C. E., Burow, M., Rollin, P. and Agerbirk, N. (2020). Glucosinolate structural diversity, identification, chemical synthesis and metabolism in plants. *Phytochemistry* 169: 112100.

Boldry, E. J., Yuan, J. M., Carmella, S. G., Wang, R., Tessier, K., Hatsukami, D. K., Hecht, S. S. and Tretyakova, N. Y. (2019). Effects of 2-phenethyl isothiocyanate on metabolism of 1,3-butadiene in smokers. *Cancer Prev. Res. (Phila)* 13(1): 91-100.

Bollard, M., Stribbling, S., Mitchell, S. and Caldwell, J. (1997). The disposition of allyl isothiocyanate in the rat and mouse. *Food Chem. Toxicol.* 35(10-11): 933-943.

Boscaro, V., Boffa, L., Binello, A., Amisano, G., Fornasero, S., Cravotto, G. and Gallicchio, M. (2018). Antiproliferative, proapoptotic, antioxidant and antimicrobial effects of Sinapis nigra L. and Sinapis alba L. extracts. *Molecules* 23(11): 3004.

Cavaiuolo, M., Cocetta, G., Spadafora, N. D., Müller, C. T., Rogers, H. J. and Ferrante, A. (2017). Gene expression analysis of rocket salad under pre-harvest and postharvest stresses: a transcriptomic resource for Diplotaxis tenuifolia. *PLoS ONE* 12(5): e0178119.

Chang, J., Wang, M., Jian, Y., Zhang, F., Zhu, J., Wang, Q. and Sun, B. (2019). Health-promoting phytochemicals and antioxidant capacity in different organs from six varieties of Chinese kale. *Sci. Rep.* 9(1): 20344.

Charron, C. S., Vinyard, B. T., Ross, S. A., Seifried, H. E., Jeffery, E. H. and Novotny, J. A. (2018). Absorption and metabolism of isothiocyanates formed from broccoli GLs: effects of BMI and daily consumption in a randomised clinical trial. *Br. J. Nutr.* 120(12): 1370–1379.

Cheng, D. L., Hashimoto, K. and Uda, Y. (2004). *In vitro* digestion of sinigrin and glucotropaeolin by single strains of *Bifidobacterium* and identification of the digestive products. *Food Chem. Toxicol.* 42(3): 351–357.

Choi, S., Park, S., Lim, Y., Kim, S., Park, J. and An, G. (2014). Metabolite profiles of GLs in cabbage varieties (Brassica oleracea var. capitata) by season, color, and tissue position introduction. *Environ. Biotechnol.* 55(3): 237–247.

Chongtham, N. et al. (2021). Mineral elements in bamboo shoots and potential role in food fortification. *J. Food Comp Anal.* 95, 103662.

Combourieu, B., Elfoul, L., Delort, A. M. and Rabot, S. (2001). Identification of new derivatives of sinigrin and glucotropaelin produced by the human digestive microflora using ^{1}H NMR spectroscopy analysis of *in vitro* incubations. *Drug Metab. Dispos.* 29(11): 1440–1445.

Conaway, C. C., Jiao, D., Kohri, T., Liebes, L. and Chung, F. L. (1999). Disposition and pharmacokinetics of phenethyl isothiocyanate and 6-phenyllhexyl isothiocyanate in F344 rats. *Drug Metab. Dispos.* 27(1): 13–20.

Davidson, R., Gardner, S., Jupp, O., Bullough, A., Butters, S., Watts, L., Donell, S., Traka, M., Saha, S., Mithen, R., Peffers, M., Clegg, P., Bao, Y., Cassidy, A. and Clark, I. (2017). Isothiocyanates are detected in human synovial fluid following broccoli consumption and can affect the tissues of the knee joint. *Sci. Rep.* 7(1): 3398.

Dekker, M. R., Verkerk, R. and Jongen, W. M. F. (2000). Predictive modelling of health aspects in the food production chain: a case study on GLs in cabbage. *Trends Food Sci. Technol.* 11(4–5): 174–181.

di Gioia, F., Pinela, J., Bailón, A. D. H., Fereira, I. C. F. R. and Petropoulos, S. A. (2019). The dilemma of "good" and "bad" GLs and the potential to regulate their content. In: *GLS: Properties, Recovery, and Applications*. Elsevier, 1–45.

Dini, I. (2018). Spices and herbs as therapeutic foods. In: *Handbook of Food Bioengineering, Food Quality: Balancing Health and Disease* Holban, A. M. and Mihai Grumezescu, A. (Eds). Academic Press, 433–469.

Embaby, H., Habiba, R., Shatta, A., Elhamamy, M., Morita, N. and Ibrahim, S. (2010). GLs and other anti-nutritive compounds in canola meals from varieties cultivated in Egypt and Japan. *Afr. J. Food Agric. Nutr. Dev.* 10(8). Available at: https://www.ajol.info/index.php/ajfand/article/view/66216.

Embaby, H. E. (2010). Effect of soaking, dehulling and cooking on antinutrients and in-vivo protein digestibility of bitter and sweet lupin sweets. *Food Sci. Biotech.* 19, 1055–1062.

Esteve, M. (2020). Mechanisms underlying biological effects of cruciferous glucosinolate-derived isothiocyanates/indoles: a focus on metabolic syndrome. *Front. Nutr.* 7: 111.

Guo, Q., Sun, Y., Tang, Q., Zhang, H. and Cheng, Z. (2020). Isolation, identification, biological estimation, and profiling of GLs in Isatis indigotica roots. *J. Liq. Chromatogr. Relat. Technol.* 43(15–16): 645–656.

Guzmán-Pérez, V., Bumke-Vogt, C., Schreiner, M., Mewis, I., Borchert, A. and Pfeiffer, A. F. (2016). Benzylglucosinolate derived isothiocyanate from Tropaeolum majus reduces gluconeogenic gene and protein expression in human cells. *PLoS ONE* 11(9): e0162397.

Halkier, B. A. and Gershenzon, J. (2006). Biology and biochemistry of GLs. *Annu. Rev. Plant Biol.* 57: 303–333.

Heaney, R. P. (2001). Factors influencing the measurement of bioavailability, taking calcium as model. *J. Nutr.* 131(4) (Suppl.): 1344S–1348S.

Holst, B. and Williamson, G. (2004). A critical review of the bioavailability of GLs and related compounds. *Nat. Prod. Rep.* 21(3): 425–447.

Kaczmarek, J. L., Liu, X., Charron, C. S., Novotny, J. A., Jeffery, E. H., Seifried, H. E., Ross, S. A., Miller, M. J., Swanson, K. S. and Holscher, H. D. (2019). Broccoli consumption affects the human gastrointestinal microbiota. *J. Nutr. Biochem.* 63: 27–34.

Kleinwächter, M., Schnug, E. and Selmar, D. (2008). The glucosinolate-myrosinase system in nasturtium (Tropaeolum majus L.): variability of biochemical parameters and screening for clones feasible for pharmaceutical utilization. *J. Agric. Food Chem.* 56(23): 11165–11170.

Kong, F. and Singh, R. P. (2008). Disintegration of solid foods in human stomach. *J. Food Sci.* 73(5): R67–R80.

Krul, C., Humblot, C., Philippe, C., Vermeulen, M., van Nuenen, M., Havenaar, R. and Rabot, S. (2002). Metabolism of sinigrin (2-propenyl glucosinolate) by the human colonic microflora in a dynamic *in vitro* large-intestinal model. *Carcinogenesis* 23(6): 1009–1016.

Lachance, J. C., Radhakrishnan, S., Madiwale, G., Guerrier, S. and Vanamala, J. K. P. (2020). Targeting hallmarks of cancer with a food-system-based approach. *Nutrition* 69: 110563.

Lian, Y., Lin, Z., Zhang, Z. and Wang, X. D. (2021). Active-targeting polymeric dual nanosensor for ratiometrically measuring proton and oxygen concentrations in mitochondria. *Anal. Chem.* 93(23): 8291–8299.

Liu, S., Huang, H., Yi, X., Zhang, Y., Yang, Q., Zhang, C., Fan, C. and Zhou, Y. (2020). Dissection of genetic architecture for glucosinolate accumulations in leaves and seeds of Brassica napus by genome-wide association study. *Plant Biotechnol. J.* 18(6): 1472–1484.

Loi, M., Villani, A., Paciolla, F., Mulè, G. and Paciolla, C. (2020). Challenges and opportunities of light-emitting diode (LED) as key to modulate antioxidant compounds in plants. A review. *Antioxidants (Basel)* 10(1): 42.

Ma, L., Liu, G., Zong, G., Sampson, L., Hu, F. B., Willett, W. C., Rimm, E. B., Manson, J. E., Rexrode, K. M. and Sun, Q. (2018a). Intake of GLs and risk of coronary heart disease in three large prospective cohorts of US men and women. *Clin. Epidemiol.* 10: 749–762.

Ma, T., Zhu, D., Chen, D., Zhang, Q., Dong, H., Wu, W., Lu, H. and Wu, G. (2018b). Sulforaphane, a natural isothiocyanate compound, improves cardiac function and remodeling by inhibiting oxidative stress and inflammation in a rabbit model of chronic heart failure. *Med. Sci. Monit.* 24: 1473–1483.

Maina, S., Misinzo, G., Bakari, G. and Kim, H. (2020). Human, animal and plant health benefits of GLs and strategies for enhanced bioactivity: a systematic review. *Molecules* 25(16): 3682.

Maldini, M., Maksoud, S. A., Natella, F., Montoro, P., Petretto, G. L., Foddai, M., De Nicola, G. R., Chessa, M. and Pintore, G. (2014). Moringa oleifera: study of phenolics and GLs by mass spectrometry. *J. Mass Spectrom.* 49(9): 900-910.

Márquez-Lema, A., Fernández-Martínez, J. M., Pérez-Vich, B. and Velasco, L. (2009). Inheritance of very high glucosinolate content in Ethiopian mustard seeds. *Plant Breed.* 128(3): 278-281.

Martínez-Ballesta, M., Moreno, D. and Carvajal, M. (2013). The physiological importance of GLs on plant response to abiotic stress in Brassica. *Int. J. Mol. Sci.* 14(6): 11607-11625.

Mawlong, I., Sujith Kumar, M., Gurung, B., Singh, K. and Singh, D. (2017). International Journal of Food Properties: A simple spectrophotometric method for estimating total GLs in mustard de-oiled cake (vol. 2017). Available at: https://www.tandfonline.com/action/journalInformation?journalCode=ljfp20.

Mennah-Govela, Y. A. and Bornhorst, G. M. (2017). Fresh-squeezed orange juice properties before and during *in vitro* digestion as influenced by orange variety and processing method. *J. Food Sci.* 82(10): 2438-2447.

Michaelsen, S., Otte, J., Simonsen, L.-O. and Sørensen, H. (1994). Absorption and degradation of individual intact GLs in the digestive tract of rodents. *Acta Agric Scand Sect - Anim Sci.* 44(1): 25-37.

Miękus, N., Marszałek, K., Podlacha, M., Iqbal, A., Puchalski, C. and Świergiel, A. H. (2020). Health benefits of plant-derived sulfur compounds, GLs, and organosulfur compounds. *Molecules* 25(17): 3804.

Mirmiran, P., Bahadoran, Z., Golzarand, M., Zojaji, H. and Azizi, F. (2014). A comparative study of broccoli sprouts powder and standard triple therapy on cardiovascular risk factors following H.pylori eradication: a randomized clinical trial in patients with type 2 diabetes. *J. Diabetes Metab. Disord.* 13(1): 64.

Mithen, R. F., Dekker, M., Verkerk, R., Rabot, S. and Johnson, I. T. (2000). The nutritional significance, biosynthesis and bioavailability of GLs in human foods. *J. Sci. Food Agric.* 80(7): 967-984.

Momtazmanesh, S., Amirimoghaddam-Yazdi, Z., Moghaddam, H. S., Mohammadi, M. R. and Akhondzadeh, S. (2020). Sulforaphane as an adjunctive treatment for irritability in children with autism spectrum disorder: a randomized, double-blind, placebo-controlled clinical trial. *Psychiatry Clin. Neurosci.* 74(7): 398-405.

Nguyen, V. P. T., Stewart, J., Lopez, M., Ioannou, I. and Allais, F. (2020). GLs: natural occurrence, biosynthesis, accessibility, isolation, structures, and biological activities. *Molecules* 25(19): 4537.

Popović, M., Maravić, A., Čikeš Čulić, V., Đulović, A., Burčul, F. and Blažević, I. (2020). Biological effects of glucosinolate degradation products from horseradish: a horse that wins the race. *Biomolecules* 10(2): 343.

Prieto, M., López, C. and Simal-Gandara, J. (2019). GLs: molecular structure, breakdown, genetic, bioavailability, properties and healthy and adverse effects. In: *Advances in Food and Nutrition Research* (vol. 90) Ferreira, I. and Barros, L. (Eds). Academic Press Inc, 305-350.

Putnik, P., Gabrić, D., Roohinejad, S., Barba, F. J., Granato, D., Mallikarjunan, K., Lorenzo, J. M. and Bursać Kovačević, D. (2019). An overview of organosulfur compounds from Allium spp.: from processing and preservation to evaluation of their bioavailability, antimicrobial, and anti-inflammatory properties. *Food Chem.* 276: 680-691.

Rhoden, A., Friedrich, F. W., Brandt, T., Raabe, J., Schweizer, M., Meisterknecht, J., Wittig, I., Ulmer, B. M., Klampe, B., Uebeler, J., Piasecki, A., Lorenz, K., Eschenhagen, T.,

Hansen, A. and Cuello, F. (2021). Sulforaphane exposure impairs contractility and mitochondrial function in three-dimensional engineered heart tissue. *Redox Biol.* 41: 101951.

Rouzard, G., Duncan, A. J., Rabot, S., Ratcliffe, B., Durao, S., Garrido, S. and Young, S. (2000). Factors influencing the release of cancer-protective isothiocyanates in the digestive tract of rats following consumption of glucosinolate-rich brassica vegetables. *Dietary Anticarcinogens and Antimutagens*: 92–95.

Sahin, N., Orhan, C., Erten, F., Tuzcu, M., Defo Deeh, P. B., Ozercan, I. H., Juturu, V. and Kazim, S. (2019). Effects of allyl isothiocyanate on insulin resistance, oxidative stress status, and transcription factors in high-fat diet/streptozotocin-induced type 2 diabetes mellitus in rats. *J. Biochem. Mol. Toxicol.* 33(7): e22328.

Sawicka, H. (2020). GLS as natural plant substances-structural and application aspects: a review. *Cell Tissue Res.* 20(2): 6919-6928. Available at: www.tcrjournals.com.

Shiina, A., Kanahara, N., Sasaki, T., Oda, Y., Hashimoto, T., Hasegawa, T., Yoshida, T., Iyo, M. and Hashimoto, K. (2015). An open study of sulforaphane-rich broccoli sprout extract in patients with schizophrenia. *Clin. Psychopharmacol. Neurosci.* 13(1): 62-67.

Singh, K., Connors, S. L., Macklin, E. A., Smith, K. D., Fahey, J. W., Talalay, P. and Zimmerman, A. W. (2014). Sulforaphane treatment of autism spectrum disorder (ASD). *Proc. Natl Acad. Sci. U. S. A.* 111(43): 15550–15555.

Sones, K., Heaney, R. K. and Fenwick, G. R. (1984). An estimate of the mean daily intake of GLs from cruciferous vegetables in the UK. *J. Sci. Food Agric.* 35(6): 712–720.

Sørensen, J. C., Frandsen, H. B., Jensen, S. K., Kristensen, N. B., Sørensen, S. and Sørensen, H. (2016). Bioavailability and in vivo metabolism of intact GLs. *J. Funct. Foods* 24: 450–460.

Steinbrecher, A. and Linseisen, J. (2009). Dietary intake of individual glucosinolates. *Ann. Nutr. Metab.* 54(2), 87–96.

Tabart, J., Pincemail, J., Kevers, C., Defraigne, J. and Dommes, J. (2018). Processing effects on antioxidant, glucosinolate, and sulforaphane contents in broccoli and red cabbage. *Eur. Food Res. Technol.* 244(12): 2085–2094.

Tacer-Caba, Z. (2019). Different sources of GLs and their derivatives. In: *Properties, Recovery, and Applications* Galanakis, C. M. (Ed.). Academic Press, 143–180.

Tahata, S., Singh, S. V., Lin, Y., Hahm, E. R., Beumer, J. H., Christner, S. M., Rao, U. N., Sander, C., Tarhini, A. A., Tawbi, H., Ferris, L. K., Wilson, M., Rose, A., Dietz, C. M., Hughes, E., Fahey, J. W., Leachman, S. A., Cassidy, P. B., Butterfield, L. H., Zarour, H. M. and Kirkwood, J. M. (2018). Evaluation of biodistribution of sulforaphane after administration of oral broccoli sprout extract in melanoma patients with multiple atypical nevi. *Cancer Prev. Res. (Phila)* 11(7): 429–438.

Traka, M. H., Melchini, A., Coode-Bate, J., Al Kadhi, O., Saha, S., Defernez, M., Troncoso-Rey, P., Kibblewhite, H., O'Neill, C. M., Bernuzzi, F., Mythen, L., Hughes, J., Needs, P. W., Dainty, J. R., Savva, G. M., Mills, R. D., Ball, R. Y., Cooper, C. S. and Mithen, R. F. (2019). Transcriptional changes in prostate of men on active surveillance after a 12-mo glucoraphanin-rich broccoli intervention–results from the Effect of sulforaphane on prostate Cancer PrEvention (Escape) randomized controlled trial. *Am. J. Clin. Nutr.* 109(4): 1133–1144.

Truong, T. and Koyama, T. (2020). Glucosinolate sinigrin improves insulin resistance to suppress glutathione consumption in type 2 diabetic mice. *Diabetes Metab. J.* 11(12).

Verkerk, R. R., Schreiner, M., Krumbein, A., Ciska, E., Holst, B., Rowland, I., Schrijver, R. D., Hansen, M., Gerhäuser, C., Mithen, R. and Dekker, M. (2009). Glucosinolates in Brassica vegetables. *Molec. Nutr. Food Res.* 53(Suppl 2), S219.

Wood, R. J. (2005). Bioavailability: definition, general aspects and fortificants. *Encyclopedia of Human Nutrition* (2nd edn.). Elsevier Ltd, Oxford.

Wu, W., Huang, H., Childs, H., Wu, Y., Yu, L. and Pehrsson, P. R. (2021). GLs in *Brassica* Vegetables: characterization and Factors that Influence distribution, Content and Intake. *Annu. Rev. Food Sci. Technol.* 12: 485-511.

Wu, X., Sun, J., Haytowitz, D. B., Harnly, J. M., Chen, P. and Pehrsson, P. R. (2017). Challenges of developing a valid dietary glucosinolate database. *J. Food Compos. Anal.* 64: 78–84.

Zhang, P., Li, T., Liu, C., Sindi, M., Cheng, X., Qi, S., Liu, X., Yan, Y., Bao, Y., Brand-Saberi, B., Yang, W., Wang, G. and Yang, X. (2021). Nano-sulforaphane attenuates PhIP-induced early abnormal embryonic neuro-development. *Ann. Anat.* 233: 151617.

Zhang, Z., Garzotto, M., Davis, E. W., Mori, M., Stoller, W. A., Farris, P. E., Wong, C. P., Beaver, L. M., Thomas, G. V., Williams, D. E., Dashwood, R. H., Hendrix, D. A., Ho, E. and Shannon, J. (2020). Sulforaphane bioavailability and chemopreventive activity in men presenting for biopsy of the prostate gland: a randomized controlled trial. *Nutr. Cancer* 72(1): 74–87.

Chapter 6

Nutraceutical potential of glucosinolates

Ella O'Grady and Greta Pileckaite, School of Food Science and Environmental Health, Technological University Dublin – City Campus, Ireland; Aline Alberti, Graduate Program in Food Science and Technology, State University of Ponta Grossa, Brazil; and Swarna Jaiswal and Amit K. Jaiswal, School of Food Science and Environmental Health, Technological University Dublin – City Campus and Environmental Sustainability and Health Institute, Technological University Dublin – City Campus, Ireland

1 Introduction

Glucosinolates (*S*-glucopyranosyl thiohydroximates; GLs) are secondary metabolites of plants, largely synthesized in the *Brassica* genus of the *Brassicaceae* family such as broccoli, cabbage, cauliflower, Brussel sprouts, and kale. GLs are anions (β-thioglycoside N-hydroxysulfates) with a side chain (R) of alkyl, aralkyl, or indolyl and a sulphur-linked β-D-glucopyranose moiety, and so far, over 130 GLs have been identified. Side chain modification of the amino acid precursors prior to the GLs formation and a wide range of secondary modifications, including oxidation, sulfation, hydroxylation, methoxylation, glucosylation, as well as substitutions with acyl conjugation on the sugar moieties, are responsible for the high number of different GLs. It can be classified into three groups, aliphatics, aromatics, or indoles, where classification comes from the R chain of the compound which is derived from one of eight amino acids of the plant. These compounds are typically stable within a plant cell, but upon damage to the cell, GLs are hydrolyzed by myrosinase resulting in its more bioactive products such as isothiocyanates (ITC).

http://dx.doi.org/10.19103/AS.2022.0101.20

GLs are hydrolyzed by the enzyme myrosinase when the plant tissue is injured, a protective response induced by the plant. In the presence of water, myrosinase cleaves a glucose group from the GLs to form an ITC, a thiocyanate, or a nitrile (Lachance et al, 2020) (unstable aglucones) (Blažević et al, 2020). ITCs are a chemical group of organosulfur compounds, -N=C=S, characterized by the presence of sulphur in place of oxygen in an ITC molecule. These compounds are typically formed by the enzymatic conversion of indole GLs and are abundant in cruciferous vegetables. Sulforaphane (SFN) is the most widely studied ITC with several studies reporting the nutraceutical benefits of this compound. SFN is synthesized from GL glucoraphanin and has been demonstrated to have potent bioactive properties. These breakdown products not only have a protective effect on the plant but also are responsible for, along with their precursor GLs, the beneficial health effects associated with the *Brassica* plants (Lachance et al, 2020).

In Chapter 5, we have provided a comprehensive account of the health benefits of GLs and their breakdown products. This chapter is focused on classes of GLs, current research on mechanisms of action, and finally nutraceutical applications of the GLs and their breakdown products.

2 Classes of glucosinolates and their breakdown products

The health benefits of GLs (please refer to Chapter 5 for a detailed account of the health benefits of GLs) are attributed to their enzymatic hydrolysis products instead of the GLs themselves. Numerous studies confirmed the relationship between myrosinase and specifier proteins, relative protein abundance of epithiospecifier protein (ESP), and thus the formation of different GL products such as ITCs, nitriles, and epithionitriles (EPTs) that affect the plant defence system (Hanschen et al., 2018).

The three main chemical classes of GLs are aliphatic GLs (derived from methionine, isoleucine, leucine, or valine), indole GLs (derived from tryptophan), and aromatic GLs (benzenics, derived from phenylalanine or tyrosine) (Fuentas et al., 2015). While other GLs from three different classes have been identified in the edible parts of *Brassica* types, methionine-derived GLs have been reported as the most significant class of GLs in *Brassica* vegetables (Cartea and Velasco, 2008). Both pathways result in the formation of GLs of different classes, although they share enzymes and mutually inhibit each other (Esfandiari et al., 2017). When plants are challenged by biotic and abiotic stimuli, the enzyme myrosinase hydrolyzes these chemicals, resulting in a variety of breakdown products. GLs and their hydrolysis products are essential components of plant defence responses to various stressors (Sánchez-Pujante et al., 2017).

The biosynthesis of aliphatic and indole GLs is well understood, while the biosynthesis of aromatic GLs is mostly unknown.

2.1 Aliphatic glucosinolates

Aliphatic GLs are formed from methionine, isoleucine, leucine, or valine (Tacer-Caba, 2019). It is also possible to classify them as methylthioalkyl, methylsulfinylalkyl, alkenyl, and hydroxyalkenyl based on their changing side chain length with changing number of carbon atoms (3, 4, or 5C) or based on the changing side chain structure of methylthioalkyl, methylsulfinylalkyl, alkenyl, and hydroxyalkenyl (Li and Quiros, 2003). A lengthy series of enzymatic conversions are required for the biosynthesis of GLs (Halkier and Gershenzon, 2006). The pathway to aliphatic GLs is divided into three stages, beginning with methionine deamination, followed by side chain elongation via sequential condensation reactions with acetyl-CoA, isomerization and decarboxylation, and finally synthesis of the core structure (Beewilder et al., 2008). Side chains may then undergo secondary transformations, such as sulfinyl groups. Methylthioalkylmalate synthases, an aconitase, and an isopropylmalate dehydrogenase catalyze elongation reactions (Textor et al., 2007). At least five cytosolic enzymatic steps are required to produce the GLs core structure, beginning with the oxidation of chain elongated amino acids to aldoximes by the cytochrome P450 encoding genes CYP79F and ending with the addition of sulfate to the penultimate intermediate, the desulfo-GLs, resulting in the production of the parent methylthioalkyl GLs (also known as thio-GLs in the following sections) (Piotrowski et al., 2004).

The final stage of GLs biosynthesis involves a variety of secondary modifications to the thio-GLs side chain (Nour-Eldin and Halkier, 2008). The methylthio sulfur moiety is oxidized to a methylsulfinyl moiety by a flavin monooxygenase, such as the flavin-monooxygenase glucosinolate *S*-oxygenase 1, to form the methylsulfinyl GLs (also known as sulfinyl GLs in the following sections) (Tacer-Caba, 2019). The *S*-oxygenated GLs can be converted directly to the corresponding hydroxyalkyl GLs or converted to alkenyl GLs by the 2-oxoglutarate-dependent dioxygenase (AOP2), which can then be converted to hydroxyalkenyl GLs (Kliebenstein et al., 2001). Finally, the benzothiazole (BZO) genes can conjugate hydroxylated GLs to benzoic acid to produce benzoyloxy GLs (Kliebenstein et al., 2007).

2.2 Indole glucosinolates

Tryptophan is the primary source of indole GLs (indolics). One of the most important indole GLs is indole-3-carbinol, which is the breakdown product of the indole GL glucobrassicin. Indolic GLs are thought to be one of the most important factors of pest and disease resistance in plants (Tacer-Caba, 2019). In addition to their direct antibacterial action, indole GLs have a role in eliciting highly conserved immunological responses in the plant world (Kim and Jander, 2007).

In Arabidopsis, indole GLs breakdown is required for bacteria-induced callose deposition and can also control pathogen-induced hypersensitive programmed cell death. Indole-3-carbinol, a byproduct of indole GLs breakdown, can function as an auxin antagonist, influencing the synthesis and location of auxin transporters. Furthermore, when challenged by a bacterial pathogen, indolic GLs were observed to be induced systemically at uninfected tissue, contributing to systemic acquired resistance (Zhou et al., 2019).

The biosynthesis of indole GLs begins with the conversion of tryptophan to indole-3-acetaldoxime by CYP79B2 and CYP79B3. The aldoxime is then catalyzed by CYP83B1 to produce an unidentified intermediate, which undergoes sulfur incorporation and thiohydroximate formation via the activities of GSTF9, GSTF10, GGP1, and SUR1 (Chhajed et al., 2020). UGT74B1 is required for thiohydroximate glucosylation, and SOT16 is responsible for the sulfation step to produce intact indole GLs, similarly to aliphatic GLs biosynthesis. CYP81Fs catalyze the hydroxylation of indole GLs, for example, CYP81F2 is responsible for the production of 4-hydroxyindole GLs (Sønderby et al., 2010). Furthermore, CYP86A7 and CYP71B26 may be involved in the hydroxylation of indole GLs, particularly at the 1-position.

Through indole GLs methyltransferases 1 and 2, hydroxyindole GLs can be further metabolized to methoxyindole derivatives (IGMT1 and IGMT2) (Pfalz et al., 2011). Furthermore, cytoplasmic protein phosphatase 2A regulatory subunit B′ (PP2A-B′) controls methylation of 4-hydroxyindol-3-ylmethyl GLs (4MI3G), which physically interacts with IGMTs and regulates IGMT activities in catalyzing *O*-methylation at the 4-position (Rahikainen et al., 2017). Recently, it was discovered that 1-hydroxyindol-3ylmethylglucosinolate can be methylated via indole GL *O*-methyl transferase 5 (IGMT5) (Pfaaz et al., 2016). Moreover, the PP2A-B′ may influence indole GLs catabolism via direct regulation of the phosphorylation of myrosinase TGG1, which is involved in GLs hydrolysis (Durian et al., 2016).

2.3 Aromatic glucosinolates

Aromatic GLs (benzenics) are mostly produced from two amino acids, phenylalanine and tyrosine (Tacer-Caba, 2019). The breakdown product of one of the aromatic GLs is benzyl isothiocyanate (BITC); glucotropaeolin and phenethyl isothiocyanates are the degradation products of another aromatic GL, gluconasturtiin. These two degradation products are important because they have the potential to act as cancer chemo-protectors (Cartea and Velasco, 2008).

Because of their potential biological activity and therapeutic characteristics, aromatic GLs are significant members of the GL family of chemicals. Aromatic

GLs like natural products (aralkyl-glucose, glucosinalbin) or some non-natural products have shown that they have numerous biological properties and prospective uses in biochemistry, genetics, and pharmaceutics (Vo et al., 2013). Compared with aliphatic and indole GLs, the aromatic GLs are quite unknown. A number of aromatic GLs have an extra sugar moiety, rhamnose or arabinose, attached with the aromatic ring by a glycosidic bond; nonetheless, this is not well understood (Liu et al., 2016).

Furthermore, studies show that 3-butenyl, 4-pentenyl, 2-phenylethyl, and BITCs have substantial inhibitory action against a variety of gram-positive and gram-negative bacterial pathogens. In comparison to aliphatic GLs of 3-butenyl and 4-pentenyl ITCs, aromatic GL-derived compounds of 2-phenylethyl and BITCs have greater bactericidal action (Jang et al., 2010). The mechanism involves the activation of oxidative stress responses and changes in cellular redox homeostasis. Interactions between ITC groups, thiol or amine groups of microbial proteins, and aromatic GL-derived ITCs had higher reactivity due to their ability to act as electron donors from the benzene ring (Sánchez-Pujante et al., 2017). Because branched-chain, aromatic, and indole GLs have a similar range of structural diversity to aliphatic GLs, they are likely to undergo similar hydroxylations, desaturations, and oxidations. The exception is that at least ten of the aromatic and indole GLs are methoxylated singly or multiple times. It has been proposed that benzoyloxyalkyl GLs form when a hydroxylalkyl GL is conjugated with benzoic acid (Mithen et al., 2000). Figure 1 shows the structure of possible GLs degradation products after enzymatic hydrolysis and their breakdown products. GLs structures are shown in green, and rearrangement upon hydrolysis is shown in pink. Some examples of fairly well-characterized GLs are listed in the later section.

2.4 Isothiocyanates

ITCs are made by hydrolyzing GLs, which are sulfur-containing compounds found in cruciferous vegetables. When GLs are hydrolyzed, they each produce a different ITC. Broccoli, for example, is high in glucoraphanin, a GL precursor of SFN, and sinigrin, a GL precursor of allyl ITC (Higdon, 2005). ITCs are rapidly conjugated to glutathione in the liver before being metabolized in the mercapturic acid pathway and excreted in the urine (Hanigan and Cooper, 2018). SFN, one of the most extensively studied ITC, was isolated from broccoli extracts as a potent inducer of mammalian cytoprotective enzymes (Zhang et al., 1992). The GL precursor of SFN, glucoraphanin, is most abundant in seeds, and 3-day-old broccoli sprouts have 20- to 50-fold higher - and much more uniform - levels than mature broccoli, typically 6 µmol/g fresh weight. In terms of bioactivity, ITCs are the most researched GL-derived bioactive dietary components. As a result, broccoli sprouts and extracts have been used in human intervention studies to deliver glucoraphanin or SFN (after complete enzymatic

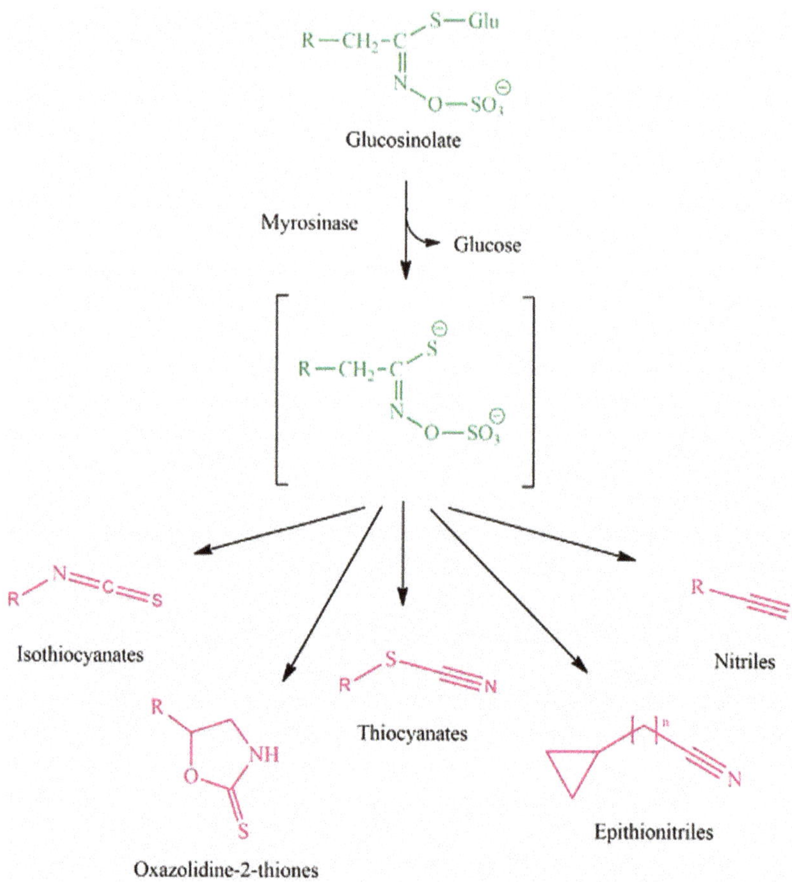

Figure 1 Structure of possible GLs degradation products after enzymatic hydrolysis and their breakdown products. GLs structures are shown in green, and rearrangement upon hydrolysis is shown in pink (Redovniković et al., 2008).

hydrolysis) (Dinkova-Kostova and Kostov, 2012). The chemo-preventive effects of ITCs on breast, lung, colorectal, and prostate cancer, the four most prevalent malignancies globally, have been highlighted in a number of studies (Novío et al., 2019). Various *in vitro* investigations have demonstrated that BITC promotes apoptosis of human breast, prostate, and pulmonary cancer cells and inhibits colon cancer cell migration and invasion (Lee et al., 2018a,b).

2.5 Thiocyanate

One of the breakdown products of GLs is thiocyanate. Thiocyanate production occurs relatively infrequently in plant tissues. Furthermore, only a few GLs may serve as precursors for the creation of suitable thiohydroximes because their chemical structures allow for the development of a stable carbocation form, which is required for thiocyanate reactions. Thiocyanate has also been used to treat hypertension in the past (Agrawal et al., 2018). Additionally, thiocyanate substances block the thyroid's iodine-concentrating process, and their goitrogenic action can be countered with iodine (Chandra, 2010). Thiocyanate inhibits the sodium iodide symporter competitively, resulting in decreased iodide uptake. Thiocyanate also appears to increase iodide efflux while inhibiting thyroid peroxidase-mediated organification (Willemin and Lumen, 2017). The role of thiocyanate in goitrogenesis is supported by both animal and human data. Goitre can develop in animals fed high glycosinolate diets in a dose-dependent manner (Marwaha et al., 2003; Eisenbrand and Gelbke, 2016). Cassava is one of the most well-known dietary sources of thiocyanate and other antioxidants. Cassava consumption has been shown to reduce iodine uptake in endemic goitre regions (Delange and Ermans, 1971).

2.6 Nitriles

Nitriles are formed from GLs in the absence of any terminal double bonds. They are also formed at a more acidic pH (pH < 4) or at an increased level of ferrous ions (Hanschen et al., 2018). Previous research on salad crops has found that they create a lot of simple nitriles at the expense of ITCs. In garden cress, nitrile production is controlled by a specifier protein, but in watercress, it is controlled by an unidentified, non-enzymatic route (Williams et al., 2009).

Several studies have demonstrated that simple nitriles are inefficient as inducers of these detoxifying enzymes and as anti-proliferative agents, in contrast to ITC (Williams et al., 2009). The mechanism of the nitrile specifier protein can change the outcome of GLs hydrolysis and create a shift from a direct to an indirect defence strategy, which is characterized as having a versatile and dynamic nature (Burow and Halkier, 2017).

2.7 Oxazolidine-2-thione

Oxazolidine-2-thione derivatives are GL-related dietary components that give some cruciferous vegetables (thyreo) toxic characteristics. They play a significant role in contemporary antibiotics, with linezolid being the first to target bacterial protein synthesis (Agerbirk et al., 2018). A single oxygen that is either absent or present in two stereochemical positions has drastic biochemical consequences (Agerbirk and Olsen, 2015).

2.8 Epithionitriles

In *Brassica* crops, EPTs are crucial but underappreciated GL hydrolysis products that are produced instead of cancer-fighting ITC (Hanschen et al., 2017). On enzymatic hydrolysis, *Brassica* plants frequently produce EPTs and nitriles instead of ITCs. The presence of ESPs is responsible for this. When plant cells are disturbed, the plant endogenous enzyme myrosinase produces the labile thiohydroximate-*O*-sulfate GL aglucon (Hanschen et al., 2017).

3 Hydrolysis of glucosinolates and the glucosinolate-myrosinase system

In the cell, myrosinase (EC 3.2.1.147 also known as thioglucoside glucohydrolase, sinigrinase, sinigrase, MYR) hydrolyzes glucosinolates (Malka and Cheng, 2017). The GLs-myrosinase system is a 'two-component' defence system, also known as the 'mustard oil bomb' (Ratzka et al., 2002). Myrosinase is found in 'myrosin cells/idioblasts,' which are distinct from the GLs and are stored in S-cells (Hunziker et al., 2019). At high turgor pressure, GLs and myrosinase are found in cells outside the vasculature, in the phloem cap region, and along the leaf margin (Burow and Halkier, 2017). When plant tissue is damaged by insects, herbivores, or cutting, myrosinase comes into contact with the GLs substrate and catalyzes the hydrolysis of the thioglucoside bond-forming glucose, hydrogen ion, and an unstable aglucone (Sturm and Wagner, 2017). This unstable intermediate then degrades instinctively to one of several products, depending on factors such as pH and the presence of various cofactors. In the case of sinigrin, for example, low pH favours the production of allyl cyanide, whereas, at neutral and alkaline pH, allyl ITC is the dominant breakdown product (Uda et al., 1986). Epithionitriles, ITCs, nitriles, and thiocyanates are the hydrolysis products of GLs (Román et al., 2018). The diversity of the product that develops itself is ultimately determined by the parent GLs and the instability of a thiohydroxamate-*O*-sulfonate aglycone, which is also affected by the pH of the environment, the presence of ferrous ions, the ESP, nitrile specific protein, or thiocyanate forming protein (Glindermann et al., 2019).

4 Analysis of glucosinolates

The biological effects of GLs and their metabolites have heightened the interest in developing precise methods for extracting, isolating, and characterizing these materials. The quantification of GLs can be divided into three categories: (1) total GLs, (2) individual GLs, and (3) breakdown products including metabolites (Verkerk and Dekker, 2008). The presence (intact) or absence (desulfo) of the sulphate group in the structure is commonly used to determine GLs, but rarely both. Although many studies have converted intact GLs into desulfo-GLs to facilitate a determination by liquid chromatography (LC), a large body of research has examined intact GLs using the same collection of detection methods but with modified sample preparation techniques (Maldini et al., 2014).

Multiple analytical methods, including high-performance liquid chromatography (HPLC), gas chromatography, and capillary electrophoresis, have been used to determine the identity and levels of GLs in *Brassica* vegetables (Smiechowska et al., 2010). The most commonly used technique has been HPLC with ultraviolet or diode-array detection, which requires a desulfation step to reduce the polarity of GLs, making them more amenable to separation by reversed-phase chromatography (Rochfort and Jones, 2011). One of the most valuable techniques for GLs analysis is mass spectrometry (MS). MS is a highly selective and sensitive technique for profiling plant extracts and identifying unknown compounds. The combination of a separation system (for example, liquid chromatography) and MS (LC-MS) has been developed as a fast, simple, precise, and sensitive technique for improving the chemical analysis of complex biological samples containing many compounds. However, in the case of two compounds with identical molecular weights (or, more precisely, mass to charge ratio m/z), it is extremely difficult to distinguish them using only LC-MS because the precursor ion is detected and analyzed. This disadvantage can be overcome by using tandem MS (LC-MS/MS), which allows differentiation by fragmentation pattern of both the original molecular ion and its fragments, increasing structural assignment accuracy and sensitivity (Almushayti et al., 2021).

5 Mechanisms of action of glucosinolates

Mechanism of action refers to a pharmacological process, where a molecule, drug, or a bioactive compound like GLs binds to an active site or another molecule in order to elute an effect. In this section, the mechanism of action summarizes the process, by which a specific compound leads to biological benefits like cell growth or suppression, as well as interactions and modulations of biological targets like proteins or nucleic acid.

5.1 Detoxification

Detoxification, often known as 'detox', is a biological process involving the removal of a toxic substance from a living organism. Naturally, detoxification is mainly carried out through the liver but also through the kidneys, lungs, gut, or skin all of which allow the body to process toxins in order to make space for the new ones and repeat the process. Detoxification is important when managing symptoms of alcohol or drug withdrawal.

From *in vitro* and *in vivo* studies, it is known that GLs work by modulating phase I and phase II enzymes, including cytochrome P450 enzyme category, through detoxification pathway. Detoxification process happens in two stages. Initially, fat-soluble toxins enter the liver, where stage I cytochrome P450 enzymes begin the detoxification process through oxidation, reduction, hydrolysis, hydration, and dehalogenation reactions. Following this, the toxic compound moves to stage II conjugation pathway, where the process of sulfation, glucuronidation, glutathione conjugation, acetylation, amino acid conjugation, and methylation is carried out (Esteve, 2020). Once both stages of detoxification are completed, waste products are eliminated through urine, bile, or stools, while useful products move back to the bloodstream for further distribution to their active site. Between stage I and stage II, reactive oxygen species (ROS) get eliminated, which can cause tissue damage. For the current reason, consumption of GLs is of vital importance, as they cleave excessive ROS from the body and lead to detoxification process.

5.2 Brain cancer

A brain tumour is a mass of cells presenting abnormal growth in the brain. GLs, and in particular SFN, are commonly linked to anticarcinogenic activity (Fig. 2). In the brain, SFN induces the nuclear factor-erythroid 2-related factor-2 (Nrf2) protein which is linked to anti-inflammation, antioxidant, and mitochondrial role in the brain. Upregulation of Nrf2 leads to activation of genes like heme oxygenase-1 (HO-1), glutathione *S*-transferase (GST), superoxide dismutase, catalase, NAD(P)H dehydrogenase (quinone) 1, and others, which are linked to improved resistance to oxidative insult. Nrf2 directly regulates the expression of inflammatory mediators like interleukin-17D, CD36, macrophage receptors

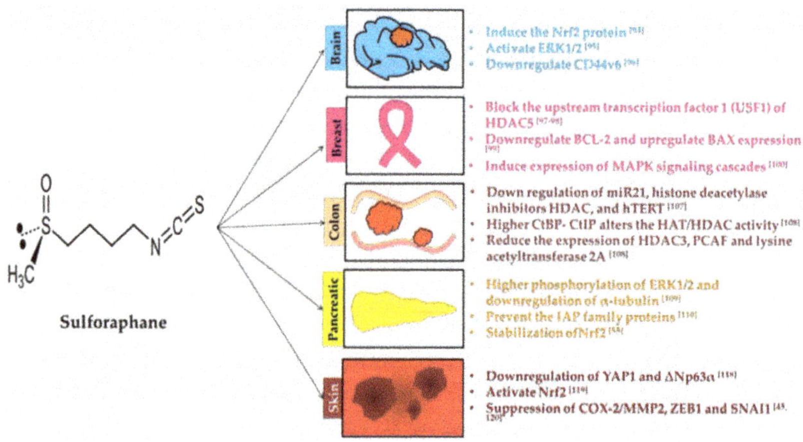

Figure 2 Processes of sulforaphane in the prevention of cancer (Soundararajan and Kim, 2018).

with collagenous structure, and G protein-coupled receptor kinase, and it has also been linked to reduced expression of pro-inflammatory cytokines such as tumour necrosis factor (TNF)-α, IL-6, IL-8, and IL-1β in microglia, macrophages, monocytes, and astrocytes (Brandes and Gray, 2020). Lastly, Nrf2 induces free-radical scavenging enzymes, which protect mitochondria from oxidative stress. It diminishes the overproduction of intracellular ROS and regulates mitochondrial biogenesis through enzymes including malic enzyme 1, isocitrate dehydrogenase 1, glucose-6-phosphate dehydrogenase, and 6-phosphogluconate dehydrogenase (Brandes and Gray, 2020). However, anti-inflammatory, antioxidant, and mitochondrial roles are interlinked and work together for anticarcinogenic activity in the brain.

SFN activates extracellular signal-regulated kinases (ERK)1/2, which is generally responsible for long-term memory; however, it also has numerous functions in tumour suppression. Most commonly, ERK is linked to RAS/RAF/MEK pathway which leads to proliferation and stimulation of mitogens like epidermal growth factor. LLRC4 anchors ERK in cytoplasm, which competitively inhibits MEK binding to ERK, leading to ERK phosphorylation, which then inhibits cell proliferation. ERK can also modulate MMP genes like MMP-9, leading to reduced invasive potential of glioblastoma cells (Hannen et al., 2017). There are more positive modulatory functions linked to ERK1/2 activation.

Lastly, SFN downregulates CD44v6, which is found during cell growth, survival, differentiation, motility, tumour growth, proliferation, and metastasis. Downregulation of CD44 blocks glioblastoma and other cancer growth and sensitizes the cells to cytotoxic drugs (Wu et al., 2020).

5.3 Breast cancer

Breast cancer occurs when the cells in a breast begin to grow in an abnormal way, eventually forming a tumour. In breast cancer, SFN blocks the upstream transcription factor 1 (USF1) of a protein-coding gene Histone Deacetylase 5, which regulates the expression of angiogenesis-related genes in endothelial cells, basal type of breast cancer cells proliferation, and therapeutic resistance (Xue et al., 2019). Therefore, blocking USF1 leads to higher breast cancer susceptibility to therapeutic drugs and lower cancer cell proliferation.

Following this, SFN downregulates B-cell lymphoma 2 (BCL-2), which naturally has a negative effect on breast cancer since it inhibits the programmed cell death. Meanwhile, SFN upregulates BCL-2-associated X protein (BAX) expression, which is often linked to apoptotic mechanisms through p53 proteins and caspase-3 in breast cancer cells with BAX protein.

Lastly, SFN induces expression of mitogen-activated protein kinase signalling cascade, which further induces cyclin D1 and $p21^{CIP1}$ gene, causing cell cycle arrest at G1 phase (the first phase of cell growth) (Soundararajan and Kim, 2018).

5.4 Colorectal cancer

Colorectal cancer, also referred to as bowel cancer, is the growth of abnormal cells in multiple places of the bowel. Colon cancer is characterized by the initial cancerous growth in the large bowel which may lead to a bowel blockage. Similar treatment methods as for breast cancer are often applied. In colon cancer, SFN downregulates miR21 gene, histone deacetylase inhibitors, and human telomerase reverse transcriptase (hTERT) cells. Naturally, miR21 gene promotes the proliferation of cancer cells (You et al., 2018), while hTERT cells participate in the cancer formation by inducing changes like gene amplifications, structural variants, promoter germline and somatic mutations, epigenetic changes, and alternative lengthening of telomere (Dratwa et al., 2020). Therefore, downregulation of miR21, histone deacetylase inhibitors, and hTERT leads to reduced proliferation and formation of cancer cells.

Following this, SFN increases amounts of C-terminal binding protein, C-terminal interacting protein, which in turn alters the enzymatic histone acetylation (HAT)/histone deacetylation (HDAC) activity. HAT/HDAC activity is of importance in colon and other types of cancer since it defines the status of loci, transcription factors, and DNA binding proteins (Zhan et al., 2020).

SFN also reduces the expression of genes: HDAC3, p300/CBP-associated protein (PCAF), and lysine acetyltransferase 2A (KAT2A). HDAC3 leads to decreased levels of histone acetylation followed by upregulation of cancer stem cell-related genes (Zhan et al., 2020); PCAF and KAT2A downregulation results in weaker repair of the colon cancer cells (Soundararajan and Kim, 2018).

5.5 Pancreatic cancer

Pancreatic cancer, which forms from abnormal cell proliferation leading to tumour in the pancreas, is often asymptomatic at the beginning. In pancreatic cancer, SFN works by producing higher phosphorylation of ERK1/2 and downregulation of α-tubulin. ERK1/2 is often linked to number of cancers including pancreatic and previously discussed brain and breast cancer. Each mechanism of action slightly differs, but overall upregulation of ERK1/2 often proves beneficial for the management of cancer cells. In pancreatic cancer, α-tubulin regulates mitosis and intracellular transport, where downregulation leads to poor cancer cell migration, invasion, and cancer metastasis, due to prevented focal adhesion for lamellipodial extension during cell migration, all of which are generally caused by acetylation reactions (Lee et al., 2018a,b). In pancreatic cancer, downregulation of α-tubulin is beneficial, as it reduces the likelihood of cancer spreading to other organs or around the body.

Following this, SFN prevents inhibitors of apoptosis protein family of proteins and stabilizes Nrf2 by demethylating promoter region, which enhances the expression of Nrf2 leading to beneficial effect in the management of pancreatic cancer (Soundararajan and Kim, 2018).

5.6 Skin cancer

Non-melanoma skin cancer is characterized by abnormal squamous and basal cell growth in the skin. It is often treated with surgery by removing the area of abnormal skin, following which skin cancer rarely spreads to other parts of the body. In skin cancer, SFN downregulates the *yes*-associated protein 1 (YAP1) and ΔNp63α. YAP coactivates TAZ and often works in combination to control the maintenance, activation, and coordination of epidermal and dermal cells during development, homeostasis, wound healing, and cancer. However, increased YAP/TAZ signalling could lead to aberrant extracellular deposition or remodelling, cancer cell proliferation, and suppression of apoptotic genes (Rognoni and Walko, 2019). ΔNp63α is a p63 protein isoform, which plays a role in the development of stratified epithelia; however, overexpression was linked to enhanced mutant Ras-driven tumourigenesis of undifferentiated cutaneous squamous cell cancer (Smirnov et al., 2019).

Following this, SFN activates Nrf2 which works through similar pathways as discussed for pancreatic and brain cancer, leading to easier management of the cancer cells. And lastly, SFN suppresses cyclooxygenase 2 (COX-2) protein/matrix metalloproteinase 2 (MMP2), Zinc Finger E-Box Binding Homeobox 1 (ZEB1) gene, and Zinc finger protein SNAI1 (SNAI1). COX-2 has been linked to stimulation of angiogenesis, apoptosis inhibition, increase in cell proliferation, immunosuppression, production of mutagens, and cell invasion. Simultaneously, COX-2 induces MMP2 which results in degradation of the extracellular matrix, tumour invasion, and vascular mimicry in melanoma (Valentina Tudor et al., 2020). Therefore, suppression of COX-2/MMP2 leads to reduced tumour invasion, production of mutagens, and immunosuppression, followed by improved apoptosis. ZEB1 drives the epithelial-mesenchymal transition which activates epigenetic reprogramming and immune invasion (Zhang et al., 2019). SNAI1 expression has been linked to tumour-infiltrating immune cells, which leads to reduced chances of survival in people with skin cancer (Fang and Ding, 2020).

The aforementioned are some of the pathways linked to anticarcinogenic activity of SFN, which are summarized in Fig. 2. However, there are a number of other pathways within cancer types discussed, as well as other types. Currently discussed SFN is the most commonly considered ITC for the treatment of cancer and other conditions; however, other GLs (including BITC or phenethyl ITC) are also beneficial in the management or treatment of cancer through a variety of mechanisms of action.

5.7 Cardiovascular disease and diabetes

The summarized pathway to cardiovascular complications, seen in Fig. 3, showcases where visceral white adipose tissue (WAT) experiences an increase in size leading to release of adipocytokine hormones, which are then linked to inflammation. Simultaneously, lipid deposit in the liver is also observed, which results in release of cytokines and other compounds leading to inflammation. This results in atherosclerosis and further cardiovascular complications like heart attack or stroke (Esteve, 2020). Alternatively, inflammation can cause other metabolic syndromes like insulin resistance. When ITC and indoles are consumed, they lead to an increase in Nrf2, peroxisome proliferator-activated receptor alpha (PPARα), lipid oxidation, glucose oxidation, WAT browning, and leptin signalling, while a decrease in nuclear factor kappa-light-chain-enhancer of activated B cells (NF-kB), sterol regulatory element-binding protein 1 (SREBP1c), and peroxisome proliferator-activated receptor gamma

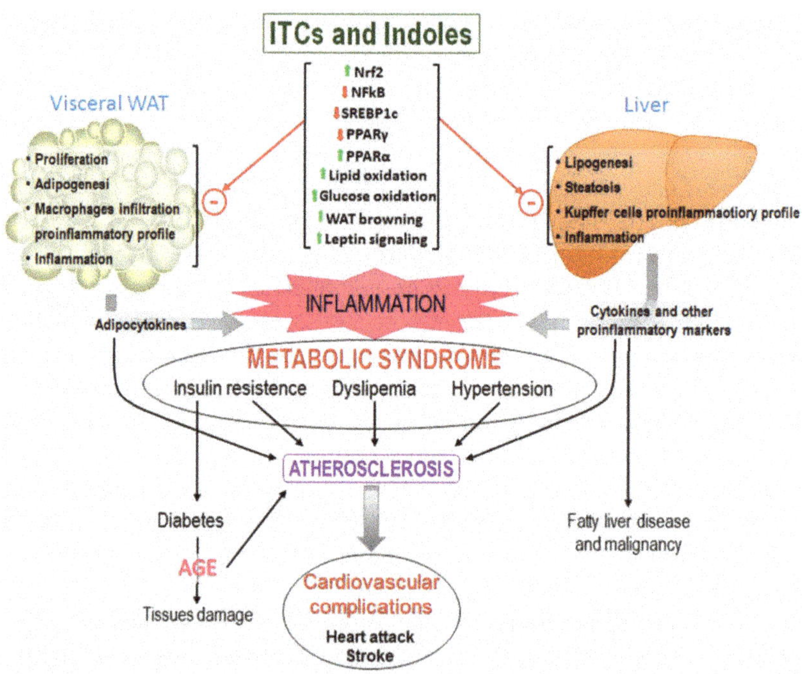

Figure 3 Mechanism of action of isothiocyanate and indole linked to cardiovascular problems (Esteve, 2020).

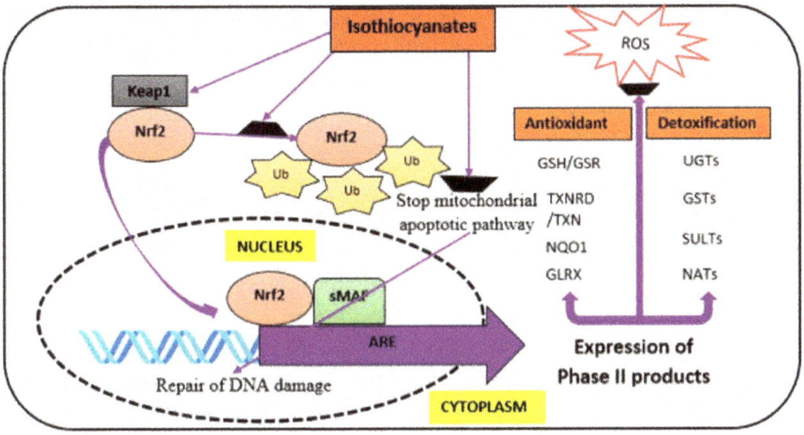

Figure 4 Summary of the mechanism of action for the neuroprotective effect of isothiocyanates (Jaafaru et al. 2018).

(PPARγ). Current changes inhibit the visceral WAT increase in size and lipid deposition in the liver, which leads to reduced inflammation and reduced risk of cardiovascular disease or diabetes (Esteve, 2020).

5.8 Central nervous system

It was found that increase in antioxidant enzymes like quinone reductase (NQO1), HO-1, both of which were discussed in anticarcinogenic pathway, and glutamate-cysteine ligase catalytic subunit results in the activation of Nrf2/antioxidant responsive element (ARE) pathway. Overall, neuroprotective effect of ITCs, and in particular SFN, is linked to the ability to activate Nrf2/ARE pathway in the central nervous system, resulting in the expression of phase II enzymes, eventually leading to inhibition of ROS (Jaafaru et al., 2018). Fig. 4 summarizes the pathways of neuroprotection by ITCs.

Following this, SFN inhibits translocation of Nrf2, allowing it to react with the active site of nucleus, where it can interact with ARE. This in turn results in the decline in production of pro-inflammatory mediators, cytokines, and oxidative markers leading to neural apoptotic pathway in the brain (Jaafaru et al., 2018). Overall, neuroprotective benefits of GLs are often linked to their antioxidant and anti-inflammatory functions.

6 Glucosinolates as nutraceuticals

Food contains a wide range of pharmacologically active chemicals that people consume. Obtaining enhanced nutritional and medicinal qualities in vegetables and fruits will become a much larger component of private and public breeding programs (Raskin and Ripoll, 2004). Technological advancements in manipulating plant metabolism and metabolites, combined with the explosive growth of the 'functional food' industry, have resulted in numerous efforts to increase the concentration of these health-promoting compounds in specific plant-based foods (Finley, 2005)

A 'nutraceutical' is a substance that is either a food or a component of a food that provides medical or health benefits, including disease prevention and treatment. Nutraceuticals include products ranging from isolated nutrients, dietary supplements, and diets to genetically engineered 'designer' foods, herbal products, and processed foods (cereals, soups, and beverages) (Dudeja and Gupta, 2017). Nutraceuticals are compounds found naturally in fruits and vegetables that can also be obtained in high concentrations through dietary supplements (Cencic and Chinwaru, 2010). GLs are one of the most notable nutraceutical compounds found in cruciferous vegetables. Because of their high potential for preventing chronic diseases, particularly cancer, they have been the subject of extensive research (García and Velázquez, 2016).

GLs are plant secondary metabolites found in varying concentrations in plant organs and throughout the plant's developmental stages. Furthermore, it is believed that GLs, along with myrosinase, are part of a defence mechanism used by plants to protect themselves from biotic and abiotic stress (Bohinc et al., 2012). Since it was discovered that SFN (4-methylsulfinylbutyl isothiocyanate), a prominent ITC in broccoli and broccoli sprouts, potently induces mammalian cytoprotective proteins/enzymes, GLs and ITC have been studied for more than a half-century (Dinkova-Kostova and Kostov, 2012). Ever since scientific research has increased its efforts to discover new pathways and/or mechanisms to improve human health and combat disease using GLs and ITC. It is widely accepted that one's diet and the occurrence of cancer are inextricably linked (Dinkova-Kostova and Kostov, 2012). Several epidemiological studies have suggested that individuals who consume fewer cruciferous vegetables have a higher risk of developing cancers such as colorectal, pancreatic, lung, breast, gastrointestinal, and ovarian cancer than those who consume more of these vegetables (Olsen et al., 2011). ITC has demonstrated a remarkable ability to act on all three stages of carcinogenesis, tumour initiation, promotion, and progression, as well as by suppressing the final stages of carcinogenesis (angiogenesis and metastasis) (Traka and Mithenm, 2008). Furthermore, GLs and their breakdown products, specifically the sulfur-containing compounds, are known for their fungicidal, bactericidal, nematocidal, and allelopathic properties, as well as their use as cancer chemo-preventive and chemotherapeutic agents (Ishida et al., 2014). GLs and ITC have also been shown in studies to reduce the risk of developing not only carcinogenesis but also cardiovascular, articular, inflammatory, and neurological (central nervous system) diseases, as well as asthma, diabetes, and cholesterol (Jaafaru et al., 2018).

6.1 Glucosinolate dietary supplements

GLs can be found in dietary supplements in a variety of forms, including broccoli and other cruciferous vegetable extracts and broccoli sprouts high in glucoraphanin (Higdon et al., 2007). Broccoli extracts, on the other hand, may be less effective because GLs must be hydrolyzed in order to be completely absorbed in the gastrointestinal tract, even though some of them can be hydrolyzed by gut microflora (Johnson, 2002). Nonetheless, products such as Endura Cell have preserved myrosinase activity in order to effectively produce ITCs.

In terms of biotechnological production, the most common method is to use plants of the *Brassica* genus to produce GLs dietary supplements. Black kale has been identified as a high source of GLs (De Nicola et al., 2014), despite this, broccoli is frequently used to produce extracts for the dietary supplement industry. However, it is important to note that GLs can differ depending on the plant's organ. It has been reported, for example, that aliphatic GLs may be found in higher concentrations in florets and leaves, whereas indolyl GLs are more likely to be found in roots (Ormirou et al., 2016). Furthermore, total GLs content has been found to vary depending on the developmental stage of the plant; for example, it has been reported that the concentration of GLs is higher in broccoli sprouts than in seeds (Bhandari et al., 2015). Additionally, the GLs content of different cultivars can vary greatly (Brown et al., 2002).

6.2 Production of dietary supplements

The procedure begins with the defatting of the plant material (i.e. broccoli florets, sprouts, or seeds), which should be pulverized to achieve a higher yield in the extraction process. Defatting can be accomplished by adding hexane and agitating the mixture for 3 h. C1-C4 alcohols (e.g. ethanol) or C3-C4 ketones (e.g. acetone) or mixtures of both can be used for extraction in an aqueous medium (Villareal-Garcia, 2020). Broccoli that is unfit for human consumption due to poor postharvest practices may be a good source of GLs. The majority of the GLs in these lost crops are well preserved due to cell compartmentalization, which keeps myrosinase separate from their substrate; however, cell rupture is likely to occur during the extraction process; thus, it is critical to inactivate the enzyme in order to preserve GLs content throughout the entire process, which can be accomplished by extracting at temperatures around 70°C. The mixture must be centrifuged to obtain an alcoholic/ketonic extract, which can then be filtered to remove insoluble matter and evaporated to remove volatiles

(Villareal-García and Jacobo-Velázquez, 2016). If it is evaporated, it must be redissolved in water, alcohols, ketones, or a combination of these. The extract can be purified further by passing it through a cation-exchange column in its acidic form, which meets regulatory requirements for food processing. The extract will now be adsorbed onto a basic resin that is regulatory-compliant. The GLs can be eluted from the column using a base such as sodium hydroxide, ammonia dissolved in water, alcohol, or a mixture (Doheny-Adams et al., 2017). The eluate can then be evaporated, freeze dried, or spray dried to produce a solid extract high in GLs. Microencapsulation can be applied to the obtained solid powder.

Other methods that have been reported include the use of more complex extraction solvent systems. For example, Fahey et al. (2003) proposed semi-preparative scale separation and purification of GLs from a variety of plant sources using high-speed countercurrent chromatography with a highly polar mixture of 1-propanol–acetonitrile–ammonium sulphate–water (1:0.5:1.2:1) before transferring to preparative scale. The authors claim that GLs extracted from broccoli seed extract are more than 95% pure (Fahey et al., 2003).

6.3 Biological activity

GLs are a group of phytochemicals that are extremely diverse and variable. According to research, they increase the activity of biotransformation enzymes in a variety of tissues (Kos et al., 2011). Glutathione peroxidases (GSH-Px), glutathione reductase (GR), GST, and superoxide dismutase (SOD) are antioxidant enzymes that play an important role in cellular oxidative stress. It was discovered that I-3-C does not induce oxidative enzymes at normal dietary levels.

In mice fed semi-purified diets containing I-3-C, however, there is a significant increase in both hepatic and intestinal GST. In rat liver, I-3-C was found to reduce GSSGR while increasing GSH-Px and SOD. Glucoraphanin also increased the activity of hepatic quinone reductase and GST in mice (Guerrera, 2005).

GLs have also shown biological activities that are linked to cancer prevention, among other things. These include (1) anti-inflammatory properties mediated by NF-kB regulation and downstream signalling, as well as inhibition of TNF-α and lipopolysaccharide-stimulated inflammatory responses (Folkard et al., 2014); (2) androgen receptor signalling downregulated in prostate cancer prevention by suppressing AR transcription and protein levels (Gibba et al., 2009); (3) epigenetic modification (Chen et al., 2016); and (4) antibacterial properties against *Helicobacter pylori*, a well-known cause of gastritis and peptic ulcers, but also linked to an increased risk of gastric cancer (Moon et al., 2010) (Fig. 5).

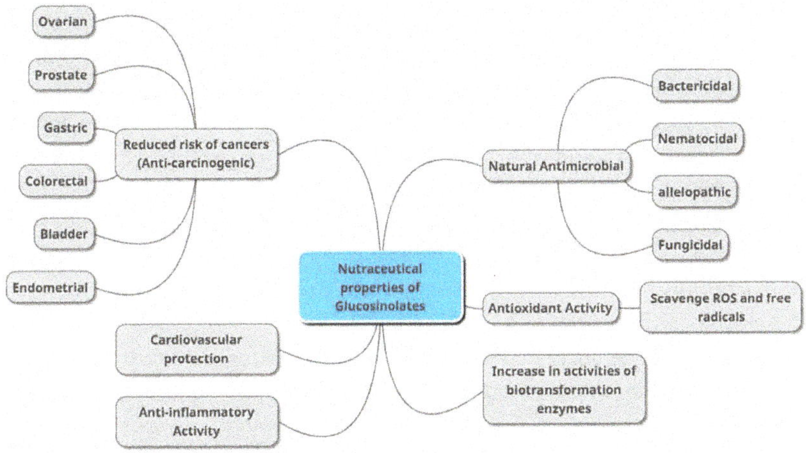

Figure 5 Overview of the nutraceutical properties of GLs.

7 Summary and future trends

GLs are a distinct class of secondary bioactives that play an important role in both plant defence and human health. The research on GLs is very dynamic and multifaceted, and there are numerous comprehensive studies and reviews on those compounds that take different approaches. Numerous studies have confirmed the correlation between myrosinase and specifier proteins, as well as the relative protein abundance of ESP, and thus the creation of various GL products such as ITCs, nitriles, and epithionitriles, which affect the plant defence system. Knowledge of the biochemical basis of GLs biosynthesis explains the diversity and relationships of various GSLs found in *Brassicaceae* species. This chapter highlights that GLs are classified into three groups based on their amino acid precursors: methionine, phenylalanine, and tryptophan. GLs biosynthesis consists of three independent stages: (1) some amino acids are elongated by one or more methylene groups, (2) the precursor amino acids are converted into parent GLs, and (3) the parent GLs are subjected to secondary modifications (oxygenations, hydroxylations, alkenylations, and methoxylations). Despite the fact that these molecules share a basic chemical structure, the range of compounds and the resulting hydrolysis products varies greatly. This complexity stems from a series of reactions that evolved from primary metabolism and necessitated subtle changes to allow for the use of new substrates.

The phenomena of GLs metabolism should be further investigated, as to determine what percentage of bioactive compounds reach their active site and elute beneficial effect, the consequences of denatured myrosinase, and the possibility of creating nutritional supplements containing GLs together with myrosinase for better activity and more potent health benefits. Currently, there are no dietary guidelines for the consumption of GLs or cruciferous vegetables, due to lack of dietary GLs intake around the world. Studies using larger sample size should be carried out in order to determine current intake, adverse effects from low intake, amount needed for beneficial effect, and the ability to make educated intake suggestions for general population.

There are numerous in *vitro* studies investigating anticarcinogenic properties of GLs; however, in *vivo* mechanisms of action for cancer protection through short- and long-term human intervention studies are yet unknown. Therefore, pharmacological pathways of GLs should be further investigated. While there is evidence that SFN reduces cancer development, more research on the underlying molecular mechanisms is needed to better understand sulforaphane's role in cancer metabolic rewiring. Remarkably, recent research has found that using SFN in conjunction with anti-cancer treatments like chemotherapy increases cancer cell sensitivity (Jabbarzadeh Kaboli et al., 2020), reduces toxic side effects (Calcabrini et al., 2020), and inhibits key survival pathways in cancer progression (Mokhtari et al., 2021). This suggests that SFN could be used not only as a potential drug candidate but also in conjunction with existing anti-cancer treatments.

8 Where to look for further information

- Galanakis, C. M. (Ed.). (2019). *Glucosinolates: Properties, recovery, and applications*. Academic Press.
- Traka, M. H. (2016). Health benefits of glucosinolates. *Advances in botanical research*, *80*, 247-279.
- Traka, M. and Mithen, R. (2008). Glucosinolates, isothiocyanates and human health. *Phytochemistry reviews*, *8*(1), 269-282.
- Lund. (2003). Non-nutritive bioactive constituents of plants: dietary sources and health benefits of glucosinolates. *International journal for vitamin and nutrition research*, *73*(2), 135-143.
- Maina, S., Misinzo, G., Bakari, G. and Kim, H. Y. (2020). Human, animal and plant health benefits of glucosinolates and strategies for enhanced bioactivity: A systematic review. *Molecules*, *25*(16), 3682.
- Andersen, K. E., Frandsen, H. B., Jensen, S. K., Bellostas, N. M., Sørensen, A. D., Sørensen, J. C. and Sørensen, H. (2010). Glucosinolates in Brassica-health risks, but also benefits. *The Norwegian Academy of Science and Letters, 2010a*, 104-124.

9 References

Agerbirk, N., Matthes, A., Erthmann, P. Ø., Ugolini, L., Cinti, S., Lazaridi, E., Nuzillard, J. M., Müller, C., Bak, S., Rollin, P. and Lazzeri, L. (2018). Glucosinolate turnover in

Brassicales species to an oxazolidin-2-one, formed via the 2-thione and without formation of thioamide. *Phytochemistry* 153: 79–93.

Agerbirk, N. and Olsen, C. E. (2015). Glucosinolate hydrolysis products in the crucifer Barbarea vulgaris include a thiazolidine-2-one from a specific phenolic isomer as well as oxazolidine-2-thiones. *Phytochemistry* 115(1): 143–151.

Agrawal, V., Ghaznavi, S. A. and Paschke, R. (2018). Environmental goitrogens. In: *Encyclopedia of Endocrine Diseases*. Elsevier, 506–511.

Almushayti, A. Y., Brandt, K., Carroll, M. A. and Scotter, M. J. (2021). Current analytical methods for determination of GLs in vegetables and human tissues. *Journal of Chromatography* 1643: 4620-4660.

Beewilder, J., Leeuwen, W. V., Dam, N. M., Bertossi, M., Grandi, V., Mizzi, L., Soloviev, M., Szabados, L., Molthoff, J. W., Schipper, B., Verbocht, H., de Vos, R. C. H., Morandini, P., Aarts, M. G. M., and Arnaud, B. (2008). The impact of the absence of aliphatic glucosinolates on insect herbivory in arabidopsis. *PLoS ONE* 3(4): e2068.

Bhandari, S. R., Jo, J. S. and Lee, J. G. (2015). Comparison of glucosinolate profiles in different tissues of nine brassica crops. *Molecules* 20(9): 15827–15841.

Blažević, I., Montaut, S., Burčul, F., Olsen, C. E., Burow, M., Rollin, P. and Agerbirk, N. (2020). Glucosinolate structural diversity, identification, chemical synthesis and metabolism in plants. *Phytochemistry* 169: 112100.

Bohinc, T., Ban, S., Ban, D. and Trdan, S. (2012). GLs in plant protection strategies: a review. *Archives of Biological Sciences* 64(3): 821-828.

Brandes, M. S. and Gray, N. E. (2020). NRF2 as a therapeutic target in neurodegenerative diseases. *ASN Neuro* 12: 1759091419899782.

Brown, A. F., Yousef, G. G., Jeffery, E. H., Klein, B. P., Wallig, M. A., Kushad, M. M. and Juvik, J. A. (2002). Glucosinolate profiles in broccoli: variation in levels and implications in breeding for cancer chemoprotection. *Journal of the American Society for Horticultural Science* 127(5): 807–813.

Burow, M. and Halkier, B. A. (2017). How does a plant orchestrate defense in time and space? Using glucosinolates in Arabidopsis as case study. *Current Opinion in Plant Biology* 38: 142–147.

Calcabrini, C., Maffei, F., Turrini, E. and Fimognari, C. (2020). Sulforaphane potentiates anticancer effects of doxorubicin and cisplatin and mitigates their toxic effects. *Frontiers in Pharmacology* 11: 567.

Cartea, M. E. and Velasco, P. (2008). Glucosinolates in Brassica foods: bioavailability in food and significance for human health. *Phytochemistry Reviews* 7(2): 213-229.

Cencic, A. and Chinwaru, W. (2010). The role of functional foods, nutraceuticals and food supplements in intestinal health. *Nutrients* 2(6): 611–625.

Chandra, A. K. (2010). Goitrogen in food: cyanogenic and flavonoids containing plant foods in the development of goiter. In: *Bioactive Foods in Promoting Health*. Elsevier Inc., 691–716.

Chen, Y., Cang, S., Han, L., Liu, C., Yang, P., Solangi, Z., Lu, Q., Liu, D. and Chiao, J. W. (2016). Establishment of prostate cancer spheres from a prostate cancer cell line after phenethyl isothiocyanate treatment and discovery of androgen-dependent reversible differentiation between sphere and neuroendocrine cells. *Oncotarget* 7(18): 26567–26579.

Chhajed, S., Mostafa, I., He, Y., Abou-Hashem, M., El-Domiaty, M. and Chen, S. (2020). Glucosinolate biosynthesis and the glucosinolate-myrosinase system in plant defense. *Agronomy* 10(11): 1786.

De Nicola, G. R., Bagatta, M., Pagnotta, E., Angelino, D., Gennari, L., Ninfali, P., Rollin, P. and Iori, R. (2014). Comparison of bioactive phytochemical content and release of isothiocyanates in selected brassica sprouts. *Food Chemistry* 141(1): 297-303.

Delange, F. and Ermans, A. M. (1971). Role of a dietary goitrogen in the etiology of endemic goiter Idjw Island. *American Journal of Clinical Nutrition* 24(11): 1354-1360.

Dinkova-Kostova, A. T. and Kostov, R. V. (2012). Glucosinolates and isothiocyanates in health and disease. *Trends in Molecular Medicine* 18(6): 337-347.

Doheny-Adams, T., Redeker, K., Kittipol, V., Bancroft, I. and Hartley, S. E. (2017). Development of an efficient glucosinolate extraction method. *Plant Methods* 13: 17.

Dratwa, M., Wysoczańska, B., Łacina, P., Kubik, T. and Bogunia-Kubik, K. (2020). Tert-Regulation and roles in cancer formation. *Frontiers in Immunology* 11: 589929.

Dudeja, P. and Gupta, R. K. (2017). Nutraceuticals. *Food Safety in the 21st Century*: 491-496.

Durian, G., Rahikainen, M., Alegre, S., Brosché, M. and Kangasjarvi, S. (2016). Protein phosphatase 2A in the regulatory network underlying biotic stress resistance in plants. *Frontiers in Plant Science* 7: 812.

Eisenbrand, G. and Gelbke, H. P. (2016). Assessing the potential impact on the thyroid axis of environmentally relevant food constituents/contaminants in humans. *Archives of Toxicology* 90(8): 1841-1857.

Esfandiari, A., Saei, A., McKenzie, M. J., Matich, A. J., Babalar, M. and Hunter, D. A. (2017). Preferentially enhancing anti-cancer isothiocyanates over glucosinolates in broccoli sprouts: how NaCl and salicylic acid affect their formation. *Plant Physiology and Biochemistry* 115: 343-353.

Esteve, M. (2020). Mechanisms underlying biological effects of cruciferous glucosinolate-derived isothiocyanates/indoles: a focus on metabolic syndrome. *Frontiers in Nutrition* 7: 111.

Fahey, J. W., Wade, K. L., Stephenson, K. K. and Chou, F. E. (2003). Separation and purification of GLs from crude plant homogenates by high-speed counter-current chromatography. *Journal of Chromatography. A* 996(1-2): 85-93.

Fang, J. and Ding, Z. (2020). SNAI1 is a prognostic biomarker and correlated with immune infiltrates in gastrointestinal cancers. *Aging* 12(17): 17167-17208.

Finley, J. W. (2005). Proposed criteria for assessing the efficacy of cancer reduction by plant foods enriched in carotenoids, GLs, polyphenols and selenocompounds. *Annals of Botany* 95(7): 1075-1096.

Folkard, D. L., Melchini, A., Traka, M. H., Al-Bakheit, A., Saha, S., Mulholland, F., Watson, A. and Mithen, R. F. (2014). Suppression of LPS-induced transcription and cytokine secretion by the dietary isothiocyanate sulforaphane. *Molecular Nutrition and Food Research* 58(12): 2286-2296.

Fuentas, F., Paredes-Gonzalez, X. and Kong, A.-N. T. (2015). Dietary glucosinolates sulforaphane, phenethyl isothiocyanatem indole-3-carbinol/3,3′-diindolylmethane: anti-oxidative stress/inflammation, Nrf2, epigenetics/epigenomics and in vivo cancer chemoprotective efficacy. *Current Pharmacology Reports* 1(3): 179-196.

García, D.-V. and Jacobo-Velázquez, D. A. (2016). GLs from broccoli: nutraceutical properties and their purification. *Journal of Nutraceuticals and Food Science*.

Gibba, A., Schwartzman, J., Deng, V. and Alumkal, J. (2009). Sulforaphane destabilizes the androgen receptor in prostate cancer cells by inactivating histone deacetylase

6. *Proceedings of the National Academy of Sciences of the United States of America* 106(39): 1663-1668.

Glindermann, C. P., Backenkohler, A., Strieker, M., Wittstock, U. and Klahn, P. (2019). Synthesis and biochemical evaluation of an artificial, fluorescent glucosinolate (GSL). *Chembiochem* 20(18): 2341-2345.

Guerrera, P. M. (2005). Traditional phytotherapy in Central Italy (Marche, Abruzzo, and Latium). *Fitoterapia* 76(1): 1-25.

Halkier, B. A. and Gershenzon, J. (2006). Biology and biochemistry of glucosinolates. *Annual Review of Plant Biology* 57: 303-333.

Hanigan, M. H. and Cooper, A. J. L. (2018). Metabolism of Gluthathione S-conjugates: multiple pathways. *Comprehensive Toxicology*: 363-406.

Hannen, R., Hauswald, M. and Bartsch, J. W. (2017). A rationale for targeting extracellular regulated kinases ERK1 and ERK2 in glioblastoma. *Journal of Neuropathology and Experimental Neurology* 76(10): 838-847.

Hanschen, F. S., Kaufmann, M., Kupke, F., Hackl, T., Kroh, L. W., Rohn, S. and Schreiner, M. (2017). Brassica vegetables as sources of epithionitriles: novel secondary products formed during cooking. *Food Chemistry* 245: 564-569.

Hanschen, F. S., Pfitzmann, M., Witzel, K., Stützel, H., Schreiner, M. and Zrenner, R. (2018). Differences in the enzymatic hydrolysis of glucosinolates increase the defense metabolite diversity in 19 Arabidopsis thaliana accessions. *Plant Physiology and Biochemistry* 124: 126-135.

Higdon, J. (2005). *Isothiocyanates*. Micronutrient Information Center, Linus Pauling Institute, Oregon State University.

Higdon, J. V., Delage, B., Williams, D. E. and Dashwood, R. H. (2007). Cruciferous vegetables and human cancer risk: epidemiologic evidence and mechanistic basis. *Pharmacological Research* 55(3): 224-236.

Hunziker, P., Halkier, B. A. and Schulz, A. (2019). Arabidopsis glucosinolate storage cells transform into phloem fibres at late stages of development. *Journal of Experimental Botany* 70(16): 4305-4317.

Ishida, M., Hara, M., Fukino, N., Kakizaki, T. and Morimitsu, Y. (2014). Glucosinolate metabolism, functionality and breeding for the improvement of Brassicaceae vegetables. *Breeding Science* 64(1): 48-59.

Jaafaru, M. S., Abd Karim, N. A., Enas, M. E., Rollin, P., Mazzon, E. and Abdull Razis, A. F. (2018). Protective effect of GLs hydrolytic products in neurodegenerative diseases (NDDs). *Nutrients* 10(5): 580.

Jabbarzadeh Kaboli, P., Afzalipour Khoshkbejari, M., Mohammadi, M., Abiri, A., Mokhtarian, R. and Vazifemand, R. (2020). Targets and mechanisms of sulforaphane derivatives obtained from cruciferous plants with special focus on breast cacner - contradictory effects and future perspectives. *Biomedicine and Pharmacotherapy* 121: 109-635.

Jang, M., Hong, E. and Kim, G. H. (2010). Evaluation of antibacterial activity of 3-butenyl, 4-pentenyl, 2-phenylethyl, and benzyl isothiocyanate in Brassica vegetables. *Journal of Food Science* 75(7): M412-M416.

Johnson, I. T. (2002). GLs in the human diet: bioavailability and implications for health. *Phytochemistry Reviews* 1(2): 183-188.

Kim, J. H. and Jander, G. (2007). Myzus persicae (green peach aphid) feeding on Arabidopsis induces the formation of a deterrent indole glucosinolate. *The Plant Journal: For Cell and Molecular Biology* 49(6): 1008-1019.

Kliebenstein, D. J., Auria, J. C., Behere, A. S., Kim, J. H., Gunderson, K. L., Breen, J. N., Lee, G., Gershenzon, J., Last, R. L. and Jander, G. (2007). Characterization of seed-specific benzoyloxyglucosinolate mutations in Arabidopsis thaliana. *Plant Journal: For Cell and Molecular Biology* 51(6): 1062-1076.

Kliebenstein, D. J., Lambrix, V. M., Reichelt, M., Gershenzon, J. and Mitchell-Olds, T. (2001). Gene duplication in the diversification of secondary metabolism: tandem 2-oxoglutearate-dependent dioxygenases control glucosinolate biosynthesis in *Arabidopsis*. *Plant Cell* 13(3): 681-693.

Kos, M., Kabouw, P., Noordam, R., Hendriks, K., Vet, L. E. M., Van Loon, J. J. A. and Dicke, M. (2011) Prey-mediated effects of glucosinolates on aphid predators. *Ecological Entomology* 36(3): 377-388.

Lachance, J. C., Radhakrishnan, S., Madiwale, G., Guerrier, S. and Vanamala, J. K. P. (2020). Targeting hallmarks of cancer with a food-system-based approach. *Nutrition* 69: 110563.

Lee, C. C., Cheng, Y. C., Chang, C. Y., Lin, C. M. and Chang, J. Y. (2018a). Alpha-tubulin acetyltransferase/MEC-17 regulates cancer cell migration and invasion through epithelial-mesenchymal transition suppression and cell polarity disruption. *Scientific Reports* 8(1): 17477.

Lee, C. F., Chiang, N. N., Lu, Y. H., Huang, Y. S., Yang, J. S., Tsai, S. C., Lu, C. C. and Chen, F. A. (2018b). Benzyl isothiocyanate (BITC) triggers mitochondria-mediated apoptotic machinery in human cisplatin-resistant oral cancer CAR cells. *BioMedicine* 8(3): 15.

Li, G. and Quiros, C. F. (2003). In planta side-chain glucosinolate modification in Arabidopsis by introduction of dioxygenase Brassica homolog BoGSL-ALK. *TAG. Theoretical and Applied Genetics. Theoretische und Angewandte Genetik* 106(6): 1116-1121.

Liu, T., Zhang, X., Yang, H., Agerbirk, N., Qiu, Y., Wang, H., Shen, D., Song, J. and Li, X. (2016). Aromatic glucosinolate biosynthesis pathway in Barbarea vulgaris and its response to plutella xylostella Infestation. *Frontiers in Plant Science* 7: 83.

Maldini, M., Maksoud, S. A., Natella, F., Montoro, P., Petretto, G. L., Foddai, M., De Nicola, G. R., Chessa, M. and Pintore, G. (2014). Moringa oleifera: study of phenolics and GLs by mass spectrometry. *Journal of Mass Spectrometry* 49(9): 900-910.

Malka, S. K. and Cheng, Y. (2017). Possible interactions between the biosynthetic pathways of indole glucosinolate and auxin. *Frontiers in Plant Science* 8(8): 2131.

Marwaha, R. K., Tandon, N., Gupta, N., Karak, A. K., Verma, K. and Kochupillai, N. (2003). Residual goitre in the postiodization phase: iodine status, thiocyanate exposure and autoimmunity. *Clinical Endocrinology* 59(6): 672-681.

Mithen, R. F., Dekker, M., Verkerk, R., Rabot, S. and Johnson, I. T. (2000). The nutritional significance biosynthesis and bioavailability of glucosinolates in human foods. *Journal of the Science of Food and Agriculture* 80(7): 967-984.

Mokhtari, R. B., Qorri, B., Baluch, N., Sparaneo, A., Fabrizio, F. P., Muscarella, L. A., Tyker, A., Kumar, S., Cheng, H. M., Szewczuk, M. R., Das, B. and Yeger, H. (2021). Next-generation multimodality of nutrigenomic cancer therapy: sulforaphane in combination with acetazolamide actively target bronchial carcinoid cancer in disabling the PI3K/Akt/mTOR survival pathway and inducing apoptosis. *Oncotarget* 12(15): 1470-1489.

Moon, J. K., Kim, J. R., Ahn, Y. J. and Shibamoto, T. (2010). Analysis and anti-Helicobacter activity of sulforaphane and related compounds present in broccoli (*Brassica oleracea L.*) sprouts. *Journal of Agricultural and Food Chemistry* 58(11): 6672-6677.

Nour-Eldin, H. H. and Halkier, B. A. (2008). Piecing together the transport pathway of aliphatic glucosinolates. *Phytochemistry Reviews* 8(1): 53-67.

Novío, S., Núñez-Iglesias, M. J. and Freire-Garabal, M. (2019). Isothiocyanates, epigenetics, and cancer prevention. In: *Epigenetics of Cancer Prevention*. Elsevier, 149-168.

Olsen, G. W., Mandel, J. S., Gibson, R. W., Wattenberg, L. W. and Schuman, L. M. (2011). A case-control study of pancreatic cancer and cigarettes, alcohol, coffee and diet. *American Journal of Public Health* 79(8): 1016-1019.

Ormirou, M. D., Papadopoulou, K. K., Papastylianou, I., Constantinou, M., Karpouzas, D. G., et al. (2016). Impact of nitrogen and sulfur fertilization on the composition of GLs in relation to sulfur assimilation in different plant organs of broccoli. *Journal of Agriculture and Food Chemistry* 57: 9408-9417.

Pfalz, M., Mikkelsen, M. D., Bednarek, P., Olsen, C. E., Halkier, B. A. and Kroymann, J. (2011). Metabolic engineering in *Nicotiana benthamiana* reveals key enzyme functions in Arabidopsis indole glucosinolate modification. *Plant Cell* 23(2): 716-729.

Pfalz, M., Mukhaimar, M., Perreau, F., Kirk, J., Hansen, C. I., Olsen, C. E., Agerbirk, N. and Kroymann, J. (2016). Methyl transfer in glucosinolate biosynthesis mediated by indole glucosinolate o-methyltransferase 5. *Plant Physiology* 172(4): 2190-2203.

Piotrowski, M., Schemenewitz, A., Lopukhina, A., Muller, A., Janowitz, T., Weiler, E. W. and Oecking, C. (2004). Desulfoglucosinolate sulfotransferases from *Arabidopsis thaliana* catalyze the final step in the biosynthesis of the glucosinolate core structure. *Journal of Biological Chemistry* 279(49): 50717-50725.

Rahikainen, M., Trotta, A., Alegre, S., Pascual, J., Vuorinen, K., Overmyer, K., Moffatt, B., Ravanel, S., Glawischnig, E. and Kangasjarvi, S. (2017). PP2A-B'gamma modulates foliar trans-methylation capacity and the formation of 4-methoxy-indol-3-yl-methyl glucosinolate in Arabidopsis leaves. *Plant Journal: For Cell and Molecular Biology* 89(1): 112-127.

Raskin, I. and Ripoll, C. (2004). Can an apple a day keep the doctor away? *Current Pharmaceutical Design* 10(27): 3419-3429.

Ratzka, A., Vogel, H., Kiebenstein, D. J., Mitxhell-Olds, T. and Kroymann, J. (2002). Disarming the mustard oil bomb. Proceedings of the National Academy of Sciences 99: 11223-11228.

Redovniković, I. R., Glivetić, T., Delonga, K. and Vorkapić-Furač, J. (2008). Glucosinolates and their potential role in plant. *Periodicum Biologorum* 110(4): 297-309.

Rochfort, S. J. and Jones, R. (2011). Glucosinolate phytochemicals from broccoli (*Brassica oleracea L. var. botrytis L.*) seeds and their potential health effects. *Nuts and Seeds in Health and Disease Prevention*: 253-261.

Rognoni, E. and Walko, G. (2019). The roles of YAP/TAZ and the hippo pathway in healthy and diseased skin. *Cells* 8(5): 411.

Román, J., Castillo, A., Cottet, L. and Mahn, A. (2018). Kinetic and structural study of broccoli myrosinase and its interaction with different glucosinolates. *Food Chemistry* 254: 87-94.

Sánchez-Pujante, P. J., Borja-Martínez, M., Pedreño, M. Á. and Almagro, L. (2017). Biosynthesis and bioactivity of glucosinolates and their production in plant in vitro cultures. *Planta* 246(1): 19-32.

Smiechowska, A., Bartoszek, A. and Namiesnik, J. (2010). Determination of GLs and their decomposition products - indoles and isothiocyanates in cruciferous vegetables. *Critical Reviews in Analytical Chemistry* 40(3): 202-216.

Smirnov, A., Anemona, L., Novelli, F., Piro, C. M., Annicchiarico-Petruzzelli, M., Melino, G. and Candi, E. (2019). p63 is a promising marker in the diagnosis of unusual skin cancer. *International Journal of Molecular Sciences* 20(22): 5781.

Sønderby, I. E., Geu-Flores, F. and Halkier, B. A. (2010). Biosynthesis of glucosinolates-gene discovery and beyond. *Trends in Plant Science* 15(5): 283-290.

Soundararajan, P. and Kim, J. (2018). Anti-carcinogenic GLs in cruciferous vegetables and their antagonistic effects on prevention of cancers. *Molecules* 23(11): 2983.

Sturm, C. and Wagner, A. E. (2017). Brassica-derived plant bioactives as modulators of chemopreventive and inflammatory signaling pathways. *International Journal of Molecular Sciences* 18(9): 1890.

Tacer-Caba, Z. (2019). Different sources of GLs and their derivatives. In: *GLS: Properties, Recovery, and Applications*. Elsevier, pp.143-180.

Textor, S., de Kraker, J. W., Hause, B., Gershenzon, J. and Tokuhisa, J. G. (2007). MAM3 catalyzes the formation of all aliphatic glucosinolate chain lengths in Arabidopsis. *Plant Physiology* 144(1): 60-71.

Traka, M. and Mithenm, R. (2008). GLs, isothiocyanates and human health. *Phytochemistry Reviews* 8: 269-282.

Uda, Y., Kurata, T. and Arakawa, N. (1986). Effects of pH and ferrous ion on the degradation of glucosinolates by myrosinase. *Agricultural and Biological Chemistry* 50: 2735-2740.

Valentina Tudor, D. V., Bâldea, I., Lupu, M., Kacso, T., Kutasi, E., Hopârtean, A., Stretea, R. and Gabriela Filip, A. (2020). COX-2 as a potential biomarker and therapeutic target in melanoma. *Cancer Biology and Medicine* 17(1): 20-31.

Verkerk, R. and Dekker, M. (2008). Glucosinolates. In Gilbert, J. and Senyuva, H. (Eds.), *Bioactive Compounds in Foods*. Blackwell Publishing Ltd, Oxford.

Villareal-García, D. (2020). GLs: nutraceutical properties and purification. *Journal of Nutraceuticals and Food Science* 5(5).

Vo, Q. V., Trenerry, C., Rochfort, S., Wadeson, J., Leyton, C. and Hughes, A. B. (2013). Synthesis and anti-inflammatory activity of aromatic glucosinolates.*Bioorganic and Medicinal Chemistry* 21(19): 5945-5954.

Willemin, M. E. and Lumen, A. (2017). Thiocyanate: a review and evaluation of the kinetics and the modes of action for thyroid hormone perrturbations. *Critical Reviews in Toxicology* 47(7): 537-563.

Williams, D. J., Critchley, C., Pun, S., Chaliha, M. and O'Hare, T. J. (2009). Differing mechanisms of simple nitrile formation on glucosinolate degradation in Lepidium sativum and Nasturtium officinale seeds. *Phytochemistry* 70(11-12): 1401-1409.

Wu, G., Song, X., Liu, J., Li, S., Gao, W., Qiu, M., Yang, C., Ma, Y. and Chen, Y. (2020). Expression of CD44 and the survival in glioma: a meta-analysis. *Bioscience Reports* 40(4).

Xue, Y., Lian, W., Zhi, J., Yang, W., Li, Q., Guo, X., Gao, J., Qu, H., Lin, W., Li, Z., Lai, L. and Wang, Q. (2019). HDAC5-mediated deacetylation and nuclear localisation of SOX9 is critical for tamoxifen resistance in breast cancer. *British Journal of Cancer* 121(12): 1039-1049.

You, C., Jin, L., Xu, Q., Shen, B., Jiao, X. and Huang, X. (2018). Expression of miR-21 and miR-138 in colon cancer and its effect on cell proliferation and prognosis. *Oncology Letters* 17(12): 2271-2277.

Zhan, W., Liao, X., Liu, J., Tian, T., Yu, L. and Li, R. (2020). USP38 regulates the stemness and chemoresistance of human colorectal cancer via regulation of HDAC3. *Oncogenesis* 9(5): 48.

Zhang, Y., Talalay, P., Cho, C. G. and Posner, G. H. (1992). A major inducer of anticarcinogenic protective enzymes from broccoli: isolation and elucidation of structure. *Proceedings of the National Academy of Sciences of the United States of America* 89(6): 2399–2403.

Zhang, Y., Xu, L., Li, A. and Han, X. (2019). The roles of ZEB1 in tumorigenic progression and epigenetic modifications. *Biomedicine and Pharmacotherapy* 110: 400–408.

Zhou, J., Kong, W., Zhao, H., Li, R., Yang, Y. and Li, J. (2019). Transcriptome-wide identification of indole glucosinolate dependent flg22-response genes in Arabidopsis. *Biochemical and Biophysical Research Communications* 520(2): 311–319.

Chapter 7

Understanding the health benefits and nutraceutical properties of organosulphur compounds in vegetables

Greta Pileckaite and Ella O'Grady, School of Food Science and Environmental Health, Technological University Dublin – City Campus, Ireland; and Swarna Jaiswal and Amit K. Jaiswal, School of Food Science and Environmental Health, Technological University Dublin – City Campus and Environmental Sustainability and Health Institute, Technological University Dublin – City Campus, Ireland

1 Introduction

Organosulphur compounds (OSCs) refer to a subclass of sulphur-containing (-SH) organic molecules (Fig. 1) naturally abundant both in nature and in the body, which include isothiocyanates, indoles, sulphoraphane (SFN) and allylic sulphur compounds (Ruhee et al., 2020). These compounds are typically identified by their unpleasant odours with certain products, such as saccharin, being the exception. Sulphur, calcium and phosphorus are some of the most abundant minerals in the human body, yet sulphur is found and exclusively derived from proteins where only 2 of the 20 essential amino acids (methionine and cystine) contain sulphur as part of their molecular composition (Nimni et al., 2007). Sulphur, an essential element, is of great importance in human health and disease due to its presence in vital biological molecules including proteins, peptides, enzymes, vitamins and hormones; however, there is no current recommended daily allowance (RDA), as it is considered prevalent in food sources, unless in cases of extreme protein deprivation where deficiency has been implicated in several diseases.

http://dx.doi.org/10.19103/AS.2022.0101.19

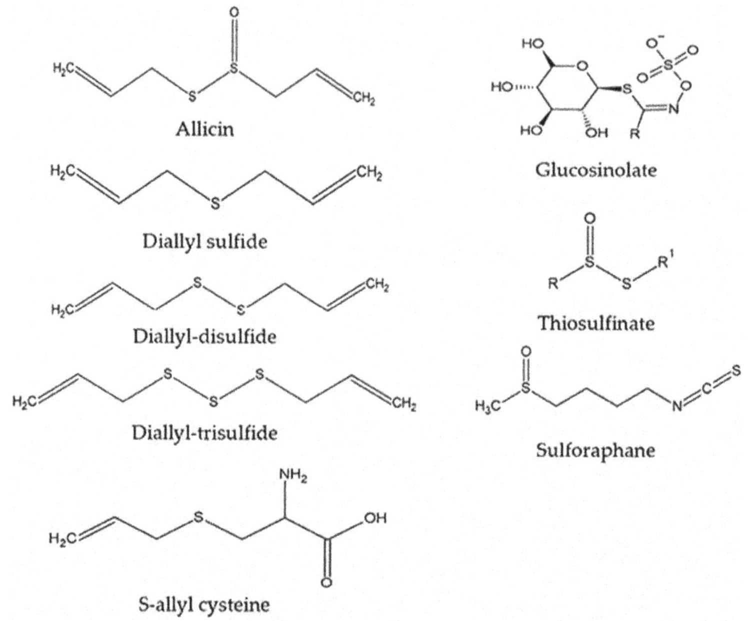

Figure 1 Generic structure of some organosulphur compounds (Ruhee et al., 2020).

OSCs have gained considerable attention recently for their role in the prevention and treatment of several non-communicable diseases such as diabetes, cancer, inflammatory and cardiovascular diseases. This is mainly linked to *S*-alk(en)yl-L-cysteine sulphoxides and *S*-methylcysteine-L-sulphoxid forms, which also help the breakdown of oestrogen toxic metabolites in the liver (Majumder and Annegowda, 2021) making them invaluable in the human diet and disease management. OSCs are naturally found in a variety of vegetables, namely the *Allium* genus, which has more than 900 known members, and cruciferous vegetables of the *Brassica* genus. *Allium* spp., encompassing *Allium sativum* L. (garlic), *Allium cepa* L. (onion), *Allium schoenoprasum* L. (chives) and *Allium ampeloprasum* L. (leek), primarily contain *S*-alk(en)yl-L-cysteine sulphoxides, whereas *Brassicaceae* spp. including *Brassica oleracea* var. *capitata* (cabbage), *Brassica oleracea* var. *botrytis* (cauliflower), *Brassica oleracea* var. *gemmifera* (Brussel sprouts) and *Brassica oleracea* var. *sabellica* (kale) are abundant in *S*-methyl cysteine-L-sulphoxides (Goncharov et al., 2021). To date, most clinical studies and research have focused on garlic due to its known medical potential; however, leeks, onions and chives have shown protective activity against cancer (Putnik et al., 2019a,b).

OSCs comprise a wide variety of sulphur-containing organic molecules such as, amongst others, sulphides, disulphides, sulphoxides and glucosinolates. The classification of OSCs is primarily according to the functional group attached to sulphur and can be further categorised according to their physicochemical properties in lipid-soluble compounds such as diallyl sulphides (DASs), which have a more potent smell, followed by water-soluble compounds like *S*-allyl cysteine (SAC) and *S*-allyl mercapto cysteine (SAMC) which have a less potent odour (Ruhee et al., 2020). OSCs are able to neutralise reactive oxygen species (ROS) and play a role in redox reactions and the formation of vitally important disulphide bridges, which are made by thiol-disulphide exchange facilitated by thioredoxin protein and are important in the post-translational modification of proteins to maintain their tertiary structure and produce the desired activity (Goncharov et al., 2021).

The aim of this chapter is to introduce OSC breakdown products, their functions, bioavailability and health benefits as well as nutraceutical applications in food, pharmaceutical and food supplement industries.

2 Bioavailability of organosulphur compounds

Bioavailability is a critical criterion for determining the relationship between food and its health benefits (Dima et al. 2020). Nutraceutical bioavailability in food is defined as the proportion of an ingested biocomponent that reaches systemic circulation (blood flow) to be dispersed to organs and tissues and manifests its bioactivity (Jafari and McClements, 2017). During cyclic or non-cyclic formation, OSCs are distinguished by a sulphur atom bonded to a cyanate group (Barba and Orlien, 2017). Allium OSCs have anticarcinogenic properties due to allyl derivatives, which inhibit carcinogenesis in the stomach, oesophagus, colon, mammary glands and lungs (Ramirez et al., 2017). The synthesis of allicin in garlic cloves is activated by damage to the cellular membrane and is unavailable in intact cells as it is toxic to plants and may harm garlic tissue. The enzyme alliinase is activated via tissue degradation, which rapidly metabolises alliin to allicin using pyridoxal 5-phosphate. Alliinase is inactivated by 40% in the human stomach due to chloric acid and digestive fluids, whereas allicin is degraded by intestinal cells (Phadatare et al. 2014).

Even though the bioavailability of OSC in the human body is limited, it has been reported that normal garlic consumption provides sufficient dosage for biological activity (e.g. anticancer effects). The majority of OSC in garlic extract is in the allyl and methyl forms (Chope et al., 2011). Early animal studies using canine and murine models exposed to aqueous garlic extracts (both mice and rats) revealed relatively high SAC bioavailability in the liver, blood plasma and kidneys (98%, 103% and 87% for rats, mice and dogs, respectively). A gram of garlic contains 2.5 mg of allicin, 60 g of SAC, 1000 100 g of diallyl trisulphide (DATS) and 570 40 g of diallyl disulphide (DADS) (Gao et al. 2013).

Allicin is synthesised and transformed in the human body during chewing, and its metabolites are transported to target tissues (organs) via the stomach, intestine and blood, while losses are channelled through breath, urine and stool (Lawson and Hughes, 1992). Research by Rahman (2007) found that pure allicin was not found in the bloodstream, urine or stool after consumption, which may be due to the bioavailability of allicin being hampered by binding to proteins and fatty acids in the plasma membrane. Aside from the high reactivity and rapid conversion to allyl mercaptan (AM), other reasons include allicin's oxidative binding to red blood cells (RBCs) and the inability to bypass the digestive tract due to binding with the luminal membrane (Marchese et al., 2016).

Garlic consumption can promote iron absorption from carbonyl iron via bioavailable compounds such as DAS, DADS, diallyl trisulphide, ajoene, SAC and by modifying DMT1 mRNA expression (Nahdi et al., 2010). It has been proposed that leeks and shallots can increase the bioavailability of iron from cereals and legumes (Luo et al. 2013).

3 Health benefits of organosulphur compounds

Discoveries in the field of natural food components with immunomodulatory properties led to the identification of novel natural compounds that could maintain homeostasis of the human immune system. Natural therapies include plant-derived food compounds that have been shown to be more beneficial than synthetic or uncontrolled (such as neutralised pathogens) compounds (Hadden, 1993). Natural immune boosters are a very promising approach to immunomodulation and may be preferable to synthetic drugs due to their lower cost, reduced (or even eliminated) toxicity and few side effects (Miękus et al., 2020). Sulphur-containing plant-based secondary metabolites have long been known to have therapeutic value. As shown in Table 1, OSCs are an integral part of the human diet and have been linked to improved health and well-being. The beneficial properties and therapeutic applications of OSCs have been reported in folk and traditional systems of medicine and practices for thousands of years (Walag et al., 2020). Sulphur accounts for about 1% of the dry weight of garlic, and the sulphur-rich compounds found in garlic correspond to its health benefits, particularly in cancer prevention and therapy (Nicastro et al., 2015).

3.1 Anti-inflammatory activity

Inflammation is a physiological state that occurs when our bodies are exposed to potentially harmful endogenous or exogenous substances, as well as injury or trauma (Kumar et al. 2014). Inflammation is the tissue's reaction to infection, irritation or a foreign substance, and it consists of removing the injurious stimuli in order to start the healing process. Following inflammation,

Table 1 Metabolic role of different organosulphur compounds in the human body

Organosulphur Compound	Type of Compound	Health-promoting Role	Reference
Alliin (*S*-allylcysteine sulphoxide)	Natural bioactive constituent, with the general formula $C_6H_{11}NO_3S$.	Increases the expression of cytokine genes such as IL-6, MCP-1 and EGR-1, which modulates the production of pro-inflammatory cytokines. It also has powerful antioxidant and free-radical scavenging properties. In addition, alliin has been shown to increase the immune response in blood.	Quintero-Fabián et al. (2013), Nair et al. (2019)
Allicin	Thiosulphinate and the precursor of various sulphur-containing compounds, with the general formula $C_6H_{10}OS_2$.	Allicin has anticancer, antibacterial, antifungal and antitumour properties. Allicin inhibits tumour cell proliferation and induces apoptosis in gastric cells by activating both the intrinsic and extrinsic pathways. Allicin has antibacterial properties against a wide variety of Gram-negative and Gram-positive bacteria (*Staphylococcus aureus*, *Escherichia coli*, *Klebsiella*, *Salmonella*, *Bacillus*, *Streptococcus*, *Proteus* and *Clostridium*). Furthermore, allicin can stimulate cytokine release, improve immune resistance and has activity against a variety of parasites.	Feng et al. (2012), Nok et al. (1996), Coppi et al. (2006)
Sulphenic acid	The first member of the organosulphur oxoacid family, with the general formula RSOH.	The enzyme alliinase catalyses the decomposition of alliin into short-lived and unstable sulphenic acid when Allium plants are chopped, damaged, chewed or crushed. It is believed to control antioxidant activity.	Gupta and Carroll (2014)
Diallyl sulphide (DAS)	Derived bioactive compound, which is a lipophilic thioether, with the general formula $C_6H_{10}S$.	Diallyl sulphide can improve the detoxification functions of liver cells, thereby preventing inflammation symptoms. It boosts the production of the enzyme glutathione *S*-transferase (GST), which binds electrophilic toxins inside the cell. Diallyl sulphide can prevent the activation of nicotine-derived nitrosamine ketone (NNK), which is linked to carcinogenesis. Diallyl sulphide preventive treatment reduces acetaminophen-induced hepatotoxicity and nephrotoxicity, indicating that it can reduce drug-induced liver damage. It has also been shown to be effective in the treatment of cardiovascular disease and colon cancer.	Shin et al. (2013), Colín-González and Santamaría (2017)

(Continued)

Table 1 (Continued)

Organosulphur Compound	Type of Compound	Health-promoting Role	Reference
Diallyl disulphide (DADS)	Derived organosulphur compound with the general formula $C_6H_{10}S_2$.	Diallyl disulphide has multiple anticarcinogenic activities, including: (1) promoting carcinogen metabolism, (2) slowing cell cycle progression, (3) inhibiting cell proliferation and (4) inducing apoptosis. Furthermore, diallyl disulphide inhibits histone deacetylase activities, which has a therapeutic effect in stopping cancer as it can modulate histone hyper-acetylation and reactivate tumour suppressor genes involved in cancer progression. Aside from these advantages, it can cause allergens in Allium plants.	Shin et al. (2013), Colín-González and Santamaría (2017)
Allitriden or diallyl trisulphide (DATS)	Derived organosulphur compound with the formula $C_6H_{10}S_3$.	Diallyl trisulphide has several health-promoting properties, including anticancer properties, being an antiviral immune booster, increasing reactive oxygen species (ROS) levels, causing platelet aggregation, lowering blood pressure, lowering cholesterol and being useful in the treatment of cardiac arrhythmias. It selectively kills cancerous cells in the breast and prostate while leaving healthy cells alone.	Kim et al. (2013), Fu et al. (2015a), You et al. (2013), Yi et al. (2005), Fu et al. (2015b), Ho et al. (2014), Li et al. (2013)

Source: Miękus et al. (2020).

the human body develops a series of defence mechanisms, including the release of histamine, bradykinin and prostaglandins (PGs) (Pan et al. 2015). Pro-inflammatory cytokines (e.g. TNF-α, IL-6 and IL-1b) are produced during the inflammation process by phagocytic cells, which stimulate cellular responses by increasing PGs, ROS and reactive nitrogen species. These reactive species have the potential to cause the onset of cardiovascular and neurodegenerative disorders such as Alzheimer's disease, atherosclerosis, cataracts, inflammation and cancer (Liao et al., 2012).

Anti-inflammatory properties are associated with OSCs; for example, in a study of inflammatory bowel disease, allicin was found to inhibit the production of cytokine messengers known to be pro-inflammatory (Lang et al., 2004), which is primarily a protective response required for survival. The normal inflammatory response entails the accumulation of white blood cells, particularly neutrophils, macrophages and plasma proteins, in areas of injury or foreign particle location, in order to selectively eliminate cells and restore homeostasis. As a result, secondary responses such as redness, pain, fever and swelling are frequently visible. Such an inflammatory response is referred to as acute inflammation, but if it lasts for an extended period of time, it is referred to as chronic or homeostatic, inflammation (Kumar et al. 2014). In addition, activating the innate immune system can cause a more chronic inflammatory response (Patel and Patel, 2015). As a result, regulating both metabolic and immune responses is critical for maintaining central homeostasis because a deviation in normal immune responses may occur if metabolic homeostasis is disrupted chronically by endogenous or exogenous inducers (Chaplin, 2010).

Allium sativum L. inhibits NF-B activity by modulating cytokine expression in lipopolysaccharide-activated human blood, making it a useful tool in anti-inflammatory processes (Keiss et al., 2003a,b). The anti-inflammatory potential of *Allium flavum* L. subsp. *flavum*, Alliaceae, was also demonstrated by inhibiting cyclooxygenase-1 (COX-1) and 12-lipoxygenase (12-LOX) activity (Simin et al., 2013). Some *Allium* species' targeted OSCs, such as allicin, have been implicated in the prevention of inflammatory processes. According to this line of research, the anti-inflammatory activity of garlic is associated with allicin inhibiting the TNF-initiated secretion of pro-inflammatory cytokines from epithelial digestive cells. Garlic oil (*via* DADS) can modulate the production of Th-cytokines in rat lymph and stimulate the immune response *ex vivo* (Zhang et al., 2015). By reducing the formation of inflammatory lipopolysaccharides, DADS, DATS and SAC repressed NF-κB and mitogen-activated protein kinase (MAPK) signalling pathways (Zhang et al., 2015). According to the same study, allyl methyl disulphide is one of the main bioactives found in garlic that successfully suppressed IL-8/IP-10 formation by TNF-α in intestinal cells. Allyl methyl disulphide has also been implicated as an IL-8 mRNA suppressor in HT-29 cells, which is responsible for IκB degradation and NF-κB p65 translocation.

This information is useful in the treatment of ulcerative colitis, Crohn's disease and other inflammatory bowel diseases (Zhang et al., 2015).

3.2 Anticarcinogenic effects

Several individual compounds isolated from garlic have been identified to have two major groups of compounds with active anticancer effects. One group consists of lipid-soluble allyl sulphur compounds such as DADS and DATS, while the other consists of water-soluble compounds such as SAC and SAMC (Thomson and Ali, 2003). The protective action of any chemopreventive agent cannot be attributed to a single mechanism, and the chemoprevention activity of OSCs can be explained using a variety of mechanisms (Moriarty et al. 2007). Several procedures have been proposed to explain the cancer-preventive effects of Allium vegetables and related OSCs, including mutagenesis inhibition, enzyme activation modulation, DNA adduct formation inhibition, free-radical scavenging, and effects on cell proliferation and tumour growth.

3.2.1 Inhibition of mutagenesis/carcinogen formation

Some OSCs may act as antimutagens or anticarcinogens by inhibiting the formation of genotoxic compounds (Dion et al. 1997). In *Salmonella typhimurium*, the mutagenic activity of aqueous and methanolic garlic extracts inhibited aflatoxin B1 (Soni et al., 1997). Dion et al. (1997) discovered that a water extract of garlic, deodorised garlic or onion and SAC were effective in reducing the *in vitro* formation of NMOR, whereas DADS, DPDS and DAS were ineffective NMOR inhibitors. OSCs can reduce the formation of mutagenic heterocyclic amines during the cooking of meat. The mutagenicity of boiled pork juice was found to be inhibited by DADS, DPDS, DAS, AM and AMS, with DADS and DPDS exhibiting a significant inhibitory effect (Tsai et al. 1996). It has been proposed that these compounds may inhibit the formation of mutagens by reducing the formation of Maillard reaction products; however, the nature of the Maillard reaction inhibition by OSCs remains unknown.

3.2.2 Effects on carcinogen-metabolising enzymes

Several OSCs inhibit the development of cancer, primarily when administered prior to, or concurrently with, the carcinogen. DAS and DADS, as well as other OSCs, have been shown to be effective cytochrome P450 2E1 inhibitors, preventing the activation of nitrosamine and other compounds activated by this cytochrome (Kwak et al., 1994). DAS and DADS have also been shown to induce other cytochrome P450s in rats, including cytochrome P450 2B and cytochrome P450 1A (Siess et al., 1997). These inhibitory or inducing effects were observed in the liver and other tissues such as gastrointestinal tract (Haber et al., 1995). The effects of OSCs on the mutagenicity of several genotoxic compounds are mediated by the modification (enhancement or inhibition) of specific P450 cytochromes involved in their activation (Guyonnet et al., 2000).

3.2.3 Effects on cell proliferation, apoptosis and cell differentiation

The anticarcinogenic properties of OSCs appear to be linked to changes in cell proliferation and apoptosis rates. The direct effect of OSCs on tumour cell growth is well documented, and research has shown that OSCs inhibit tumour growth in animals (Le Bon and Seiss, 2000). Direct inhibition of cancer cell growth in culture has been demonstrated in various tumoural cells, with human tumour cell lines from the colon, lung and skin shown to be inhibited by DADS (Sundaram and Milner, 1996). The ability to decrease the proportion of cells in the G1 phase and increase the proportion of cells in the G2/M phase was linked to the antiproliferative effect in human colon tumour cells (Knowles and Milner, 1998). As evidenced by morphological changes and DNA fragmentation, DADS and DATS are also capable of inducing apoptosis in the same cells. DADS also inhibited the growth and differentiation of mouse erythroleukaemia cells (Lea and Ayyala, 1997); this differentiation could be mediated by the induction of acetylation.

3.2.4 Enhancement of the immune system

Garlic extracts and OSCs have been shown to modulate both specific and non-specific antitumour immunity. It has been reported that the pretreatment of Ehrlich ascites cells with garlic extract inhibits the development of malignant ascites, with mice appearing to develop antitumour immunity (Fujiwara and Natata, 1967). DAS protected Balb/c mice from NDMA-induced immunosuppression of humoral and cellular responses (Jeong and Lee, 1998). An aged garlic extract (AGE) was found to stimulate immunity, including macrophage activity, natural killer and killer cells, and LAK cells, as well as increasing cytokine production (Lamm and Riggs, 2000), suggesting one mechanism for the role of garlic in cancer prevention could be immune system stimulation.

3.3 Antioxidant activity

Excessive ROS production and a decrease in antioxidant content leads to the development of oxidative stress, which causes oxidative damage to molecular cellular structures (e.g. carbohydrates, nucleic acids, lipids and proteins) and changes their functions, ultimately leading to cell death (Munné-Bosch and Pintó-Marijuan, 2017). Many diseases, including cancer, neurodegenerative disorders, liver damage, ageing, atherosclerosis, hypertension, ischaemia/ perfusion, diabetes, chronic obstructive pulmonary disease and asthma, are thought to be linked to oxidative stress (Osipova et al., 2021). Antioxidants of synthetic and natural origin are widely used to prevent the effects of oxidative stress. Low concentrations of antioxidants can neutralise ROS *via* a variety of pathways (Green and Shuaib, 2006; Ziakas et al., 2006). As biologically active components of natural plant products can alleviate and prevent a variety of pathological conditions, they are constantly tested in experimental and clinical trials (Stefanucci et al., 2020).

3.4 Antimicrobial activity

Allium vegetables have antibacterial, antifungal, antiviral and antiprotozoal properties, due to thiosulphinates and other sulphur-containing compounds found in these vegetables. The main antimicrobial agents are allicin breakdown products such as DADS, DATS, DAS and ajoene, the antimicrobial strength of which is greater than that of allicin (Corzo-Martinez et al., 2007). Thiosulphinates are generated after the activation of this enzyme following vegetable tissue injury, hence the enzyme alliinase plays an important role in antimicrobial action. Alliinase inactivation caused by prolonged heating results in the loss of antimicrobial activity (Lawson, 1996). Generally, when compared with crude extracts of *Allium* spp., every other sulphur compound has weaker antimicrobial activity (Lawson, 1998).

Allicin's antibacterial activity is bacteriostatic rather than bactericidal. It has been proposed that allicin's antibacterial action results in the rapid and complete inhibition of RNA biosynthesis, as well as a partial inhibition of DNA and protein synthesis. Other thiosulphinates, in addition to allicin, have been found to exhibit antibacterial activity. Allium species have been shown to inhibit both Gram-positive and Gram-negative bacteria, as well as toxin production. The most researched vegetable, i.e. garlic, has been shown to be effective against *Pseudomonas*, *Proteus*, *Escherichia coli* (*E. coli*), *Staphylococcus aureus* (*S. aureus*), *Klebsiella*, *Salmonella*, *Micrococcus*, *Bacillus subtilis* (*B. subtilis*), *Mycobacterium* and *Clostridium difficile* (*C. difficile*) strains. It may also inhibit beneficial intestinal microflora, but potentially harmful Enterobacteriaceae species are more sensitive to garlic compounds, particularly allicin (Corzo-Martinez et al., 2007). Garlic and onion have been the focus of most research into the antibacterial activity of thiosulphinates, with garlic exhibiting more effective bacterial inhibition than onion as its sulphur compound content is four times that of onion (Salem et al., 2010).

Many researchers have also reported that Allium extracts have antifungal properties. Yin and Tsao (1999) discovered that garlic bulb extract had the greatest inhibitory effect on *Aspergillus niger* (*A. niger*), *Aspergillus flavus* (*A. flavus*) and *Aspergillus fumigatus* (*A. fumigatus*), with MIC values of 35, 75 and 104 mg/mL.

Freshly crushed garlic has long been known to have antiparasitic properties, with the Chinese traditionally using an alcoholic extract of garlic cloves to treat intestinal diseases. There are only a few reports on the antiparasitic activity of Allium vegetables or their sulphur compounds. Allicin (30 mg/mL) was found to effectively inhibit the growth of some bacteria, including *Giardia lamblia, Leishmania major, Leptomonas colosoma* and *Crithidia fasciculate* (Ankri and Mirelman, 1999).

In pre-clinical and clinical studies, garlic and its active OSCs have been shown to alleviate a variety of viral infections. Pre-clinical data from a study carried out by Rouf et al. colleagues (2020) show that garlic and its OSCs have activity against various human, animal and plant pathogenic viruses by preventing viral entry into host cells, inhibition of viral RNA polymerase, reverse transcriptase, DNA synthesis and transcription of the immediate–early gene 1 (IEG1), as well as by downregulating the extracellular-signal-regulated kinase (ERK)/MAPK signalling pathway. The immunomodulatory effects of garlic and its OSCs have also been linked to the alleviation of viral infection. In addition, clinical studies have shown that garlic has a prophylactic effect in the prevention of common viral infections in humans by enhancing the immune response (Rouf et al., 2020).

3.5 Improving immune function

Garlic, or its constituents, exert an immunomodulatory effect via modulating cytokine production, which is a mediator of inflammation. Nuclear factor-KB (NF-KB) is a transcription factor that plays an important role in the expression of genes that regulate the immune response. NF-KB plays a crucial role in the activation and regulation of key molecules linked to inflammatory diseases and cancer (Li and Verma, 2002) and increases the expression of some cytokine genes. The inhibition of NF-KB by garlic products was controlled indirectly by the modulation of pro- and anti-inflammatory cytokines (Keiss et al., 2003a,b). Allicin exhibits anti-inflammatory properties and plays a key role in intestinal inflammation. Allicin has been shown to inhibit the secretion of pro-inflammatory cytokines and chemokines from intestinal epithelial cells both spontaneously and in response to TNF (Lang et al., 2004). Josling (2001) discovered that an allicin-containing supplement can protect against the common cold virus. Over a 12-week period, 146 volunteers were randomly assigned to receive either a placebo or an allicin-containing garlic supplement, one capsule daily. Colds were significantly lower in the active treatment group compared with the placebo group. Furthermore, the placebo group had significantly more days challenged virally and symptoms that lasted significantly longer.

3.6 Neurogenerative disorders

The roles of long-term and short-term responses of phytochemical OSCs in neurodegenerative disease have been extensively researched. Garlic powders are generally thought to be beneficial to human health and may also have neuroprotective properties. Recent research has shown that given their antioxidative and neuroprotective properties, OSCs can help with neurological disorders (Dwivedi et al., 2020). *S*-allyl cysteine, the active ingredient, was tested in models of neurodegenerative diseases such as stroke, ischaemia and Alzheimer's and Parkinson's disease. As a result, OSCs can be successfully used to manage inflammation-related neurodegenerative diseases such as Alzheimer's disease, stroke, and Parkinson's disease (Abdulzahra and Hussein, 2014). Recent research indicates that a garlic OSC has a short-term effect on the proliferation of neural progenitor cells (NPCs) and hippocampal neurogenesis. The herbal OSC, extracted DADS, significantly reduced NPC proliferation, while other OSC-extracted components had no effect on neurodegenerative disorders. In adults, hippocampal neurogenesis can be influenced by a variety of signals that modulate the effects of neurotransmitters, growth factors and neurotrophic factors, as well as environmental cues (Yun et al., 2014). Herbal supplements such as OSCs have been shown to both positively and negatively modulate hippocampal neurogenesis in adults (Jiang, 2005). Several studies on the neuroprotective effects of garlic or AGE in Alzheimer's disease, cerebral ischaemia and apoptotic cell death have been published (Chauhan, 2006). The long-term goals of OSCs in neurodegenerative disorders need to be addressed against a longitudinal cohort, as preliminary studies show that OSCs have a positive effect in several neurodegenerative disorders and play a role in neuroprotection (Patil et al., 2016).

4 Nutraceutical applications

Nutraceutical refers to a natural compound, part or derivative of a food, such as omega-3 oil, or nutrient-rich whole food like spirulina or cumin, used in the food or pharmaceutical sector to obtain a beneficial function or health benefit. Nutraceutical applications of OSCs involve the use of natural sulphur-containing compounds, e.g. allicin present in garlic, to obtain a health benefit or beneficial function in the food, supplement and pharmaceutical sectors through the use of antioxidant, antibacterial and other pharmacological properties.

4.1 Organosulphur compounds' uses in the food industry

OSCs are commonly utilised in the food industry as food dyes, food preservatives, sweeteners or for use in water softening. However, as food additives, sulphites must always be noted on the ingredient list as they are considered to be one of 14 allergens.

Natural and synthetic food dyes are assigned an E number between E100–E200 depending on their chemical structure and colour appearance. Aromatic sulphonic acid derivatives of benzene, naphthalene, anthracene and anthraquinone, such as *p*-cresidine sulphonic acid, are used to produce synthetic water-soluble food dyes some of which include Tartrazine, Sunset Yellow FCF, Azorubine, Amaranth, Ponceau 4R, Allura Red AC, Patent Blue V, Indigo Carmine, Brilliant Blue FCF, Green S, Brilliant Black BN, Brown HT and Litholrubine BK, all represented in Fig. 2 corresponding to their colour and chemical structure.

Sulphur dioxide and synthetic or natural sulphites are approved by the European Food Safety Authority (EFSA) and are commonly used preservatives in the food industry with assigned E numbers E220–E228, including sulphur dioxide (E220), sodium sulphite (E221), sodium hydrogen sulphite (E222), sodium metabisulphite (E223), potassium metabisulphite (E224), calcium sulphite (E226), calcium hydrogen sulphite (E227) and potassium hydrogen sulphite (E228). There are legal limits for their use in food products, including dried fruits, breakfast sausages and burger meat where they maintain an appealing colour *via* the inhibition of enzymatic activity of polyphenol oxidase and other enzymes, and extended shelf life through reduced growth of fungi and bacteria. All added sulphites, with E numbers are considered allergens and must be indicated on the food label in bold script. Sulphur dioxide and many of the sulphites are naturally occurring, however, in the food industry, they are often artificially synthesised to obtain a higher yield and stronger antibacterial or alternative properties.

Natural sources of 'sulphur dioxide' include fires, volcanos and burning of fossil fuels such as coal, oil or petroleum. In the food industry, it is used in its gaseous, liquid or dry form as sulphite, bisulphite and metabisulphite salts. Sulphur dioxide is obtained by burning sulphur from its natural sources, following which the gas can be used for the disinfection of fruits. While the sulphur dioxide source is burning, fruits like grapes get exposed to the fumes prior to dehydration and transport. Liquid sulphur dioxide is also known for its use in wineries; however, it is often more expensive due to the requirement for a special steel container during the wine ageing stage (Prabhakar and Mallika, 2014). 'Calcium sulphite' is another naturally occurring sulphite used in food preservation. Its main source is naturally occurring gypsum, appearing as a colourless powder often extracted by deep mining, which is the only

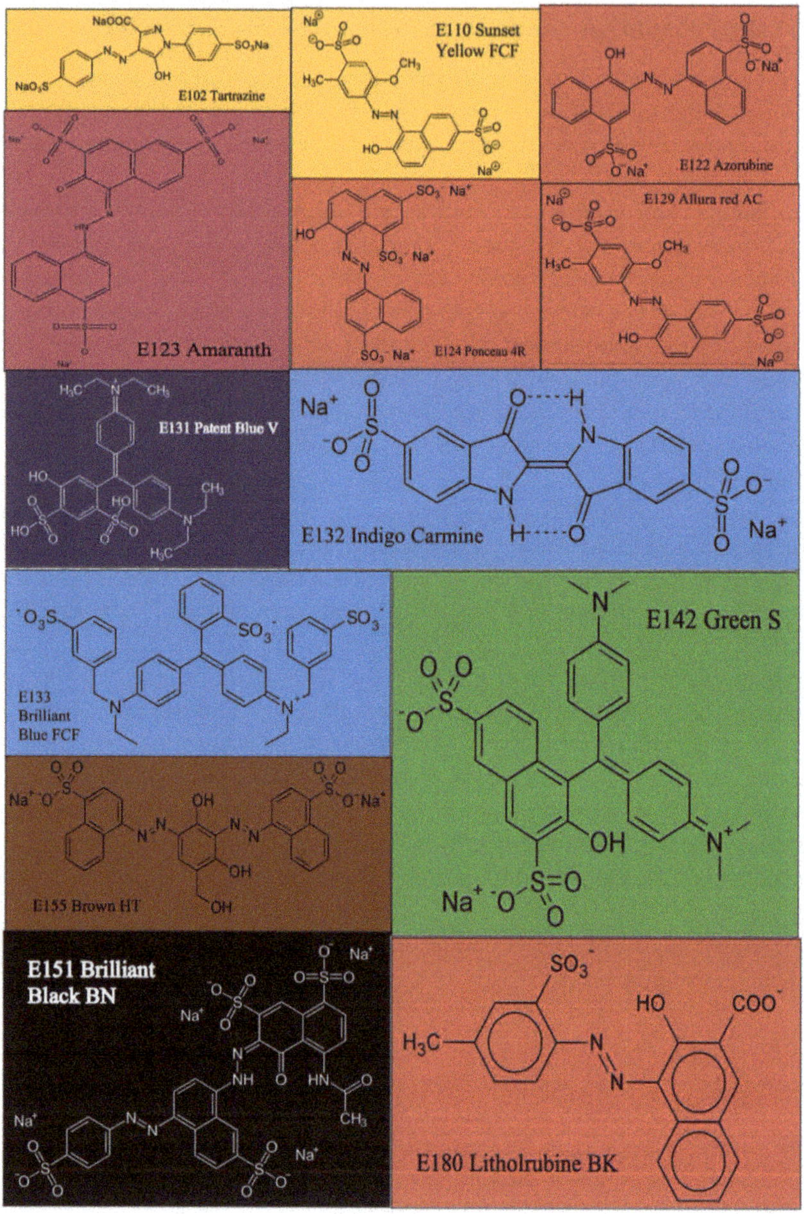

Figure 2 Approved food dyes produced using *p*-cresidine sulphonic acid.

acceptable way of producing calcium sulphite for use in the food industry. Synthetically manufactured calcium sulphite is strictly prohibited. Approved natural calcium sulphite is used in flour, bread, cereals, canned vegetables, juices, jellies and other food products as it serves as an anticaking agent, pH regulator and thickener.

Natural and synthetic sweeteners in the food industry are used to impact the sweet taste of food and have an assigned E number between E700–E999, where some of the sweeteners are produced using OSCs. Most artificial sweeteners have zero or relatively no glycaemic index, making them a safe alternative for people with diabetes or obesity. E954 is assigned to the artificial sweetener 'saccharin' appearing as a white crystalline solid. It adds no nutritional value to the food and is 400 times sweeter than sucrose, however, when used in large amounts it can impart a bitter or metallic aftertaste. It is derived from a sulphonamide derivative of toluene and is often used in food products such as fizzy drinks, sweets, jams or biscuits (Mahmood and Al-Juboori, 2020).

'Acesulfame potassium' (Ace K), E900, is another artificial sweetener closely related to saccharin and, as it is calorie-free sugar, it does not add nutritional value to the food. It appears as a white crystalline powder similar to regular sugar and is around 200 times sweeter than sucrose. Similar to saccharin, it also exhibits a slightly bitter aftertaste especially when used in higher quantities. Ace K is heat stable and is commonly used in combination with other sweeteners such as sucralose or aspartame in soft drinks, frozen desserts, sweets and even tabletop sweeteners to give them a more sucrose-like taste. Ace K is produced using a potassium salt and a sulphonate compound, e.g. sulphur trioxide, to make *N*-sulphoacetoacetamide. This is later neutralised with potassium hydroxide, making the final product approved for use in food production by the Food and Drug Authority (FDA) in America and EFSA in Europe, within an acceptable daily intake of 15 mg/kg of body weight (Zeece, 2020). Generally, Ace K is safe within recommended limits as further health data is still limited. Some studies suggest that the consumption of Ace K has an effect on neurocognitive function by decreasing intracellular ATP production in neuronal cells (Yalamanchi et al., 2016); however, *in vitro* and *in vivo* studies are often carried out using higher amounts than those used in the human diet, therefore, further investigation is required in human trials within recommended concentrations.

4.2 Uses in the pharmaceutical industry

Sulpha drug is a synthetic antibiotic produced from the sulphonamide functional group of the molecule. In 2018, the sulpha drug market value was US$20.5 billion with a gross domestic product (GDP) of 5.27% (Menafn, 2019), signifying increasing popularity and need in the pharmaceutical market. Sulpha drugs are mainly used for the treatment of skin infections, followed by gastrointestinal tract (GIT) infections, urinary tract infections (UTIs) and respiratory tract infections (RTIs) (MordorIntelligence, 2021), with the main companies in the market including, amongst others, F. Hoffmann-La Roche AG, Pfizer Inc., Teva Pharmaceutical Industries Ltd., AA Pharma Inc., Lexine Technochem Pvt. Ltd., Abbott Laboratories and Mylan Pharmaceuticals. Currently, the sulphonamide market does not have a dominant player and is therefore a competitive and fragmented market.

In 1928, bacteriologist Alexander Fleming discovered penicillin by noticing an area with no bacterial growth on an agar plate, which was linked to the presence of mould from the *Penicillium* genus (Gaynes, 2017). The chemical structure of penicillin is shown in Fig. 3, where the presence of sulphur makes it the first discovered and most well-known sulpha drug in history. The active part of penicillin was not isolated until 1939, when Howard Florey assembled a team of scientists who managed to purify an active strain of penicillin, which proved successful in animal trials for the treatment of microbial infection in mice. The production and purification process took scientists 11 years, following which in 1941, the first dose of purified penicillin was administered to humans for the treatment of serious infection, which sparked the beginning of its use and discovery of other sulpha drugs.

However, the isolation and purification of penicillin would not have happened, without the discovery of prontosil in 1932, when German pathologist and bacteriologist Gerhard Domagk proved its antimicrobial activity and won a Nobel Prize in 1939. It was discovered that prontosil is metabolised into sulphanilamide and shows obvious health benefits, and this link resumed interest in the isolation of penicillin and further investigation of other sulphanilamide drugs (Gaynes, 2017).

Currently, there are numerous sulphonamides and sulphaguanidine drugs on the market. 'Pediazole' is a combination of the antibiotic erythromycin and an antibacterial sulpha drug sulphisoxazole used for the treatment of ear infection in children. Other sulpha drugs are summarised in Table 2, depending on their function including antimicrobial, antidiabetic or antidiuretic activity.

'Sulphacetamide' is a short-acting antibacterial drug mainly used to treat skin infections such as acne rosacea, seborrheic dermatitis or dandruff. A combination of 10% sodium sulphacetamide and 5% sulphur is the most

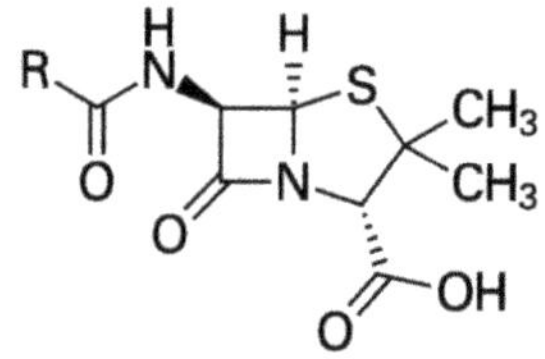

Figure 3 Chemical composition of penicillin.

Table 2 Examples of sulpha drugs corresponding to their function

	Antimicrobial	
Short acting	Intermediate acting	Long acting
Sulphacetamide	Sulphadoxine	Sulphadimethoxine
Sulphadiazine	Sulphamethoxazole	Sulphamethoxypyridazine
Sulphadimidine	Sulphamoxole	Sulphametoxydiazine
Sulphafurazole		
	Antidiabetic	**Diuretics**
First Generation	**Second Generation**	
Acetohexamide	Glibornuride	Acetazolamide
Carbutamide	Glyclopyramide	Bumetanide
Chlorpropamide	Gliquidone	Chlorthalidone
Tolbutamide	Glibenclamide	Clopamide
Tolazamide	Glipizide	

common formulation used in lotion, cream, cleansers or foam wash (Grobel and Murphy, 2018). 'Sulphadiazine' is another short-acting sulpha antibiotic often prescribed for treatment of UTIs, in particular *Toxoplasma gondii*. Medication is administered orally or intravenously, following which adequate hydration is needed to prevent the possibility of developing sulphadiazine crystalluria and obstructive uropathy, as well as other kidney-related issues (Miller et al., 2012). 'Sulphadimidine' is mostly employed in veterinary medicine, as a short-acting antibacterial agent with low potency and a half-life of 1.5–5 h. It has proven activity against *Streptococcus pyogenes*, *Streptococcus pneumoniae*, *Haemophilus influenzae*, *Escherichia coli*, *Neisseria meningitidis* and other bacterial strains (Greenwood, 2010). Lastly, 'sulphafurazole' is effective against a wide range of Gram-positive and Gram-negative microorganisms in a short-acting fashion, where effects can be observed after 2–3 h following an oral dose of 2 g (Greenwood, 2010). It is often employed in the treatment of UTI, but also for chlamydia and nocardia infections.

Sulphadoxine is a well-known intermediate-acting antimicrobial sulphonamide drug with a plasma half-life of 6 days and maximum activity observed after 3–4 h, following oral consumption. It is mainly used in combination with pyrimethamine to treat malaria caused by *Plasmodium falciparum* or *Plasmodium vivax* (Aderibigbe and Mukaya, 2016). 'Sulphamethoxazole' has a plasma half-life of about 10 h and is used on its own or in combination with trimethoprim to treat UTIs, RTIs and GIT infections caused by *Mycobacterium fortuitum* or *Mycobacterium marinum* (Wallace et al. 2015). However, sulphamethoxazole has a low solubility in water and is rapidly eliminated from blood, followed by some possible side effects including skin rashes, fever, hepatotoxicity, haematological disorder and lymphadenopathy (Singh et al., 2017).

'Sulphadimethoxine' is a long-acting antimicrobial sulpha drug used in veterinary medicine to treat UTIs, RTIs, enteric and soft tissue infections. 'Sulphamethoxypyridazine' is another long-acting sulphonamide drug with a half-life of 38 h, previously prescribed for vaginal irritation and the treatment of dermatitis herpetiformis. However, sulphamethoxypyridazine, sulphametoxydiazine and other long-acting antimicrobial sulpha drugs are currently not allowed for single daily dose therapy in the USA, due to a variety of side effects and hypersensitivity reactions (Zinner and Mayer, 2015).

Early sulphonylureas known as first-generation 'antidiabetic' sulpha drugs include tolbutamide, carbutamide, acetohexamide, tolazamide and chlorpropamide, followed by more potent second-generation sulphonylureas including gliclazide, glimepiride and others which are summarised in Table 2. First- and second-generation antidiabetic sulpha drugs work through a similar mechanism of action, by acting directly on the beta cells in the pancreas. They bind to the cytosolic surface of the sulphonylurea receptor and close ATP-sensitive potassium channels, followed by opening calcium channels and depolarising the plasma membrane, all of which results in induced insulin secretion (Reusch, 2015). Current drugs are employed in the treatment of type 2 diabetes, as the stimulation of insulin production is needed to efficiently metabolise carbohydrates.

Diuretic medication increases the production of urine in order to remove water and salts from the body. They are commonly employed to manage health conditions such as high blood pressure, liver disease, heart disease or some kidney disease. 'Acetazolamide', also called diamox, is used to treat health conditions such as glaucoma and altitude sickness. Acetazolamide creates mild metabolic acidosis by losing Na^+ and K^+ into tubular fluid, which results in alkaline urine; meanwhile, H^+ remains in the plasma, resulting in mild acidosis. Overall, this reaction results in stimulation of carbonic anhydrase activity, leading to the diuretic action of acetazolamide (Waller and Sampson, 2018). 'Bumetanide', traded as bumex, is used to treat high blood pressure and swelling from heart, liver or kidney problems. Bumetanide is often compared with furosemide, however, bumex is 40 times more potent. Bumetanide passively diffuses to its site of action, which results in rapid diuresis and relief of pulmonary oedema (Kandasamy and Carlo, 2017). 'Chlorthalidone' is a preferred initial treatment for high blood pressure and chronic kidney disease, but it also treats swelling induced by heart failure, liver failure, nephrotic syndrome, diabetes insipidus and renal tubular acidosis (Peixoto and Bakris, 2015).

There are numerous effective sulpha drugs with different mechanisms of action as well as effects. The market is constantly growing and it is often the case that the combination of products could have a synergistic health benefit.

In the general population, around 3–8% of patients report an allergic reaction to sulphonamide drugs, with the most common symptom being rash, but some cases have reported more serious reactions such as Stevens-Johnson Syndrome. All sulphonamides contain an NH_2-SO_2 moiety, but only sulphonamide antimicrobials contain arylamine at the N_4 position and a nitrogen-containing ring at the N_1 position (Giles et al., 2019). Substitutions at N_4 and N_1 positions are linked to the allergic reactions to sulphonamide drugs, meaning that most nonantimicrobial sulpha drugs do not cause those adverse reactions.

4.3 Uses in nutritional supplements

Glucosamine sulphate (GS) is a natural compound found in the fluid and tissue around human joints, known as cartilage, where it cushions the joints. For various supplement brands, GS is advertised as a natural nutritional supplement, closely mirroring the body's natural glucosamine to improve joint mobility and health. The consumption of GS stimulates the production of glycosaminoglycans and incorporation of sulphur into cartilage. At times, people lose the ability to manufacture glucosamine, which results in cartilage loss, leading to osteoarthritis. Through clinical trials, it was found that GS could be used as an option to help people suffering from osteoarthritis (Murray, 2020a,b).

For use in supplements, GS is extracted from the shells of prawns, crabs and other shellfish, making it unsuitable for people on specific diets, i.e. vegetarian or vegan, as well as people allergic to shellfish. However, some alternative replacement options are available, where GS is extracted from vegetable sources such as fermented corn, as in the production of Glucosamine Sulphate 2KCl supplements by Healthspan.

Lastly, GS is often well tolerated with very few side effects. Unlike in the food industry with the use of food additives, or in the sulpha drug industry, no allergic reactions have been linked to the sulphur present as part of the GS supplement, as sulphur is an essential mineral within the human body. Only minor side effects have been noted from the consumption of GS, and those include stomach upset, heartburn, diarrhoea, nausea and indigestion (Murray, 2020a,b). Studies have shown that the consumption of GS supplements is safe for the overall population, and would be recommended for people suffering from osteoarthritis.

5 Conclusion and future trends

OSCs are linked to anti-inflammatory, anticarcinogenic, antioxidant and antimicrobial activity, as well as improvements in immune function and neurodegenerative disease. In the food industry, OSCs are used as food dyes, preservatives and sweeteners, however, they must be indicated as additives, due to potential allergic reactions. In the pharmaceutical sector, sulphonamide drugs are mainly used for treating skin infections, gastrointestinal infections, UTIs and RTIs, however, they must be prescribed with caution, as 3-8% of patients report allergic reactions to sulphonamide drugs, linked to the subunits at N_4 and N_1 positions. In nutritional supplements, OSC GS is advertised as improving joint mobility and health and was not linked to any allergic reactions, as sulphur is an essential mineral in the body, following which, the supplement was recommended for people suffering from osteoarthritis.

Future OSC research should focus on the mechanisms and underpinnings of structure-activity relationships, as well as the toxicological effects of isolated bioactive compounds from edible Allium species. In the nutraceutical industry, the development of sustainable and safe OSC production should be researched, as well as applications in other diseases. However, sulphur is difficult to work with and introduce into molecules with high precision preventing scientists from developing OSCs with structural geometry, thus improved development methods and precision would allow research into new medication to protect against diseases such as cancer, heart disease, diabetes or neurodegenerative disorders, all of which could be improved by consuming naturally occurring OSCs in vegetables such as garlic, onions, chives and leeks.

6 Where to look for further information

- Walag, A. M. P., Ahmed, O., Jeevanandam, J., Akram, M., Ephraim-Emmanuel, B. C., Egbuna, C., Semwal, P., Iqbal, M., Hassan, S. and Uba, J. O. (2020). Health Benefits of Organosulfur Compounds. In *Functional Foods and Nutraceuticals* (pp. 445-472). Springer, Cham.Key journals/conferences
- Guillamón, E., Andreo-Martínez, P., Mut-Salud, N., Fonollá, J. and Baños, A. (2021). Beneficial Effects of Organosulfur Compounds from *Allium cepa* on Gut Health: A Systematic Review. *Foods*, *10*(8), 1680.
- Ruhee, R. T., Roberts, L. A., Ma, S. and Suzuki, K. (2020). Organosulfur compounds: A review of their anti-inflammatory effects in human health. *Frontiers in Nutrition*, *7*, 64.
- Miękus, N., Marszałek, K., Podlacha, M., Iqbal, A., Puchalski, C. and Świergiel, A. H. (2020). Health benefits of plant-derived sulfur compounds, glucosinolates, and organosulfur compounds. *Molecules*, *25*(17), 3804.
- Kothari, D., Lee, W. D. and Kim, S. K. (2020). Allium flavonols: Health benefits, molecular targets, and bioavailability. *Antioxidants*, *9*(9), 888.
- Ansary, J., Forbes-Hernández, T. Y., Gil, E., Cianciosi, D., Zhang, J., Elexpuru-Zabaleta, M., Simal-Gandara, J., Giampieri, F. and Battino, M. (2020). Potential health benefit of garlic based on human intervention studies: A brief overview. *Antioxidants*, *9*(7), 619.

- Zinner, S. and Mayer, K. (2015). Sulfonamides and trimethoprim. Mandell, Douglas, and Bennett's principles and practice of infectious diseases, 8(1): 410-418.

7 References

Abdulzahra, M. D. and Hussein, F. M. (2014). The antibacterial effect of Ginger and Garlic extracts on some pathogenic bacteria isolated from patients with otitis media. *International Research Journal of Medical Sciences* 2(5): 1-5.

Aderibigbe, B. and Mukaya, H. (2016). Nanobiomaterials architectured for improved delivery of antimalaria drugs. *Nanoarchitectonics for Smart Delivery and Drug Targeting* 1(7): 169-200.

Ankri, S. and Mirelman, D. (1999). Antimicrobial properties of allicin from garlic. *Microbes and Infection* 1(2): 125-129.

Barba, F. J. and Orlien, V. (2017). Processing, bioaccessibility and bioavailability of bioactive sulfur compounds: Facts and gaps. *Journal of Food Composition and Analysis* 61: 1-3.

Chaplin, D. D. (2010). Overview of the immune response. *Journal of Allergy and Clinical Immunology* 125(2): S3-S23

Chauhan, N. B. (2006). Effect of aged garlic extract on APP processing and tau phosphorylation in Alzheimer's transgenic model Tg2576. *Journal of Ethnopharmacology* 108(3): 385-394.

Chope, G. A., Cools, K. and Terry, L. A. (2011). Alliums (onions, garlic, leek and shallot). *Health Promoting Properties of Fruit and Vegetables* 1: 5-27.

Colín-González, A. L. and Santamaría, A. (2017). Garlic, gastrointestinal protection and oxidative stress. *Gastrointestinal Tissue* 2017: 275-288.

Coppi, A., Cabinian, M., Mirelman, D. and Sinnis, P. (2006). Antimalarial activity of allicin, a biologically active compound from garlic cloves. *Antimicrobial Agents and Chemotherapy* 50(5): 1731-1737.

Corzo-Martinez, M., Corzo, N. and Villamiel, M. (2007). Biological properties of onions and garlic. *Trends in Food Science and Technology* 18(12): 609-625.

Dima, C., Assadpour, E., Dima, S. and Jafari, S. M. (2020). Bioactive-loaded nanocarriers for funvtional foods: From designing to bioavailbility. *Current Opinion in Food Science* 33: 21-29.

Dion, M. E., Agler, M. and Milner, J. A. (1997). S-allyl cysteine inhibits nitrosomorpholine formation and bioactivation. *Nutrition and Cancer* 28(1): 1-6.

Dwivedi, A., Singh, S., Kumar, S. and Mittal, P. C. (2020). Organosulfur phytochemicals against metabolic and neurodegenerative disease: Benefits and risks. *Bio-Functional Leads for Drug Development* 11: 179-194.

Feng, Y., Zhu, X., Wang, Q., Jiang, Y., Shang, H., Cui, L. and Cao, Y. (2012). Allicin enhances host pro-inflammatory immune responses and protects against acute murine malaria infection. *Malaria Journal* 11: 268.

Food Safety Authority (FSAI). (2017). *Sulphur Dioxide and Sulphites*. Available at: https://www.fsai.ie/faq/additives/sulphur_dioxide_sulphites.html (Accessed on 4 November 2021).

Fu, E., Tsai, M. C., Chin, Y. T., Tu, H. P., Fu, M. M., Chiang, C. Y. and Chiu, H. C. (2015a). The effects of diallyl sulfide upon P *Orphyromonas gingivalis* lipopolysaccharide stimulated proinflammatory cytokine expressions and nuclear factor-kappa B activation in human gingival fibroblasts. *Journal of Periodontal Research* 50(3): 380–388.

Fu, R., Zhang, Y., Peng, T., Guo, Y. and Chen, F. (2015b). Phenolic composition and effects on allergic contact dermatitis of phenolic extracts *Sapium sebiferum* (L.) Roxb. leaves. *Journal of Ethnopharmacology* 162: 176–180.

Fujiwara, M. and Natata, T. (1967). Induction of tumour immunity with tumour cells treated with extract of garlic (*Allium sativum*). *Nature* 216(5110): 83–84.

Gao, S., Basu, S., Yang, G., Deb, A. and Hu, M. (2013). Oral bioavailability challenges of natural products used in cancer chemoprevention. *Progress in Chemistry* 25(9): 1553–1574.

Gaynes, R. (2017). The discovery of penicillin–New insights after more than 75 years of clinical use. *Emerging Infectious Diseases* 23(5): 849–853.

Giles, A., Foushee, J., Lantz, E. and Gumina, G. (2019). Sulfonamide allergies. *Pharmacy* 7(3): 132.

Goncharov, N. V., Belinskaia, D. A., Ukolov, A. I., Jenkins, R. O. and Avdonin, P. V. (2021). Organosulfur compounds as nutraceuticals. In: *Nutraceuticals, Efficacy, Safety and Toxicity* (pp. 911-924). Academic Press, Cambridge, Massachusetts, USA. DOI: 10.1016/C2019-0-02045-6.

Green, A. R. and Shuaib, A. (2006). Therapeutic strategies for the treatment of stroke. *Drug Discovery Today* 11(15–16): 681–693.

Greenwood, D. (2010). Sulfonamides. *Antibiotic and Chemotherapy* 9(29): 337–343.

Grobel, H. and Murphy, S. (2018). Acne vulgaris and acne rosacea. *Integrative Medicine* 4: 759–770.

Gupta, V. and Carroll, K. S. (2014). Sulfenic acid chemistry, detection and cellular lifetime. *Biochimica et Biophysica Acta* 1840(2): 847–875.

Guyonnet, D., Belloir, C., Suschetet, M., Siess, M. H. and Le Bon, A. M. (2000). Liver subcellular fractions from rats treated by organosulfur compounds from Allium modulate mutagen activation. *Mutation Research* 466(1): 17–26.

Haber, D., Siess, M. H., Canivenc-Lavier, M. C., Le Bon, A. M. and Suschetet, M. (1995). Differential effects of dietary diallyl sulfide and diallyl disulfide on rat intestinal and hepatic drug-metabolizing enzymes. *Journal of Toxicology and Environmental Health* 44(4): 423–434.

Hadden, J. W. (1993). Immunostimulants. *Trends in Pharmacological Sciences* 14(5): 169–174.

Ho, C. Y., Weng, C. J., Jhang, J. J., Cheng, Y. T., Huang, S. M. and Yen, G. C. (2014). Diallyl sulfide as a potential dietary agent to reduce TNF-α-and histamine-induced proinflammatory responses in A 7r5 cells. *Molecular Nutrition and Food Research* 58(5): 1069–1078.

Jafari, S. M. and McClements, D. J. (2017). Nanotechnology approaches for increasing nutrient bioavailability. *Advances in Food and Nutrition Research* 81: 1–30.

Jantan, I., Ahmad, W. and Bukhari, S. N. A. (2015). Plant-derived immunomodulators: An insight on their preclinical evaluation and clinical trials. *Frontiers in Plant Science* 6: 655.

Jeong, H. G. and Lee, Y. W. (1998). Protective effects of diallyl disulfide on *N*-nitrosodimethylamine-induced immunosuppression in mice. *Cancer Letters* 134(1): 73–79.

Jiang, A. (2005). Na^+/Cl^- coupled GABA transporter, GAT-1 from *Caenorhabditis elegans*. *Journal of Biological Chemistry* 280: 2065–2077.

Josling, P. (2001). Preventing the common cold with a garlic supplement: A double-blind, placebo-controlled survey. *Advances in Therapy* 18(4): 189–193.

Kandasamy, J. and Carlo, W. (2017). Pharmacologic therapies IV. *Assisted Ventilation of the Neonate* 6: 366–379.

Keiss, H.-P., Dirsch, V. M., Hartung, T., Trueman, L., Auger, J., Kahane, R. and Vollmar, A. J. (2003a). Garlic (*Allium sativum* L.) modulates cytokine expression in lipopolysaccharide-activated human blood thereby inhibiting NF-κB activity. *The Journal of Nutrition* 133(7): 2171-2175.

Keiss, H.-P., Dirsch, V. M., Hartung, T., Trueman, L., Auger, J., Rémi, K. and Vollmar, A. M. (2003b). Garlic (*Allium sativum L.*) modulates cytokine expression in lipopolysaccharide-activated human blood therapy inhibiting NF-κB activity. *The Journal of Nutrition* 133(7): 2171–2175.

Kim, S. R., Jung, Y. R., An, H. J., Kim, D. H., Jang, E. J., Choi, Y. J., Moon, K. M., Park, M. H., Park, C. H., Chung, K. W., Bae, H. R., Choi, Y. W., Kim, N. D. and Chung, H. Y. (2013). Anti-wrinkle and anti-inflammatory effects of active garlic components and the inhibition of MMPs via NF-F-κB signaling. *PLoS ONE* 8(9): e73877.

Knowles, L. M. and Milner, J. A. (1998). Depressed p34cdc2 kinase activity and G2/M phase arrest induced by diallyl disulfide in HCT-15 cells. *Nutrition and Cancer* 30(3): 169–174.

Kumar, V., Abbas, A. K., Fausto, N. and Aster, J. C. (2014). Robins and cortan pathologic basis of disease. *Inflammation and Repair* 9: 69–112.

Kwak, M. K., Kim, S. G., Kwak, J. Y., Novak, R. F. and Kim, N. D. (1994). Inhibition of cytochrome P4502E1 expression by organosulfur compounds allylsulfide, allylmercaptan and allylmethylsulfide in rats. *Biochemical Pharmacology* 47(3): 531-539.

Lamm, D. L. and Riggs, D. R. (2000). The potential application of *Allium sativum* (garlic) for the treatment of bladder cancer. *Urologic Clinics of North America* 27(1): 157–162.

Lang, A., Lahav, M., Sakhnini, E., Barshack, I., Fidder, H. H., Avidan, B., Bardan, E., Hershkoviz, R., Bar-Meir, S. and Chowers, Y. (2004). Allicin inhibits spontaneous and TNF-α induced secretion of proinflammatory cytokines and chemokines from intestinal epithelial cells. *Clinical Nutrition* 23(5): 1199–1208.

Lawson, L. D. (1996). The composition and chemistry of garlic cloves and processed garlic. In: Koch, H. P. and Lawson, L. D. (Eds). *Garlic: The Science and Therapeutic Application of Allium sativum L., and Related Species.* Williams and Wilkins Press, Baltimore.

Lawson, L. D. (1998). Garlic: A review of its medicinal effects and indicated active compounds. In: Lawson, L. D. and Bauer, R. (Eds). *Phytomedicines of Europe: Chemistry and Biological Activity, ACS Symposium Series.* Am Chem S, Washington, DC.

Lawson, L. D. and Hughes, B. G. (1992). Characterization of the formation of allicin and other thiosulfinates from garlic. *Planta Medica* 58(4): 345–350.

Le Bon, A. M. and Siess, M. H. (2000). Organosulfur compounds from allium and the chemoprevention of cancer. *Drug Metabolism and Drug Interactions* 17(1–4): 51–79.

Lea, M. A. and Ayyala, U. S. (1997). Differentiating and growth inhibitory effects of diallyl disulfide on cancer cells. *International Journal of Oncology* 11(1): 181–185.

Lea, M. A., Randolph, V. M. and Patel, M. (1999). Increased acetylation of histones induced by diallyl disulfide and structurally related molecules. *International Journal of Oncology* 15(2): 347–352.

Li, Q. and Verma, I. M. (2002). NF-kappaB regulation in the immune system. *Nature Reviews. Immunology* 2(10): 725–734.

Li, Y. N., Huang, F., Liu, X. L., Shu, S. N., Huang, Y. J., Cheng, H. J. and Fang, F. (2013). Allium sativum-derived allitridin inhibits Treg amplification in cytomegalovirus infection. *Journal of Medical Virology* 85(3): 493–500.

Liao, J. C., Deng, J. S., Chiu, C. S., Huang, S. S., Hou, W. C., Lin, W. C. and Huang, G. J. (2012). Chemical compositions, anti-inflammatory, antiproliferative and radical-scavenging activities of *Actinidia callosa var. ephippioides*. *The American Journal of Chinese Medicine* 40(5): 1047–1062.

Luo, Y., Xie, W., Hao, Z., Jin, X. and Wang, Q. (2013). Use of shallot (*Allium ascalonicum*) and leek (*Allium tuberosum*) to improve thein vitroavailaible iron and zinc from cereals and legumes. *Journal of Food* 12(12): 195–198.

Mahmood, A. A. R. and Al-Juboori, S. B. (2020). A review: Saccharin discovery, synthesis, and applications. *Ibn AL-Haitham Journal for Pure and Applied Sciences* 33(2): 43–61.

Majumder, P. and Annegowda, H. (2021). *Fruit and Vegetable By-Products: Novel Ingredients for a Sustainable Society–Valorization of Agri-Food Wastes and By-Products* (pp. 133–156). Academic Press, Cambridge, Massachusetts, USA. DOI: 10.1016/C2020-0-01248-X.

Marchese, A., Barbieri, R., Sanches-Silva, A., Daglia, M., Nabavi, S. F., Jafari, N. J., Izadi, M., Ajami, M. and Nabavi, S. M. (2016). Antifungal and antibacterial activities of allicin: A review. *Trends in Food Science and Technology* 52: 49–56.

Menafn (2019). Sulfa drugs market size, share, demand, opportunity, outlook, trends, revenue, future growth opportunities till 2026. Available at: https://menafn.com/1098751811/Sulfa-Drugs-Market-Size-Share-Demand-Opportunity-Outlook-Trends-Revenue-Future-Growth-Opportunities-Till-2026 (Accessed on 4 November 2021).

Miękus, N., Marszalek, K., Podlacha, M., Iqbal, A., Puchalski, C. and Swiergiel, A. H. (2020). Health benefits of plant-derived sulfur compounds, glucosinolates and organosulfur compounds. *Molecules* 25(17): 3804.

Miller, R., Lipman, M. and Morris, A. (2012). Pulmonary infections in patients with human immunodeficiency virus disease. *Clinical Respiratory Medicine* 4: 346–373.

MordorIntelligence (2021). Sulphonamides market-growth, trends, COVID-19 impact, and forecasts. Available at: https://www.mordorintelligence.com/industry-reports/sulphonamides-market (Accessed on 4 November 2021).

Moriarty, R. M., Naithani, R. and Surve, B. (2007). Organosulfur compounds in cancer chemoprevention. *Mini Reviews in Medicinal Chemistry* 7(8): 827–838.

Munné-Bosch, S. and Pintó-Marijuan, M. (2017). Free radicals, oxidative stress and antioxidants. *Encyclopaedia of Applied Plant Sciences* 1: 16–19.

Murray, M. (2020a). Glucosamine. *Textbook of Natural Medicine* 5: 629–633.

Murray, M. (2020b). Osteoarthritis. *Textbook of Natural Medicine* 5: 1622–1632.

Nahdi, A., Hammami, I., Brasse-Lagnel, C., Pilard, N., Hamdaoui, M. H., Beaumont, C. and El May, M. (2010). Influence of garlic or its main active component diallyl disulfide on iron bioavailability and toxicity. *Nutrition Research* 30(2): 85–95.

Nair, A., Chattopadhyay, D. and Saha, B. (2019). *Plant-Derived Immunomodulators*. In *New Look to Phytomedicine* (pp. 435–499). Elsevier, Amsterdam.

Nicastro, H. L., Ross, S. A. and Milner, J. A. (2015). Garlic and onions: Their cancer prevention properties. *Cancer Prevention Research* 8(3): 181–189.

Nimni, M. E., Han, B. and Cordoba, F. (2007). Are we getting enough sulfur in our diet? *Nutrition and Metabolism* 4(24): 24.

Nok, A. J., Williams, S. and Onyenekwe, P. C. (1996). Allium sativum-induced death of African trypanosomes. *Parasitology Research* 82(7): 634–637.

Osipova, V., Polovinkina, M., Gracheva, Y., Shpakovsky, D., Osipova, A. and Berberova, N. (2021). Antioxidant activity of some organosulfur compounds *in vitro*. *Arabian Journal of Chemistry* 14(4): 103068.

Pan, S., Lakshmil, A. and Priyanka, P. (2015). Anti-inflammatory activity of aqueous of *Allium satium* leaves. *Asian Journal of Pharmaceutical and Clinical Research* 8(3): 78–80.

Patel, H. and Patel, V. H. (2015). Inflammation and metabolic syndrome – An overview. *Current Research in Nutrition and Food Science Journal* 3(3): 263–268.

Patil, T. R., Patil, S., Patil, A. and Patil, S. T. (2016). Garlic and neurodegenerative disorders: A review. *International Journal of Pharmacognosy and Phytochemical Research* 8(10): 1634–1644.

Peixoto, A. and Bakris, G. (2015). Approach to the patient with hypertensive nephrosclerosis. In: *Chronic Renal Disease* (pp. 455–469). Wiley Blackwell, Hoboken, NJ.

Phadatare, A. G., Viswanathan, V. and Mukne, A. (2014). Novel strategies for optimized delivery of select components of *Allium sativum*. *Pharmacognosy Research* 6(4): 334–340.

Prabhakar, K. and Mallika, E. (2014). Permitted preservatives – Sulfur dioxide. *Encyclopedia of Food Microbiology* 2: 108–112.

Putnik, P., Gabrić, D., Roohinejad, S., Barba, F. J., Granato, D., Lorenzo, J. M. and Bursać Kovačević, D. B. (2019a). Bioavailability and food production of organosulfur compounds from edible Allium species. In: Innovative thermal and non-thermal processing, bioaccessibility and bioavailability of nutrients and bioactive compounds (pp. 293-308). Woodhead Publishing. Cambridge, United Kingdom. DOI: 10.1016/B978-0-12-814174-8.00010-X.

Putnik, P., Gabric, D., Roohinejad, S., Barba, F. J., Granato, D., Mallikarjunan, K., Lorenzo, J. M. and Kovacevic, D. B. (2019b). An overview of organosulfur compounds from *Allium* spp.: From processing and preservation to evaluation of their bioavailability, antimicrobial, and anti-inflammatory properties. *Food Chemistry* 276: 680–691.

Quintero-Fabián, S., Ortuño-Sahagún, D., Vázquez-Carrera, M. and López-Roa, R. I. (2013). Allium, a garlic (Allium sativum) compound, prevents LPS-induced inflammation in 3T3-L1 adipocytes. *Mediators of Inflammation* 2013: 381815.

Rahman, M. S. (2007). Allicin and other functional active components in garlic: Health benefits and bioavailability. *International Journal of Food Properties* 10(2): 245–268.

Ramirez, D. A., Locatelli, D. A., González, R. E., Cavagnaro, P. F. and Camargo, A. B. (2017). Analytical methods for bioactive sulfur compounds in *Allium*: An integrated review and future directions. *Journal of Food Composition and Analysis* 61: 4–19.

Reusch, C. (2015). Feline diabetes mellitus. *Canine and Feline Endocrinology* 4: 258–314.

Rouf, R., Uddin, S. J., Sarker, D. K., Islam, M. T., Ali, E. S., Shilpi, J. A., Nahar, L., Tiralongo, E. and Sarker, S. D. (2020). Antiviral potential of garlic (*Allium sativum*) and its organosulfur compounds: A systematic update of pre-clinical and clinical data. *Trends in Food Science and Technology* 104: 219–234.

Ruhee, R. T., Roberts, L. A., Ma, S. and Suzuki, K. (2020). Organosulfur compounds: A review of their anti-inflammatory effects in human health. *Frontiers in Nutrition* 7: 64.

Salem, A. M., Amin, R. A. and Afifi, G. S. A. (2010). Studies on antimicrobial and antioxidant efficiency of some essential oils in minced beef. *Journal of American Science* 6(12): 691.

Shin, I. S., Hong, J., Jeon, C. M., Shin, N. R., Kwon, O. K., Kim, H. S., Kim, J. C., Oh, S. R. and Ahn, K. S. (2013). Diallyl-disulfide, an organosulfur compound of garlic, attenuates airway inflammation via activation of the Nrf-2/HO-1 pathway and NF-kappaB suppression. *Food and Chemical Toxicology* 62: 506–513.

Siess, M. H., Le Bon, A. M., Canivenc-Lavier, M. C. and Suschetet, M. (1997). Modification of hepatic drug-metabolizing enzymes in rats treated with alkyl sulfides. *Cancer Letters* 120(2): 195–201.

Simin, N., Orcic, D., Cetojevic-Simin, D., Mimica-Dukic, N., Anackov, G., Beara, I., Mitic-Culafic, D. and Bozin, B. (2013). Phenolic profile, antioxidant, anti-inflammatory, and cytotoxic activities of small yellow onion. *LWT: Food Science and Technology* 54(1): 139–146.

Singh, A., Sharma, A., Khan, I., Gothwal, A., Gupta, L. and Gupta, U. (2017). Oral drug delivery potential of dendrimers. In: Nanostructures for Oral Medicine (pp. 231-261). Elsevier, Academic Press, Cambridge, Massachusetts, USA. DOI: 10.1016/B978-0-323-47720-8.00010-9.

Soni, K. B., Lahiri, M., Chackradeo, P., Bhide, S. V. and Kuttan, R. (1997). Protective effect of food additives on aflatoxin-induced mutagenicity and hepatocarcinogenicity. *Cancer Letters* 115(2): 129–133.

Stefanucci, A., Zengin, G., Llorent-Martinez, E. J., Dimmito, M. P., Della Valle, A., Pieretti, S., Ak, G., Sinan, K. I. and Mollica, A. (2020). Chemical characterization, antioxidant properties and enzyme inhibition of Rutabaga roots' pulp and peel (*Brassica napus* L.). *Arabian Journal of Chemistry* 13(9): 7078–7086.

Sundaram, S. G. and Milner, J. A. (1996). Diallyl disulfide inhibits the proliferation of human tumour cells in culture. *Biochimica et Biophysica Acta* 1315(1): 15–20.

Thomson, M. and Ali, M. (2003). Garlic [*Allium sativum*]: A review of its potential use as an anti-cancer agent. *Current Cancer Drug Targets* 3(1): 67–81.

Tsai, S. J., Jenq, S. N. and Lee, H. (1996). Naturally occurring diallyl disulfide inhibits the formation of carcinogenic heterocyclic aromatic amines in boiled pork juice. *Mutagenesis* 11(3): 235–240.

Walag, A. M. P., Akram, M., Ahmed, O., Jeevanadam, J., Ephriam-Emmanuel, B., Egbuna, C., Semwal, P. and Iqbal, M. (2020). Health benefits of organosulfur compounds. In: Egbuna, C. and Dable Tupas, G. (ed). *Functional Foods and Nutraceuticals*. Springer Nature Switzerland AG, Geneva, Switzerland.

Walang, A. M. P., Ahmed, O., Jeevanadam, J., Muhammad, A., Ephraim-Emmanuel, B. C., Egbuna, C., Semwal, P., Iqbal, M., Hassan, S. and Uba, J. O. (2020). Health benefits of organosulfur compounds. *Functional Foods and Nutraceuticals* 21: 445–460.

Wallace, R., Philley, J. and Griffith, D. (2015). Antimycobacterial agents. Mandell, Douglas, and Bennett's principles and practice of infectious diseases. The Lancet Infectious Diseases 8(1): 463–478.

Waller, D. and Sampson, A. (2018). Diuretics. *Medical Pharmacology and Therapeutics* 5: 219–229.

Yalamanchi, S., Srinath, R. and Dobs, A. (2016). *Acesulfame-K: Encyclopedia of Food and Health*, 1–5.

Yamada, A., Fahey, J. W., Fukumoto, A., Nakayama, M., Inoue, S., Zhang, S., Tauchi, M., Suzuki, H., Hyodo, I. and Yamamoto, M. (2009). Dietary sulforaphane-rich broccoli sprouts reduce colonization and attenuate gastritis in *Helicobacter pylori*-infected mice and humans. *Cancer Prevention Research* 2(4): 353–360.

Yi, X., Feng, F., Xiangm, Z. and Ge, L. (2005). The effects of allitridin on the expression of transcription factors T-bet and GATA-3 in mice infected by murine cytomegalovirus. *Journal of Medicinal Food* 8(3): 332–336.

Yin, M. C. and Tsao, S. M. (1999). Inhibitory effect of seven *Allium* plants upon three *Aspergillus* species. *International Journal of Food Microbiology* 49(1-2): 49-56.

You, S., Nakanishi, E., Kuwata, H., Chen, J., Nakasone, Y., He, X., He, J., Liu, X., Zhang, S., Zhang, B. and Hou, D. X. (2013). Inhibitory effects and molecular mechanisms of garlic organosulfur compounds on the production of inflammatory mediators. *Molecular Nutrition and Food Research* 57(11): 2049-2060.

Yun, H. M., Ban, J. O., Park, K. R., Lee, C. K., Jeong, H. S., Han, S. B. and Hong, J. T. (2014). Potential therapeutic effects of functionally active compounds isolated from garlic. *Pharmacology and Therapeutics* 142(2): 183–195.

Zeece, M. (2020). Flavours. In: *Introduction to the Chemistry of Food* (pp. 213–250), Academic Press, Cambridge, Massachusetts, USA. DOI: 10.1016/B978-0-12-809434-1.00006-2.

Zhang, Y., Wang, Y., Zhang, F., Wang, K., Liu, G., Yang, M., Luan, Y., Zhao, Z., Zhang, J., Cao, X. and Zhang, D. (2015). Allyl methyl disulfide inhibits IL-8 and IP-10 secretion in intestinal epithelial cells via the NF-κB signalling pathway. *International Immunopharmacology* 27(1): 156–163.

Ziakas, G. N., Rekka, E. A., Gavalas, A. M., Eleftheriou, P. T. and Kourounakis, P. N. (2006). New analogues of butylated hydroxytoluene as anti-inflammatory and antioxidant agents. *Bioorganic and Medicinal Chemistry* 14(16): 5616–5624.

Zinner, S. and Mayer, K. (2015). Sulfonamides and trimethoprim. *Mandell, Douglas, and Bennett's Principles and Practice of Infectious Diseases* 8(1): 410–418.

Part 3

Phytochemicals and the prevention of disease

Chapter 8

Advances in understanding the role of plant phytochemicals in preventing cancer

Gulsun Akdemir Evrendilek, Bolu Abant Izzet Baysal University, Turkey

1 Introduction

The global incidence of cancer has increased dramatically as a result of demographic factors, increased urbanization, and changes in diet and lifestyles (Surh, 2003). It is estimated that 19.3 million new cases of cancer (covering 36 different types of cancer, including nonmelanoma-type skin cancer) and approximately 10 million deaths occurred in 2020. Among the different types of cancers, female breast cancer is the most common, followed by lung, colorectal, prostate, and stomach. Lung cancer has the highest mortality rate, followed by colorectal, liver, stomach, and female breast cancer (Sung et al., 2021). Although various cancer prevention strategies are in place, there is no complete and effective strategy to overcome all types of cancers. Identification

http://dx.doi.org/10.19103/AS.2022.0101.08

of carcinogens, reduced exposure to these triggers, minimization of exposure to risk factors, and changes in lifestyle are some of the suggested ways to reduce cancer rates (Surh, 2003).

Recent attention has been focused on the relationship between incidences of cancer and a healthy diet after it was reported that lifestyle modification can play an important role in preventing more than two-thirds of human cancers (Surh, 2003). According to statistical and epidemiological data, approximately 10–70% of mortality caused by cancer can be related to diet and nutritional factors (Doll and Peto, 1981). Various substances and constituents found in food can trigger tumor development, growth and spread, and play a role in converting normal cells to malignant cells, while others can reduce the risk of cancer. Health-promoting activities of foods are mostly related to regular consumption of fruits and vegetables due to their bioactive compounds such as phytochemicals. Phytochemicals present in a plant-based diet possessing antimutagenic and anticarcinogenic properties are defined as non-nutritive components.

The phytochemicals produced through primary or secondary metabolism by different parts of plants have an essential function in plant defense mechanisms against microorganisms, insects, animals, and abiotic stress and are classified as constitutive metabolites (Molyneux et al., 2007; Santhi and Sengottuvel, 2016). Phytochemicals possess great variability in their chemical structure, thus it is more appropriate to determine their anticarcinogenic activity based on their mechanism rather than their structural properties. More importantly, signal-transduction pathways are a better way to determine their anticarcinogenic properties (Sung et al., 2021; Surh, 2003).

2 Importance of plant-based foods

Cancer is defined as the abnormal growth of cells with the potential to invade and metastasize to other parts of the body. Several factors are involved in the initiation of cancer but changes in genes regulating normal body functions are vital. Radiotherapy and chemotherapy are the most common and widely used treatments, but it is essential to stop cancer in its initiation stage (Golemis et al., 2018; Sung et al., 2021). Reversing the initial phase of carcinogenesis or preventing the potential for premalignant cells by using synthetic, natural, or biological agents known as chemoprevention (Ball et al., 2019; Chapa and Mejía-Teniente, 2016) is becoming more popular with increased knowledge of cancer biology, determination, and characterization of molecular targets, helping prevent various cancers such as colon, prostate, and breast (Golemis et al., 2018; Ranjan et al., 2019; Sung et al., 2012; Vogel et al., 2010). The effects of chemoprevention are defined as primary, secondary, and tertiary. The general population and those at high risk of cancer are subjected to primary chemoprevention, while patients diagnosed with premalignant lesions indicating invasive cancer are subjected to secondary chemoprevention, and patients that need to prevent the recurrence of cancer are included in tertiary chemoprevention (Ball et al., 2019; De Flora and Ferguson, 2005). While dietary phytochemical and non-steroidal anti-inflammatory drugs (NSAID) are classified as primary chemopreventive agents, prevention of recurrence of cancer with tamoxifen is practiced as tertiary chemoprevention, especially in breast cancer (Ball et al., 2019; Bishayee and Sethi, 2016; De Flora and Ferguson, 2005; Sung et al., 2021; Wong et al., 2016).

Both vegetables and fruit are important sources of cancer-preventive substances, containing various types of phytochemicals. Among them, ginger, tomatoes, garlic, onion, soybeans, turmeric, and cruciferous vegetables such as cabbage, Brussels sprouts, broccoli, and cauliflower are important sources of phytochemicals. Different substances are classified as phytochemicals, including micronutrients (secondary metabolites) such as trace minerals, antioxidant vitamins and their precursors, and macronutrients (primary metabolites) such as proteins, carbohydrates, fiber, and fat. Primary metabolites are essential for plants as they have a direct relationship with plant growth and metabolism, whereas secondary metabolites biosynthetically originated from the primary metabolites do not have a vital function for survival, but are important for some functions such as competition, protection, and interaction of species (Errayes et al., 2020; Ko and Moon, 2015; Pichersky and Gang, 2000). Based on their chemical structures and biosynthetic origins, these chemicals can be classified into three major groups - phenolic compounds, terpenoids, and nitrogen-/sulfur-containing compounds (Irchhaiya et al., 2015). Phenolic substances classified as secondary metabolites are present in a wide variety of foods such as fruits, cereals, vegetables, legumes, horticultural crops, and chocolate and in beverages such as coffee and tea (Shahidi and Ambigaipalan, 2015). Polyphenols are classified as flavonoids and non-flavonoids based on having at least one aromatic ring and one or more hydroxyl group(s) (Fig. 1). Flavonoids, as the most widespread and diverse group of polyphenols, are further subdivided into flavonols (quercetin, myricetin, kaempferol, rutin, etc.), flavanones (hesperidin, hesperetin, naringenin, etc.), isoflavones (daidzein, genistein, etc.), anthocyanins (delphinidin, cyanidin, pelargonidin, malvidin, etc.), and flavones (apigenin, luteolin, tangeretin, etc.) (Gonzales et al., 2014), depending on the degree of methoxylation, hydroxylation, glycosylation, and prenylation. Non-flavonoids, on the other hand, contain a wide range of polyphenols such as lignans, stilbenes (resveratrol), and phenolic acids (hydroxybenzoic acids and hydroxycinnamic acids, and hydrolyzable tannins) (Faridi Esfanjani et al., 2018).

Recent studies have focused on the effect of phytochemicals as their anticarcinogenic effects have been proven with clinical studies since they are

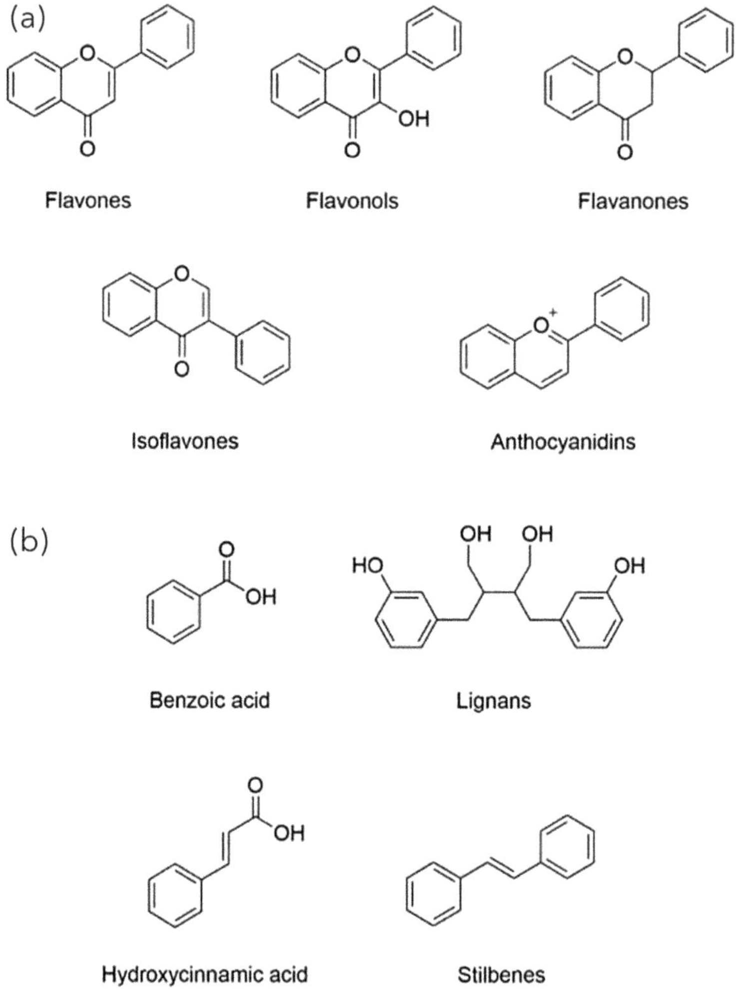

Figure 1 Chemical structure of (a) common flavonoids and (b) non-flavonoid-type phenolic compounds.

used in the treatment of carcinomatous-related diseases and have revealed various anti-cancer properties, including apoptotic cell death activity and anti-proliferation (Errayes et al., 2020). Approximately 1000 different phytochemicals present in plant-based foods are reported to have cancer-preventive properties (Surh, 2003).

3 Mechanisms of chemoprevention

Tumor formation and development is a multistage and complex process that includes tumor initiation, promotion, and progression. Initiation is described as a rapid and irreversible process with extracellular and intracellular chain events, including exposure or uptake of a carcinogenic agent as well as its transport to organs and tissues, where activation and detoxification in addition to covalent interaction of reactive species with target-cell DNA causing genotoxic damage may occur (Surh, 2003). Tumor development, a reversible process, takes a longer time and causes accumulation of proliferating preneoplastic. In contrast to initiation, tumor promotion is a relatively lengthy and reversible process in which cell accumulation takes place. Invasive tumor growth and development of metastatic potential occur in the last stage, called 'progression', also defined as the final stage of neoplastic transformation (Ranjan et al., 2019; Surh, 2003).

Based on their function, chemopreventive agents are categorized as suppressing and blocking agents (Fig. 2). Suppressing agents act as an inhibitor to stop transformation of the initiated malignant cells at the promotion or progression stage while blocking agents prevent carcinogens from reaching the target sites by either undergoing metabolic activation or interaction with critical macromolecules of RNA, DNA, and proteins (Gescher et al., 1998; Manson et al., 2000). Blocking or reversing initiation and promotion at the premalignant stage as well as retarding or halting the development and progression of precancerous cells into malignant cells can be prevented by chemopreventive phytochemicals (Manson et al., 2000). This preventive function of chemoprotective compounds involves various cellular molecules

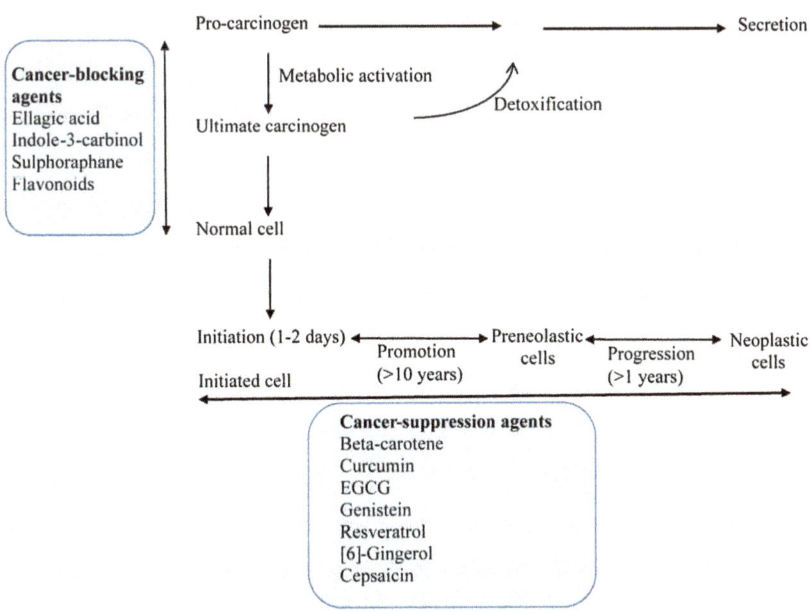

Figure 2 Blocking or suppressing of multistage carcinogenesis by dietary phytochemicals. Source: Surh (2003).

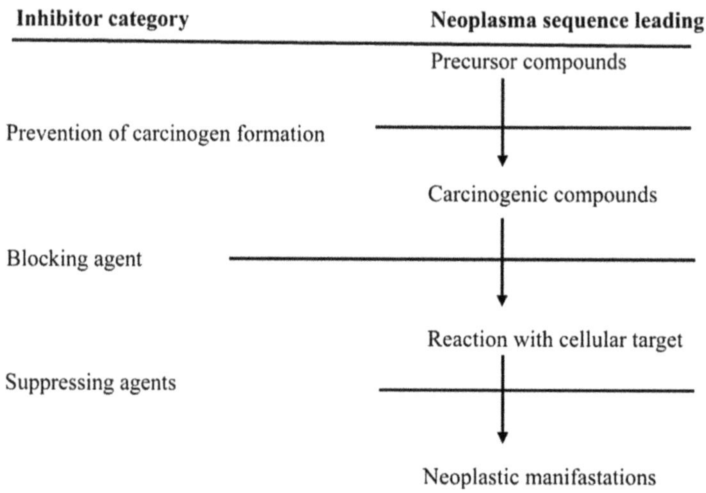

Figure 3 Classification of chemoprotective agents based on their protective effects. Source: Surh (1999).

(Milner et al., 2001; Wattenberg, 1985). Therefore, prevention of tumor development by chemopreventives involves a combination of various distinct sets of intercellular effects (Surh, 1999) (Fig. 3).

Chemopreventive phytochemicals have several functions for the prevention of cancer development, including repairing DNA damage, activation/ detoxification by xenobiotic, metabolizing enzymes, cell proliferation, cell-cycle progression, apoptosis and differentiation, metastasis and angiogenesis, hormonal and growth-factor activity, and functional activation and expression of oncogenes or tumor-suppressor genes (Surh, 2003).

4 Phytochemicals and their mode of action

Both *in vivo* and *in vitro* studies show that plant-originated bioactive compounds have a significant effect on the treatment and prevention of cancer (Bishayee and Sethi, 2016; Gullett et al., 2010; Wong et al., 2016). Terpenoids, alkaloids, and organosulfur compounds are the most effective and encouraging groups of phytochemicals in cancer therapy and prevention (Kaur et al., 2018; Krajka-Kuźniak et al., 2015; Thoppil and Bishayee, 2011) with some compounds having a specific effect on cancer prevention or tumor development. Camptothecin, a naturally occurring quinoline alkaloid extracted from *Camptotheca acuminate*, is effective in inhibiting DNA topoisomerase I. Inhibition of DNA topoisomerase I is also performed by topotecan and irinotecan drugs and semisynthetic derivatives. The antitumor properties of celastrol, a terpenoid, are reported as including the regulation of various transcription factors, angiogenesis, proteases, apoptotic processes, and cell cycle (Kashyap et al., 2018; Shanmugam et al., 2016; Zhang et al., 2019). Studies of one of the most reviewed soy phytoestrogens, genistein, have been performed with this isoflavone alone and in combination with chemotherapy for lung and prostate cancer therapies. Having a role in redox homeostasis, it is effective in the inhibition of hormone-dependent cancers and has an important function in altering the expression of various estrogen receptors, transcription factors, and tumor suppressors in cancerous cells (Pool et al., 2018; Shanmugam et al., 2016). Clinical studies report that the cancer-preventive therapeutic effects of antioxidant phytochemicals and polyphenols are not definite and mostly exhibit multifunctional activities (Lagoa et al., 2020).

5 Cancer-preventive effects of antioxidant and anti-inflammatory activities

Polyphenols such as resveratrol, epigallocatechin-3-gallate (EGCG), and curcumin have important roles in different stages of metastasis and carcinogenesis by possessing both anti-inflammatory and antioxidant activities (Gullett et al., 2010; Kasi et al., 2016; Sinha et al., 2017), presenting inflammatory signaling and oxidative stress modulation in healthy cells by decreasing ROS production, reducing damage in DNA and other cellular components, inhibiting nuclear factor-κB (NF-κB) pathways, and activating nuclear factor erythroid 2-related factor 2 (Nrf2)-mediated antioxidant response (Krajka-Kuźniak et al., 2015; Samadi et al., 2015; Suphim et al., 2010).

Resveratrol, EGCG, curcumin, lycopene, and genistein act as NF-κB suppressor phytochemicals in cancer prevention and therapy by having a critical role in the cancer initiation stage where activation of NF-κB in a tumor-growing environment induces activation and transcription of several anti-apoptotic and pro-proliferative genes activated by inflammation (Gullett et al., 2010; Taniguchi and Karin, 2018). Phase I xenobiotic metabolizing enzymes such as cytochrome P450 inhibition and activation of phase II detoxifying and antioxidant enzymes by various phytochemicals and mechanisms providing minimization of carcinogen active sites are important chemopreventive activities (Kaur et al., 2018; Lagoa et al., 2020). Some compounds such as indoles, isothiocyanates, and polyphenols, such as EGCG, tannic, and protocatechuic acids, accelerate the expression of phase II and antioxidant genes via the Nrf2 pathway (Krajka-Kuźniak et al., 2015; Lambert and Elias, 2010; Shanmugam et al., 2016). Reactive oxygen species (ROS) at moderate levels trigger proliferation and migration causing metastasis and tumor development in malignant cells (Sznarkowska et al., 2016). Inhibition of ROS sources by phytochemicals, especially in mitochondria, provides preventive effects and therapeutic strategies for cancer treatment (Lagoa et al., 2011; Sznarkowska et al., 2016). Moreover, genistein in free form, as well as loaded into NPs, decreases hydrogen peroxide production in human colon cancer cells, demonstrating its antiproliferative activity (Pool et al., 2018).

6 Angiogenesis suppression activities of phytochemicals

Angiogenesis is defined as invasion, tumor progression, and metastasis. Vascular endothelial growth factor (VEGF) is the most important pro-angiogenic factor due to its ability to produce other factors capable of stimulating the proliferation of endothelial cells and blood vessel formation. Moreover, hypoxia-inducible factor-1 (HIF-1), having a vital role in the process of angiogenesis, coordinates the expression of VEGF as it is involved in redox and metabolic remodeling cancer cells studies (Kaur et al., 2018). It has been proven that EGCG, resveratrol, and curcumin are able to inhibit angiogenesis, HIF-1α expression, and VEGF in different cancer models (Bishayee and Darvesh, 2012). As has been shown by some *in vivo* studies, phytochemicals, in addition to angiogenesis, work as VEGF signal inhibitor mechanisms (Lagoa et al., 2020).

7 Functions of phytochemicals in cell death pathway regulation

Various phytochemicals, including their derivatives, can exert their prooxidant properties similar to the effect of polyphenols, camptothecins, and etoposide, which are able to increase the concentration of ROS, leading to cell death (Kaur et al., 2018; Suphim et al., 2010). Different factors such as pH, oxygen concentration, and transition metal level, as well as concentration of polyphenols, have a significant effect on prooxidant behavior as only high micromolar concentration provides cancer cell necrosis (Kaur et al., 2018; Rengasamy et al., 2019).

The ratio between Bax and Bcl-2 proteins increased in several cancer models in apoptosis triggered by EGCG (Khan et al., 2014; Lagoa et al., 2020). Both Bcl-2 and Bcl-xL are the antiapoptotic proteins that are overexpressed in several malignancies that inhibit the mitochondrial release of cytochrome C (Maji et al., 2018). Apoptosis by intrinsic and extrinsic pathways assisted by increased concentration of caspase-8 and caspase-9 in cancer cells can be induced by gallic acid (Kaur et al., 2018). Celastrol is another phytochemical compound presenting *in vivo* and *in vitro* apoptosis-inducing capacity regulated by several different pathways such as Bcl-2 and NF-κB signal mechanism inhibition and p53 and Bax (Aqil et al., 2016; Zhang et al., 2019). Due to its ability to promote cell proliferation and inhibition of apoptosis by regulating various gene expressions, such as Bcl-2 and Bcl-xL, and antagonizing p53 and NF-κB, the regulatory network is a popular antitumor agent (Taniguchi and Karin, 2018).

Thus, nanoformulations of quercetin and EGCG are reported to be effective in preventing NF-κβ activation and inducing apoptosis in cancer models and inflammation (Chakraborty et al., 2012; Siddiqui et al., 2014). Besides p53 pathway, alteration of genes included in apoptosis, androgen pathway, and cell-cycle control mechanisms have been observed in studies with resveratrol. However, its effects on microRNA profiling and epigenetic remodeling are yet to be studied (Huminiecki and Horbańczuk, 2018; Xie et al., 2016).

Curcumin at low concentrations exhibiting apoptosis-inducing or antiproliferative properties has the ability to inhibit NF-κB signal mechanism, cause partial mitochondrial depolarization, and decrease Bcl-xL levels with no detectable oxidative stress in cholangiocarcinoma (Suphim et al., 2010). Curcumin at the same concentration also causes an increase in Bax (proapoptotic proteins) and p53 (related to cell death) whereas at higher concentrations, it is unable to alter these proteins. However, promoted ROS generation results in a collapse in mitochondria and cell death. Although catechin-type flavonoid of EGCG activity against melanoma cells is stronger than that of the curcumin, it (proportionally at high and prooxidant concentration) exhibits cytotoxicity (Branco et al., 2015). Polyphenols, having cytotoxicity or antiproliferative effects, have the ability to modulate cell signaling pathways related to the promotion and initiation of carcinogenesis in addition to invasive and migration phenotypes at low concentrations, but higher concentrations are required for efficient antitumor activity (Kaur et al., 2018; Rengasamy et al., 2019). However, higher concentrations of these compounds similar to hepatotoxicity of catechins may raise safety concerns. Although its role in cellular transformation and cancer progression is still being debated, autophagy is a complicated process with an important role in the degradation of damaged organelles and proteins in eukaryotic cells (Sznarkowska et al., 2016). Different phytochemicals such as curcumin, thymoquinone, celastrol, resveratrol, ursolic acid, and γ-tocotrienol have been implicated in the induction of autophagy in various cancer models (Deng et al., 2019). In addition to autophagy, different cell death mechanisms such as non-canonical programmed cell death mechanisms may become an effective therapeutic approach (Diederich and Cerella, 2016).

8 Combined use of phytochemicals with antineoplastic agents

Due to toxicity concerns and the desire to reduce the therapeutic dose, combinations of standard chemotherapeutics with anticancer phytochemicals are gaining popularity. Application of low-dose doxorubicin or etoposide combined with quercetin or apigenin polyphenols in lymphoid and myeloid leukemia cells has reduced ATP levels, lowered the induction of apoptosis, especially caspase-3, -8, and -9 activation, and increased S and/or G2/M phase cell cycle in malignant cell lines (Mahbub et al., 2015). It has been proven that the combination of the two treatments provides a synergistic decrease in glutathione levels and an increase in DNA damage, whereas a single application only affected these parameters on a moderate level. A combination of emodin, rhein, and *cis*-stilbene polyphenols with chemotherapeutic drugs, on the other hand, exhibited a lower anticancer effect. It is estimated that the synergistic effect is related to intrinsic apoptosis and, particularly with the activation of caspase-9, is because p53 cell lines are found to be null, indicating that applied polyphenols might affect the mitochondrial target in cancer cells as also observed upon inactivation of Jurkat cells by application of quercetin followed by menadione (Baran et al., 2014). Applications of polyphenols alone or in combination with etoposide or doxorubicin presented no toxic effect on non-cancerous cells, and more importantly, these combined applications have a protective effect on these cells from the adverse side effects of chemotherapeutic agents (Mahbub et al., 2015). Protecting non-cancerous cells from the adverse effects of chemo- and radiotherapies, in addition to inhibiting multidrug resistance proteins, breast cancer resistance protein, and P-glycoprotein, is another advantage of antioxidants and natural phytochemicals (Asensi et al., 2011; El-Ashmawy et al., 2017; Li et al., 2017; Mercader and Pomilio, 2012). There is a correlation between the therapeutic and chemopreventive capacity of phytochemicals with tumor suppressor genes and oncogenes (Muddineti et al., 2017).

9 Importance of delivery systems for phytochemicals

Although the anticancer activities of phytochemicals are highly promising, there are several concerns regarding the translation of beneficial effects. The poor solubility of polyphenols in the aqueous phase prevents them from being retained in circulation. Due to their low absorption in the gastrointestinal system, high metabolism, rapid clearance, and chemical degradation, they may not reach blood or tumor tissues. Although they have considerable antitumor activity, the level of micromolar concentration of EGCG and quercetin in blood and the very short half-time (several minutes) of resveratrol is insufficient to provide a cytoprotective effect (Jung et al., 2015; Lagoa et al., 2017). Efforts are therefore being made to develop delivery systems to increase the solubility and stability of bioactive phytochemicals, improve oral bioavailability, and increase their target specificity to tumor cells. Due to their increased solubility and easier intestinal uptake, lipid-based systems have been found to be successful. Applications of chylomicrons, supported by berberine, increase the drug absorption rate in plasma. Different delivery systems for resveratrol are also found to be superior (Elsheikh et al., 2018; Peng et al., 2018). Proliposomes formulated for gingerol are more effective than that of both the free phenolic compound to inhibit HepG2 cancer cells (Wang et al., 2018) and celastrol, a lipid-based carrier, tested in rabbits (Freag et al., 2018). Encapsulation of curcumin with NPs protein exhibits an adverse effect on the viability of MCF-7 cells with increased bioavailability in rat cells (Liu et al., 2017).

Nanoformuled form of curcumin, registered as Lipocurc® (liposomal, IV), has increased bioavailability and provides better treatment in cancer patients. However, its faster systematic elimination causes problems in treatment (Greil et al., 2018). Another formulation, Meriva®, a phytosomal complex of natural curcuminoids with lecithin, exhibits higher oral absorption and is reported to be effective for the metabolization of demethoxycurcumin (Cuomo et al., 2011) with diminishing oxidative stress and systemic inflammation in patients receiving chemotherapy (Panahi et al., 2014). Some adverse effects of radio- and chemotherapy have been reported (Belcaro et al., 2014), and a gemcitabine response increase in pancreatic cancer patients (Pastorelli et al., 2018). Quercetin nanoemulsion, supported by low doses of flavonoid and pemetrexed, presents an increase in membrane permeability and oral bioavailability antifolate against A549 tumors in mice (Pangeni et al., 2018). Inhibition of colon cancer cells *in vivo* is possible with the introduction of quercetin polymeric micelle orally and by injection (Chang et al., 2018). Studies of nanoparticles of phytochemicals are found to be more effective than that of their free form, depending on the type of cancer. Formation of nanoparticles with encapsulated EGCG in polylactic acid–polyethylene glycol (PLA–PEG) enhanced its antiangiogenic and chemopreventive proapoptotic potential (Siddiqui et al., 2009).

Nanoparticles formed by iron oxide core material and chitosan are able to release phytic acid slowly following a pseudo-second-order model kinetics, effective in impairing HT-29 colon cancer cells' viability (Barahuie et al., 2017; Soares et al., 2016). Compared to its free form, cyclodextrin-added carbon nanotubes also present cytotoxicity to MCF-7 cells and Hela cells *in vitro* (Liu et al., 2018). At concentrations above 10 μM, EGCG shows a growth inhibitory effect for melanoma cells, but its cytotoxic effect becomes clear after it is encapsulated as nanoparticles (Siddiqui et al., 2014).

Resveratrol-containing PLA–PEG nanoparticles are effective in reducing tumor growth, colony-forming capacity, and the number of cells in addition to the production of ROS and uptake of fluorodeoxyglucose, resulting in cell apoptosis in *in vitro* studies (Jung et al., 2015). Hyaluronic acid decorated PCL nanoparticles containing naringenin causes overexpressing in tumor cells (Parashar et al., 2018). Zein involved nanoparticles with oil-based fraction in the core of the particle, having exemestane and resveratrol, present high efficacy against breast cancer in mouse models and at cellular level (Elzoghby et al., 2017). A combination of resveratrol with other antioxidants gives encouraging results in *in vivo* studies against breast cancer promoted by 7,12-dimethylbenz[a] anthracene (Jain et al., 2017; Rehman et al., 2017). An increase in cytotoxicity with slow release is observed by the encapsulation of lycobetaine as liposomes and nanoemulsions (Chen et al., 2018). Encapsulated curcumin also shows *in vitro* cytotoxic action against A549 lung cancer cells (Ibrahim et al., 2018). Application of phosphatidylcholine liposomes on silica-coated iron oxide nanoparticles accompanied by doxorubicin show cytotoxicity against MCF-7 human breast adenocarcinoma and U87 glioblastoma cells under magnetic field application (Sharifabad et al., 2016).

10 Routes of administration for phytochemicals

Brain-targeted treatments have seen some encouraging results against cancer cells (Da Fonseca et al., 2013). Nasal introduction of kaempferol to rat brain as a mucoadhesive nanoemulsion is reported to be compatible with mucosa (Colombo et al., 2018). While low-dose free resveratrol is not detected in cerebrospinal fluid after 3 h, its equivalent encapsulated dose administrated by nasal introduction measured in micro gram per milliliter concentrations in less than 1 h, indicating that the microparticle carrier system uses a nose-CNS system, not a blood-brain barrier (Trotta et al., 2018). Aglycone flavonoids, due to their lipophilicity, can pass the blood-brain barrier (Youdim et al., 2003) by passive transcellular diffusion and downregulation or inhibition of breast cancer resistance protein (Kaur and Badhan, 2017). It is desirable to treat brain cancer by facilitating barrier penetration (Kaur and Badhan, 2017) and nanoparticles are effective in extending the retention times of phytochemicals in different regions of the brain (Tsai et al., 2011).

In addition to the brain, colon-targeted delivery systems have been found to be effective for resveratrol interventions. Encapsulated Ca-pectinate beads provide increased resveratrol capacity to prevent inflammation in ulcerative colitis model in rats and inhibit sphingosine kinase 1 in colon cancer implications (Abdin, 2013). Similarly, the protective action of alginate-chitosan microspheres carrying icariin (prenylated kaempferol) is presented by the accumulation of the vehicle in mouse colon (Wang et al., 2016). Alginate – as anionic polysaccharides – and pectin are commonly used oral vehicles to target the colon due to their bioadhesive properties and safety. Quercetin magnetic nanoparticles coated with PLGA have been shown to be effective against lung cancer in mice (Verma et al., 2013), while a combination of low-dose etopside and berberine with albumin-supported nanoparticles is suggested for deep lung deposition for two drugs having different release trends (Elgohary et al., 2018). Camptothecin-loaded PLGA nanoparticles are also found to be effective in preventing the development of adenovirus-induced vaginal cancer in mice (das Neves et al., 2015).

Compared to other delivery systems, transdermal delivery, due to having some advantages such as more efficient drug penetration, is particularly preferred for incidences of skin cancer (Liu et al., 2018). Sesamol-releasing gelatin-oleic acid nanoparticles applied through mice skin *in vitro* are effective against MCF-7 breast cancer (El Masry et al., 2018). Liposome-loaded curcumin, in combination with iontophoresis applied to porcine skin, is effective against human epidermoid cancer cells (Jose et al., 2017). EGCG results in a reduction in proliferation and invasion/migration potential for melanoma, as well as metastasis to distant tissues (Silva et al., 2017).

11 Combining phytochemicals with other applications

A combination of different applications for cancer treatment is more advantageous to gain the benefit of cumulative effects as having a different mode of action with reduced side effects is currently in demand. The chemosensitization effect of nanoparticles prepared by encapsulation of EGCG and theaflavin in PLGA is reported to be higher than that of the free compounds against HeLa cervical carcinoma, A549 lung carcinoma, THP-1 leukemia cells, and mice bearing Ehrlich ascites carcinoma cells by increasing the anticancer activity of cisplatin (Singh et al., 2015). Nanoparticles designed for simultaneous delivery of cisplatin and wortmannin have shown a superior therapeutic effect against platinum-sensitive and platinum-resistant ovarian cancer than the use of these compounds separately (Karve et al., 2012). Tailored nanoparticles with a combination of cisplatin and wortmannin are reported to be superior to that of free drug and single-drug-loaded nanoparticles against platinum-resistant and platinum-sensitive ovarian cancer cells (Zhang et al., 2018). Moreover, the superior effects of β-elemene plus celastrol (Zhang et al., 2019) and pemetrexed plus resveratrol (Abdelaziz et al., 2019), 10-hydroxycamptothecin in gold-zein nanoparticle complex (Mashhadi Malekzadeh et al., 2017), and gallic acid (Usman et al., 2018), in addition to nanoparticles/gadolinium contrast agents (Wang et al., 2018), use of apoferritin as curcumin carrier (Conti et al., 2016) against various forms of cancer cells are also reported.

12 Conclusion

Understanding cancer mechanisms is the key to inhibiting cancer growth and metastasis. Effects of dietary polyphenols as new chemopreventive and chemotherapeutic agents, either alone or in combination with medicine, epigenetic inhibitors, and/or other polyphenols might be a stronger and more effective anticancer approach. Phytochemicals, mostly depending on the applied dose, may themselves have an adverse effect, thus minimization of side effects should also be taken into consideration. Besides general targeting, strategies for specific delivery mechanisms, efficient and responsive carriers, and specially designed target molecules may provide better opportunities for specific treatments. Phytochemicals have huge potential for cancer treatment, but further studies will provide a better understanding as to whether phytochemicals will take the place of traditional cancer treatments with reduced side effects.

13 Where to look for further information

13.1 The following articles provide a good overview of the subject

- Kim, Y. S., Young, M. R., Bobe, G., Colburn, N. H., Milner, J. A. 2009. Bioactive food components, inflammatory targets, and cancer prevention. *Cancer Prevention Res. 2*(3), 200–208.
- Siddiqui, I. A., Adhami, V. M., Ahmad, N., Mukhtar, H. 2010. Nanochemoprevention: sustained release of bioactive food components for cancer prevention. *Nutr. Cancer* 62(7), 883–890.
- Ferguson, L. R., Philpott, M. 2007. Cancer prevention by dietary bioactive components that target the immune response. *Curr. Cancer Drug Targets* 7(5), 459–464.
- Kris-Etherton, P. M., Hecker, K. D., Bonanome, A., Coval, S. M., Binkoski, A. E., Hilpert, K. F., Griel, A. E. and Etherton, T. D. 2002. Bioactive compounds in foods: their role in the prevention of cardiovascular disease and cancer. *Am. J. Med.* 113(9), 71–88.
- Watson, R. R., Preedy, V. R. (Eds.). (2010). *Bioactive Foods and Extracts: Cancer Treatment and Prevention*. CRC Press.
- Patil, B. S., Jayaprakasha, G. K., Chidambara Murthy, K. N., Vikram, A. 2009. Bioactive compounds: historical perspectives, opportunities, and challenges. *J. Agric. Food Chem.* 57(18), 8142–8160.

13.2 Key research in this area can be found at the following organizations

- American Institute for Cancer Research.
- American Cancer Society.

14 References

Abdelaziz, H. M., Elzoghby, A. O., Helmy, M. W., Samaha, M. W., Fang, J. Y. and Freag, M. S. 2019. Liquid crystalline assembly for potential combinatorial chemo-herbal drug delivery to lung cancer cells. *Int. J. Nanomedicine* 14, 499–517. https://doi.org/10.2147/IJN.S188335.

Abdin, A. A. 2013. Targeting sphingosine kinase 1 (SphK1) and apoptosis by colon-specific delivery formula of resveratrol in treatment of experimental ulcerative colitis in rats. *Eur. J. Pharmacol.* 718(1-3), 145-153. https://doi.org/10.1016/j.ejphar.2013.08.040.

Aqil, F., Kausar, H., Agrawal, A. K., Jeyabalan, J., Kyakulaga, A. H., Munagala, R. and Gupta, R. 2016. Exosomal formulation enhances therapeutic response of celastrol against lung cancer. *Exp. Mol. Pathol.* 101(1), 12-21. https://doi.org/10.1016/j.yexmp.2016.05.013.

Asensi, M., Ortega, A., Mena, S., Feddi, F. and Estrela, J. M. 2011. Natural polyphenols in cancer therapy. *Crit. Rev. Clin. Lab. Sci.* 48(5-6), 197-216. https://doi.org/10.3109/10408363.2011.631268.

Ball, S., Arevalo, M., Juarez, E., Payne, J. D. and Jones, C. 2019. Breast cancer chemoprevention: An update on current practice and opportunities for primary care physicians. *Prev. Med.* 129, 105834. https://doi.org/10.1016/j.ypmed.2019.105834.

Barahuie, F., Dorniani, D., Saifullah, B., Gothai, S., Hussein, M. Z., Pandurangan, A. K., Arulselvan, P. and Norhaizan, M. E. 2017. Sustained release of anticancer agent phytic acid from its chitosan-coated magnetic nanoparticles for drug-delivery system. *Int. J. Nanomed.* 12, 2361-2372. https://doi.org/10.2147/IJN.S126245.

Baran, I., Ionescu, D., Filippi, A., Mocanu, M. M., Iftime, A., Babes, R., Tofolean, I. T., Irimia, R., Goicea, A., Popescu, V., Dimancea, A., Neagu, A. and Ganea, C. 2014. Novel insights into the antiproliferative effects and synergism of quercetin and menadione in human leukemia Jurkat T cells. *Leuk. Res.* 38(7), 836-849. https://doi.org/10.1016/j.leukres.2014.04.010.

Belcaro, G., Hosoi, M., Pellegrini, L., Appendino, G., Ippolito, E., Ricci, A., Ledda, A., Dugall, M., Cesarone, M. R., Maione, C., Ciammaichella, G., Genovesi, D. and Togni, S. 2014. A controlled study of a lecithinized delivery system of curcumin (Meriva®) to alleviate the adverse effects of cancer treatment. *Phytother. Res.* 28(3), 444-450. https://doi.org/10.1002/ptr.5014.

Bishayee, A. and Darvesh, A. S. 2012. Angiogenesis in hepatocellular carcinoma: A potential target for chemoprevention and therapy. *Curr. Cancer Drug Targets* 12(9), 1095-1118. https://doi.org/10.2174/156800961120910 95.

Bishayee, A. and Sethi, G. 2016. Bioactive natural products in cancer prevention and therapy: Progress and promise. *Semin. Cancer Biol.* 40-41, 1-3. https://doi.org/10.1016/j.semcancer.2016.08.006.

Branco, Cdos S., de Lima, É. D., Rodrigues, T. S., Scheffel, T. B., Scola, G., Laurino, C. C. F. C., Moura, S. and Salvador, M. 2015. Mitochondria and redox homoeostasis as chemotherapeutic targets of *Araucaria angustifolia* (Bert.) O. Kuntze in human larynx HEp-2 cancer cells. *Chem. Biol. Interact.* 231, 108-118. https://doi.org/10.1016/j.cbi.2015.03.005.

Chakraborty, S., Stalin, S., Das, N., Choudhury, S. T., Ghosh, S. and Swarnakar, S. 2012. The use of nano-quercetin to arrest mitochondrial damage and MMP-9 upregulation during prevention of gastric inflammation induced by ethanol in rat. *Biomaterials* 33(10), 2991-3001. https://doi.org/10.1016/j.biomaterials.2011.12.037.

Chang, C. E., Hsieh, C. M., Huang, S. C., Su, C. Y., Sheu, M. T. and Ho, H. O. 2018. Lecithin-stabilized polymeric micelles (LSBPMS) for delivering quercetin: Pharmacokinetic studies and therapeutic effects of quercetin alone and in combination with doxorubicin. *Sci. Rep.* 8(1), 17640. https://doi.org/10.1038/s41598-018-36162-0.

Chapa, A. M. and Mejía-Teniente, L. 2016. Capsaicin: From plants to a cancer-suppressing agent. *Molecules* 21(8), 931. https://doi.org/10.3390/molecules21080931.

Chen, T., Gong, T., Zhao, T., Fu, Y., Zhang, Z. and Gong, T. 2018. A comparison study between lycobetaine-loaded nanoemulsion and liposome using nRGD as therapeutic adjuvant for lung cancer therapy. *Eur. J. Pharm. Sci.* 111, 293–302. https://doi.org/10.1016/j.ejps.2017.09.041.

Colombo, M., Figueiró, F., de Fraga Dias, A., Teixeira, H. F., Battastini, A. M. O. and Koester, L. S. 2018. Kaempferol-loaded mucoadhesive nanoemulsion for intranasal administration reduces glioma growth in vitro. *Int. J. Pharm.* 543(1–2), 214–223. https://doi.org/10.1016/j.ijpharm.2018.03.055.

Conti, L., Lanzardo, S., Ruiu, R., Cadenazzi, M., Cavallo, F., Aime, S. and Crich, S. G. 2016. L-ferritin targets breast cancer stem cells and delivers therapeutic and imaging agents. *Oncotarget* 7(41), 66713–66727. https://doi.org/10.18632/oncotarget.10920.

Cuomo, J., Appendino, G., Dern, A. S., Schneider, E., McKinnon, T. P., Brown, M. J., Togni, S. and Dixon, B. M. 2011. Comparative absorption of a standardized curcuminoid mixture and its lecithin formulation. *J. Nat. Prod.* 74(4), 664–669. https://doi.org/10.1021/np1007262.

Da Fonseca, C. O., Teixeira, R. M., Silva, J. C., De Saldanha, D. A., Gama Fischer, J., Meirelles, O. C., Landeiro, J. A. and Quirico-Santos, T. 2013. Long-term outcome in patients with recurrent malignant glioma treated with perillyl alcohol inhalation. *Anticancer Res.* 33(12), 5625–5631.

das Neves, J., Nunes, R., Machado, A. and Sarmento, B. 2015. Polymer-based nanocarriers for vaginal drug delivery. *Adv. Drug Deliv. Rev.* 92, 53–70. https://doi.org/10.1016/j.addr.2014.12.004.

De Flora, S. and Ferguson, L. R. 2005. Overview of mechanisms of cancer chemopreventive agents. *Mutat. Res.* 591(1–2), 8–15. https://doi.org/10.1016/j.mrfmmm.2005.02.029.

Deng, S., Shanmugam, M. K., Kumar, A. P., Yap, C. T., Sethi, G. and Bishayee, A. 2019. Targeting autophagy using natural compounds for cancer prevention and therapy. *Cancer* 125(8), 1228–1246. https://doi.org/10.1002/cncr.31978.

Diederich, M. and Cerella, C. 2016. Non-canonical programmed cell death mechanisms triggered by natural compounds. *Semin. Cancer Biol.* 40–41, 4–34. https://doi.org/10.1016/j.semcancer.2016.06.001.

Doll, R. and Peto, R. 1981. The causes of cancer: Quantitative estimates of avoidable risks of cancer in the United States today. *J. Natl. Cancer Inst.* 66(6), 1191–1308. https://doi.org/10.1093/jnci/66.6.1192.

El-Ashmawy, N. E., Khedr, E. G., Ebeid, E. M., Salem, M. L., Zidan, A. A. and Mosalam, E. M. 2017. Enhanced anticancer effect and reduced toxicity of doxorubicin in combination with thymoquinone released from poly-N-acetyl glucosamine nanomatrix in mice bearing solid Ehrlish carcinoma. *Eur. J. Pharm. Sci.* 109, 525–532.

Elgohary, M. M., Helmy, M. W., Abdelfattah, E.-Z. A., Ragab, D. M., Mortada, S. M., Fang, J. Y. and Elzoghby, A. O. 2018. Targeting sialic acid residues on lung cancer cells by inhalable boronic acid-decorated albumin nanocomposites for combined chemo/herbal therapy. *J. Control. Release* 285, 230–243. https://doi.org/10.1016/j.jconrel.2018.07.014.

El Masry, S. R., Hathout, R. M., Abdel-Halim, M. and Mansour, S. 2018. *In vitro* transdermal delivery of sesamol using oleic acid chemically-modified gelatin nanoparticles as a potential breast cancer medication. *J. Drug Deliv. Sci. Technol.* 48, 30–39. https://doi.org/10.1016/j.jddst.2018.08.017.

Elsheikh, M. A., Elnaggar, Y. S. R., Hamdy, D. A. and Abdallah, O. Y. 2018. Novel cremochylomicrons for improved oral bioavailability of the antineoplastic phytomedicine berberine chloride: Optimization and pharmacokinetics. *Int. J. Pharm.* 535(1-2), 316-324. https://doi.org/10.1016/j.ijpharm.2017.11.023.

Elzoghby, A. O., El-Lakany, S. A., Helmy, M. W., Abu-Serie, M. M. and Elgindy, N. A. 2017. Shell-crosslinked zein nanocapsules for oral codelivery of exemestane and resveratrol in breast cancer therapy. *Nanomedicine (Lond)* 12(24), 2785-2805. https://doi.org/10.2217/nnm-2017-0247.

Errayes, A. O., Abdussalam-Mohammed, W. and Darwish, M. O. 2020. Review of phytochemical and medical applications of *Annona muricata* fruits. *J. Chem. Rev.* 2(1), 70-79. https://doi.org/10.33945/SAMI/JCR.2020.1.5.

Faridi Esfanjani, A., Assadpour, E. and Jafari, S. M. 2018. Improving the bioavailability of phenolic compounds by loading them within lipid-based nanocarriers. *Trends Food Sci. Technol.* 76, 56-66.

Freag, M. S., Saleh, W. M. and Abdallah, O. Y. 2018. Self-assembled phospholipid-based phytosomal nanocarriers as promising platforms for improving oral bioavailability of the anticancer celastrol. *Int. J. Pharm.* 535(1-2), 18-26. https://doi.org/10.1016/j.ijpharm.2017.10.053.

Gescher, A., Pastorino, U., Plummer, S. M. and Manson, M. M. 1998. Suppression of tumour development by substances derived from the diet–Mechanisms and clinical implications. *Br. J. Clin. Pharmacol.* 45(1), 1-12. https://doi.org/10.1046/j.1365-2125.1998.00640.x.

Golemis, E. A., Scheet, P., Beck, T. N., Scolnick, E. M., Hunter, D. J., Hawk, E. and Hopkins, N. 2018. Molecular mechanisms of the preventable causes of cancer in the United States. *Genes Dev.* 32(13-14), 868-902. https://doi.org/10.1101/gad.314849.118.

Gonzales, G. B., Raes, K., Coelus, S., Struijs, K., Smagghe, G. and Van Camp, J. 2014. Ultra(high)-pressure liquid chromatography-electrospray ionization-time-of-flight-ion mobility-high definition mass spectrometry for the rapid identification and structural characterization of flavonoid glycosides from cauliflower waste. *J. Chromatogr. A* 1323, 39-48. https://doi.org/10.1016/j.chroma.2013.10.077.

Greil, R., Greil-Ressler, S., Weiss, L., Schönlieb, C., Magnes, T., Radl, B., Bolger, G. T., Vcelar, B. and Sordillo, P. P. 2018. A phase 1 dose-escalation study on the safety, tolerability and activity of liposomal curcumin (Lipocurc™) in patients with locally advanced or metastatic cancer. *Cancer Chemother. Pharmacol.* 82(4), 695-706. https://doi.org/10.1007/s00280-018-3654-0.

Gullett, N. P., Ruhul Amin, A. R. M., Bayraktar, S., Pezzuto, J. M., Shin, D. M., Khuri, F. R., Aggarwal, B. B., Surh, Y. J. and Kucuk, O. 2010. Cancer prevention with natural compounds. *Semin. Oncol.* 37(3), 258-281. https://doi.org/10.1053/j.seminoncol.2010.06.014.

Huminiecki, L. and Horbańczuk, J. 2018. The functional genomic studies of resveratrol in respect to its anti-cancer effects. *Biotechnol. Adv.* 36(6), 1699-1708. https://doi.org/10.1016/j.biotechadv.2018.02.011.

Ibrahim, S., Tagami, T., Kishi, T. and Ozeki, T. 2018. Curcumin marinosomes as promising nano-drug delivery system for lung cancer. *Int. J. Pharm.* 540(1-2), 40-49. https://doi.org/10.1016/j.ijpharm.2018.01.051.

Irchhaiya, R., Kumar, A., Yadav, A., Gupta, N., Kumar, S., Gupta, N., Kumar, S., Yadav, V., Prakash, A. and Gurjar, H. 2015. Metabolites in plants and its classification. *World J Pharm. Pharm. Sci WJPPS* 4, 287-305.

Jain, S., Garg, T., Kushwah, V., Thanki, K., Agrawal, A. K. and Dora, C. P. 2017. α-Tocopherol as functional excipient for resveratrol and coenzyme Q10-loaded SNEDDS for improved bioavailability and prophylaxis of breast cancer. *J. Drug Target.* 25(6), 554–565. https://doi.org/10.1080/1061186X.2017.1298603.

Jose, A., Labala, S. and Venuganti, V. V. K. 2017. Co-delivery of curcumin and STAT3 siRNA using deformable cationic liposomes to treat skin cancer. *J. Drug Target.* 25(4), 330–341. https://doi.org/10.1080/1061186X.2016.1258567.

Jung, K. H., Lee, J. H., Park, J. W., Quach, C. H. T., Moon, S. H., Cho, Y. S. and Lee, K. H. 2015. Resveratrol-loaded polymeric nanoparticles suppress glucose metabolism and tumor growth in vitro and in vivo. *Int. J. Pharm.* 478(1), 251–257. https://doi.org/10.1016/j.ijpharm.2014.11.049.

Karve, S., Werner, M. E., Sukumar, R., Cummings, N. D., Copp, J. A., Wang, E. C., Li, C., Sethi, M., Chen, R. C., Pacold, M. E. and Wang, A. Z. 2012. Revival of the abandoned therapeutic wortmannin by nanoparticle drug delivery. *Proc. Natl. Acad. Sci. U. S. A.* 109(21), 8230–8235. https://doi.org/10.1073/pnas.1120508109.

Kashyap, D., Sharma, A., Tuli, H. S., Sak, K., Mukherjee, T. and Bishayee, A. 2018. Molecular targets of celastrol in cancer: Recent trends and advancements. *Crit. Rev. Oncol. Hematol.* 128, 70–81. https://doi.org/10.1016/j.critrevonc.2018.05.019.

Kasi, P. D., Tamilselvam, R., Skalicka-Woźniak, K., Nabavi, S. F., Daglia, M., Bishayee, A., Pazoki-Toroudi, H. and Nabavi, S. M. 2016. Molecular targets of curcumin for cancer therapy: An updated review. *Tumour Biol.* 37(10), 13017–13028. https://doi.org/10.1007/s13277-016-5183-y.

Kaur, M. and Badhan, R. K. S. 2017. Phytochemical mediated-modulation of the expression and transporter function of breast cancer resistance protein at the blood-brain barrier: An in-vitro study. *Brain Res.* 1654(A), 9–23. https://doi.org/10.1016/j.brainres.2016.10.020.

Kaur, V., Kumar, M., Kumar, A., Kaur, K., Dhillon, V. S. and Kaur, S. 2018. Pharmacotherapeutic potential of phytochemicals: Implications in cancer chemoprevention and future perspectives. *Biomed. Pharmacother.* 97, 564–586. https://doi.org/10.1016/j.biopha.2017.10.124.

Khan, N., Bharali, D. J., Adhami, V. M., Siddiqui, I. A., Cui, H., Shabana, S. M., Mousa, S. A. and Mukhtar, H. 2014. Oral administration of naturally occurring chitosan-based nanoformulated green tea polyphenol EGCG effectively inhibits prostate cancer cell growth in a xenograft model. *Carcinogenesis* 35(2), 415–423. https://doi.org/10.1093/carcin/bgt321.

Ko, E. Y. and Moon, A. 2015. Natural products for chemoprevention of breast cancer. *J. Cancer Prev.* 20(4), 223–231. https://doi.org/10.15430/JCP.2015.20.4.223.

Krajka-Kuźniak, V., Paluszczak, J., Szaefer, H. and Baer-Dubowska, W. 2015. The activation of the Nrf2/ARE pathway in HepG2 hepatoma cells by phytochemicals and subsequent modulation of phase II and antioxidant enzyme expression. *J. Physiol. Biochem.* 71(2), 227–238. https://doi.org/10.1007/s13105-015-0401-4.

Lagoa, R., Graziani, I., Lopez-Sanchez, C., Garcia-Martinez, V. and Gutierrez-Merino, C. 2011. Complex I and cytochrome C are molecular targets of flavonoids that inhibit hydrogen peroxide production by mitochondria. *Acta BBA: Bioenerg.* 1807(12), 1562–1572. https://doi.org/10.1016/j.bbabio.2011.09.022.

Lagoa, R., Samhan-Arias, A. K. and Gutierrez-Merino, C. 2017. Correlation between the potency of flavonoids for cytochrome C reduction and inhibition of cardiolipin-induced peroxidase activity. *Biofactors (Oxf. Engl.)* 43(3), 451–468. https://doi.org/10.1002/biof.1357.

Lagoa, R., Silva, J., Rodrigues, J. R. and Bishayee, A. 2020. Advances in phytochemical delivery systems for improved anticancer activity. *Biotechnol. Adv.* 38, 107382. https://doi.org/10.1016/j.biotechadv.2019.04.004.

Lambert, J. D. and Elias, R. J. 2010. The antioxidant and pro-oxidant activities of green tea polyphenols: A role in cancer prevention. *Arch. Biochem. Biophys.* 501(1), 65–72. https://doi.org/10.1016/j.abb.2010.06.013.

Li, L., Ni, J., Li, M., Chen, J., Han, L., Zhu, Y., Kong, D., Mao, J., Wang, Y., Zhang, B., Zhu, M., Gao, X. and Fan, G. 2017. Ginsenoside Rg3 micelles mitigate doxorubicin-induced cardiotoxicity and enhance its anticancer efficacy. *Drug Deliv.* 24(1), 1617–1630. https://doi.org/10.1080/10717544.2017.1391893.

Liu, C., Yang, X., Wu, W., Long, Z., Xiao, H., Luo, F., Shen, Y. and Lin, Q. 2017. Elaboration of curcumin-loaded rice bran albumin nanoparticles formulation with increased in vitro bioactivity and in vivo bioavailability. *Food Hydrocoll.* 77, 834–842. https://doi.org/10.1016/j.foodhyd.2017.11.027.

Liu, X., Xu, D., Liao, C., Fang, Y. and Guo, B. 2018. Development of a promising drug delivery for formononetin: Cyclodextrin-modified single-walled carbon nanotubes. *J. Drug Deliv. Sci. Technol.* 43, 461–468. https://doi.org/10.1016/j.jddst.2017.11.018.

Mahbub, A. A., Le Maitre, C. L., Haywood-Small, S. L., Cross, N. A. and Jordan-Mahy, N. 2015. Polyphenols act synergistically with doxorubicin and etoposide in leukaemia cell lines. *Cell Death Discov.* 1, 15043. https://doi.org/10.1038/cddiscovery.2015.43.

Maji, S., Panda, S., Samal, S. K., Shriwas, O., Rath, R., Pellecchia, M., Emdad, L., Das, S. K., Fisher, P. B. and Dash, R. 2018. Bcl-2 antiapoptotic family proteins and chemoresistance in cancer. *Adv. Cancer Res.* 137, 37–75. https://doi.org/10.1016/bs.acr.2017.11.001.

Manson, M. M., Gescher, A., Hudson, E. A., Plummer, S. M., Squires, M. S. and Prigent, S. A. 2000. Blocking and suppressing mechanisms of chemoprevention by dietary constituents. *Toxicol. Lett.* 112–113, 499–505. https://doi.org/10.1016/s0378-4274(99)00211 8.

Mashhadi Malekzadeh, A., Ramazani, A., Tabatabaei Rezaei, S. J. and Niknejad, H. 2017. Design and construction of multifunctional hyperbranched polymers coated magnetite nanoparticles for both targeting magnetic resonance imaging and cancer therapy. *J. Colloid Interface Sci.* 490, 64–73. https://doi.org/10.1016/j.jcis.2016.11.014.

Mercader, A. G. and Pomilio, A. B. 2012. (Iso)flav(an)ones, chalcones, catechins, and theaflavins as anticarcinogens: Mechanisms, anti-multidrug resistance and QSAR studies. *Curr. Med. Chem.* 19(25), 4324–4347. https://doi.org/10.2174/092986712802884277.

Milner, J. A., McDonald, S. S., Anderson, D. E. and Greenwald, P. 2001. Molecular targets for nutrients involved with cancer prevention. *Nutr. Cancer* 41(1–2), 1–16. https://doi.org/10.1080/01635581.2001.9680606.

Molyneux, R. J., Lee, S. T., Gardner, D. R., Panter, K. E. and James, L. F. 2007. Phytochemicals: The good, the bad and the ugly? *Phytochemistry* 68(22–24), 2973–2985. https://doi.org/10.1016/j.phytochem.2007.09.004.

Muddineti, O. S., Kumari, P., Ghosh, B., Torchilin, V. P. and Biswas, S. 2017. D-α-tocopheryl succinate/phosphatidyl ethanolamine conjugated amphiphilic polymer-based nanomicellar system for the efficient delivery of curcumin and to overcome multiple drug resistance in cancer. *ACS Appl. Mater. Interfaces* 9(20), 16778–16792. https://doi.org/10.1021/acsami.7b01087.

Panahi, Y., Saadat, A., Beiraghdar, F. and Sahebkar, A. 2014. Adjuvant therapy with bioavailability-boosted curcuminoids suppresses systemic inflammation and improves quality of life in patients with solid tumors: A randomized double-blind placebo-controlled trial. *Phytother. Res.* 28(10), 1461–1467. https://doi.org/10.1002/ptr.5149.

Pangeni, R., Panthi, V. K., Yoon, I. S. and Park, J. W. 2018. Preparation, characterization, and in vivo evaluation of an oral multiple nanoemulsive system for co-delivery of pemetrexed and quercetin. *Pharmaceutics* 10(3), 158. https://doi.org/10.3390/pharmaceutics10030158.

Parashar, P., Rathor, M., Dwivedi, M. and Saraf, S. A. 2018. Hyaluronic acid decorated naringenin nanoparticles: Appraisal of chemopreventive and curative potential for lung cancer. *Pharmaceutics* 10(1), 33. https://doi.org/10.3390/pharmaceutics10010033.

Pastorelli, D., Fabricio, A. S. C., Giovanis, P., D'Ippolito, S., Fiduccia, P., Soldà, C., Buda, A., Sperti, C., Bardini, R., Da Dalt, G., Rainato, G., Gion, M. and Ursini, F. 2018. Phytosome complex of curcumin as complementary therapy of advanced pancreatic cancer improves safety and efficacy of gemcitabine: Results of a prospective phase II trial. *Pharmacol. Res.* 132, 72–79. https://doi.org/10.1016/j.phrs.2018.03.013.

Peng, R. M., Lin, G. R., Ting, Y. and Hu, J. Y. 2018. Oral delivery system enhanced the bioavailability of stilbenes: Resveratrol and pterostilbene. *Biofactors (Oxf. Engl.)* 44(1), 5–15. https://doi.org/10.1002/biof.1405.

Pichersky, E. and Gang, D. R. 2000. Genetics and biochemistry of secondary metabolites in plants: An evolutionary perspective. *Trends Plant Sci.* 5(10), 439–445. https://doi.org/10.1016/S1360-1385(00)01741-6.

Pool, H., Campos-Vega, R., Herrera-Hernández, M. G., García-Solis, P., García-Gasca, T., Sánchez, I. C., Luna-Bárcenas, G. and Vergara-Castañeda, H. 2018. Development of genistein-pegylated silica hybrid nanomaterials with enhanced antioxidant and antiproliferative properties on HT29 human colon cancer cells. *Am. J. Transl. Res.* 10(8), 2306–2323.

Ranjan, A., Ramachandran, S., Gupta, N., Kaushik, I., Wright, S., Srivastava, S., Das, H., Srivastava, S., Prasad, S. and Srivastava, S. K. 2019. Role of phytochemicals in cancer prevention. *Int. J. Mol. Sci.* 20(20), 4981. https://doi.org/10.3390/ijms20204981.

Rehman, F. U., Shah, K. U., Shah, S. U., Khan, I. U., Khan, G. M. and Khan, A. 2017. From nanoemulsions to self-nanoemulsions, with recent advances in self-nanoemulsifying drug delivery systems (SNEDDS). *Expert Opin. Drug Deliv.* 14(11), 1325–1340. https://doi.org/10.1080/17425247.2016.1218462.

Rengasamy, K. R. R., Khan, H., Gowrishankar, S., Lagoa, R. J. L., Mahomoodally, F. M., Khan, Z., Suroowan, S., Tewari, D., Zengin, G., Hassan, S. T. S. and Pandian, S. K. 2019. The role of flavonoids in autoimmune diseases: Therapeutic updates. *Pharmacol. Ther.* 194, 107–131. https://doi.org/10.1016/j.pharmthera.2018.09.009.

Samadi, A. K., Bilsland, A., Georgakilas, A. G., Amedei, A., Amin, A., Bishayee, A., Azmi, A. S., Lokeshwar, B. L., Grue, B., Panis, C., Boosani, C. S., Poudyal, D., Stafforini, D. M., Bhakta, D., Niccolai, E., Guha, G., Vasantha Rupasinghe, H. P., Fujii, H., Honoki, K., Mehta, K., Aquilano, K., Lowe, L., Hofseth, L. J., Ricciardiello, L., Ciriolo, M. R., Singh, N., Whelan, R. L., Chaturvedi, R., Ashraf, S. S., Shantha Kumara, H. M. C.,

Nowsheen, S., Mohammed, S. I., Keith, W. N., Helferich, W. G. and Yang, X. 2015. A multi-targeted approach to suppress tumor-promoting inflammation. *Semin. Cancer Biol.* 35 (Suppl.), S151-S184. https://doi.org/10.1016/j.semcancer.2015.03.006.

Santhi, K. and Sengottuvel, R. 2016. Qualitative and quantitative phytochemical analysis of Moringa concanensis Nimmo. *Int. J. Curr. Microbiol. Appl. Sci.* 5(1), 633-640. https://doi.org/10.20546/ijcmas.2016.501.064.

Shahidi, F. and Ambigaipalan, P. 2015. Phenolics and polyphenolics in foods, beverages and spices: Antioxidant activity and health effects - A review. *J. Funct. Foods Nat. Antioxid.* 18, 820-897. https://doi.org/10.1016/j.jff.2015.06.018.

Shanmugam, M. K., Lee, J. H., Chai, E. Z. P., Kanchi, M. M., Kar, S., Arfuso, F., Dharmarajan, A., Kumar, A. P., Ramar, P. S., Looi, C. Y., Mustafa, M. R., Tergaonkar, V., Bishayee, A., Ahn, K. S. and Sethi, G. 2016. Cancer prevention and therapy through the modulation of transcription factors by bioactive natural compounds. *Semin. Cancer Biol.* 40-41, 35-47. https://doi.org/10.1016/j.semcancer.2016.03.005.

Sharifabad, M. E., Mercer, T. and Sen, T. 2016. Drug-loaded liposome-capped mesoporous core-shell magnetic nanoparticles for cellular toxicity study. *Nanomedicine (Lond)* 11(21), 2757-2767. https://doi.org/10.2217/nnm-2016-0248.

Siddiqui, I. A., Adhami, V. M., Bharali, D. J., Hafeez, B. B., Asim, M., Khwaja, S. I., Ahmad, N., Cui, H., Mousa, S. A. and Mukhtar, H. 2009. Introducing nanochemoprevention as a novel approach for cancer control: Proof of principle with green tea polyphenol epigallocatechin-3-gallate. *Cancer Res.* 69(5), 1712-1716. https://doi.org/10.1158/0008-5472.CAN-08-3978.

Siddiqui, I. A., Bharali, D. J., Nihal, M., Adhami, V. M., Khan, N., Chamcheu, J. C., Khan, M. I., Shabana, S., Mousa, S. A. and Mukhtar, H. 2014. Excellent anti-proliferative and pro-apoptotic effects of (-)-epigallocatechin-3-gallate encapsulated in chitosan nanoparticles on human melanoma cell growth both in vitro and in vivo. *Nanomedicine* 10(8), 1619-1626. https://doi.org/10.1016/j.nano.2014.05.007.

Silva, J., Videira, P. and Lagoa, R. 2017. Bioactivity gradients of cytoprotective and anticancer catechins in skin: Simulation studies for the design of controlled release systems. In: *Presented at the 2017 IEEE 5th Port. Meeting on Bioengineering (ENBENG)*, vol. 2017, pp. 1-4. https://doi.org/10.1109/ENBENG.2017.7889467.

Singh, M., Bhatnagar, P., Mishra, S., Kumar, P., Shukla, Y. and Gupta, K. C. 2015. PLGA-encapsulated tea polyphenols enhance the chemotherapeutic efficacy of cisplatin against human cancer cells and mice bearing Ehrlich ascites carcinoma. *Int. J. Nanomedicine* 10, 6789-6809. https://doi.org/10.2147/IJN.S79489.

Sinha, D., Biswas, J., Nabavi, S. M. and Bishayee, A. 2017. Tea phytochemicals for breast cancer prevention and intervention: From bench to bedside and beyond. *Semin. Cancer Biol.* 46, 33-54. https://doi.org/10.1016/j.semcancer.2017.04.001.

Soares, P. I. P., Sousa, A. I., Ferreira, I. M. M., Novo, C. M. M. and Borges, J. P. 2016. Towards the development of multifunctional chitosan-based iron oxide nanoparticles: Optimization and modelling of doxorubicin release. *Carbohydr. Polym.* 153, 212-221. https://doi.org/10.1016/j.carbpol.2016.07.109.

Sung, B., Prasad, S., Yadav, V. R. and Aggarwal, B. B. 2012. Cancer cell signaling pathways targeted by spice-derived nutraceuticals. *Nutr. Cancer* 64(2), 173-197. https://doi.org/10.1080/01635581.2012.630551.

Sung, H., Ferlay, J., Siegel, R. L., Laversanne, M., Soerjomataram, I., Jemal, A. and Bray, F. 2021. Global cancer statistics 2020: GLOBOCAN estimates of incidence and mortality worldwide for 36 cancers in 185 countries. *CA Cancer J. Clin.* 71(3), 209-249. https://doi.org/10.3322/caac.21660.

Suphim, B., Prawan, A., Kukongviriyapan, U., Kongpetch, S., Buranrat, B. and Kukongviriyapan, V. 2010. Redox modulation and human bile duct cancer inhibition by curcumin. *Food Chem. Toxicol.* 48(8-9), 2265-2272. https://doi.org/10.1016/j.fct.2010.05.059.

Surh, Y. J. 1999. Molecular mechanisms of chemopreventive effects of selected dietary and medicinal phenolic substances. *Mutat. Res.* 428(1-2), 305-327. https://doi.org/10.1016/s1383-5742(99)00057-5.

Surh, Y. J. 2003. Cancer chemoprevention with dietary phytochemicals. *Nat. Rev. Cancer* 3(10), 768-780. https://doi.org/10.1038/nrc1189.

Sznarkowska, A., Kostecka, A., Meller, K. and Bielawski, K. P. 2016. Inhibition of cancer antioxidant defense by natural compounds. *Oncotarget* 8(9), 15996-16016. https://doi.org/10.18632/oncotarget.13723.

Taniguchi, K. and Karin, M. 2018. NF-κB, inflammation, immunity and cancer: Coming of age. *Nat. Rev. Immunol.* 18(5), 309-324. https://doi.org/10.1038/nri.2017.142.

Thoppil, R. J. and Bishayee, A. 2011. Terpenoids as potential chemopreventive and therapeutic agents in liver cancer. *World J. Hepatol.* 3(9), 228-249. https://doi.org/10.4254/wjh.v3.i9.228.

Trotta, V., Pavan, B., Ferraro, L., Beggiato, S., Traini, D., Des Reis, L. G., Scalia, S. and Dalpiaz, A. 2018. Brain targeting of resveratrol by nasal administration of chitosan-coated lipid microparticles. *Eur. J. Pharm. Biopharm.* 127, 250-259. https://doi.org/10.1016/j.ejpb.2018.02.010.

Tsai, Y. M., Chien, C. F., Lin, L. C. and Tsai, T. H. 2011. Curcumin and its nano-formulation: The kinetics of tissue distribution and blood-brain barrier penetration. *Int. J. Pharm.* 416(1), 331-338. https://doi.org/10.1016/j.ijpharm.2011.06.030.

Usman, M. S., Hussein, M. Z., Kura, A. U., Fakurazi, S., Masarudin, M. J. and Saad, F. F. A. 2018. Synthesis and characterization of protocatechuic acid-loaded gadolinium-layered double hydroxide and gold nanocomposite for theranostic application. *Appl. Nanosci.* 8(5), 973-986. https://doi.org/10.1007/s13204-018-0752-6.

Verma, N. K., Crosbie-Staunton, K., Satti, A., Gallagher, S., Ryan, K. B., Doody, T., McAtamney, C., MacLoughlin, R., Galvin, P., Burke, C. S., Volkov, Y. and Gun'ko, Y. K. 2013. Magnetic core-shell nanoparticles for drug delivery by nebulization. *J. Nanobiotechnology* 11, 1. https://doi.org/10.1186/1477-3155-11-1.

Vogel, V. G., Costantino, J. P., Wickerham, D. L., Cronin, W. M., Cecchini, R. S., Atkins, J. N., Bevers, T. B., Fehrenbacher, L., Pajon, E. R., Wade, J. L., Robidoux, A., Margolese, R. G., James, J., Runowicz, C. D., Ganz, P. A., Reis, S. E., McCaskill-Stevens, W., Ford, L. G., Jordan, V. C., Wolmark, N. and National Surgical Adjuvant Breast and Bowel Project 2010. Update of the national surgical adjuvant breast and bowel project study of tamoxifen and raloxifene (STAR) P-2 Trial: Preventing breast cancer. *Cancer Prev. Res. PA* 3(6), 696-706. https://doi.org/10.1158/1940-6207.CAPR-10-0076.

Wang, Q., Wei, Q., Yang, Q., Cao, X., Li, Q., Shi, F., Tong, S. S., Feng, C., Yu, Q., Yu, J. and Xu, X. 2018. A novel formulation of [6]-gingerol: Proliposomes with enhanced oral bioavailability and antitumor effect. *Int. J. Pharm.* 535(1-2), 308-315. https://doi.org/10.1016/j.ijpharm.2017.11.006.

Wang, Q. S., Wang, G. F., Zhou, J., Gao, L. N. and Cui, Y. L. 2016. Colon targeted oral drug delivery system based on alginate-chitosan microspheres loaded with icariin in the treatment of ulcerative colitis. *Int. J. Pharm.* 515(1–2), 176–185. https://doi.org/10.1016/j.ijpharm.2016.10.002.

Wattenberg, L. W. 1985. Chemoprevention of cancer. *Cancer Res.* 45(1), 1–8.

Wong, V. K.-W., Law, B. Y.-K., Yao, X. J., Chen, X., Xu, S. W., Liu, L. and Leung, E. L.-H. 2016. Advanced research technology for discovery of new effective compounds from Chinese herbal medicine and their molecular targets. *Pharmacol. Res.* 111, 546–555. https://doi.org/10.1016/j.phrs.2016.07.022.

Xie, J., Yang, Z., Zhou, C., Zhu, J., Lee, R. J. and Teng, L. 2016. Nanotechnology for the delivery of phytochemicals in cancer therapy. *Biotechnol. Adv.* 34(4), 343–353. https://doi.org/10.1016/j.biotechadv.2016.04.002.

Youdim, K. A., Dobbie, M. S., Kuhnle, G., Proteggente, A. R., Abbott, N. J. and Rice-Evans, C. 2003. Interaction between flavonoids and the blood-brain barrier: In vitro studies. *J. Neurochem.* 85(1), 180–192. https://doi.org/10.1046/j.1471-4159.2003.01652.x.

Zhang, M., Hagan, C. T., Min, Y., Foley, H., Tian, X., Yang, F., Mi, Y., Au, K. M., Medik, Y., Roche, K., Wagner, K., Rodgers, Z. and Wang, A. Z. 2018. Nanoparticle co-delivery of wortmannin and cisplatin synergistically enhances chemoradiotherapy and reverses platinum resistance in ovarian cancer models. *Biomaterials* 169, 1–10. https://doi.org/10.1016/j.biomaterials.2018.03.055.

Zhang, Q., Tian, X. and Cao, X. 2019. Transferrin-functionalised microemulsion co-delivery of β-elemene and celastrol for enhanced anti-lung cancer treatment and reduced systemic toxicity. *Drug Deliv. Transl. Res.* 9(3), 667–678. https://doi.org/10.1007/s13346-019-00623-4.

Chapter 9

Advances in understanding the role of plant phytochemicals in preventing cardiovascular disease

Nicholas J. Sadgrove and Monique S. J. Simmonds, Royal Botanic Gardens – Kew, UK

1 Introduction

Cardiovascular disease (CVD) is the biggest challenge to human health in the modern era (Flora and Nayak, 2019) and has become steadily more prevalent during the twentieth century. Today it is understood that CVD has its origins partly in high-protein atherogenic diets. These diets create epigenetic modifications that are passed down and accumulated in successive generations (Komal et al., 2021), increasing genetic susceptibility to CVD in family lineages over time (Zhang et al., 2018).

The epidemiology of CVD is described in terms of diagnosis and mortality rates. The average global rate of diagnosis is still rising. In developed nations, where the highest percentage of citizens with CVD occurs, it has plateaued. In contrast, in developing nations, whilst the average percentage of citizens with CVD is lower, the number of new cases is rising rapidly. In terms of mortality, those living with CVD in developed nations are living longer. In the

http://dx.doi.org/10.19103/AS.2022.0101.09

UK and USA, mortality from CVD peaked in the 1960s then started to decline, continuing to the present day (Jones and Greene, 2013). This improvement in survival is due to the availability of effective treatments (Bhatnagar et al., 2016).

The availability of modern pharmaceuticals is an important contributor to patient survival. However, with improved health education, people are also learning to make better dietary choices to prevent rather than live with the disease. The dietary choices of younger generations in developed countries, in particular, are changing and are projected to reduce rates of CVD diagnosis further as these groups age. Now is as good a time as ever to reintroduce healthier plant-based foods or nutraceuticals into the marketplace, given that a new health paradigm (focused more on prevention through diet and other lifestyle changes) is taking shape amongst consumers.

A recent systematic review concluded that the three plant foods with the strongest evidence base in CVD prevention (prophylaxis) and treatment are tomato, cranberry, and pomegranate (Rouhi-Boroujeni et al., 2017). These species are rich in phytochemicals that are protective against CVD, such as lycopene in tomato (Thies et al., 2016), anthocyanins in cranberry (Cassidy et al., 2016) and phloroglucinol in pomegranate (Chang et al., 2012). However, there are substantially more plant-food candidates with a strong theoretical case for protection against CVD that still require a stronger epidemiological evidence base to establish a clear benefit. With a greater focus on prevention through diet, the modern health paradigm regards fruits and vegetables as important in CVD prophylaxis (Liu et al., 2020) although, for now, nuts continue to be at the wrong end of the food pyramid of healthy foods, mainly because of their fat content.

2 Cardiovascular disease and diet

CVD is one of the many lifestyle diseases afflicting people in both the developed and the developing world. While its prevalence was initially considered a disease of the 'western world', a recent rise in incidence in other countries, such as China (Cheng, 2012), Iran (Mirmirani et al., 2017), and sub-Saharan Africa (Bigna and Noubiap, 2019), has occurred, lagging behind the trend in the western world. One of several theories to explain the increase of CVD in countries is the adoption of a 'western diet' and lifestyle by those societies as they become wealthier and more urbanised (Casas et al., 2018).

The western diet is characterised by highly satiating foods due to their high fat, sugar, or carbohydrate content. Foods are often processed to remove components with less sensory appeal and refined for maximum satiety, characterised by feeling full at a faster rate. Consequently, western diets can be low in minerals, vitamins, fibres, and plant phytochemicals that normally help our bodies to cope with the sugar/starch and fat load (Carrera-Bastos et al., 2011).

Lifestyles for many western consumers involve regular eating and less regular exercise as well as other factors such as stress which then contribute to CVD risk. Regular caloric loading eliminates periods of detoxification between meals (intermittent fasting), which promotes accumulation of highly reactive by-products of sugar metabolism (starches become sugars in digestion) (Johnson et al., 2017). These toxic by-products are known as reactive oxygen species (ROS) (Panth et al., 2016) and advanced glycation end products (AGEs) (Hegab et al., 2012). They promote fatty liver and inflammatory effects in tissues, which creates insulin resistance, from which further problems such as CVD develop (Hegab et al., 2012; Panth et al., 2016).

When it was realized that plant phytochemicals could protect against many of the negative effects of western diets and lifestyles, researchers started seeking ways to put these phytochemicals back into the diet, as a nutraceutical additive, as a food supplement, or as a 'superfood' like kale. Dietary interventions against CVD such as these target high blood pressure, high cholesterol (hyperlipidaemia), uraemia, diabetes, and obesity.

An example of a modern dietary intervention is to fortify common foods with plant phytochemicals for their prophylactic effects in preventing CVD. There are e.g. many kinds of butter or spreads available in the market that advertise a high sterol content, a plant phytochemical class that has been proven to lower cholesterol. However, fortification may be insufficient on its own, considering that consumers who use prophylactic supplements may also continue to indulge in unhealthy items under the false assumption they are protected by a 'phytochemical safety net'.

The industry does not always invest wholly in the fortification strategy either. Often the phytochemical content of a fortified food item is less than needed to be of true benefit (Sadgrove, 2021; Sadgrove and Jones, 2019). For this and many other reasons, researchers are increasingly arguing that fortification and supplementation are less effective than altering the diet to consume more of the actual foods that contain these phytochemicals (Chen et al., 2019). Isolated phytochemicals sold as nutraceuticals are often not as effective as the whole plant (Koss-Mikolajczyk et al., 2019; Yuan et al., 2017). In contrast, to supplement use or fortification, where people may continue to make bad dietary choices, adding healthier foods to the diet involves substitution of good for bad food choices. There are also many examples where plant phytochemicals are not as available in a food supplement when compared to the original food source (Sadgrove and Jones, 2019; Sadgrove, 2021). Whilst there is still a place for supplementation and fortification, by far the best choice is to modify one's diet to include plant-based ingredients that have protective effects against CVD.

3 Cardiovascular disease and its comorbidities

3.1 Characteristics of cardiovascular diseases

Cardiovascular degeneration has three major characteristics that form the basis of several diseases. They include either:

- hardening of arteries;
- contrasting softening of artery walls; and
- blood clotting.

Hardening of the arteries is known broadly as atherosclerotic cardiovascular disease, which involves a plaque build up inside arteries. The plaque is built from fat, cholesterol, and calcium, creating a hard insoluble layer that is permanent unless surgically rectified. Over time the plaque layers make the arteries narrower and limit circulation of oxygen-rich blood to organs, creating heart problems or numbness in the extremities. The diseases that are associated with atherosclerosis include coronary heart disease (e.g. angina, heart attacks, and heart failure; i.e. myocardial infarction) and peripheral arterial disease (e.g. numbness, ulcers, and cramping) (Golledge and Norman, 2010).

The second major characteristic of CVD, softening or weakening of the arteries, generally leads to the formation of aneurysms. An aneurysm is a bulge in the artery wall that can burst and cause potentially fatal internal bleeding. The diseases caused by aneurysm include strokes in general, transient ischaemic attack (mini stroke), and aortic disease (aortic aneurysm). While the character of an aneurysm is opposite to the hardening of arteries of atherosclerosis, the two are not mutually exclusive. Atherosclerotic cardiovascular disease and aneurysm commonly occur together and theories centre on whether one causes the other if they are the same disease or whether they are caused by the same lifestyle factors (Golledge and Norman, 2010).

The third cause, blood clotting or 'thrombosis' (Jackson, 2011), frequently occurs with both atherosclerotic disease and aneurysm. Thrombosis can create a disease that is seemingly like the effects of a ruptured aneurysm or atherosclerotic condition, creating symptomatic overlap between the first two causes of CVD. Hence, actual thrombotic diseases include heart attack, strokes in general, transient ischaemic attack (mini stroke), and peripheral arterial disease (e.g. painful, discoloured, and cold limbs).

Although the three major causes of CVD can create similar disease symptoms, they require dramatically different treatment measures. This has significant implications for the effectiveness of therapies between ethnic groups because the causes of CVD differ between different races and ethnic groups (Agyemang et al., 2009). Racial disparities in CVD pathogenesis may be partly related to differing phases of globalization across countries that have altered diets and lifestyles at different points in time for each race or ethnic group. This is an important consideration because it takes time to create disease in populations, in fact several generations. The chances of developing CVD increase in family lineages that pass on and accumulate epigenetic modifications caused by obesogenic diets (Komal et al., 2021).

3.2 Comorbidities

There are several diseases that are either associated or directly linked to the pathogenesis of CVD. These diseases are termed comorbidities and include insulin resistance, metabolic syndrome (Lind et al., 2021), diabetes (Einarson et al., 2018), osteoporosis or loss of bone mineral density (Wen et al., 2018), subclinical magnesium deficiency (Di Nicolantonio et al., 2018), kidney disease or uraemia (Fujii et al., 2016), hypertension and dyslipidaemia (Petrie et al., 2018), inflammatory bowel disease (Bigeh et al., 2020), and chronic obstructive pulmonary disease (Rabe et al., 2018). One group of authors argues that all comorbidities generate systemic inflammation and propose that all types of inflammation can lead to heart disease (Bigeh et al., 2020).

There is also a strong link between gastrointestinal bacterial dysbiosis and CVD (Jin et al., 2018). Dysbiosis may be defined as a condition involving an imbalance of gut bacteria, triggering a wide range of digestive disturbance symptoms. These include milder complaints, such as bloating, constipation, cramping, diarrhoea, and indigestion, and more severe conditions, such as 'small intestinal bacterial overgrowth', inflammatory bowel disease, chronic inflammation, and progression into cancers.

A healthy cardiovascular and renal system has increasingly been seen as dependent upon the crosstalk in the gut-kidney-heart 'triangle', and it is becoming evident that gut microbiota is a strong participant in this crosstalk. Hence, disruption to the harmony of the gut negatively affects the renal or cardiovascular system, and conversely, disruption to the renal or cardiovascular system negatively affects the gut, creating a feedback loop that has come to be known as the 'vicious cycle' (Onal et al., 2019).

The gut microbiome in CVD is also related to the modulation of local and systemic inflammation profiles. Inflammation can be caused by leakage of bacterial lipopolysaccharides into and across the intestinal mucosal or epithelial barrier (Onal et al., 2019). In cases of more severe disturbance to the intestinal epithelial barrier function, live bacteria escape the gut lumen and translocate into the systemic circulation, contributing to atherosclerotic symptoms and myocardial infarction (Zhou et al., 2018). As a result, one of the key aspects of pre- or probiotic use in prophylaxis or treatment of CVD is the strengthening of the intestinal epithelial barrier via the nurturing of commensal gut bacteria (Ohland and Macnoughton, 2010).

Whilst many diseases are confirmed comorbidities of CVD, the connection between disease states is complex. Whilst there can be a causal link between disease states, they might be merely coincident, or they may share a common pathogenesis pathway. As a result, comorbidities that are demonstrated to be interdependent triggers of each other (A leads to B, and B leads to A) are sometimes referred to as being on an 'axis'. Research suggests there is a gut-brain axis (Mukhtar et al., 2019), a gut-renal axis (Yang et al., 2017), and a neuro-immune axis (Bonaz et al., 2017). The gut-renal axis is one of the more significant interactions in the context of CVD because it has the power to control hypertension (Yang et al., 2017). There is mounting evidence that these three axes are all significant in the development of CVD (Onal et al., 2019).

3.3 Plants as prebiotics

Although both pre- and probiotics have been considered as promising prophylactic or therapeutic measures to combat CVD, the use of plant-based phytochemicals is limited to prebiotic effects. Probiotics are usually made by growing 'good' bacteria in controlled conditions, whereas prebiotics are plant sources that are considered to stimulate the growth and activity of 'good' bacteria in the gut, boosting their population density relative to other bacteria (La Fata et al., 2017).

Plant prebiotics can include vitamins, specialised metabolites, and polymers. Plant-based polymers are currently the main focus of current prebiotic research, occurring either as non-digestible yet water-soluble carbohydrates or polyphenol polymers (Lamuel-Raventos and Onge, 2017). Plant foods that are rich in polyphenol polymers, such as ellagitannins and procyanidins, are considered good prebiotics. Such foods include nuts, particularly almonds (Liu et al., 2014), as well as berries and grapes.

Water-soluble carbohydrate polymers are commonly referred to as 'dietary fibre' and have many chemical versions, such as fructooligosacharide from yacon (Padilla-González et al., 2020b), galactomannan from fenugreek (Hamden et al., 2010), glucomannan from konjac (Al-Chazzewi and Tester, 2010), acemannan from *Aloe vera* (Quezada et al., 2017), or β-glucan recovered from wheat germ (Aktas et al., 2015).

Beneficial effects from prebiotics are not limited to maintaining a healthy gut microbial community. Derivatives produced during microbial digestion of prebiotics are also linked to non-digestive health benefits in the context of CVD. The main products of prebiotic digestion include phenyl-γ-valerolactones from procyanidins (Angelino et al., 2020), urolithins from ellagitannins (Piwowarski et al., 2014), and short-chain fatty acids (propionic and butyric acids) from carbohydrate polymers (Liu et al., 2014).

Phenyl-γ-valerolactones are associated with improved metabolic health and cognitive function, reversal or prevention of inflammation (Angelino et al., 2020), and platelet modulation (Montagnana et al., 2018), which may reduce thrombotic events in CVD. Urolithins are associated with systemic anti-inflammatory effects (Piwowarski et al., 2014) and improved lipid metabolism (Kang et al., 2016). Short-chain fatty acids have numerous effects, such as appetite suppression, attenuation of insulin resistance, improved colonic health and gut barrier function, and attenuated weight gain (Chambers et al., 2018).

4 Assessing protective and therapeutic effects of phytochemicals

Targeting the risk factors for CVD or its comorbidities is not only preventative (prophylactic) but can be used in treatment. Common pharmaceutical strategies to treat or prevent CVD include the use of lipid-lowering drugs, antihypertensives, antiplatelet, and anticoagulation therapies (Flora and Nayak, 2019). The effectiveness of these pharmaceuticals can be measured by monitoring blood or urine biomarkers. When researchers assess the viability of a plant-based dietary component in attenuating risk factors for CVD, they will often use the same biomarkers.

The most common effects described for plants that are protective against CVD include basic antioxidant effects. However, they are now seen to include antiplatelet (Khan et al., 2018), anti-hyperlipidaemic, vasorelaxant, antithrombotic, anti-uraemic, and diuretic effects (Michel et al., 2020). More recently identified positive effects include angiotensin-converting enzyme (ACE) inhibition (Nileeka Balasuriya and Vasantha Rupasinghe, 2011; Kim et al., 2019), ferroptosis inhibition (Xie et al., 2016), inhibition of 'A Disintegrin and Metalloproteinases' (ADAM) (Malemud, 2019), such as ADAM17 (Kawai et al., 2021), and matrix metalloproteinase (MMP) inhibition such as MMP-9 (Chaturvedi and Kaczmarek, 2014; Ende and Gebhardt, 2004). The range of protective mechanisms against CVD is summarised in Table 1.

The health-promoting properties of different foods are mechanistically diverse and sometimes poorly explained or understood in the scientific community. As an example, studies of phytochemical content of a particular food often fail to account for the fact that health conferring properties are restricted to organs that may be removed during processing or lost during food preparation. As an example, vegetables that are cooked by boiling will lose most of the water-soluble antioxidants and retain more of the lipophilic ones. Grains that have high antioxidant capacity e.g. lose their potency when they are refined and the germ is removed, turning them from wholemeal to white, leaving only depleted starch. Phytochemical studies of foods must be interpreted in the context of food processing methods, whether at home or in the factory.

Table 1 Examples of biochemical mechanisms that are measured by scientists to predict or explain cardiovascular protection of plants (and pharmaceuticals), description of the mechanism and an example of a natural product that meets the criteria of efficacy

Mechanism	Description	Natural product example
Antioxidant	Water-soluble or lipophilic compounds that quench the free radicals produced in respiratory metabolism. Powerful and penetrating plant-based antioxidants quench radicals generated in metabolism and attenuate the inflammatory effects, which treat the symptoms of CVD and help to reduce plaque, thrombosis, strokes, and abnormal blood glucose levels.	Turmeric (curcumin), most flavonoids, anthocyanins, and phenols such as chlorogenic acid (Brewer, 2011). Another example is sulforaphane, by switching on NF-E2-related factor 2, a gene region encoding for hundreds of antioxidant proteins (Bai et al., 2015).
Anti-hyperlipidaemic	Elevated blood lipid promotes atherosclerotic plaque formation. Upregulation of the low-density lipoprotein receptor (LDLR) has lipid-lowering effects (Vijayakumar et al., 2010).	Diosgenin from greater yams (*Dioscorea alata*) ameliorates hyperlipidaemia (Harijono et al., 2016) and several mechanisms proposed. Diosgenin is also part of the steroidal fraction of seeds from *Trigonella foenum-graecum* (fenugreek) and may be responsible for the upregulated low-density lipoprotein receptor activity.
Anti-inflammatory (COX-2, NOX, interleukin-1, TNF-α, or IFN-γ inhibition; PPAR-α and FPAR-γ activation)	Systemic inflammation is associated with all cardiovascular diseases and is regarded as a trigger for venereal calcification.	Citral from *Cymbopogon citratus* activates PPAR-α, PPAR-γ and suppresses COX-2 (Katsukawa et al., 2010); *E*-cinnamaldehyde from *Cinnamomum zeylanicum* attenuated LPS and IFN-γ activation of macrophages, confirmed by reduced expression of NOX and TNF-α (Gunawardena et al., 2015).
Vasorelaxant/dilator	Vasorelaxation tends to correlate with anti-hypertension. Vasorelaxation helps to dilate blood vessels and reduce pressure against the vessel walls and improves blood circulation to body tissues. Furthermore, aneurysms and rupturing of the blood vessels can occur with constricted arteries.	Allicin from garlic has vasorelaxant effects by generation of H_2S and promotion of the nitric oxide and prostaglandin pathways (Cui et al., 2020).
Anti-thrombotic	Platelet modulation to prevent vascular thrombosis by 'primary clot preventing'. Three mechanisms are explored: collagen/adenosine-5'-diphosphate (Col/ADP) and collagen/epinephrine (Col/Epi) (Montagnana et al., 2018) and thromboxane inhibition (Chang et al., 2012). Cycloxygenase-1 inhibition is also linked to 'blood thinning' (Undas et al., 2007).	Aspirin is a derivative of salicylic acid (from *Salix alba*) as a blood thinner (Undas et al., 2007) and phenyl-γ-valerolactones, which are by-products of procyanidin metabolism and inhibit Col/ADP but not Col/Epi (Montagnana et al., 2018).

Anti-uraemic	Development of uraemia indicates kidney damage as urea accumulates in the blood (Fujii et al., 2016). In the case of uraemia, it is better to prevent rather than cure.	Plant compounds can retard renal fibrosis by altering the expression of the TGF-β/Smad and NF-κB signaling pathways. Betulinic acid from the bark of *Betula pendula* is nephroprotective and reduces uraemia in a rat chronic kidney disease model (Sharma et al., 2017).
Diuretic	Diuretics block calcium-activated chloride transport into platelets, reducing the incidences of epinephrine-induced vascular thrombosis (Spalding et al., 1998).	Dandelion (*Taraxacum officinale*) extract (tea) is a natural diuretic (Clare et al., 2009).
ACE inhibition	Angiotensin-1-converting enzyme (ACE) converts the hormone angiotensin-1 into the active vasoconstrictor angiotensin-2 (Xu et al., 2021).	Inhibitors of ACE prevent high blood pressure. Pomegranate juice is associated with reduced serum ACE activity (Aviram and Dornfeld, 2001).
Ferroptosis Inhibition	Ferroptosis is an iron-dependent process of cell death caused by the fatal accumulation of lipid peroxides that store in and compromise mitochondrial function (Li et al., 2021).	Ferroptosis can be inhibited by iron chelators and lipophilic antioxidants (Galluzzi et al., 2018). Flavonoids are well-known iron chelators (Wang et al., 2021) and vitamin E is a well-known lipophilic antioxidant.
ADAM inhibition	A disintegrin and metalloproteases (ADAMs) regulate cell proliferation, cell growth, and inflammation. ADAM17 is also known as TNF- α-converting enzyme (TACE) (Kawai et al., 2021), but its inflammatory effects are caused by widespread proteolysis (Menghini et al., 2013).	Rosmarinic acid *from Salvia officinalis* (previously *Rosmarinus officinalis*) inhibits ADAM17 expression (Huang et al., 2021).
MMP-9 inhibition	MMP-9 directly mediates extracellular matrix proteins involved in inflammation and fibrosis. MMP-9 is upregulated in the advancing edges of the atherosclerotic plaque (Orbe et al., 2003). There is some overlap between therapies for MMP-9 inhibition and therapies for arthritis.	*Epi*-lupeol from frankincense (*Boswellia frereana*) inhibited the expression of MMP-9 (Blain et al., 2010).
Cholesterol efflux promotion	Cholesterol efflux from macrophages can be increased by promoting *ABCA1* gene expression and inhibiting metalloproteinases that control *ABCA1* protein degradation.	Falcarindiol from carrots promotes *ABCA1* expression and inhibits metalloproteins that cause degradation of this protein (Wang et al., 2017).

Another factor to consider is bioavailability: the proportion of a chemical that finally enters circulation and is able to have an active effect. The bioavailability of phytochemicals can vary widely, depending on factors such as chemical structure and form of dietary intake (Selby-Pham et al., 2017). However, it is important to note that even phytochemicals that have poor or zero bioavailability can still make foods healthier. For example, starches that are digested quickly can cause unhealthy spikes in blood glucose content, which triggers comorbidities of CVD such as diabetes. Phytochemicals that are eaten together with these starches can slow the release of sugar from starches, even if the phytochemicals themselves do not get absorbed into the body. These phytochemicals, therefore, reduce the glycaemic index of foods. They are either digested by gut microbes or excreted, but generally do not enter portal circulation unless they are transformed into another molecule.

Whilst plant-based vitamins, minerals, and metal chelates are considered an important aspect of CVD prevention and treatment, establishing a clear beneficial effect in preventing or alleviating conditions such as CVD can be extraordinarily difficult (Jenkins et al., 2021). However, 'no evidence of efficacy' is not the same as 'evidence of no efficacy'. CVD occurs over the long term, so it is hard to measure the net effects of prophylactic initiatives.

As an example, whilst beneficial effects from vitamin and mineral supplementation in the context of CVD have a strong theoretical basis, definitive *in vivo* evidence of their effects is still lacking, and there is a consensus that individuals who already have a balanced diet may have nothing to gain. However, it is evident that vitamins and minerals are beneficial to those individuals who experience a deficiency (Ingles et al., 2020), whether clinical or sub-clinical (Di Nicolantonio et al., 2018). Research has not fully explored the synergistic or antagonistic interactions of vitamins and minerals with the biological effects of plant phytochemicals (Koss-Mikolajczyk et al., 2019).

Another complication is that the phytochemical composition of foods often changes according to factors such as geography (Padilla-González et al., 2020a), weather, light availability or frequency (Bian et al., 2015), and cultivar or chemotype (Sadgrove et al., 2014, 2020; Sadgrove and Jones, 2014). This is of particular concern because, in the face of climate change, the current understanding of the phytochemistry of plants may become outdated, particularly if the chemical composition of species is affected by changing weather patterns (Ahmed and Stepp, 2016). The effects of abiotic stress on food phytochemicals vary according to species. It is important to note variations in phytochemical content caused by climate change, which may be positive as well as negative e.g. some species will produce more phytochemical compounds under abiotic stress (Sadgrove, 2020), even when the crop yields are negatively affected (Raza et al., 2019).

5 Types of phytochemical compounds: flavonoids, phenols, organosulphur compounds, alkaloids, lignans, sterols, tannins and soluble fibres

Plant phytochemicals that are becoming significant in the context of preventing or alleviating CVD include flavonoids, phenols, organosulphur compounds, lignans, sterols, phloroglucinols, and soluble fibres.

5.1 Flavonoids

The flavonoids class is most frequently cited in the context of positive health outcomes in relation to conditions such as CVD (Micek et al., 2021). There are many different types of flavonoids, some bonded to sugars (glycosides) and the others in their free form as 'aglycones'. The sugar-bound flavonoids, the glycosides, have a different pharmacokinetic profile compared to the aglycones. The disaccharides (two sugars) as well as tri-, tetra-, and other types of saccharides are generally not absorbed until they are deglycosylated in digestion (sugars removed). This makes it possible for the aglycone to be absorbed into portal circulation (the circulation of nutrient-rich blood between the gut and the liver). The monosaccharides (one sugar) are partly absorbed in the small intestine via the hexose transporter pathway that is designed to process free sugar. The remaining monosaccharide is then cleaved in digestion and absorbed as an aglycone (Sadgrove and Jones, 2019).

Flavonoid aglycones are represented by a diversity of types, but the most common forms are flavones, flavanones, flavonols, isoflavones, isoflavanones, flavanols (catechins), chalcones, anthocyanins, and procyanidins (Palma-Tenango et al., 2017). Flavonoids are well known for their anti-inflammatory (Ginwala et al., 2019), antioxidant (Pietta, 2000), and antithrombotic effects (Bojić et al., 2019). As with all nutraceuticals, the beneficial effects of flavonoid supplementation are difficult to prove definitively. However, one meta-analysis was able to demonstrate a 14% reduction rate of stroke in men associated with supplementation of 20 mg of a flavonol per day (Wang et al., 2014).

Many flavonoids are already available in the market as nutraceuticals. An example is biochanin A (Yu et al., 2019) which is present in high concentration in red clover (*Trifolium pratense*) and is also found in vegetables from the *Brassica* and legume family. Most of the positive health effects ascribed to biochanin A use are against the comorbidities of CVD. In the gut, biochanin A is metabolically converted by microbes into genistein and daidzein. Daidzein is converted to equol, a flavanol with particular protective effects in the context of CVD (Mayo et al., 2019). Biochanin A is also a selective agonist of the oestrogen receptor (Yu et al., 2019), which may have positive implications for age-related cardiovascular disorders. Other prominent flavonoids include the isoflavones genistein and daidzein from soy, celery, and other legumes; quercetin from onion, citrus fruits, broccoli, red grapes, apples, and cherries; and kaempferol from broccoli and radishes (Yu et al., 2019).

One of the more impressive innovations in health has been the reintroduction of anthocyanin-rich foods to the modern diet. Anthocyanins, either as aglycones or glycosides, are the colour compounds in e.g. flower petals, in fruits such as blueberries, and in autumn leaves. They are e.g. associated with the colour of purple potatoes (Montilla et al., 2011) and purple cauliflower (Yan et al., 2019) or with the black colour of black beans (Takeoka et al., 1997) as well as black carrot (Akhtar et al., 2017). Anthocyanins are relatively unstable, although some types are more stable than others. It has therefore been challenging to make oral anthocyanins supplements with the exception of freeze-dried fruits which are able to retain the compound. Efforts have been directed instead into enriching anthocyanin in common foods.

One example is the appearance of purple cauliflowers on supermarket shelves in recent years. The colour derives from an anthocyanin that is encoded by activation of the gene BoMYB2 (Yan et al., 2019). The common white head cauliflower, known as curd, was created by selective breeding. It is thought to have originated in southern Italy during the Middle Ages by cross-breeding the Sicilian purple variety with a cabbage, producing a larger white head, making it popular in cultivation as a food crop (Smith and King, 2000). The return of the purple anthocyanin to the curd phenotype represents an important success in the use of genetics to guide breeding without having to resort to genetic modification (Chiu and Li, 2012). Genetics merely guided selective breeding of silent genes to optimise the expression of the anthocyanin pigment, making the modern cauliflower healthier and more resilient.

5.2 Phenols

One of the best-known phenols in the context of metabolic health is the stilbene resveratrol, a phytoalexin in grapes produced in response to the threat of disease (Jeandet et al., 1995), and the main point of interest in the 'French Paradox' (Catalgol et al., 2012). The French Paradox was defined in 1992 as the unexpected cardiovascular resilience in the French population, despite a high-fat diet, with red wine drinking put forward as a potential explanation (Renaud and De Lorgeril, 1992). Because resveratrol is one of the main phenols in red wine, subsequent research has focused on this stilbene and demonstrated a plethora of effects that strongly suggest prophylaxis of CVD (Wang et al., 2012) with clinical evidence supporting this (Zordoky et al., 2015).

Other significant phenols include chlorogenic acid from coffee, epigallocatechin gallate from green tea, curcumin from turmeric, caffeic acid from propolis or olives, rosmarinic acid from rosemary, and gallic acid from gallnuts. The main mechanism of these phenols, including resveratrol, is the attenuation of systemic inflammation, platelet aggregation, oxidation, and glycation (Ali et al., 2020).

5.3 *Organosulphur compounds*

The organosulphur compounds that have come to be recognised as beneficial in health are generally very small molecules that include at least one sulphur atom. They are expressed in onions, garlic, and cruciferous vegetables and are regarded as prophylactic for CVD (Vazquez-Prieto and Miatello, 2010). The two best-known organosulphur classes are the isothiocyanates and the sulfoxide derivatives of the amino acid cysteine: alliin and its derivative allicin. Alliin and allicin have been implicated in the amelioration of cardiovascular disorders and may attenuate gut dysbiosis as well as having efficacy against chronic kidney disease (Ribeiro et al., 2021), thus supporting this part of the gut–kidney–heart triangle.

The isothiocyanates are derived from glucosinolate compounds by enzymatic hydrolysis of the thioglucoside bond by myrosinase (Vanduchova et al., 2019). This generally occurs when the tissue of cruciferous vegetables is damaged, perhaps via the chewing motion of a grazing herbivore, removing the physical barrier between precursor and enzyme and creating a sudden burst of flavour in the process. The most studied isothiocyanates in terms of CVD are sulforaphane, allyl isothiocyanate, benzyl isothiocyanate, phenethyl isothiocyanate, and goitrin (Yeger and Mokhtari, 2020).

5.4 *Alkaloids*

While some alkaloids have been implicated in adverse cardiovascular and cerebrovascular effects, such as the performance-enhancement alkaloids known as 'ephedra' (Andrews et al., 2005), several alkaloids have also been recognised as beneficial against risk factors for CVD. Berberine alkaloids e.g. are associated with improved serum lipid and glucose profiles (Liu et al., 2008). The alkaloid colchicine, long known to be effective against gout and osteoarticular pain (familial Mediterranean fever), is also being recognised as effective against CVD (Andreis et al., 2021). Colchicine was originally isolated from the bulb-like corm of *Colchicum autumnale* and has been used for millennia in medicine, with records from the ancient Egyptian Ebers Papyrus (1500 BCE) (Dasgeb et al., 2018). Alkaloids from the husk fibre of coconut (*Cocus nucifera*), used in traditional Nigerian medicine, have e.g. been found to lower HDL cholesterol in mice (Joshua and Muyiwa, 2019).

5.5 Lignans

Lignans are the result of a dimerization of two phenol derivatives of cinnamic acid. They are widely distributed in the plant kingdom in seeds, grains, legumes, fruits, and vegetables. The most-researched lignans are secoisolariciresinol, matairesinol, pinoresinol, and lariciresinol. These sometimes occur naturally in glucoside forms such as secoisolariciresinol diglucoside from flax seed (Peterson et al., 2010). Dietary lignans are metabolised by the gut microbiota into 'enterolignans', which are suggested to be therapeutic against hypertension and hypercholesterolemia (Witkowska et al., 2018).

5.6 Sterols

The three sterols that are most frequently examined in the context of cholesterol studies are sitosterol, stigmasterol, and campesterol. It is well known that consumption of these sterols lowers the blood serum concentration of low-density lipoprotein cholesterol (LDL-C, also known as non-HDL-C), which is implicated in cardiovascular disease. High-density lipoprotein is called 'good cholesterol' because it is the lipoprotein that circulates cholesterol back to the liver for excretion, whereas LDL-C is the class of lipoprotein that transports cholesterol to the tissues, making it 'bad cholesterol'.

The suggested mechanism by which phytosterols lower the LDL-C serum level is by reducing the amount of cholesterol absorbed from the intestines (Cabral and Klein, 2017). However, there is controversy over the possible cancelling out of the LDL-C lowering benefit conferred by phytosterols, by promoting CVD via other mechanisms (unrelated to LDL-C levels). One group of authors argues that 38% of coronary artery disease cases are not explained by LDL-C and that phytosterols have a net beneficial effect (Helgadottir et al., 2021). However, another group of authors argues that the link between phytosterols and non-cholesterol-affiliated CVD is speculative and not supported by the data (Plat et al., 2021). To further confound this, a meta-analysis of phytosterols in CVD prevention was unable to demonstrate a strong preventative relationship (Genser et al., 2012).

The hypothesis that phytosterols can increase the risk of CVD originates from the observation of people living with sitosterolemia (also known as phytosterolemia), which is a rare autosomal recessive disease that is characterised by a 50-fold increase in phytosterol absorption (Kaur and Myrie, 2020). Individuals affected by this condition demonstrate an increased incidence of CVD. While this condition is rare, many researchers have extrapolated this effect to the general population. Nevertheless, examination of intermediate markers of cardiovascular disease has demonstrated both positive and neutral (no) effects, but reports of negative effects are rare and restricted to candidates living with sitosterolemia (Cabral and Klein, 2017). In 2013, the European Atherosclerosis Society published a recommendation that phytosterols are suitable for lowering LDL-C and, because of the absence of adverse signs, the consumption of phytosterol-fortified foods can be recommended (https://www.eas-society.org/page/phytosterol_comment).

5.7 Tannins

Tannins are polymerised phenols that frequently (but not always) have a sugar molecule as a nucleus. Tannins are widely distributed in the plant kingdom, e.g. concentrated in tree bark and some fruits such as grapes and pomegranate. Common hydrolysable tannins include ellagitannins, constructed of ellagic acid and sometimes gallic acid; gallotannins, constructed of gallic acid; and phlorotannins, constructed of phloroglucinols.

In digestion, the tannins are reduced to simpler phenols, such as ellagic acid, gallic acid, or phloroglucinol. Some of these phenols are absorbed and the remainder catabolised by gut microbes into smaller forms. For example, ellagic acid is reduced to urolithin and gallic acid to pyrogallol (Septembre-Malaterre et al., 2017). Phloroglucinol enters portal circulation efficiently, meaning catabolism is less common. Aside from digestion of phlorotannins, phloroglucinol is also produced from catabolism of quercetin (Kawabata et al., 2019). In portal circulation, phloroglucinol is quickly metabolised in phase two liver processes, giving it a short half-life (Dollo et al., 1999).

The beneficial effects of phenols have been described earlier. The anti-platelet activity of phloroglucinol is related to the inhibition of thromboxane A_2 production, among other effects (Chang et al., 2012). Tannins and phenols from pomegranate have demonstrated platelet aggregation inhibition at physiologically relevant concentrations (Mattiello et al., 2009).

5.8 Soluble fibres

As mentioned earlier, soluble fibres work as prebiotics. However, they are also modulators of the glycaemic index of foods, stabilising blood sugar levels (Scazzina et al., 2013). They do this by delaying gastric emptying, inhibiting the action of digestive enzymes, and slowing the rate of digestion of starch (Hamden et al., 2010; Jayachandran et al., 2018). As previously mentioned, the most common soluble fibres include fructooligosaccharide (Padilla-González et al., 2020b), galactomannan (Hamden et al., 2010), glucomannan (Al-Chazzewi and Tester, 2010), acemannan (Quezada et al., 2017), or β-glucan (Aktas et al., 2015).

6 Wild crop relatives as sources of phytochemicals

Modern domestic crops are the result of thousands of years of selective breeding in human agricultural systems. Generally, selection has been in favour of bigger yields and sweeter-tasting vegetables and fruits. Over time this increasing sweetness has been achieved by the disappearance of 'bitters', the phytochemicals that are now seen to have a role in preventing or treating CVD. Phytochemicals generally do not add flavour and may even reduce the sweetness of vegetables and even make them slightly bitter. This has favoured selecting cultivars with reduced phytochemical content (Gasparini et al., 2020). This trend is now reversing with a greater interest in wild crop relatives. This is not just because of their higher phytochemical content but because of their wider resilience to biotic and abiotic stresses compared to modern high-yielding varieties, particularly in the face of climate change and the need to reduce reliance on fertilizers, herbicides, and insecticides (Dempewolf et al., 2014; Ahmed and Stepp, 2016; Gasparini et al., 2020). This section looks at a number of examples of modern versus wild relatives of common crops.

6.1 Carrot

Differences between the wild carrot, *Daucus carota* L. subsp. *carota* (Fig. 1a), and the domestic carrot, *Daucus carota* L. subsp. *sativa* (Fig. 1b), were evident during the sixth-century AD when images were added to the pharmacopoeia *De Materia Medica* originally created by Pedanius Dioscorides in AD 65 (Janick, 2014). It was in the second-century AD when the domestic carrot was recognised as a more palatable food source than the wild version (Akhtar et al., 2017). Genetic studies suggest the origin of the domestic carrot might be in central Asia (Que et al., 2019).

Although the modern carrot is regarded as a healthy food source (Just et al., 2009), alongside other modern vegetables, olive oil, and nuts (Estruch et al., 2018), the amount of some specific classes of phytochemical has generally declined. The modern carrot is a rich source of carotenoids, which are precursors to Vitamin A (Just et al., 2009), often referred to as 'provitamin A'. In contrast, a wild carrot from Scotland was demonstrated to be rich in antioxidant glycosides of the flavonoid luteolin (Kumarasamy et al., 2005). This has raised the possibility of carrot hybridisation to retain carotenoid content but enrich overall phytochemical content (Just et al., 2009; Que et al., 2019). However, hybridisation has proved difficult, due to sterility of some hybrids and the predominance of the metabolic profile of wild carrots in successful hybrids (Que et al., 2019).

The species *D. carota* includes over six subspecies. The domestic carrot (subsp. *sativa*) contains a diverse concentration range of the bioactive polyacetylenes falcarinol (82–518 $\mu g.g^{-1}$ dry weight, DW) and falcarindiol (236–1553 $\mu g.g^{-1}$ DW) (Roman et al., 2011). These polyacetylenes are associated with health benefits which include anti-inflammatory and anti-platelet-aggregatory effects that are beneficial in the context of CVD. Furthermore, falcarinol is a potent anti-inflammatory compound that is beneficial in cases of gastrointestinal inflammation and has positive implications for CVD (Stefanson and Bakovic, 2018).

The closest relative of the domestic carrot, *D. carota* subsp. *carota*, contains similar quantities of polyacetylenes, but generally the concentration of falcarindiol is higher (583 $\mu g.g^{-1}$ DW), which is the metabolite responsible for imparting bitterness (Roman et al., 2011). This metabolite also promotes cholesterol efflux from macrophages (Wang et al., 2017), which is the first step in reverse cholesterol transport, limiting its circulation to peripheral tissues and attenuating atherosclerotic plaque accumulation.

The polyacetylenes in carrots are not evenly distributed throughout organs or spatially in the root. Their expression is activated by insect attack and possibly by sunlight exposure. They are therefore more concentrated in the skin, in the basal (upper) part of the root in phloem tissue, and in blemishes such as black spots (Roman et al., 2011). However, it is common for blemishes in the skin and the basal part to be removed in food preparation, in part for cosmetic reasons, reducing their availability and potential benefits.

Wild carrot subspecies are generally more bitter than domestic carrot and this may be attributed to the higher concentration of falcarindiol. Two of the subspecies that produce similar size carrots to the domestic carrot are *D. carota* subsp. *maximus* Desf. (syn. *D. maximus*) and *D. carota* subsp. *gummifer* (Syme) Hook.f. These have polyacetylene concentrations up to 2000 $\mu g.g^{-1}$ at fresh weight, many times higher than domestic carrots. The bitterness of these large wild carrots has prevented their domestication. Furthermore, *D. maximus* requires harvesting while the roots are immature because, as it matures, it becomes woody and tough, and thus unpalatable. This poses challenges to introducing them into diets.

6.2 Potato

The current domestic potato *Solanum tuberosum* came from South America and was introduced into Europe by the Spanish in the sixteenth century (Bradshaw and Ramsay, 2009). It was previously believed that *S. tuberosum* was produced from selective breeding and domestication over various locations in South America, but a genetic study of wild crop relatives produced evidence in favour of a single location in Northern Peru (Spooner et al., 2005).

Potatoes belong to the genus Solanaceae, which is known colloquially as the nightshade or potato family. Although the taxonomy of wild potatoes is

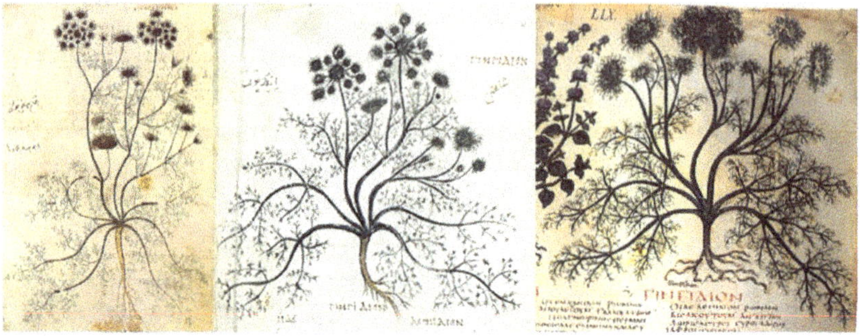

(a) - ***Daucus carota*** **L. subsp.** ***carota***

(b) - ***Daucus carota*** **L. subsp.** ***sativa***

Figure 1 Differences between (a) wild carrot (*Daucus carota* subsp. *carota*) and (b) domestic carrot (*Daucus carota* subsp. *sativa*). Source: Images taken from an online database https://tulip.hort.purdue.edu/herbalimages/search.html.

constantly under revision, there are at least 219 recognised species of *Solanum* in South America and southern North America (Bradshaw and Ramsay, 2009). When the potato is eaten on its own, either cooked in water or oil, the starches are rapidly digested, and the blood sugar level quickly rises. For this reason, it is regarded as a high glycaemic index food (Henry et al., 2005). As previously mentioned, diabetes or insulin resistance are risk factors for CVD.

Various studies have demonstrated that the purple or red-fleshed wild relatives of *S. tuberosum* have high antioxidant content (Campos et al., 2006). These species and subspecies are rich in anthocyanins which are inversely associated with CVD (Cassidy et al., 2016). It has been suggested that domestic potato could be improved by cross breeding with native Andean *Solanum* to improve its phytochemical content (Andre´ et al., 2009).

The antioxidant content is even higher in tuber species that belong to other genera (Campos et al., 2006). One promising South American species is *Smallanthus sonchifolius*, which produces a tuber that is rich in chlorogenic acid esters and fructooligosaccharides (Padilla-González et al., 2020b). Both are known for their role in controlling blood glucose spikes, but they follow different mechanisms. The fructooligosaccharides are soluble dietary fibres that act as prebiotics and slow the rate of sugar release from starches in digestion, by slowing the activity of digestive enzymes and also by slowing gastric emptying (Chen et al., 2016). Chlorogenic acids lower the risk of CVD by anti-diabetic and anti-lipidemic effects achieved via the activation of AMPK (5′adenosine monophosphate-activated protein kinase) (Ong et al., 2013), which is an enzyme that plays a role in cellular energy homeostasis, mostly by activation of glucose and fatty acid uptake followed by oxidation. Many wild crop relatives of domestic potato and other tubers may be a healthier choice in the context of CVD.

6.3 Wheat

Arguably the most famous of all crops is the 'golden grain' common wheat (*Triticum aestivum*) (Hazard et al., 2020). It is currently believed that the ancestral route of modern *T. aestivum* started with the wild crop relative *Triticum urartu*, a diploid that crossed with another diploid goat grass species *Aegilops speltoides*. The cross created *Triticum dicoccoides* (wild emmer), a tetraploid wild crop relative that is widespread in modern Israel (Copper, 2015). This tetraploid crossed with another diploid species *Aegilops tauschii* (goat grass) on two occasions and produced two modern commercial crops, *Triticum dicoccum*, a tetraploid, and *Triticum spelta*, a hexaploid. Over the course of crop selective breeding, the modern wheat hexaploid *T. aestivum* and modern wheat tetraploid *Triticum durum* were developed (Peng et al., 2011).

Approximately 95% of wheat production worldwide is from *T. aestivum*. Most studies highlight wheat as a beneficial dietary component (Shewry and Hey, 2015; Wieser et al., 2020). However, despite its role as a staple food crop, there is an ongoing debate about how far it may promote or negatively impact human health (Di Nicolantonio et al., 2018). However, it is becoming evident that it is not wheat that is the problem but the refining of white wheat flour and removal of fibre, which then make cereal products a potential risk factor for dyslipidaemia and metabolic syndrome (Amin and Gilani, 2013). The fibre content is lowered when the grain is processed to remove the germ (outer skin of the grain), which makes the meal whiter, producing white flour and white bread instead of the brown wholemeal versions that are also available.

Some phytochemical parameters in wheat can be measured to predict its long-term effect on health. For example, the amylose to amylopectin ratio is important in determining starch digestibility and thus the glycaemic index (Singh et al., 2010). Amylose is slower to digest, due to limited branching links on the carbohydrate chain. Hence, a higher amylose content equates to a lower glycaemic index meal (Van Amelsvoort and Westrate, 1992), even after the grain is processed to remove the germ.

For nearly 20 years, a distinction has been made between the relative health benefits of refined compared to whole grain diets with the latter seen as more beneficial (Liu, 2002). Dietary guidelines in Europe, America, Australia, and elsewhere have subsequently recommended that wholegrain foods be substituted in place of refined grain foods, making up one-half or two-thirds of grain-food intake (Williams, 2012). An alternative approach has been to genetically modify wheat so that the white starch is also rich in soluble fibre, though this approach is limited by continuing consumer and regulatory caution over the use of genetic modification (Hazard et al., 2020).

A well-known wheat alternative is spelt (*Triticum spelta*), which is regarded as a 'natural' alternative to modern wheat. The content of soluble fibre is higher in wholegrain spelt compared to wheat but, after refinement and whitening, the two flours can be almost identical in fibre content, neutralising the advantage (Escarnot et al., 2012). A similar conundrum is faced with other domestic forms of wheat that have high polyphenol content compared to *T. aestivum* because the phenols are restricted to the wheat germ and are rare in the endosperm used to make flour (Brandolini et al., 2013). This has led to alternative solutions such as the use of hulled wheats (Beta, 2021).

Today's wheats are recognised as either tetraploid (e.g. *T. dicoccum*, *T. dicoccoides*, and *T. durum*) or hexaploid (e.g. *T. aestivum* and *T. spelta*). In this regard, wild crop relatives of wheat, or less used crops, such as durum or einkorn wheat, are generally tetraploids. Significantly, the tetraploid species have been shown to have higher amylose content than their hexaploid counterparts, particularly durum and Polish wheats (e.g. *T. durum* and *T. polonicum*) (Rodriguez-Quijano et al., 2003). Hexaploids tend to yield amylose content within the range of 18–35% (Labuschagne et al., 2007), whereas tetraploids often produce values above 40% (Rodriguez-Quijano et al., 2003). Unfortunately, a key issue in adoption of these healthier grains is that high amylose flours are less easy to use in bread making, potentially reducing their availability and appeal (Lee et al., 2001).

6.4 Brassicas: broccoli, cabbage, kale, cauliflower, mustard and Brussel sprouts

Cruciferous vegetables are a very promising prophylactic food source in the context of CVD. Those varieties and species that belong to the genus *Brassica* are regarded as important in the Mediterranean diet and are also inversely correlated to subclinical atherosclerosis in older women (Blekkenhorst et al., 2018). However, the relatively short history of many cultivars and hybrids that are considered as F1 hybrids means they have not faced natural selective pressures, making them potentially vulnerable to climate change. Genotype-guided selective breeding approaches have been recommended to recover resilience genes from wild species of *Brassica* (Lv et al., 2020).

The wild cabbage (*Brassica oleracea* L. subsp. *oleracea*) is found on the sea cliffs of western and southern Europe. It is from this leafy biennial, or a common ancestor, that the modern cabbage (*B. oleracea* subsp. *capita* L.), kale and collard (*B. oleracea* subsp. *acephala* DC.), cauliflower and broccoli (*B. oleracea* subsp. *botrytis* L.), brussels sprouts (*B. oleracea* subsp. *gemmifera* DC.), and kohl-rabi (*B. oleracea* subsp. *caulo-rapa* DC.) originate. From as early as the 1920s it was noticed that wild cabbage was heterozygous for resistance (Walker, 1928), meaning resistance varied from specimen to specimen but that it could be more easily selected in breeding. This has provided the basis for improving resistance in modern varieties of e.g. cauliflower (Singh et al., 2018).

Pathogen resistance is often associated with plant phytochemicals, or phytoalexins, that are also of value in human nutrition and CVD prophylaxis. The phytoalexin resveratrol has already been mentioned (Jeandet et al., 1995). *Brassica* species also express isothiocyanates such as sulforaphane. Sulforaphane is an effective natural activator of the Nrf2 pathway, which cascades into the expression of antioxidant proteins (Kubo et al., 2017). This pathway is ubiquitously expressed in mammalian tissue, suggesting sulforaphane has the potential to reverse inflammation at the whole-body level. Research has demonstrated enhancement to gut barrier efficiency (Canxia et al., 2018). Theoretically, gut barrier function is improved via a two-fold mechanism involving attenuated dysbiosis and activation of the Nrf2 pathway.

Brassica is also a viable source of vitamin K1, a methyl naphthaquinone compound that was discovered in the 1930s by a Danish chemist who named it 'Koagulations-Vitamin'. As the Danish name suggests, vitamin K is an 'anti-haemorraghic factor' (Lippi and Franchini, 2011). Another version of the vitamin, denoted K2, is produced by gut bacterial fermentation and is accumulated in the liver which gives it a prebiotic function in preventing heart disease (Haugsgjerd et al., 2020). There is also a strong theoretical rationale for an inverse correlation between vitamin K1 intake and prevention of atherosclerotic disease because it increases the activity of matrix-Gla protein, a potent inhibitor of vascular calcification. While the link has been conclusively demonstrated in animal studies, it has not yet been proven in humans (Villines et al., 2005; Haugsgjerd et al., 2020; Palmer et al., 2020).

7 Herbs and spices as sources of phytochemicals

Herbs, spices, and their blends (known in Indian cuisine as masalas) are much more than simply flavourings and colourants. These colourful and flavour-dense plants get their properties from a high phytochemical content while their textures often come from fibres with prebiotic effects. For example, fenugreek is a rich source of galactomannan, a soluble fibre with prebiotic effects (Mathern et al., 2009). Fenugreek is also rich in a water-soluble antidiabetic compound, 4-hydroxy isoleucine (Jetté et al., 2009), and various flavonoids (Shang et al., 1998). For this reason, fenugreek seed powder has long been added to hot water, allowed to cool, and drunk as a remedy in middle eastern or north African cultures for respiratory complaints or for 'strength' (Khalki et al., 2012). Turmeric e.g. contains the polyphenol curcumin which is associated with CVD prophylaxis, including when combined with black pepper which improves curcumin bioavailability (Shamsi et al., 2017; Dastani et al., 2019). Quinones in black cumin are considered effective for numerous ailments, including pathologies that are considered risk factors for CVD (Amin and Hosseinzadeh, 2016). Seeds contain saponins, flavonoids, alkaloids as well as essential oils, fixed terpenes, free fatty acids, and high mineral content (Mukhtar et al., 2021).

There are several traditional foods or medicinal plant species that have been found to have therapeutic properties for CVD *in vitro* but are not feasible *in vivo* because blood serum concentrations cannot possibly be matched. These include medicinal plants that inhibit ACE from countries such as Australia and Iran (Deo et al., 2016; Sharifi et al., 2013; Rad et al., 2019). When searching in the literature for plants demonstrated *in vitro* as potentially useful against CVD, the dose requirement must be taken into consideration. Alternatively, it may be necessary to focus on studies that present *in vivo* results in animal models, such as those summarised by Michel et al (2020).

8 Future trends in research

Furthering our understanding of the range of phytocompounds in plants as well as the levels of intake required to have benefits is clearly a top priority that will involve breeders and academics working closely together. Another important factor is associated with keeping the diversity of plants available for future study, especially as seeds from many wild heritage varieties as well as underutilised food plants are not always available for study. Advances in genomics provides us with a better understanding of the genes and biosynthetic processes that can assist select and breed for varieties with higher levels of CVD beneficial compounds. Such research should not only assists conserve plant genetic resource, but also supports the health and nutrition of wealthy societies as well as poorer communities that are often the custodians of wild relatives of crop plants. There is considerable evidence that supports the notion that plant-based diets have cardiovascular benefits. Future research should also focus on how this knowledge feeds into policy strategies that can impact different populations.

9 Where to look for further information

There is no one database that brings together all the data about the levels of specific CVD compounds in crop plants, the availability of cultivars or of which are the key species to study. There are different organisations that can provide information about heritage varieties and these will vary among countries. For example, the Consultative Group on International Agricultural Research (CGIAR) is a global partnership that unites organizations engaged in research for a food secure future and provides details about sourcing different cultivars (https://www.cgiar.org/). In the UK information about heritage grains can be found on https://www.heritagegraintrust.org. The organisation Bioversity also provides information about different cultivars, heritage varieties and wild relatives of crop plants (http://www.cropwildrelatives.org/). Another very good source of information including trade data is the Food and Agriculture Food Organisation of the United Nations. (https://www.fao.org/home/en). Search engines such as Pub Med are good for searching for papers that cover the most recent findings on plant-derived compounds on different aspects of CVD (https://pubmed.ncbi.nlm.nih.gov/). Other information that is more specific to CVD and plant-based diets can be found on health websites, such as the British Heart Foundation (https://www.bhf.org.uk).

10 Conclusion

Cardiovascular disease and its comorbidities are a leading cause of illness and death worldwide. Research is highlighting the complex causes involved such as the role of gastrointestinal bacterial dysbiosis as a factor in CVD as well as the key role of diet both in causing and also in preventing or treating illness. This research has helped to highlight the important prophylactic and therapeutic role of plant phytochemicals in relation to CVD. Dietary flavonoids e.g. have been found to have diverse effects such as buffering against inflammation, oxidation, dysbiosis, diabetes, and hyperlipidaemia. Plant foods may also have prebiotic effects or attenuate comorbidities and risk factors for CVD.

The theoretical basis for the beneficial effects of plant phytochemicals (based on identifying potential mechanisms of action in the body) is now well supported by *in vitro* and animal studies, although *in vivo* and epidemiological results remain mixed and require further research to establish a strong evidence base. This is perhaps unsurprising given the multiple potential factors involved in any observed effect (particularly over a long period in the case of a chronic disease such as CVD) as well as the complex route phytochemicals must travel from plant to consumption to any effect in the body. Results are confounded by factors such as wide variations in phytochemical content in particular foods (determined by factors such as cultivar, environmental conditions, and cultivation practices), the effects of preparation and processing, and bioavailability.

The weight of evidence in favour of the beneficial effects of plant phytochemicals is leading to a range of initiatives to promote their role in diets. These range from supplementation to altering diets in favour of both familiar and more exotic foods with high phytochemical content. There are also moves to enhance and restore phytochemical content in modern cultivars by exploiting wild crop varieties.

11 References

Agyemang, C., Addo, J., Bhopal, R., Aikins, Ade G. and Stronks, K. 2009. Cardiovascular disease, diabetes and established risk factors among populations of sub-Saharan African descent in Europe: A literature review. *Globalization and Health* 5, 7.

Ahmed, S. and Stepp, J. R. 2016. Beyond yields: Climate change effects on specialty crop quality and agroecological management. *Elementa: Science of the Anthropocene* 4, 92.

Akhtar, S., Rauf, A., Imran, M., Qamar, M., Riaz, M. and Mubarak, M. S. 2017. Black carrot (Daucus carota L.), dietary and health promoting perspectives of its polyphenols: A review. *Trends in Food Science and Technology* 66, 36–47.

Aktas, K., Bilgiçli, N. and Levent, H. 2015. Influence of wheat germ and β-glucan on some chemical and sensory properties of Turkish noodle. *Journal of Food Science and Technology* 52(9), 6055–6060.

Al-Chazzewi, F. H. and Tester, R. F. 2010. Effect of konjac glucomannan hydrolysates and probiotics on the growth of the skin bacterium Propionibacterium acnes in vitro. *International Journal of Cosmetic Science* 32(2), 139–142.

Ali, S. S., Ahmad, W. A. N. W., Budin, S. B. and Zainalabidin, S. 2020. Implication of dietary phenolic acids on inflammation in cardiovascular disease. *Reviews in Cardiovascular Medicine* 21(2), 225–240.

Amin, B. and Hosseinzadeh, H. 2016. Black cumin (*Nigella sativa*) and its active constituent, thymoquinone: An overview on the analgesic and anti-inflammatory effects. *Planta Medica* 82(1–2), 8–16.

Amin, F. and Gilani, A. H. 2013. Fiber-free white flour with fructose offers a better model of metabolic syndrome. *Lipids in Health and Disease* 12, 44.

Andre´, C. M., Oufir, M., Hoffman, L., Hausman, J.-F., Rogez, H., Larondelle, Y. and Evers, D. 2009. Influence of environment and genotype on polyphenol compounds and *in vitro* antioxidant capacity of native Andean potatoes (*Solanum tuberosum* L.). *Journal of Food Composition and Analysis* 22(6), 517–524.

Andreis, A., Imazio, M. and De Ferrari, G. M. 2021. Colchicine for the treatment of cardiovascular diseases: Old drugs, new targets. *Journal of Cardiovascular Medicine* 22(1), 1–8.

Andrews, R., Chawla, P. and Brown, D. L. 2005. Cardiovascular effects of ephedra alkaloids: A comprehensive review. *Progress in Cardiovascular Diseases* 47(4), 217–225.

Angelino, D., Caffrey, A., Moore, K., Laird, E., Moore, A. J., Gill, C. I. R., Mena, P., Westley, K., Pucci, B., Boyd, K., Mullen, B., Mccarroll, K., Ward, M., Strain, J. J., Cunningham, C., Molloy, A. M., Mcnulty, H. and del Rio, D. 2020. Phenyl-c-valerolactones and healthy ageing: Linking dietary factors, nutrient biomarkers, metabolic status and inflammation with cognition in older adults (the VALID project). *Nutrition Bulletin* 45(4), 415–423.

Aviram, M. and Dornfeld, L. 2001. Pomegranate juice consumption inhibits serum angiotensin converting enzyme activity and reduces systolic blood pressure. *Atherosclerosis* 158(1), 195–198.

Bai, Y., Wang, X., Zhao, S., Ma, C., Cui, J. and Zheng, Y. 2015. Sulforaphane protects against cardiovascular disease via Nrf2 activation. *Oxidative Medicine and Cellular Longevity* 2015, 407580.

Beta, T. (Ed.). 2021. *Improving the nutritional and nutraceutical properties of wheat and other cereals*. Burleigh Dodds Science Publishing, Cambridge.

Bhatnagar, P., Wickramasinghe, K., Wilkins, E. and Townsend, N. 2016. Trends in the epidemiology of cardiovascular disease in the UK. *Heart* 102(24), 1945–1952.

Bian, Z. H., Yang, Q. C. and Liu, W. K. 2015. Effects of light quality on the accumulation of phytochemicals in vegetables produced in controlled environments: A review. *Journal of the Science of Food and Agriculture* 95(5), 869–877.

Bigeh, A., Sachez, A., Maestas, C. and Gulati, M. 2020. Inflammatory bowel disease and the risk for cardiovascular disease: Does all inflammation lead to heart disease? *Trends in Cardiovascular Medicine* 30(8), 463–469.

Bigna, J. J. and Noubiap, J. J. 2019. The rising burden of non-communicable diseases in sub-Saharan Africa. *The Lancet. Global Health* 7(10), e1295–e1296.

Blain, E. J., Ali, A. Y. and Duance, V. C. 2010. Boswellia frereana (frankincense) suppresses cytokine-induced matrix metalloproteinase expression and production of pro-inflammatory molecules in articular cartilage. *Phytotherapy Research* 24(6), 905–912.

Blekkenhorst, L. C., Bondonno, C. P., Lewis, J. R., Woodman, R. J., Devine, A., Bondonno, N. P., Lim, W. H., Zhu, K., Beilin, L. J., Thompson, P. L., Prince, R. L. and Hodgson, J. M. 2018. Cruciferous and total vegetable intakes are inversely associated with subclinical atherosclerosis in older adult women. *Journal of the American Heart Association* 7(8), e008391.

Bojić, M., Maleš, Ž, Antolić, A., Babić, I. and Tomičić, M. 2019. Antithrombotic activity of flavonoids and polyphenols rich plant species. *Acta Pharmaceutica* 69(4), 483–495.

Bonaz, B., Sinniger, V. and Pellissier, S. 2017. The vagus nerve in the neuro-immune axis: Implications in the pathology of the gastrointestinal tract. *Frontiers in Immunology* 8, 1452.

Bradshaw, J. E. and Ramsay, G. 2009. Potato origin and production. In: Singh, J. and Kaur, L. (Eds.), *Advances in Potato Chemistry and Technology*. Elsevier - Academic Press, New York, NY.

Brandolini, A., Castoldi, P., Plizzari, L. and Hidalgo, A. 2013. Phenolic acids composition, total polyphenols content and antioxidant activity of *Triticum monococcum*, *Triticum turgidum* and *Triticum aestivum*: A two-years evaluation. *Journal of Cereal Science* 58(1), 123–131.

Brewer, M. S. 2011. Natural antioxidants: Sources, compounds, mechanisms of action, and potential applications. *Comprehensive Reviews in Food Science and Food Safety* 10(4), 221–247.

Cabral, C. E. and Klein, M. R. S. T. 2017. Phytosterols in the treatment of hypercholesterolemia and prevention of cardiovascular diseases. *Arquivos Brasileiros de Cardiologia* 109(5), 475–482.

Campos, D., Noratto, G., Chirinos, R., Arbizu, C., Roca, W. and Cisneros-Zevallos, L. 2006. Antioxidant capacity and secondary metabolites in four species of Andean tuber crops: Native potato (Solanum sp.), mashua (Tropaeolum tuberosum Ruiz & Pavon), Oca. *Journal of the Science of Food and Agriculture* 86(10), 1481–1488.

Canxia, H., Huang, L., Lei, P., Liu, X., Li, B. and Shan, Y. 2018. Sulforaphane normalizes intestinal flora and enhances gut barrier in mice with BBN-induced bladder cancer. *Molecular Nutrition and Food Research* 62, 1800427.

Carrera-Bastos, P., Fontes-Villalba, M., O'Keefe, J. H., Lindeberg, S. and Cordain, L. 2011. The western diet and lifestyle and diseases of civilization. *Research Reports in Clinical Cardiology* 2, 15-35.

Casas, R., Castro-Barquero, S., Estruch, R. and Sacanella, E. 2018. Nutrition and cardiovascular health. *International Journal of Molecular Sciences* 19(12), 3988.

Cassidy, A., Bertoia, M., Chiuve, S., Flint, A., Forman, J. and Rimm, E. B. 2016. Habitual intake of anthocyanins and flavanones and risk of cardiovascular disease in men. *The American Journal of Clinical Nutrition* 104(3), 587-594.

Catalgol, B., Batirel, S., Taga, Y. and Ozer, N. K. 2012. Resveratrol: French paradox revisited. *Frontiers in Pharmacology* 3, 141.

Chambers, E. S., Preston, T., Frost, G. and Morrison, D. J. 2018. Role of gut microbiota-generated short-chain fatty acids in metabolic and cardiovascular health. *Current Nutrition Reports* 7(4), 198-206.

Chang, M. C., Chang, H. H., Chan, C. P., Chou, H. Y., Chang, B. E., Yeung, S. Y., Wang, T. M. and Jeng, J. H. 2012. Antiplatelet effect of phloroglucinol is related to inhibition of cyclooxygenase, reactive oxygen species, ERK/p38 signaling and thromboxane A2 production. *Toxicology and Applied Pharmacology* 263(3), 287-295.

Chaturvedi, M. and Kaczmarek, L. 2014. MMP-9 inhibition: A therapeutic strategy in ischemic stroke. *Molecular Neurobiology* 49(1), 563-573.

Chen, C. C., Zeng, Y., Xu, J., Zheng, H., Liu, J., Fan, R., Zhu, W., Yuan, L., Qin, Y., Chen, S., Zhou, Y., Wu, Y., Wan, J., Mi, M. and Wang, J. 2016. Therapeutic effects of soluble dietary fiber consumption on type 2 diabetes mellitus. *Experimental and Therapeutic Medicine* 12(2), 1232-1242.

Chen, F., Du, M., Blumberg, J. B., Chui, K. K. H., Ruan, M., Rogers, G., Shan, Z., Zeng, L. and Zhang, F. F. 2019. Association among dietary supplement use, nutrient intake, and mortality among U.S. adults: A cohort study. *Annals of Internal Medicine* 170(9), 604-613.

Cheng, T. O. 2012. Cardiovascular health, risks and diseases in contemporary China. *International Journal of Cardiology* 154(2), 233-242.

Chiu, L. W. and Li, L. 2012. Characterization of the regulatory network of BoMYB2 in controlling anthocyanin biosynthesis in purple cauliflower. *Planta* 236(4), 1153-1164.

Clare, B. A., Conroy, R. S. and Spelman, K. 2009. The diuretic effect in human subjects of an extract of Taraxacum officinale folium over a single day. *Journal of Alternative and Complementary Medicine* 15(8), 929-934.

Copper, R. 2015. Re-discovering ancient wheat varieties as functional foods. *Journal of Traditional and Complementary Medicine* 5(3), 138-143.

Cui, T., Liu, W., Chen, S., Yu, C., Li, Y. and Zhang, J. Y. 2020. Antihypertensive effects of allicin on spontaneously hypertensive rats via vasorelaxation and hydrogen sulfide mechanisms. *Biomedicine and Pharmacotherapy* 128, 110240.

Dasgeb, B., Kornreich, D., McGuinn, K., Okon, L., Brownell, I. and Sackett, D. L. 2018. Colchicine: An ancient drug with novel applications. *British Journal of Dermatology* 178(2), 350-356.

Dastani, M., Bigdelu, L., Hoseinzadeh, M., Rahimi, H. R., Karimani, A., Mohammadpour, A. H. and Salari, M. 2019. The effects of curcumin on the prevention of atrial and ventricular arrhythmias and heart failure in patients with unstable angina: A randomized clinical trial. *Avicenna Journal of Phytomedicine* 9(1), 1–9.

Dempewolf, H., Eastwood, R. J., Guarino, L., Khoury, C. K., Müller, J. V. and Toll, J. 2014. Adapting agriculture to climate change: A global initiative to collect, conserve, and use crop wild relatives. *Agroecology and Sustainable Food Systems* 38(4), 369–377.

Deo, P., Hewawasam, E., Karakoulakis, A., Claudie, D. J., Nelson, R., Simpson, B. S., Smith, N. M. and Semple, S. J. 2016. In vitro inhibitory activities of selected Australian medicinal plant extracts against protein glycation, angiotensin converting enzyme (ACE) and digestive enzymes linked to type II diabetes. *BMC Complementary and Alternative Medicine* 16(1), 435.

Di Nicolantonio, J. J., O'Keefe, J. H. and Wilson, W. 2018. Subclinical magnesium deficiency: A principle driver of cardiovascular disease and a public health crisis. *Open Heart* 5(1), e000668.

Dollo, G., Chevanne, F., Le Corre, P., Chemtob, C. and Le Verge, R. 1999. Bioavailability of phloroglucinol in man. *Journal de Pharmacie de Belgique* 54(3), 75–82.

Dou, F., Liu, Y. T., Liu, L., Wang, J., Sun, T., Mu, F., Guo, Q., Guo, C., Jia, N., Liu, W., Ding, Y. and Wen, A. 2019. Aloe-emodin ameliorates renal fibrosis via inhibiting PI3K/Akt/mTOR signaling pathway *in vivo* and *in vitro*. *Rejuvenation Research* 22(3), 218–229.

Einarson, T. R., Acs, A., Ludwig, C. and Panton, U. H. 2018. Prevalence of cardiovascular disease in type 2 diabetes: A systematic literature review of scientific evidence from across the world in 2007–2017. *Cardiovascular Diabetology* 17(1), 83.

Ende, C. and Gebhardt, R. 2004. Inhibition of matrix metalloproteinase-2 and -9 activities by selected flavonoids. *Planta Medica* 70(10), 1006–1008.

Escarnot, E., Jacquemin, J.-M., Agneessens, R. and Paquot, M. 2012. Comparative study of the content and profiles of macronutrients in spelt and wheat, a review. *Base* 16, 243–256.

Estruch, R., Ros, E., Salas-Salvadó, J., Covas, M. I., Corella, D., Arós, F., Gómez-Gracia, E., Ruiz-Gutiérrez, V., Fiol, M., Lapetra, J., Lamuela-Raventos, R. M., Serra-Majem, L., Pintó, X., Basora, J., Muñoz, M. A., Sorli, J. V., Martínez, J. A., Fitó, M., Gea, A., Hernán, M. A., Martínez-González, M. A. and PREDIMED Study Investigators 2018. Primary prevention of cardiovascular disease with a Mediterranean diet supplemented with extra-virgin olive oil or nuts. *New England Journal of Medicine* 378(25), e34.

Flora, G. D. and Nayak, M. K. 2019. A brief review of cardiovascular diseases, associated risk factors and current treatment regimes. *Current Pharmaceutical Design* 25(38), 4063–4084.

Fujii, H., Goto, S. and Fukagawa, M. 2016. Role of uremic toxins for kidney, cardiovascular, and bone dysfunction. *Toxins* 10(5), 202.

Galluzzi, L., Vitale, I., Aaronson, S. A., Abrams, J. M., Adam, D., Agostinis, P., Alnemri, E. S., Altucci, L., Amelio, I., Andrews, D. W., Annicchiarico-Petruzzelli, M., Antonov, A. V., Arama, E., Baehrecke, E. H., Barlev, N. A., Bazan, N. G., Bernassola, F., Bertrand, M. J. M., Bianchi, K., Blagosklonny, M. V., Blomgren, K., Borner, C., Boya, P., Brenner, C., Campanella, M., Candi, E., Carmona-Gutierrez, D., Cecconi, F., Chan, F. K., Chandel, N. S., Cheng, E. H., Chipuk, J. E., Cidlowski, J. A., Ciechanover, A., Cohen, G. M., Conrad, M., Cubillos-Ruiz, J. R., Czabotar, P. E., D'Angiolella, V., Dawson, T. M., Dawson, V. L., De Laurenzi, V., De Maria, R., Debatin, K. M., Deberardinis, R. J., Deshmukh, M., di Daniele, N., di Virgilio, F., Dixit, V. M., Dixon, S. J., Duckett, C. S., Dynlacht, B. D., El-Deiry, W. S., Elrod, J. W., Fimia, G. M., Fulda, S., Garcia-Saez, A. J., Garg, A. D., Garrido, C., Gavathiotis, E., Golstein, P., Gottlieb, E., Green,

D. R., Greene, L. A., Gronemeyer, H., Gross, A., Hajnoczky, G., Hardwick, J. M., Harris, I. S., Hengartner, M. O., Hetz, C., Ichijo, H., Jaattela, M., Joseph, B., Jost, P. J., Juin, P. P., Kaiser, W. J., Karin, M., Kaufmann, T., Kepp, O., Kimchi, A., Kitsis, R. N., Klionsky, D. J., Knight, R. A., Kumar, S., Lee, S. W., Lemasters, J. J., Levine, B., Linkermann, A., Lipton, S. A., Lockshin, R. A., Lopez-Otin, C., Lowe, S. W., Luedde, T., Lugli, E., Macfarlane, M., Madeo, F., Malewicz, M., Malorni, W., Manic, G., Marine, J. C., Martin, S. J., Martinou, J. C., Medema, J. P., Mehlen, P., Meier, P., Melino, S., Miao, E. A., Molkentin, J. D., Moll, U. M., Muñoz-Pinedo, C., Nagata, S., Nuñez, G., Oberst, A., Oren, M., Overholtzer, M., Pagano, M., Panaretakis, T., Pasparakis, M., Penninger, J. M., Pereira, D. M., Pervaiz, S., Peter, M. E., Piacentini, M., Pinton, P., Prehn, J. H. M., Puthalakath, H., Rabinovich, G. A., Rehm, M., Rizzuto, R., Rodrigues, C. M. P., Rubinsztein, D. C., Rudel, T., Ryan, K. M., Sayan, E., Scorrano, L., Shao, F., Shi, Y., Silke, J., Simon, H. U., Sistigu, A., Stockwell, B. R., Strasser, A., Szabadkai, G., Tait, S. W. G., Tang, D., Tavernarakis, N., Thorburn, A., Tsujimoto, Y., Turk, B., Vanden Berghe, T., Vandenabeele, P., Vander Heiden, M. G., Villunger, A., Virgin, H. W., Vousden, K. H., Vucic, D., Wagner, E. F., Walczak, H., Wallach, D., Wang, Y., Wells, J. A., Wood, W., Yuan, J., Zakeri, Z., Zhivotovsky, B., Zitvogel, L., Melino, G. and Kroemer, G. 2018. Molecular mechanisms of cell death: Recommendations of the Nomenclature Committee on Cell Death 2018. *Cell Death and Differentiation* 25(3), 486–541.

Gasparini, K., Moreira, J. D. R., Peres, L. E. P. and Zsögön, A. 2020. De novo domestication of wild species to create crops with increased resilience and nutritional value. *Current Opinion in Plant Biology* 60, 102006.

Genser, B., Silbernagel, G., Backer, G. D., Bruckert, E., Carmena, R., Chapman, M. J., Deanfield, J., Descamps, O. S., Rietzschel, E. R., Dias, K. C. and März, W. 2012. Plant sterols and cardiovascular disease: A systematic review and meta-analysis. *European Heart Journal* 33(4), 444–451.

Ginwala, R., Bhavsar, R., Chigbu, D. I., Jain, P. and Khan, Z. K. 2019. Potential role of flavonoids in treating chronic inflammatory diseases with a special focus on the anti-inflammatory activity of apigenin. *Antioxidants (Basel)* 8(2), 35.

Golledge, J. and Norman, P. E. 2010. Atherosclerosis and abdominal aortic aneurysm: Cause, response or common risk factors? *Arteriosclerosis, Thrombosis, and Vascular Biology* 30(6), 1075–1077.

Gunawardena, D., Karunaweera, N., Lee, S., Van Der Kooy, F., Harman, D. G., Raju, R., Bennett, L., Gyengesi, E., Sucher, N. J. and Münch, G. 2015. Anti-inflammatory activity of cinnamon (*C. zeylanicum* and *C. cassia*) extracts – Identification of *E*-cinnamaldehyde and o-methoxy cinnamaldehyde as the most potent bioactive compounds. *Food and Function* 6(3), 910–919.

Hamden, K., Jaouadi, B., Carreau, S., Bejar, S. and Elfeki, A. 2010. Inhibitory effect of fenugreek galactomannan on digestive enzymes related to diabetes, hyperlipidemia, and liver-kidney dysfunctions. *Biotechnology and Bioprocess Engineering* 15(3), 407–413.

Harijono, H., Estiasih, T., Ariestiningsih, A. and Wardani, N. 2016. The effect of crude diosgenin extract from purple and yellow greater yams (*Dioscorea alata*) on the lipid profile of dyslipidemia rats. *Emirates Journal of Food and Agriculture* 28(7), 506–512.

Haugsgjerd, T. R., Egeland, G. M., Nygård, O. K., Vinknes, K. J., Sulo, G., Lysne, V., Igland, J. and Tell, G. S. 2020. Association of dietary vitamin K and risk of coronary heart disease in middle-age adults: The Hordaland Health Study Cohort. *BMJ Open* 10(5), e035953.

Hazard, B., Trafford, K., Lovegrove, A., Griffiths, S., Uauy, C. and Shewry, P. R. 2020. Strategies to improve wheat for human health. *Nature Food* 1(8), 475–480.

Hegab, Z., Gibbons, S., Neyses, L. and Mamas, M. A. 2012. Role of advanced glycation end products in cardiovascular disease. *World Journal of Cardiology* 4(4), 90–102.

Helgadottir, A., Thorleifsson, G. and Stefansson, K. 2021. Increased absorption of phytosterols is the simplest and most plausible explanation for coronary artery disease risk not accounted for by non-HDL cholesterol in high cholesterol absorbers. *European Heart Journal* 42(3), 283–284.

Henry, C. J. K., Lightowler, H. J., Strik, C. M. and Storey, M. 2005. Glycaemic index values for commercially available potatoes in Great Britain. *British Journal of Nutrition* 94(6), 917–921.

Huang, L., Chen, J., Quan, J. and Xiang, D. 2021. Rosmarinic acid inhibits proliferation and migration, promotes apoptosis and enhances cisplatin sensitivity of melanoma cells through inhibiting ADAM17/EGFR/AKT/GSK3beta axis. *Bioengineered* 12(1), 3065–3076.

Ingles, D. P., Rodriguez, J. B. C. and Garcia, H. 2020. Supplemental vitamins and minerals for cardiovascular disease prevention and treatment. *Current Cardiology Reports* 22(4), 22.

Jackson, S. P. 2011. Arterial thrombosis–Insidious, unpredictable and deadly. *Nature Medicine* 17(11), 1423–1436.

Janick, J. 2014. Dioscoridean herbals: Comparisons of images provide a clue to their relationships. *Acta Horticulturae* 1125(1125), 235–248.

Jayachandran, M., Chen, J., Chung, S. S. M. and Xu, B. 2018. A critical review on the impacts of β-glucans on gut microbiota and human health. *The Journal of Nutritional Biochemistry* 61, 101–110.

Jeandet, P., Bessis, R., Shaghi, M. and Meunier, P. 1995. Production of the phytoalexin resveratrol by grapes as a response to *Botrytis* attack Under natural conditions. *Journal of Phytopathology* 143(3), 135–139.

Jenkins, D. J. A., Spence, J. D., Giovannucci, E. L., Kim, Y. I., Josse, R. G., Vieth, R., Sahye-Pudaruth, S., Paquette, M., Patel, D., Blanco Mejia, S., Viguiliouk, E., Nishi, S. K., Kavanagh, M., Tsirakis, T., Kendall, C. W. C., Pichika, S. C. and Sievenpiper, J. L. 2021. Supplemental vitamins and minerals for cardiovascular disease prevention and treatment: JACC focus seminar. *Journal of the American College of Cardiology* 77(4), 423–436.

Jetté, L., Harvey, L., Eugeni, K. and Levens, N. 2009. 4-Hydroxyisoleucine: A plant-derived treatment for metabolic syndrome. *Current Opinion in Investigational Drugs* 10(4), 353–358.

Jin, M., Qian, Z., Yin, J., Xu, W. and Zhou, X. 2018. The role of intestinal microbiota in cardiovascular disease. *Journal of Cellular and Molecular Medicine* 23(4), 2343–2350.

Johnson, R. J., Sánchez-Lozada, L. G., Andrews, P. and Lanaspa, M. A. 2017. Perspective: A historical and scientific perspective of sugar and its relation with obesity and diabetes. *Advances in Nutrition* 8(3), 412–422.

Jones, D. S. and Greene, J. A. 2013. The decline and rise of coronary heart disease: Understanding public health catastrophism. *American Journal of Public Health* 103(7), 1207–1218.

Joshua, B. O. and Muyiwa, A. 2019. Effects of alkaloids of *Cocos nucifera* husk fibre on cardiovasular disease indices in albino mice. *Cardiovascular Pharmacology* 8, 253.

Just, B. J., Santos, C. A. F., Yandell, B. S. and Simon, P. W. 2009. Major QTL for carrot color are positionally associated with carotenoid biosynthetic genes and interact epistatically in a domesticated x wild carrot cross. *TAG. Theoretical and Applied Genetics. Theoretische und Angewandte Genetik* 119(7), 1155–1169.

Kang, I., Buckner, T., Shay, N. F., Gu, L. and Chung, S. 2016. Improvements in metabolic health with consumption of ellagic acid and subsequent conversion into urolithins: Evidence and mechanisms. *Advances in Nutrition* 7(5), 961–972.

Katsukawa, M., Nakata, R., Takizawa, Y., Hori, K., Takahashi, S. and Inoue, H. 2010. Citral, a component of lemongrass oil, activates PPARα and γ and suppresses COX-2 expression. *Biochimica et Biophysica Acta* 1801(11), 1214–1220.

Kaur, R. and Myrie, S. B. 2020. Association of dietary phytosterols with cardiovascular disease biomarkers in humans. *Lipids* 55(6), 569–584.

Kawabata, K., Yoshioka, Y. and Terao, J. 2019. Role of intestinal microbiota in the bioavailability and physiological functions of dietary polyphenols. *Molecules* 24(2), 370.

Kawai, T., Elliott, K. J., Scalia, R. and Eguchi, S. 2021. Contribution of ADAM17 and related ADAMs in cardiovascular diseases. *Cellular and Molecular Life Sciences* 78(9), 4161–4187.

Khalki, L., Bennis, M., Sokar, Z. and Ba-M'hamed, S. 2012. The developmental neurobehavioral effects of fenugreek seeds on prenatally exposed mice. *Journal of Ethnopharmacology* 139(2), 672–677.

Khan, H., Jawad, M., Kamal, M. A., Baldi, A., Xiao, J., Nabavi, S. M. and Daglia, M. 2018. Evidence and prospective of plant derived flavonoids as antiplatelet agents: Strong candidates to be drugs of future. *Food and Chemical Toxicology* 119, 355–367.

Kim, H. R., Kim, W. K. and Ha, A. W. 2019. Effects of phytochemicals on blood pressure and neuroprotection mediated via brain renin-angiotensin system. *Nutrients* 11(11), 2761.

Komal, S., Zhang, L. R. and Han, S. N. 2021. Potential regulatory role of epigenetic RNA methylation in cardiovascular diseases. *Biomedicine and Pharmacotherapy* 137, 111376.

Koss-Mikolajczyk, I., Baranowska, M., Todorovic, V., Albini, A., Sansone, C., Andreoletti, P., Cherkaoui-Malki, M., Lizard, G., Noonan, D., Sobajic, S. and Bartoszek, A. 2019. Prophylaxis of non-communicable diseases: Why fruits and vegetables may be better chemopreventive agents than dietary supplements based on isolated phytochemicals? *Current Pharmaceutical Design* 25(16), 1847–1860.

Kubo, E., Chhunchha, B., Singh, P., Sasaki, H. and Singh, D. P. 2017. Sulforaphane reactivates cellular antioxidant defense by inducing Nrf2/ARE/Prdx6 activity during aging and oxidative stress. *Scientific Reports* 7(1), 14130.

Kumarasamy, Y., Nahar, L., Byres, M., Delazar, A. and Sarker, S. D. 2005. The assessment of biological activities associated with the major constituents of the methanol extract of 'wild carrot' (Daucus carota L.) seeds. *Journal of Herbal Pharmacotherapy* 5(1), 61–72.

La Fata, G., Rastall, R. A., Lacroix, C., Harmsen, H. J. M., Mohajeri, M. H., Webber, P. and Steinert, R. E. 2017. Recent development of prebiotic research–Statement from an expert workshop. *Nutrients* 9(12), 1376.

Labuschagne, M. T., Geleta, N. and Osthoff, G. 2007. The influence of environment on starch content and amylose to amylopectin ratio in wheat. *Starch - Stärke* 59(5), 234-238.

Lamuel-Raventos, R. M. and Onge, M.-P. S. 2017. Prebiotic nut compounds and human microbiota. *Critical Reviews in Food Science and Nutrition* 57(14), 3154-3163.

Lee, M.-R., Swanson, B. G. and Baik, B.-K. 2001. Influence of amylose content on properties of wheat starch and breadmaking quality of starch and gluten blends. *Cereal Chemistry Journal* 78(6), 701-706.

Li, N., Jiang, W., Wang, W., Xiong, R., Wu, X. and Geng, Q. 2021. Ferroptosis and its emerging roles in cardiovascular diseases. *Pharmacological Research* 166, 105466.

Lind, L., Sundström, J., Ärnlöv, J., Risérus, U. and Lampa, E. 2021. A longitudinal study over 40 years to study the metabolic syndrome as a risk factor for cardiovascular diseases. *Scientific Reports* 11(1), 2978.

Lippi, G. and Franchini, M. 2011. Vitamin K in neonates: Facts and myths. *Blood Transfusion* 9(1), 4-9.

Liu, S. 2002. Intake of refined carbohydrates and whole grain foods in relation to risk of type 2 diabetes mellitus and coronary heart disease. *Journal of the American College of Nutrition* 21(4), 298-306.

Liu, W., Liu, P., Tao, S., Deng, Y., Li, X., Lan, T., Zhang, X., Guo, F., Huang, W., Chen, F., Huang, H. and Zhou, S. F. 2008. Berberine inhibits aldose reductase and oxidative stress in rat mesangial cells cultured under high glucose. *Archives of Biochemistry and Biophysics* 475(2), 128-134.

Liu, X., Guasch-Ferré, M., Drouin-Chartier, J. P., Tobias, D. K., Bhupathiraju, S. N., Rexrode, K. M., Willett, W. C., Sun, Q. and Li, Y. 2020. Changes in nut consumption and subsequent cardiovascular disease risk Among US men and women: 3 large prospective cohort studies. *Journal of the American Heart Association* 9(7), e013877.

Liu, Z., Lin, X., Huang, G., Zhang, W., Rao, P. and Ni, L. 2014. Prebiotic effects of almonds and almond skins on intestinal microbiota in healthy adult humans. *Anaerobe* 26, 1-6.

Lv, H., Miyaji, N., Osabe, K., Akter, A., Mehraj, H., Shea, D. J. and Fujimoto, R. 2020. The importance of genetic and epigenetic research in the *brassica* vegetables in the face of climate change. In: Kole, C. (Ed.). *Genomic Designing of Climate-Smart Vegetable Crops*. Switzerland AG: Springer Nature.

Malemud, C. J. 2019. Inhibition of MMPs and ADAM/ADAMTS. *Biochemical Pharmacology* 165, 33-40.

Mathern, J. R., Raatz, S. K., Thomas, W. and Slavin, J. L. 2009. Effect of fenugreek fiber on satiety, blood glucose and insulin response and energy intake in obese subjects. *Phytotherapy Research* 23(11), 1543-1548.

Mattiello, T., Trifirò, E., Jotti, G. S. and Pulcinelli, F. M. 2009. Effects of pomegranate juice and extract polyphenols on platelet function. *Journal of Medicinal Food* 12(2), 334-339.

Mayo, B., Vázquez, L. and Flórez, A. B. 2019. Equol: A bacterial metabolite from the daidzein isoflavone and its presumed beneficial health effects. *Nutrients* 11(9), 2231.

Menghini, R., Fiorentino, L., Casagrande, V., Lauro, R. and Federici, M. 2013. The role of ADAM17 in metabolic inflammation. *Atherosclerosis* 228(1), 12-17.

Micek, A., Godos, J., Del Rio, D., Galvano, F. and Grosso, G. 2021. Dietary flavonoids and cardiovascular disease: A comprehensive dose-response meta-analysis. *Molecular Nutrition and Food Research* 65(6), e2001019.

Michel, J., Rani, N. Z. A. and Husain, K. 2020. A review on the potential use of medicinal plants from Asteraceae and Lamiaceae plant family in cardiovascular diseases. *Frontiers in Pharmacology* 11, 852.

Mirmirani, P., Bahadoran, Z., Vakili, A. Z. and Azizi, F. 2017. Western dietary pattern increases risk of cardiovascular disease in Iranian adults: A prospective population-based study. *Applied Physiology, Nutrition, and Metabolism* 42(3), 326-332.

Montagnana, M., Danese, E., Angelino, D., Mena, P., Rosi, A., Benati, M., Gelati, M., Salvagno, G. L., Favaloro, E. J., Del Rio, D. and Lippi, G. 2018. Dark chocolate modulates platelet function with a mechanism mediated by flavan-3-ol metabolites. *Medicine (Baltimore)* 97(49), e13432.

Montilla, E. C., Hillebrand, S. and Winterhalter, P. 2011. Anthocyanins in purple sweet potato (Ipomoea batatas L.) varieties. *Fruit, Vegetable and Cereal Science and Biotechnology* 5, 19-24.

Mukhtar, H., Mumtaz, M. W., Tauqueer, T. and Raza, S. A. 2021. Composition of *Nigella sativa* seeds. In: Ramadan, M. F. (Ed.). *Black Cumin (Nigella sativa) Seeds: Chemistry, Technology, Functionality, and Applications: Food Bioactive Ingredients*. Springer, Cham.

Mukhtar, K., Nawaz, H. and Abid, S. 2019. Functional gastrointestinal disorders and gut-brain axis: What does the future hold? *World Journal of Gastroenterology* 25(5), 552-566.

Nileeka Balasuriya, B. W. and Vasantha Rupasinghe, H. P. 2011. Plant flavonoids as angiotensin converting enzyme inhibitors in regulation of hypertension. *Functional Foods in Health and Disease* 1(5), 172-188.

Ohland, C. L. and Macnoughton, W. K. 2010. Probiotic bacteria and intestinal epithelial barrier function. *American Journal of Physiology. Gastrointestinal and Liver Physiology* 298(6), G807-G819.

Onal, E. M., Afsar, B., Covic, A., Vaziri, N. D. and Kanbay, M. 2019. Gut microbiota and inflammation in chronic kidney disease and their roles in the development of cardiovascular disease. *Hypertension Research* 42(2), 123-140.

Ong, K. W., Hsu, A. and Tan, B. K. H. 2013. Anti-diabetic and anti-lipidemic effects of chlorogenic acid are mediated by ampk activation. *Biochemical Pharmacology* 85(9), 1341-1351.

Orbe, J., Fernandez, L., Rodríguez, J. A., Rábago, G., Belzunce, M., Monasterio, A., Roncal, C. and Páramo, J. A. 2003. Different expression of MMPs/TIMP-1 in human atherosclerotic lesions. Relation to plaque features and vascular bed. *Atherosclerosis* 170(2), 269-276.

Padilla-González, G. F., Amrehn, E., Frey, M., Gómez-Zeledón, J., Kaa, A., Da Costa, F. B. and Spring, O. 2020a. Metabolomic and gene expression studies reveal the diversity, distribution and spatial regulation of the specialized metabolism of Yacón (Smallanthus sonchifolius, Asteraceae). *International Journal of Molecular Sciences* 21(12), 4555.

Padilla-González, G. F., Sadgrove, N. J., Ccana-Ccapatinta, G. V., Leuner, O. and Fernandez-Cusimamani, E. 2020b. Feature-based molecular networking to target the isolation of new caffeic acid esters from Yacon (Smallanthus sonchifolius, Asteraceae). *Metabolites* 10(10), 407.

Palma-Tenango, M., Soto-Hernández, M. and Aguirre-Hernández, E. 2017. Flavonoids in agriculture. In: Justino, C. (Ed.). *Flavonoids - From Biosynthesis to Human Health*. IntechOpen.

Palmer, C. R., Blekkenhorst, L. C., Lewis, J. R., Ward, N. C., Schultz, C. J., Hodgson, J. M., Croft, K. D. and Sim, M. 2020. Quantifying dietary vitamin K and its link to cardiovascular health: A narrative review. *Food and Function* 11(4), 2826–2837.

Panth, N., Paudel, K. R. and Parajuli, K. 2016. Reactive oxygen species: A key hallmark of cardiovascular disease. *Advances in Medicine* 2016, 9152732.

Peng, J., Sun, D. and Nevo, E. 2011. Wild emmer wheat, *Triticum dicoccoides', occupies a pivotal position in wheat domestication process. Australian Journal of Crop Science* 5, 1127–1143.

Peterson, J., Dwyer, J., Adlercreutz, H., Scalbert, A., Jacques, P. and McCullough, M. L. 2010. Dietary lignans: Physiology and potential for cardiovascular disease risk reduction. *Nutrition Reviews* 68(10), 571–603.

Petrie, J. R., Gunzik, T. J. and Touyz, R. M. 2018. Diabetes, hypertension, and cardiovascular disease: Clinical insights and vascular mechanisms. *Canadian Journal of Cardiology* 34(5), 575–584.

Pietta, P. G. 2000. Flavonoids as antioxidants. *Journal of Natural Products* 63(7), 1035–1042.

Piwowarski, J. P., Granica, S. and Kiss, A. K. 2014. Influence of gut microbiota-derived ellagitannins' metabolites urolithins on pro-inflammatory activities of human neutrophils. *Planta Medica* 80(11), 887–895.

Plat, J., Strandberg, T. E. and Gylling, H. 2021. Intestinal cholesterol and phytosterol absorption and the risk of coronary artery disease. *European Heart Journal* 42(3), 281–282.

Que, F., Hous, X. L., Wang, G. L., Xu, Z. S., Tan, G. F., Li, T., Wang, Y. H., Khadr, A. and Xiong, A. S. 2019. Advances in research on the carrot, an important root vegetable in the Apiaceae family. *Horticulture Research* 6, 69.

Quezada, M. P., Salinas, C., Gotteland, M. and Cardemil, L. 2017. Acemannan and fructans from Aloe vera (*Aloe barbadensis* Miller) plants as novel prebiotics. *Journal of Agricultural and Food Chemistry* 65(46), 10029–10039.

Rabe, K. F., Hurst, J. R. and Suissa, S. 2018. Cardiovascular disease and COPD: Dangerous liaisons? *European Respiratory Review* 27(149), 180057.

Rad, S. Z. K., Javadi, B., Hayes, A. W. and Karimi, G. 2019. Potential angiotensin converting enzyme (ACE) inhibitors from Iranian traditional plants described by Avicenna's Canon of Medicine. *Avicenna Journal of Phytomedicine* 9(4), 291–309.

Raza, A., Razzaq, A., Mehmood, S. S., Zou, X., Zhang, X., Lv, Y. and Xu, J. 2019. Impact of climate change on crops adaptation and strategies to tackle its outcome: A review. *Plants* 8(2), 34.

Renaud, S. and De Lorgeril, M. 1992. Wine, alcohol, platelets, and the French paradox for coronary heart disease. *Lancet* 339(8808), 1523–1526.

Ribeiro, M., Alvarenga, L., Cardozo, L. F. M. F., Chermut, T. R., Sequera, J., Moreira, L. D. S., Teixeira, K. T. R., Shiels, P. G., Stenvinkel, P. and Mafra, D. 2021. From the distinctive smell to therapeutic effects: Garlic for cardiovascular, hepatic, gut, diabetes and chronic kidney disease. *Clinical Nutrition* 40(7), 4807–4819. DOI: 10.1016/j.clnu.2021.03.005.

Rodriguez-Quijano, M., Lucas, R. and Carrillo, J. M. 2003. Waxy proteins and amylose content in tetraploid wheats *Triticum dicoccum* Schulb, *Triticum durum* L. and *Triticum polonicum* L. *Euphytica* 134(1), 97–101.

Roman, M., Debrowolski, J. C., Baranska, M. and Baranski, R. 2011. Spectroscopic studies on bioactive polyacetylenes and other plant components in wild carrot root. *Journal of Natural Products* 74(8), 1757–1763.

Rouhi-Boroujeni, H., Heidarian, E., Rouhi-Boroujeni, H., Deris, F. and Rafieian-Kopaei, M. 2017. Medicinal plants with multiple effects on cardiovascular diseases: A systematic review. *Current Pharmaceutical Design* 23(7), 999–1015.

Sadgrove, N. J. 2020. Comparing essential oils from Australia's 'Victorian Christmas Bush' (*Prostanthera lasianthos* Labill., Lamiaceae) to closely allied new species: Phenotypic plasticity and taxonomic variability. *Phytochemistry* 176, 112403.

Sadgrove, N. J. 2021. Honest nutraceuticals, cosmetics, therapies, and foods (NCTFs): Standardization and safety of natural products. *Critical Reviews in Food Science and Nutrition* 62(16), 4326–4341. DOI: 10.1080/10408398.2021.1874286.

Sadgrove, N. J., Gonçalves-Martins, M. and Jones, G. L. 2014. Chemogeography and antimicrobial activity of essential oils from *Geijera parviflora* and *Geijera salicifolia* (Rutaceae): Two traditional Australian medicinal plants. *Phytochemistry* 104, 60–71.

Sadgrove, N. J. and Jones, G. L. 2014. Cytogeography of essential oil chemotypes of *Eremophila longifolia* F. Muell (Schrophulariaceae). *Phytochemistry* 105, 43–51.

Sadgrove, N. J. and Jones, G. L. 2019. From Petri dish to patient: Bioavailability estimation and mechanism of action for antimicrobial and immunomodulatory natural products. *Frontiers in Microbiology* 10, 2470.

Sadgrove, N. J., Padilla-González, G. F., Telford, I. R. H., Greatrex, B. W., Jones, G. L., Andrew, R., Bruhl, J. J., Langat, M. K., Melnikovova, I. and Fernandez-Cusimamani, E. 2020. *Prostanthera* (Lamiaceae) as a 'cradle of incense': Chemophenetics of rare essential oils from both new and forgotten Australian 'mint bush' species. *Plants* 9(11), 1570.

Scazzina, F., Siebenhandl-Ehn, S. and Pellegrini, N. 2013. The effect of dietary fibre on reducing the glycaemic index of bread. *British Journal of Nutrition* 109(7), 1163–1174.

Selby-Pham, S. N. B., Miller, R. B., Howell, K., Dunshea, F. and Bennett, L. E. 2017. Physiochemical properties of dietary phytochemicals can predict their passive ansorption in the human small intestine, Nature. *Scientific Reports* 7(1), 1931.

Septembre-Malaterre, A., Remize, F. and Poucheret, P. 2017. Fruits and vegetables, as a source of nutritional compounds and phytochemicals: Changes in bioactive compounds during lactic fermentation. *Food Research International* 104, 86–99.

Shamsi, S., Tran, H., Tan, R. S., Tan, Z. J. and Lim, L. Y. 2017. Curcumin, piperine, and capsaicin: A comparative study of spice-mediated inhibition of human cytochrome P450 isozyme activities. *Drug Metabolism and Disposition: The Biological Fate of Chemicals* 45(1), 49–55.

Shang, M., Cai, S., Han, J., Li, J., Zhao, Y., Zheng, J., Namba, T., Kadota, S., Tezuka, Y. and Fan, W. 1998. Studies on flavonoids from Fenugreek (*Trigonella foenumgraecum* L.). *Zhonggou Zhong Yao Za Zhi* 23(10), 614–616.

Sharifi, N., Souri, E., Ziai, S. A., Amin, G. and Amanlou, M. 2013. Discovery of new angiotensin converting enzyme (ACE) inhibitors from medicinal plants to treat hypertension using an *in vitro* assay. *Daru* 21(1), 74.

Sharma, A., Thakur, R., Lingaraju, M. C., Kumar, D., Mathesh, K., Telang, A. G., Singh, T. U. and Kumar, D. 2017. Betulinic acid attenuates renal fibrosis in rat chronic kidney disease model. *Biomedicine and Pharmacotherapy* 89, 796–804.

Shewry, P. R. and Hey, S. J. 2015. The contribution of wheat to human diets and health. *Food and Energy Security* 4(3), 178–202.

Singh, B. K., Singh, B. and Singh, P. M. 2018. Breeding cauliflower: A review. *International Journal of Vegetable Science* 24(1), 58–84.

Singh, J., Dartois, A. and Kaur, L. 2010. Starch digestibility in food matrix: A review. *Trends in Food Science and Technology* 21(4), 168–180.

Smith, L. B. and King, G. J. 2000. The distribution of BoCAL-a alleles in *Brassica oleracea* is consistent with a genetic model for curd development and domestication of the cauliflower. *Molecular Breeding* 6(6), 603–613.

Spalding, A., Vaitkevicius, H., Dill, S., Mackenzie, S., Schmaier, A. and Lockette, W. 1998. Mechanism of epinephrine-induced platelet aggregation. *Hypertension* 31(2), 603–607.

Spooner, D. M., McLean, K., Ramsay, G., Waugh, R. and Bryan, G. J. 2005. A single domestication for potato based on multilocus amplified fragment length polymorphism genotyping. *PNSA: Proceedings of the National Academy of Sciences of the United States of America* 102(41), 14694–14699.

Stefanson, A. L. and Bakovic, M. 2018. Falcarinol is a potent inducer of heme Oxygenase-1 and was more effective than sulforaphane in attenuating intestinal inflammation at diet-achievable doses. *Oxidative Medicine and Cellular Longevity* 2018, 3153527.

Takeoka, G. R., Dao, L. T., Full, G. H., Wong, R. Y., Harden, L. A., Edwards, R. H. and Berrios, J. D. J. 1997. Characterization of black bean (*Phaseolus vulgaris* L.) anthocyanins. *Journal of Agricultural and Food Chemistry* 45(9), 3395–3400.

Thies, F., Mills, L. M., Moir, S. and Masson, L. F. 2016. Cardivoascular benefits fo lycopene: Fantasy or reality? *Proceedings of the Nutrition Society* 76, 122–129.

Undas, A., Brummel-Ziedins, K. E. and Mann, K. G. 2007. Antithrombotic properties of aspirin and resistance to aspirin: Beyond strictly antiplatelet actions. *Blood* 109(6), 2285–2292.

Van Amelsvoort, J. M. M. and Westrate, J. A. 1992. Amylose-amylopectin ratio in a meal affects postprandial variables in male volunteers. *American Journal of Clinical Nutrition* 55(3), 712–718.

Vanduchova, A., Anzenbacher, P. and Anzenbacherova, E. 2019. Isothiocyanate from broccoli, sulforaphane, and its properties. *Journal of Medicinal Food* 22(2), 121–126.

Vazquez-Prieto, M. A. and Miatello, R. M. 2010. Organosulfur compounds and cardiovascular disease. *Molecular Aspects of Medicine* 31(6), 540–545.

Vijayakumar, M. V., Pandey, V., Mishra, G. C. and Bhat, M. K. 2010. Hypolipidemic effect of fenugreek seeds is mediated through inhibition of fat accumulation and upregulation of LDL receptor. *Obesity* 18(4), 667–674.

Villines, T. C., Hatzigeorgiou, C., Feuerstein, I. M., O'Malley, P. G. and Taylor, A. J. 2005. Vitamin K1 intake and coronary calcification. *Coronary Artery Disease* 16(3), 199–203.

Walker, J. C. 1928. A survey of the resistance of subspecies of Brassica oleracea to yellows (fusarium con-GLUTINANS). *Journal of Agricultural Research* 37, 233–241.

Wang, H., Yang, Y. J., Qian, H. Y., Zhang, Q., Xu, H. and Li, J. J. 2012. Resveratrol in cardiovascular disease: What is known from current research? *Heart Failure Reviews* 17(3), 437–448.

Wang, L., Palme, V., Schilcher, N., Ladurner, A., Heiss, E. H., Stangl, H., Bauer, R., Dirsch, V. M. and Atanasov, A. G. 2017. The dietary constituent falcarindiol promotes cholesterol efflux from THP-1 macrophages by increasing ABCA1 gene transcription and protein stability. *Frontiers in Pharmacology* 8, 596.

Wang, X., Li, Y., Han, L., Li, J., Liu, C. and Sun, C. 2021. Role of flavonoids in the treatment of iron overload. *Frontiers in Cell and Developmental Biology* 9, 685364.

Wang, Z. M., Zhao, D., Nie, Z. L., Zhao, H., Zhou, B., Gao, W., Wang, L. S. and Yang, Z. J. 2014. Flavonol intake and stroke risk: A meta-analysis of cohort studies. *Nutrition* 30(5), 518–523.

Wen, L., Chen, J., Duan, L. and Li, S. 2018. Vitamin K-dependent proteins involved in bone and cardiovascular health (review). *Molecular Medicine Reports* 18(1), 3–15.

Wieser, H., Koehler, P. and Scherf, K. A. 2020. The two faces of wheat. *Frontiers in Nutrition* 7, 517313. DOI: 10.3389/fnut.2020.517313.

Williams, P. G. 2012. Evaluation of the e aluation of the evidence between consumption of r vidence between consumption of refined gr efined grains and health ains and health outcomes. *Nutrition Reviews* 70(2), 80–99.

Witkowska, A. M., Waśkiewicz, A., Zujko, M. E., Szcześniewska, D., Stepaniak, U., Pająk, A. and Drygas, W. 2018. Are total and individual dietary lignans related to cardiovascular disease and its risk factors in postmenopausal women? A nationwide study. *Nutrients* 10(7), 865.

Xie, Y., Song, X., Sun, X., Huang, J., Zhong, M., Lotze, M. T., Zeh, H. J., Kang, R. and Tang, D. 2016. Identification of baicalein as a ferroptosis inhibitor by natural product library screening. *Biochemical and Biophysical Research Communications* 473(4), 775–780.

Xu, Y., Rong, J. and Zhang, Z. 2021. The emerging role of angiotensinogen in cardiovascular diseases. *Journal of Cellular Physiology* 236(1), 68–78.

Yan, C., An, G., Zhu, T., Zhang, W., Zhang, L., Peng, L., Chen, J. and Kuang, H. 2019. Independent activation of the BoMYB2 gene leading to purple traits in *Brassica oleracea. TAG. Theoretical and Applied Genetics. Theoretische und Angewandte Genetik* 132(4), 895–906.

Yang, J., Jose, P. A. and Zeng, C. 2017. Gastrointestinal-renal axis: Role in the regulation of blood pressure. *Journal of the American Heart Association* 6(3), e005536.

Yeger, H. and Mokhtari, R. B. 2020. Perspective on dietary isothiocyanates in the prevention, development and treatment of cancer. *Journal of Cancer Metastasis and Treatment* 2020, 26.

Yu, C., Zhang, P., Lou, L. and Wang, Y. 2019. Perspectives regarding the role of biochanin A in humans. *Frontiers in Pharmacology* 10, 793.

Yuan, H., Ma, Q., Cui, H., Liu, G., Zhao, X., Li, W. and Piao, G. 2017. How can synergism of traditional medicines benefit from network pharmacology? *Molecules* 22(7), 1135.

Zhang, W., Song, M., Qu, J. and Liu, G. H. 2018. Epigenetic modifications in cardiovascular aging and diseases. *Circulation Research* 123(7), 773–786.

Zhou, X., Li, J., Guo, J., Geng, B., Ji, W., Zhao, Q., Li, J., Liu, X., Liu, J., Guo, Z., Cai, W., Ma, Y., Ren, D., Miao, J., Chen, S., Zhang, Z., Chen, J., Zhong, J., Liu, W., Zou, M., Li, Y. and Cai, J. 2018. Gut-dependent microbial translocation induces inflammation and cardiovascular events after ST-elevation myocardial infarction. *Microbiome* 6(1), 66.

Zordoky, B. N. M., Robertson, I. M. and Dyck, J. R. B. 2015. Preclinical and clinical evidence for the role of resveratrol in the treatment of cardiovascular diseases. *Biochimica et Biophysica Acta* 1852(6), 1155–1177.

Part 4

Analysing and optimising phytochemical compounds in fruits and vegetables

Chapter 10

Advances in screening/analysis of phytochemical compounds in fruits and vegetables

Alessandro Nogueira, State University of Ponta Grossa, Brazil; Amit K. Jaiswal, School of Food Science and Environmental Health, Technological University Dublin - City Campus and Environmental Sustainability and Health Institute, Technological University Dublin - City Campus, Ireland; and Aline Alberti, State University of Ponta Grossa, Brazil

1 Introduction

Consumers have changed their perception of food. Aspects related to the health and sustainability of products are being increasingly sought after by consumers. The presence of phytochemical compounds in food products interferes with their nutritional and sensory quality. Fruit and vegetables are a source of these compounds, and therefore, studies have been conducted aiming to characterise phytochemical sources, their structures, and activities. Until recently, most of the plant metabolite studies relied on the quantitation of selected bioactive compounds. Bioactive compounds can be grouped into categories such as carotenoids, glucosinolates, phenolics, fatty acids, and non-digestible carbohydrates. Phenolics are by far the most studied phytochemicals in fruits and vegetables. Sometimes the analysis is restricted to a subclass, or a phenolic profile is performed.

Since plant extracts usually occur as a complex mixture combination of the bioactive compounds with different polarities, their analysis remains a big challenge for the process of identification and characterisation. To obtain pure compounds, it is a common practice to use some different separation

http://dx.doi.org/10.19103/AS.2022.0101.11

techniques followed by the determination of structure and biological activity. For natural products analysis, many factors including sample preparation for eliminating interfering compounds, selection of the analytical method instruments, optimisation of the conditions for the confirmation of known compounds, or the identification of unknown components need to be taken into account according to different analytical requirements.

Advances in the techniques for the identification and quantification of chemical compounds have been made throughout the last few years. With the advent of ultra performance liquid chromatography (UPLC) (or ultra-high performance liquid chromatography, UHPLC), improvements in mass spectroscopy with the development of new mass analysers have made the analysis more sensitive and accurate. At the same time, nuclear magnetic resonance (NMR) has been shown to be an adequate technique for not only identifying but quantifying bioactive compounds. Improvements in NMR and infrared spectroscopy have been made in order to overcome their shortcomings. With all these analysis methods available, choosing the most suitable one can be difficult. For this purpose, throughout this chapter, we will present the advantages and limitations of the most advanced techniques available, as well as some successful examples in the investigation of phytochemical compounds.

2 Extraction methods

Before screening/analysis of phytochemical compounds, sample preparation is regarded as an essential procedure because extracting the desired chemical components from the samples is the first step in the analytical process. The suitable sample preparation techniques can reduce matrix effects which have an impact on the accuracy, precision, and robustness of the methods. Soxhlet extraction, solid-phase extraction, liquid-liquid extraction, ultrasonic extraction, microwave-assisted extraction, supercritical fluid extraction, and pressurised liquid extraction are usual sample preparation methods (Ghomari et al., 2019; Mirzadeh et al., 2020; Mohd Hazli et al., 2019; Khoei and Chekin, 2016; Kus et al., 2018; Poojary and Passamonti, 2015; Turner and Waldebäck, 2013).

Among the methods used to obtain bioactive compounds, solid-liquid extraction is one of the oldest techniques in unit operation and is widely used in the chemical and food industries. This operation takes place through the diffusion of the liquid solvent in a certain solid matrix and selective dissolution of one or more solutes that are diffused out of the matrix (Meireles, 2008). Thus, in solid-liquid extraction processes, increased diffusivity and solubility positively contribute to obtaining bioactive compounds. The efficiency of the method will depend on the solvent properties, matrix particle size, solvent/solute ratio, extraction temperature and time, solution pH, presence of interfering compounds, and the choice between conventional and non-conventional techniques (Zhang et al., 2018). Overall, extracting a solute from porous particles to the solvent during a diffusion process involves several steps, such as diffusion of the solvent into the porous solid, dissolution of the solute in the solvent, diffusion of dissolved solute to the surface of the particle, and diffusion of dissolved solute from the particle surface to the solvent (Setford et al., 2017).

In addition to the extraction method, the choice of solvent is very important for a better and faster extraction. Usual solvents like methanol, acetone, ethanol, ethyl acetate, and hexane are not fully efficient to extract a range of non-polar and polar molecules hence, we require solvent mixtures, high temperature-pressure conditions, or the search for alternative solvents (Renard, 2018). Green and efficient alternatives such as the use of deep eutectic solvents (DES) in the extraction have been emerging, presenting favourable results when compared to conventional solvents. A study by Bubalo et al. (2016) showed that the yields of phenolic compounds from grape skin obtained with DES were higher than those obtained with water and conventional solvent using the maceration extraction process. DES are blends of original products readily available in nature. They offer many advantages, including low cost, readily available components, simple preparation, and low toxicity (Radošević et al., 2016). Guo et al. (2019) obtained strong evidence of anthocyanin extraction and anthocyanin stabilisation capacity with acid-based DES. Acid-based DES presents as a great alternative for ecological extraction and can be widely used in extraction, being a benefit for the pharmaceutical and food industries.

Simple extraction using organic solvents is still commonly used followed by centrifugation and filtration (0.45/0.22 μm) (Alghamdi et al., 2018; Ceccanti et al., 2021; Kaneria et al., 2018; Maulidiani et al., 2019). Methanol, acetone, and ethanol are the most used solvents (Alberti et al., 2014; Alghamdi et al., 2018; Ma et al., 2021; Naveen and Kavitha, 2021; Parveen et al., 2021; Pascale et al., 2018; Santos et al., 2017). Depending on the phytochemical class to be analysed, a different solvent can be used. For flavonoids, phenolics, tannins extraction with ethanol, methanol, and aqueous solvent are the most common and appropriate. In the same way, chloroform can extract alkaloids, terpenoids, tannins while ethyl acetate alkaloids, flavonoids, phytosterols, and proteins (Kaneria et al., 2018; Naveen and Kavitha, 2021). However, these methods may require more time to extract the desired molecules, when compared to non-conventional ones. In addition, they are techniques that can provide low extraction yields or consume large amounts of solvent, as occurs with the use of Soxhlet (Poojary and Passamonti, 2015; Roohinejad et al., 2016; Zhang et al., 2018). Non-conventional techniques, especially ultrasound-assisted and microwave-assisted extraction, and supercritical fluid extraction are proposed as alternatives to extract target compounds from various plant matrices. They have the potential to minimise the use of toxic solvents and allow for room temperature extractions. Thus, the quality of the extract is improved, also increasing the yield of active substances obtained in the extraction (Tiwari, 2015; Zhang et al., 2018).

Among the innovative technologies for extracting compounds, the extraction by pressurised liquid appears as a cleaner alternative, since the process requires less solvent, and less polluting solvents can be applied. The technique consists of placing the solid sample in a cell, where the high pressure applied makes the solvent used to be heated above the boiling temperature. Diatomaceous earth or glass beads can be placed in the cell to prevent clogging. The temperature can vary from 20°C to 200°C and the pressure between 35 bar and 200 bar. The time for each cycle can vary from 10 min to 30 min (Richter et al., 1996; Turner and Waldebäck, 2013). Temperature directly influences the extraction of compounds, proportionally, possibly by increasing the diffusion and solubilities of the analytes. The combination of solvents seems to favour the extraction of phenolic compounds, such as water and ethanol solutions, which make the medium moderately polar. This mixture also obtained greater extraction of anthocyanins, at temperatures of 60–80°C (Machado et al., 2015). Pressurised liquid extraction, when compared to conventional techniques, uses less solvent, about 2.25 times less, besides reduced extraction time (Santos et al., 2012).

The microwave-assisted extraction (MAE) has been used owing to increasing the yield in shorter periods and at the same time using less solvent. Microwave energy is non-ionising radiation that results in molecular motion by the migration of ions and rotation of dipoles. The higher extraction yield is related to the molecular interactions between the electric component of the microwave field with the dipolar molecules and the ionic species present in the extraction mixtures (Mirzadeh et al., 2020). MAE method is used to extract the flavonoids, isoflavonoids, other phenolics compounds, essential oils, and saponins (Karimi and Jaafar, 2011).

Ultrasound-assisted extraction (UAE) is a green and economically viable alternative to conventional techniques for bioactive compounds. The use of ultrasound has shown several benefits such as faster mass transfer (shorter extraction time), reduced temperature, selective extraction, low solvent consumption, low installation cost, and easy operation (Chemat et al., 2011, 2017). As a disadvantage, it presents non-homogeneous heating in the extraction (Lianfu and Zelong, 2008). In a liquid medium, sample cell wall disruption occurs when the implosion of cavitation bubbles generates microturbulence, high-speed collisions between particles, and internal diffusion. UAE technique has been applied in the food industry for purposes such as oil extraction (Khoei and Chekin, 2016). Phenolic compounds (Nadeem et al., 2018; Pinela et al., 2019) also act in the reduction of microbiological load (Cervantes-Elizarrarás et al., 2017).

Hyphenated extraction methods can also be an alternative for better yields. The use of new solvents (deep eutectic solvent and ionic liquids) with emergent technologies is becoming popular (Sukor et al., 2021). UAE and MAE extraction of lycopene from tomato paste were optimised using response surface methodology. The MAE overcomes the shortcomings of the UAE and appears to be a more attractive extraction method in the future (Lianfu and Zelong, 2008).

3 Methods of phytochemical analysis

Photometric measurements are recommended for a simple and inexpensive analysis of the total phytochemical compounds. However, when more precise information and the profile of compounds in a sample are desired, other techniques are more suitable, such as chromatography (Kusznierewicz et al., 2021). The current chromatographic methods are highly efficient in the separation and analysis of natural products. A variety of chromatographic techniques have been employed including thin-layer chromatography, gas chromatography (GC), and high or ultra-high-performance liquid chromatography (HPLC/UHPLC). Reversed-phase liquid chromatography HPLC/UHPLC on octadecylsilane (C18 or ODS) is doubtlessly the most widely used in phytochemical analysis. Coupled with chromatographic methods (liquid or gas) mass analysers can greatly expand the accuracy of the phytochemical analysis. In addition, NMR and infrared spectroscopy have emerged as technologies to enhance the evaluation of bioactive compounds from fruit and vegetable samples. The advantages and limitations of these techniques are summarised in Table 1.

3.1 High-performance thin-layer chromatography (HPTLC)

Thin-layer chromatography is a technique applied for the fingerprinting of phytochemicals. It is an old chromatographic technique widely used as a cost-effective method for fast analysis. The high-performance version of this technique (HPTLC), combined with the automated sample application and densitometric scanning, is sensitive and suitable for use in qualitative and quantitative analysis. Due to the polar nature of some bioactive compounds, the use of this technique results in chromatograms that could be of rather poor quality. Nevertheless, Kusznierewicz et al. (2021) proposed a reversed-phase betalain pigment separation using a plate coated with octadecyl-modified silica (C18). The use of the appropriate stationary phase resulted in a high-quality betalain profile.

3.2 Liquid chromatography (LC)

The analysis of individual phytochemicals from complex matrices requires efficient separation methods before detection. Developments in HPLC

Table 1 The advantage and limitations of the techniques for analysis of phytochemical compounds

Method	Advantage	Limitation
HPTLC	Fast analysis Low cost High sensitivity and specificity	Low accuracy in quantification Bad efficiency in separation
Liquid chromatography	Low cost High specificity High repeatability Multiple detectors	Large solvent consuming Long analysis time (HPLC) Low sensitivity
Gas chromatography	Less solvent consuming High sensitivity	Limited to volatile compounds Derivatisation may be needed Long analysis time
Mass spectroscopy	High sensitivity High resolution High accuracy	Limited libraries High cost Bad stability
Infrared spectrometry	Fast analysis Less/no solvent consuming No sample preparation Low cost	Low accuracy in quantification Low specificity Long time for calibration
NMR	Less solvent consuming Fast analysis High resolution No derivatisation needed	High cost Low accuracy in quantification Convoluted spectra

Source: Lammertyn et al. (2000); Bras et al. (2005); Bharti and Roy (2012); Wu et al. (2013); La Barbera et al. (2017); Zhang et al. (2017); Alghamdi et al. (2018); Bakhytkyzy et al. (2018) Kaneria et al. (2018); Kusznierewicz et al. (2021); Naveen and Kavitha (2021).

systems, including 2 μm particles, very pH-stable stationary phases, and fused-core particles, have considerably improved their performance in terms of resolution, reproducibility, peak capacity, and ruggedness (La Barbera et al., 2017; Wu et al., 2013). The higher peak capacity can also be achieved by the two-dimensional (2D) (LC × LC) liquid chromatographic techniques to resolve samples of great complexity (Wu et al., 2013). Then UPLC (or UHPLC) started to be used in the analysis of bioactive compounds and became popular. Even with the advantages of UHPLC and the shorter gradient times, traditional HPLC (3–5 μm particle size) is still widely employed.

According to the chemical and physical properties of the compounds, different detection techniques including photodiode array detection (DAD), refractive index, fluorescence detection, and evaporative light scattering detection are commonly used for the identification and quantification of different compounds (Bakhytkyzy et al., 2018; Luca et al., 2021; Nowicka et al., 2019; Yuenyong et al., 2021).

The photodiode array detector is the most common detector used in LC. Benvenutti et al. (2019) performed a phenolic profile in apple pomace and apple beverages using the HPLC-DAD system. In the same way, Ghomari et al. (2019) studied the yield of different phenolic extraction methods from olive leaves. The maceration of olive leaves in ethanol followed by water resulted in the higher total content and varieties of phenolic compounds and flavonoids identified by the HPLC method. After developing a method for determining 13 bioactive phenolic compounds in guava (*Psidium guajava* L.) by HPLC-DAD, Santos et al. (2017) concluded that guava is a rich source of polyphenols and can be used as a functional food, additive, and also for pharmaceutical, medical, and commercial purposes.

In some cases, HPLC can be used to fractionate complex mixtures, and then it can be directly injected into a modern detection system. LC, specially UHPLC, coupled to mass spectrometry (MS) or NMR seems to be the future method for the analysis of bioactive compounds.

3.3 Gas chromatography (GC)

Phenolic compounds, ketones, aldehydes, alcohols, heterocyclic compounds, esters, and carboxylic acids are among the substances that can be analysed by GC coupled to MS (Alghamdi et al., 2018; Kaneria et al., 2018), although less polar (or polar) compounds are the most identified by this technique. Oleic acid (antioxidant and anti-breast cancer), n-hexadecanoic acid (antioxidant, antioxidant, and antibacterial), octadecatrienoic acid (anti-inflammatory), tridecene (against respiratory irritations), and behenyl alcohol (antiviral agent) were identified in *Ximenia americana* extracts by Shettar et al. (2017). A single quadrupole GC-mass spectrometer was used in the analysis and identification was done by comparison with a library spectra database. In the same way, Naveen and Kavitha (2021) identified, in *Clerodendrum phlomidis* extract, 4-hydroxy-3-nitrocoumarin (antiallergic action) stigmasterol (anti-inflammatory activity, good membrane-stabilising action in human red blood cells, and antibacterial action against methicillin-resistant *Staphylococcus aureus*), gamma-sitosterol (hypolipidemic property), 1,2-bis(trimethylsilyl) benzene (antibacterial, antimicrobial, and antioxidant potential), and cyclotrisiloxane (antibacterial, antioxidant, anti-tumour, and cancer prevention agent potential).

The high sensitivity of GC coupled to MS allows it to be used in the identification of food adulteration (Vadivel et al., 2018). For this, some phytochemical compounds could be considered chemical markers to differentiate adulterated samples.

3.4 Mass spectrometry (MS)

Over these past decades, MS has become an indispensable method for the structure determination of compounds of interest, such as phenolics, carotenoids, alkaloids, peptides, proteins, and oligosaccharides. The improvements in sample processing, LC-MS instrumentation, and software for data analysis have added to our ability to structurally characterise biomolecules.

MS technology has a variety of ionisation, mass analysing, and detection methods. Ionisation methods are related to the class of compounds available for measurement, while the mass analyser and detector, the quality and reliability of the method. When the ions are separated according to their nominal mass, mass spectrometers are composed of only one magnetic analysis sector and are called low-resolution mass spectrometers. Mass spectrometers, which consist of an analysis sector formed by a magnetic field coupled to an electrical sector, can separate ions according to their exact masses, and for this reason, they are called high-resolution mass spectrometers. LC hyphenated with high-resolution MS provides a powerful tool for qualitative and quantitative analysis and has been increasingly used in recent years (La Barbera et al., 2017; Kusznierewicz et al., 2021; Maulidiani et al., 2019; Zielinski et al., 2016). Thus the ionisation competition and Coulombic repulsion between ions are reduced and consequently the complexity of the sample. In general, the limit of quantification of LC-MS/MS ranged between 0.01 and 100 ng/mL.

Phytochemical identification can be difficult due to the low accessibility of high-quality standards that may be used as references. MS measurement with a precision of four decimal places allows us to predict the molecular formula; therefore, the identification process can be carried out by comparing the experimental data to the results from previous studies in the literature. MS data gathered from multiple studies concerning phytochemicals can be implemented as a mass list in a local database using identification software. Kusznierewicz et al. (2021) using LC-Q-Orbitrap HRMS and Compound Discoverer software demonstrated a high potential for application to identifying various phytochemical classes, including betalains and their degradation products.

Mass analysers are classified into Fourier transform ion cyclotron resonance (FTICR), Orbitrap, and time of flight (TOF), according to the physical principle of operation (Wu et al., 2013; Zubarev and Makarov, 2013). They could be combined with quadrupole (Q) and quadrupole traps to allow the analysis of intact ions coming from analytes and their fragmentation pattern. Tandem MS (MS/MS or MS^2) is the combination of two or more MS experiments (Mittal, 2015). In the first mass spectrometer, the ion to be fragmented is isolated and passes through a cell filled with the collision gas, and the resulting fragment ions are analysed in the second mass spectrometer (van Agthoven et al., 2019). The commonly used tandem mass spectrometers include the triple quadrupole (QqQ or TQ), Q-TOF, Q-Linear Ion Trap (Q-Trap), LTQ Orbitrap, and IT-TOF although the mass analysers suitable for high mass accuracy determination are the image current-based FTICR, Orbitrap, and ion beam-based TOF instruments (La Barbera et al., 2017).

In FT-ICR (or FTMS), ions are trapped in a strong magnetic field combined with a weak electric field. The data of image current from coherently excited trapped ions is converted using Fourier transform into the frequency domain and then mass spectra. The field uniformity of superconducting magnets and high accuracy of frequency measurements lead to unique mass resolution and mass accuracy. The ability to discern ions with closely located mass-to-charge ratios (m/z), and to determine their charge, allowed researchers to measure their masses with relative accuracy. This ability became increasingly important over the last decades with the development of electrospray (ESI) and matrix-assisted laser desorption/ionisation (MALDI) which resulted in a better acceptance of MS for analysis of biological samples (Zubarev and Makarov, 2013).

There is a superior intrinsic relation between the FT-ICR and Orbitrap analysers concerning the maximum mass resolution and accuracy (van Agthoven et al., 2019). The MS/MS experiments performed on a quadrupole analyser coupled to the FT-ICR mass spectrometer provide a better limit of detection when compared to GC-MS (quadrupole), even for fatty acids analysis (Souza et al., 2019).

High resolution is necessary for more accurate mass measurement (La Barbera et al., 2017). Although image current data treatment in Orbitrap is similar to FTICR, only electrical fields are employed to trap ions. Therefore, Orbitrap is a benchtop instrument, justifying the wider spread of Orbitrap instruments over FTICR ones.

The high mass accuracy facilitates peak identification through databases such as ChemSpider and m/z cloud. The retention time, m/z, MS and MS/MS spectrum, and the information found in the open-access mass spectra databases were used by Lee et al. (2021) for the identification of 130 chemical compounds in Malaysia purple corn. A wide range of phytochemicals, such as phenolics, flavonoids, esters, fatty acids, and alkaloids, was identified and revealed that purple corn (*Zea mays*) can be considered a functional food. UHPLC- Q Exactive Orbitrap MS/MS was used by Zheng et al. (2020) for the identification and analysis of the chemical components of *Camellia reticulata* semen. The evaluation of different retention times and fragmentation characteristics, as well as comparative analysis with the literature, resulted in the identification of 35 chemical constituents, including 21 flavonoids and 14 other compounds, among them limonin, hesperidin, nobiletin, synephrine, tangeretin, 3,5,6,7,8,3′,4′-heptamethoxyflavone, and 5-hydroxide-6,7,8,3′,4′-pentamethoxyflavone. Qualitative phytochemical screening of persimmon extracts (organic acids, phenolics, flavonoids, fatty acids, and triterpenes) of six varieties of persimmon (Maulidiani et al., 2019) was performed using the same system (UPLC-ESI-Orbitrap-MS/MS). The identification of 61 compounds was based on their spectral characteristics (negative ion mode) and comparison with the literature reports and online database.

In TOF analysers, the TOF of the ion depends on the square root of its m/z, while resolution depends on the capacity of the instrument of eliminating the initial spread of kinetic energy of the injected ions. HRMS/MS analysers, such as quadrupole-time-of-flight (QTOF), have the advantage of providing accurate mass acquisition of both precursor and fragment ions. High efficiency is reached at a very high kinetic energy since TOF instrument detection is accomplished by secondary electron multipliers. In general, the highest resolution available in TOF analysers is several times lower than the highest resolution in Orbitrap and FTI-CR instruments (La Barbera et al., 2017). Even so, Glauser et al. (2013) comparing Q-TOF and Orbitrap concluded that both systems reached very similar results and suggested that they can be equivalently used for plant metabolomics.

The identification of the most active phenolics in *Memecylon edule* shoots was possible using LC-ESI-ion trap MS (Falleh et al., 2011). The *M. edule* shoots are rich in bioactive proanthocyanidins, particularly in propelargonidins. The LC-ESI-ion trap MS provides useful information due to its ability to collect MS^n data. Combining the pseudo molecular ion mass with the MS^n fragmentation and the comparison with standards allow compound identification. Since phenolic compounds are weakly acidic compounds, better sensitivity and selectivity have been described when the mass spectrometer was operated in negative ionisation mode (Bendif et al., 2020; Friedrich et al., 2000). The same study demonstrated that ESI-MS with trap collision-induced dissociation in the negative ion mode is a suitable technique to evaluate the sequence of monomeric units in proanthocyanidins.

It is well known that the beneficial health effects of tea include reducing the risks of cancer, type 2 diabetes, and cardiovascular diseases. The beneficial effect is related to the phytochemical compounds such as gallic acid, 5-galloylquinic acid, theobromine, gallocatechin, epigallocatechin, epicatechin, epigallocatechin gallate, and epicatechin gallate which are identified and quantified through LC-DAD-ESI-Q-TOF-MS/MS (Zielinski et al., 2016). In another study, the same LC-HRMS/MS analysis revealed 42 piperamides in Piper species (*Piper nigrum, Piper retrofractum, Piper borbonense, Piper guineense, and Piper cubeba*), with 22 of them being quantified by LC-DAD (Luca et al., 2021).

The advantages of the Matrix-assisted laser desorption/ionisation time-of-flight mass spectrometry (MALDI-TOF-MS) were first reported by Karas (1996) in 1987, including speed of analysis, high sensitivity, wide applicability combined with a good tolerance towards contaminants, and the ability to analyse complex mixtures. HPLC-MALDI-TOF MS has been demonstrated to be a rapid technique for the identification of flavonol glycosides. The fragment ions provided characteristic information for the structural elucidation. The multiple ions in the positive mode give more information on individual flavonol glycoside structures than the negative mode (Wang and Sporns, 2000).

In general, for identification of phytochemicals (untargeted analysis), LTQ Orbitrap, Q-TOF, and IT-TOF are preferable with high sensitivity, resolution, and accuracy. LTQ Orbitrap or IT-TOF MS provide high mass accuracy MS^n ($n > 2$) and are especially useful when a multi-stage fragmentation is necessary. Also, the high-resolution tandem mass spectrometer can provide isotopic abundances and the elemental composition of the fragment ions for the elucidation of the structure of the compounds.

In addition to the different types of analysers, the application range of liquid chromatography coupled to MS is expanded by the combination of different scanning modes. The MS data can be acquired in full scan, selected ion monitoring, product ion scanning, constant neutral loss scanning, precursor ion scanning, selected reaction monitoring (SRM), or multiple reactions monitoring (MRM) (Wu et al., 2013). Quantitative determination (targeted compounds) of phytochemical compounds is a relatively common application of LC-MS. The combination of the specific parent mass and the unique fragment ion is used to selectively monitor the compound of interest, which highly improves the sensitivity and specificity of analysis. Triple quadrupole (QqQ or TQ) and Q-Trap systems using MRM or SRM mode are the most commonly used instruments.

3.5 Infrared spectroscopy (IR)

Infrared spectroscopy (IR) is an important analytical technique able to analyse practically any sample in almost any state. It is non-destructive, precise, mechanically simple, and can be used for routine quantitative and qualitative analysis. When compared with HPLC, infrared spectroscopy has shown good results for phytochemical composition (Sato et al., 2008). In addition, conventional methods of analysis for bioactive content typically utilise time-consuming separation techniques, which are not necessary for IR methods (Zhang et al., 2017).

Chemical bonds such as C–H, O–H, and N–H usually have a molecular vibration, resulting in detectable overtones and combination bands in the near-infrared (NIR) area of 10 000–4000 cm^{-1}. The specific patterns displayed in the NIR region indicate the chemical and physical properties of the studied material. The mid-infrared (MIR) region (4000–400 cm^{-1}) is generally regarded as a reproducible region of the electromagnetic spectrum, in which very tiny differences in sample structure can be reliably measured (Arslan et al., 2018; Naveen and Kavitha, 2021). MIR spectra gave models with considerably inferior prediction power than NIR-based models. The difference in the penetration depth of NIR and MIR may affect the accuracy of the qualitative and quantitative analyses. NIR radiation can penetrate further into the sample (millimetre magnitude) than MIR (micrometre magnitude). Therefore, MIR data is more affected by problems due to particle size and sample homogeneity (Brás et al., 2005; Lammertyn et al., 2000).

Carbas et al. (2020) revealed that infrared spectroscopy is an accurate and rapid method for quantification of phytochemical composition, individual phenolic compounds, and *in vitro* antioxidant activity of *Phaseolus vulgaris* L. flour. In MIR analysis, the prediction models were obtained using the first derivative after normalisation, and in NIR, the first derivative of the spectra after normalisation.

Fourier transform infrared (FT-IR) and Fourier transform near-infrared (FT-NIR) spectroscopy are instrumental methods based on the measurement of the vibration of a molecule excited by infrared radiation at a specific wavenumber range. FT-IR technology allows taking advantage of distinct infrared intervals, namely NIR and MIR, coupled with different instrumentation (Carbas et al., 2020). Machado et al. (2017) demonstrated that FTIR spectroscopy (within the NIR and MIR ranges) associated with chemometrics allowed to differentiate fresh samples from those frozen for long periods based on the phytochemical content.

These spectroscopic analytical techniques are associated with multivariate data analysis for qualitative and quantitative analysis. Commonly, qualitative assessment of the spectra can be performed by principal components analysis, while partial least squares regression allows the enhancement of calibration models based on spectral and analytical data (Brás et al., 2005). However, to obtain reliable predictions, exhaustive calibrations should be performed in the same conditions for sample processing to remain consistent throughout an individual study. In addition, secondary validation can be employed as a way to identify potential artefacts and further evaluate the quality of predictions (Rubert-Nason et al., 2013).

3.6 Nuclear magnetic resonance (NMR)

For a long time, chemical analysis using liquid chromatography coupled to MS has been widely used for the screening and analysis of phytochemicals. However, the technique requires stringent confirmation with pure standard or highly informative complementary techniques, such as NMR. The ^1H (proton) NMR technique is related to the ^1H core energy levels that result when the sample is subjected to an external magnetic field. The NMR spectrum provides a series of acute signals whose frequencies can be related to the chemical nature of the hydrogen atoms and whose intensities are proportional to the number of hydrogens that produce the signal (Guillén and Ruiz, 2001).

NMR is an analytical technique extensively used for structural identification (Bharti and Roy, 2012). Nevertheless, innovations in the technique and data analysis enhanced the capability of NMR for discriminating similar compounds contained in complex mixtures (Felli and Brutscher, 2009; Wang et al., 2016; Kupče and Claridge, 2018). The area of an NMR signal is linearly proportional to the number of NMR active nuclei that generate the signal, and the quantification can be achieved by calculating the integral of the NMR signal since the response is independent of the molecule (Bharti and Roy, 2012). The correlation between the resonance frequency of a signal and the type of nuclei associated with that signal make NMR spectroscopy an important technique for structural determination and quantification of different compounds (Kim et al., 2011).

Phytochemical analysis by NMR technique has been under development in the past few years (Gallo et al., 2020; Musio et al., 2020; Praveen et al., 2021; Tsiokanos et al., 2021). NMR has been used for the identification and characterisation of bioactive compounds, for biomarker studies, as well as for the evaluation of the effects of these compounds in different health disorders. Despite its lower sensitivity, it is also a non-destructive technique, provides broad coverage, and has simple sample preparation, good reproducibility, and robustness. In addition, NMR can be used for the determination of the absolute concentration of individual components in a mixture along with their structural information (Bharti and Roy, 2012). In comparison, liquid chromatography needs an authenticated standard and a large volume of solvent is required. Praveen et al. (2021) developed a fast ^{1}H NMR method for the quantification of curcuminoids (curcumin, demethoxycurcumin, and bisdemethoxycurcumin). The curcuminoids were directly extracted using acetone-D_6 as a solvent, without purification steps. The standard error of NMR analysis was less than 2%, and when compared to chromatographic analysis (HPLC), the standard deviation was less than 1.5%.

The analysis of NMR spectra of complex mixtures can be difficult. For highly overlapped spectra, integration of peak area requires a good knowledge of curve-fitting methods. Softwares and deconvolution programmes are commonly used for the analysis of molecules in overlapped spectra. Moreover, NMR can be combined with other methods. Wang et al. (2016) using 2D NMR techniques including correlation spectroscopy, rotating frame nuclear overhauser effect spectroscopy (ROESY), heteronuclear single-quantum coherence (HSQC), and heteronuclear multiple-band coherence (HMBC) experiments managed to determine the unstable malonyl ginsenosides Rb1, Rb2, Rc, and Rd from fresh flowers of *Panax ginseng*.

A versatile, non-selective, NMR-based approach, termed high-resolution magic angle spinning (HR-MAS) NMR spectroscopy, has allowed the characterisation of foodstuff providing information on a wide variety of compounds without extraction procedures. The samples are submitted to fast-spinning about the so-called 'magic angle' (54.74°) to eliminate the contribution of dipolar coupling and differences in magnetic susceptibility (Santos et al., 2015).

Using ^{1}H HR-MAS NMR spectroscopy, the metabolic changes during the development of astringent ('Giombo') and non-astringent ('Fuyu') cultivars of persimmon (*Diospyros kaki*) can be performed. Among bioactive compounds, gallic acid was detected throughout the growth of 'Giombo', while for 'Fuyu', signals of polyphenols disappeared over time (Santos et al., 2018).

Gallo et al. (2020) employing interlaboratory comparisons demonstrated that NMR spectroscopy can provide statistically equivalent signals even with different magnetic field strength, manufacturer, and hardware configuration. In the same way, Musio et al. (2020) did extraordinary work to encourage the use of NMR for quantitative analysis and in the future be an internationally accepted analytical protocol.

The major drawbacks (the need for deuterated solvents, higher cost of analysis, etc) have been overcome by the advent of newer technologies, such as solvent suppression techniques and the introduction of post-column solid phase extraction (SPE) with cryogenic probes. This advent enabled its use in phytochemical analysis, promoting the use of LC-NMR and LC-NMR-MS systems.

NMR can also be used with LC-MS (LC–NMR–MS) and offers a more accurate structure analysis of the compounds in a mixture. The enhancements in high-resolution magnetic field strength, probe technology, software for controlling spectrometers, and data acquisition have resulted in the improvement of sensitivity and utility of LC-NMR instrumentation. Liquid chromatography coupled to NMR spectroscopy (LC-NMR) has emerged as a promising tool for the characterisation of minor components. However, some issues have to be overcome for this method to become popular. The larger solvent volumes, as compared to conventional NMR methods, can be very expensive. Besides, solvents used in normal phase chromatography are also rarely available in the deuterated form. Gradient elution mode is usually avoided, due to incompatibility in NMR operations from changes in solvent resonance positions (Sahu et al., 2019).

4 Case study

In this section, the use of advanced methods for the analysis of phenolic compounds from apples will be discussed. The health beneficial effects of phenolic compounds on health are well known; however, they also contribute to the sensorial and technological aspects of foods and beverages. In cider, (alcoholic apple beverage), the phenolic compounds are linked to bitterness and astringency and contribute to the formation of the cider's body (Benvenutti et al., 2019; Lea and Drilleau, 2003). The colour is essentially due to the phenolic compound and its oxidation products (Le Deun et al., 2015) and can greatly influence the overall assessment of the product. In addition, some phenolics, such as the hydroxycinnamic acids, are precursors of volatile constituents that may contribute to the aroma of the cider.

Phenolic compounds in apples are influenced by the ripening stage, variety, and cultivation conditions. New apple varieties are constantly in development with a focus on disease resistance, higher production yield, or another specific desired feature. Taking into account the relevance of phenolic compounds in the production of apple beverages and even in the sensory quality of fresh apples, the analysis of phenols is extremely important. A scheme of phenolic extraction from apples is demonstrated in Fig. 1.

Special attention must be paid to sample preparation. The enzymatic browning reactions and consequently the oxidation of phenolic compounds

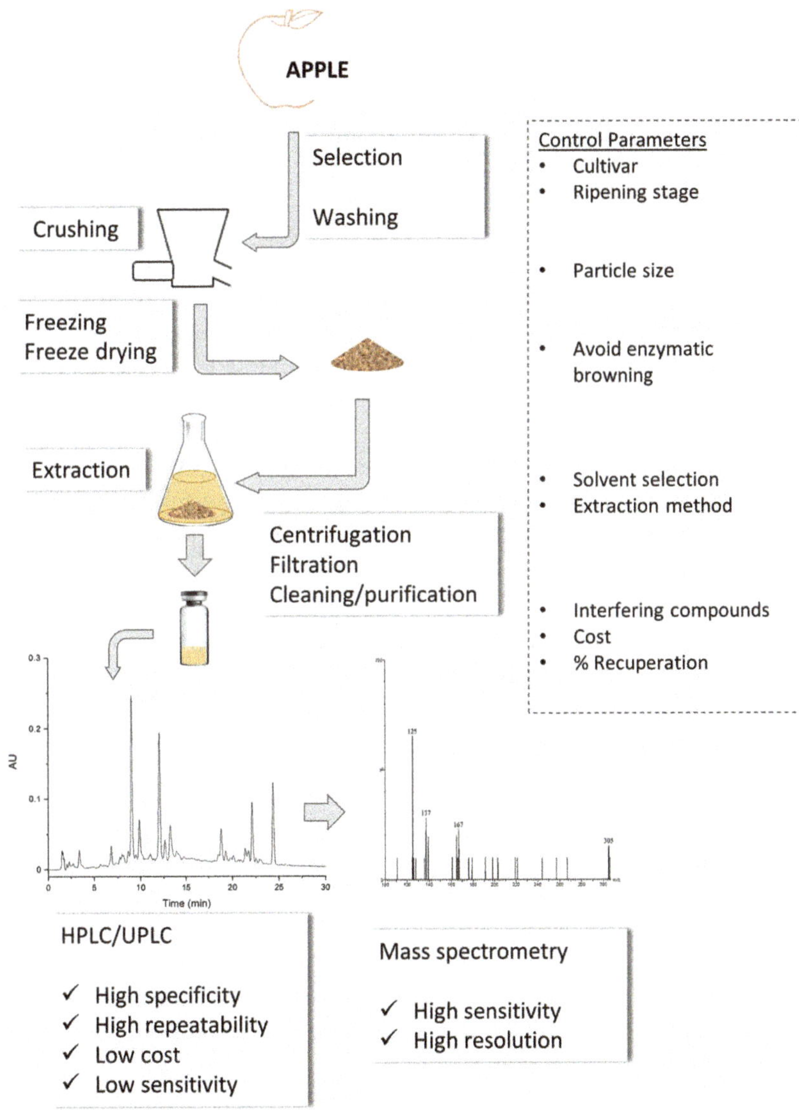

Figure 1 Steps for phenolic extraction from apples and control parameters of the process.

are extremely fast. Thus, immediate freezing of the apple after crushing using liquid nitrogen and freeze-drying are absolutely necessary. The extraction optimisation of phenolic compounds from apples was performed by Alberti et al. (2014) using methanol and acetone. Due to chemical differences between individual phenolic compounds, solvents with different polarities may be necessary to obtain better yields in the extraction of phenolic compounds. For this, a sequential extraction with methanol and acetone solutions can be the best choice. However, the use of solvents such as methanol and acetone are not recommended if the intention is to obtain extracts for use in food or medicine due to their toxicity. For this purpose, the extraction must be carried out using other solvents, despite the possibility of not obtaining the same yields. The extract purification could be performed using chromatographic methods, such as solid-phase extraction cartridges.

Then, the phenolic profile can be performed in a HPLC system coupled with a diode array detector and a C18 column. Authentic standards are used for the identification and quantification of phenolics; however, some of them are expensive or not available for sale. An alternative is the use of other advanced methods for screening and analysis such as liquid chromatography with a mass analyser. One example is the use of UHPLC coupled to a TQ-MS. A C18 column could be used to previously fractionate the sample. Since phenolic compounds from apples are already known, quantification could be performed in multiple reaction monitoring (MRM) scanning mode. In MRM mode, TQ-MS detects a specific precursor ion, isolates that ion for collision-induced fragmentation, and, finally, detects a specific product ion following fragmentation. The high sensitivity of MRM-based MS and the wide dynamic range of triple-quadrupole spectrometers provide a tool for the analysis of complex samples from fruits and vegetables. The short duration of the analysis (without extraction) makes the methodology suitable for screening studies and for use in assisting plant breeders.

5 Summary and future trends

No single best method exists for phytochemicals; the methods available are complementary and a choice should be based on the chemical structure and stability of the analyte. Qualitative analysis of phytochemical constituents from different matrices includes the confirmation of known compounds and elucidation of the structures of unknowns. NMR is an emerging technology suitable for this purpose. Nowadays, LC-DAD-MS is still routinely applied in qualitative analysis, however, a tandem mass spectrometer coupled with UHPLC is used for unequivocal identification of natural products and their metabolites present at trace concentrations.

The different analytical methods and techniques in their respective sensitivities allow the detection of different classes of compounds. Particularly, the analysis of bioactive compounds from fruit and vegetables which are complex matrices, has greatly benefited from the use of modern hyphenated techniques, especially from the combination of chromatography separation and high-resolution MS. This combination results in extraordinary tools which are being increasingly used and in constant development.

However, some shortcomings remain in the phytochemical evaluation from complex samples as fruits and vegetables. Therefore, the development of even more advanced and suitable analytical techniques is essential for future phytochemical research.

6 Where to look for further information

6.1 Further reading

- La Barbera, G., Capriotti, A. L., Cavaliere, C., Montone, C. M., Piovesana, S., Samperi, R., Chiozi, R. Z. and Laganà, A. (2017). Liquid chromatography-high resolution mass spectrometry for the analysis of phytochemicals in vegetal-derived food and beverages. *Food Research International* 100: 28–52. DOI: 10.1016/j.foodres.2017.07.080.
- McGoverin, C. M., Weeranantanaphan, J., Downey, G. and Manley, M. (2010). Review: The application of near infrared spectroscopy to the measurement of bioactive compounds in food commodities. *Journal of Near Infrared Spectroscopy* 18(2): 87–111. DOI: 10.1255/jnirs.874.
- Praveen, A., Prasad, D., Mishra, S., Nagarajan, S. and Chaudhari, S. R. (2021). Facile NMR approach for profiling curcuminoids present in turmeric. *Food Chemistry* 341: 128646. DOI: 10.1016/j.foodchem.2020.128646.
- Wu, H., Guo, J., Chen, S., Liu, X., Zhou, Y., Zhang, X. and Xu, X. (2013). Recent developments in qualitative and quantitative analysis of phytochemical constituents and their metabolites using liquid chromatography-mass spectrometry. *Journal of Pharmaceutical and Biomedical Analysis* 72: 267–291. DOI: 10.1016/j.jpba.2012.09.004.
- Zubarev, R. A. and Makarov, A. (2013). Orbitrap Mass Spectrometry. *Analytical Chemistry* 85(11): 5288–5296. DOI: 10.1021/AC4001223.

7 References

Alberti, A., Zielinski, A. A. F., Zardo, D. M., Demiate, I. M., Nogueira, A. and Mafra, L. I. (2014). Optimisation of the extraction of phenolic compounds from apples using response surface methodology. *Food Chemistry* 149: 151–158. DOI: 10.1016/j.foodchem.2013.10.086.

Alghamdi, S. S., Khan, M. A., El-Harty, E. H., Ammar, M. H., Farooq, M. and Migdadi, H. M. (2018). *Comparative Phytochemical Profiling of Different Soybean (Glycine max (L.) Merr) Genotypes Using*. GC-MS. DOI: 10.1016/j.sjbs.2017.10.014.

Arslan, M., Xiaobo, Z., Tahir, H. E., Zareef, M., Xuetao, H. and Rakha, A. (2018). Total polyphenol quantitation using integrated NIR and MIR spectroscopy: a case study of Chinese dates (*Ziziphus jujuba*). *Phytochemical Analysis* 30(3): 357-363. DOI: 10.1002/pca.2818.

Bakhytkyzy, I., Nuñez, O. and Saurina, J. (2018). Determination of flavanols by liquid chromatography with fluorescence detection: application to the characterization of cranberry-based pharmaceuticals through profiling and fingerprinting approaches. *Journal of Pharmaceutical and Biomedical Analysis* 156: 206–213. DOI: 10.1016/j.jpba.2018.04.031.

Bendif, H., Peron, G., Miara, M. D., Sut, S., Dall'Acqua, S., Flamini, G. and Maggi, F. (2020). Total phytochemical analysis of *Thymus munbyanus* subsp. coloratus from Algeria by HS-SPME-GC-MS, NMR and HPLC-MS n studies. *Journal of Pharmaceutical and Biomedical Analysis* 186: 113330. DOI: 10.1016/j.jpba.2020.113330.

Benvenutti, L., Bortolini, D. G., Nogueira, A., Zielinski, A. A. F. and Alberti, A. (2019). Effect of addition of phenolic compounds recovered from apple pomace on cider quality. *LWT* 100: 348–354. DOI: 10.1016/J.LWT.2018.10.087.

Bharti, S. K. and Roy, R. (2012). Quantitative 1H NMR spectroscopy. *TrAC Trends in Analytical Chemistry* 35: 5–26. DOI: 10.1016/j.trac.2012.02.007.

Brás, L. P., Bernardino, S. A., Lopes, J. A. and Menezes, J. C. (2005). Multiblock PLS as an approach to compare and combine NIR and MIR spectra in calibrations of soybean flour. *Chemometrics and Intelligent Laboratory Systems* 75(1): 91–99. DOI: 10.1016/J.CHEMOLAB.2004.05.007.

Bubalo, M. C., Ćurko, N., Tomašević, M., Kovačević Ganić, K. and Radojčić Redovniković, I. (2016). Green extraction of grape skin phenolics by using deep eutectic solvents. *Food Chemistry* 200: 159–166. DOI: 10.1016/j.foodchem.2016.01.040.

Carbas, B., Machado, N., Oppolzer, D., Queiroz, M., Brites, C., Rosa, E. A. S. and Barros, A. I. R. N. A. (2020). Prediction of phytochemical composition, in vitro antioxidant activity and individual phenolic compounds of common beans using MIR and NIR spectroscopy. *Food and Bioprocess Technology*. Springer 13(6): 962–977. DOI: 10.1007/S11947-020-02457-2.

Ceccanti, C., Rocchetti, G., Lucini, L., Giuberti, G., Landi, M., Biagiotti, S. and Guidi, L. (2021). Comparative phytochemical profile of the elephant garlic (*Allium ampeloprasum* var. holmense) and the common garlic (*Allium sativum*) from the Val di Chiana area (Tuscany, Italy) before and after in vitro gastrointestinal digestion. *Food Chemistry* 338: 128011. DOI: 10.1016/J.FOODCHEM.2020.128011.

Cervantes-Elizarrarás, A., Piloni-Martini, J., Ramírez-Moreno, E., Alanís-García, E., Güemes-Vera, N., Gómez-Aldapa, C. A., Zafra-Rojas, Q. Y. and Cruz-Cansino, N. D. (2017). Enzymatic inactivation and antioxidant properties of blackberry juice after thermoultrasound: optimization using response surface methodology. *Ultrasonics Sonochemistry* 34: 371–379. DOI: 10.1016/j.ultsonch.2016.06.009.

Chemat, F., Zill-E-Huma and Khan, M. K. (2011). Applications of ultrasound in food technology: processing, preservation and extraction. *Ultrasonics Sonochemistry* 18(4): 813–835. DOI: 10.1016/j.ultsonch.2010.11.023.

Chemat, F., Rombaut, N., Sicaire, A. G., Meullemiestre, A., Fabiano-Tixier, A. S. and Abert-Vian, M. (2017). Ultrasound assisted extraction of food and natural products. Mechanisms, techniques, combinations, protocols and applications: a review. *Ultrasonics Sonochemistry* 34: 540–560. DOI: 10.1016/j.ultsonch.2016.06.035.

Falleh, H., Oueslati, S., Guyot, S., Dali, A. B., Magné, C., Abdelly, C. and Ksouri, R. (2011). LC/ESI-MS/MS characterisation of procyanidins and propelargonidins responsible for the strong antioxidant activity of the edible halophyte *Mesembryanthemum edule* L. *Food Chemistry* 127(4): 1732–1738. DOI: 10.1016/j.foodchem.2011.02.049.

Felli, I. C. and Brutscher, B. (2009). Recent advances in solution NMR: fast methods and heteronuclear direct detection. *Chemphyschem* 10(9–10): 1356–1368. DOI: 10.1002/CPHC.200900133.

Friedrich, W., Eberhardt, A. and Galensa, R. (2000). Investigation of proanthocyanidins by HPLC with electrospray ionization mass spectrometry. *European Food Research and Technology*. Springer 211(1): 56–64. DOI: 10.1007/S002170050589.

Gallo, V., Ragone, R., Musio, B., Todisco, S., Rizzuti, A., Mastrorilli, P., Pontrelli, S., Intini, N., Scapicchio, P., Triggiani, M., Pascazio, A., Cobas, C., Mari, S., Garino, C., Arlorio, M., Acquotti, D., Airoldi, C., Arnesano, F., Assfalg, M., Barison, A., Benevelli, F., Borioni, A., Cagliani, L. R., Casadei, L., Marincola, F. C., Colson, K., Consonni, R., Costantino, G., Cremonini, M. A., Davalli, S., Duarte, I., Guyader, S., Hamon, E., Hegmanns, M., Lamanna, R., Longobardi, F., Mallamace, D., Mammi, S., Markus, M., Menezes, L. R. A., Milone, S., Molero-Vilchez, D., Mucci, A., Napoli, C., Rossi, M. C., Sáez-Barajas, E., Savorani, F., Schievano, E., Sciubba, F., Sobolev, A., Takis, P. G., Thomas, F., Villa-Valverde, P. and Latronico, M. (2020). A contribution to the harmonization of non-targeted NMR methods for data-driven food authenticity assessment. *Food Analytical Methods* 13(2): 530–541. DOI: 10.1007/s12161-019-01664-8.

Ghomari, O., Sounni, F., Massaoudi, Y., Ghanam, J., Drissi Kaitouni, L. B., Merzouki, M. and Benlemlih, M. (2019). Phenolic profile (HPLC-UV) of olive leaves according to extraction procedure and assessment of antibacterial activity. *Biotechnology Reports* 23: e00347. DOI: 10.1016/J.BTRE.2019.E00347.

Glauser, G., Veyrat, N., Rochat, B., Wolfender, J. L. and Turlings, T. C. (2013). Ultra-high pressure liquid chromatography-mass spectrometry for plant metabolomics: a systematic comparison of high-resolution quadrupole-time-of-flight and single stage Orbitrap mass spectrometers. *Journal of Chromatography. A* 1292: 151–159. DOI: 10.1016/J.CHROMA.2012.12.009.

Guillén, M. D. and Ruiz, A. (2001) High resolution 1H nuclear magnetic resonance in the study of edible oils and fats. *Trends in Food Science and Technology* 12(9): 328–338. DOI: 10.1016/S0924-2244(01)00101-7.

Guo, N., Ping-Kou, J. Y. W., Jiang, Y. W., Wang, L. T., Niu, L. J., Liu, Z. M. and Fu, Y. J. (2019). Natural deep eutectic solvents couple with integrative extraction technique as an effective approach for mulberry anthocyanin extraction. *Food Chemistry* 296: 78–85. DOI: 10.1016/j.foodchem.2019.05.196.

Kaneria, M. J., Rakholiya, K. D., Marsonia, L. R., Dave, R. A. and Golakiya, B. A. (2018). Nontargeted metabolomics approach to determine metabolites profile and antioxidant study of Tropical Almond (*Terminalia catappa* L.) fruit peels using GC-QTOF-MS and LC-QTOF-MS. *Journal of Pharmaceutical and Biomedical Analysis* 160: 415–427. DOI: 10.1016/J.JPBA.2018.08.026.

Karas, M. (1996). Matrix-assisted laser desorption ionization MS: a progress report. *Biochemical Society Transactions* 24(3): 897–900. DOI: 10.1042/BST0240897.

Karimi, E. and Jaafar, H. Z. E. (2011). Molecules HPLC and GC-MS determination of bioactive compounds in microwave obtained extracts of three varieties of *Labisia pumila* Benth. *Molecules* 16(8): 6791–6805. DOI: 10.3390/molecules16086791.

Khoei, M. and Chekin, F. (2016). The ultrasound-assisted aqueous extraction of rice bran oil. *Food Chemistry* 194: 503–507. DOI: 10.1016/J.FOODCHEM.2015.08.068.

Kim, H. K., Choi, Y. H. and Verpoorte, R. (2011). NMR-based plant metabolomics: where do we stand, where do we go? *Trends in Biotechnology* 29(6): 267–275. DOI: 10.1016/j.tibtech.2011.02.001.

Kupče, Ē. and Claridge, T. D. W. (2018). Molecular structure from a single NMR super-sequence. *Chemical Communications* 54(52): 7139–7142. DOI: 10.1039/C8CC03296C.

Kus, P. M., Okínczyc, P., Jakovljevíc, M., Jokić, S. and Jerković, I. (2018). Development of supercritical CO2 extraction of bioactive phytochemicals from black poplar (*Populus nigra* L.) buds followed by GC-MS and UHPLC-DAD-QqTOF-MS. *Journal of Pharmaceutical and Biomedical Analysis* 158: 15–27. DOI: 10.1016/j.jpba.2018.05.041.

Kusznierewicz, B., Mróz, M., Koss-Mikołajczyk, I. and Namieśnik, J. (2021). Comparative evaluation of different methods for determining phytochemicals and antioxidant activity in products containing betalains-verification of beetroot samples. *Food Chemistry* 362: 130132. DOI: 10.1016/j.foodchem.2021.130132.

La Barbera, G., Capriotti, A. L., Cavaliere, C., Montone, C. M., Piovesana, S., Samperi, R., Zenezini Chiozzi, R. and Laganà, A. (2017). Liquid chromatography-high resolution mass spectrometry for the analysis of phytochemicals in vegetal-derived food and beverages. *Food Research International* 100(1): 28–52. DOI: 10.1016/j.foodres.2017.07.080.

Lammertyn, J., Peirs, A., De Baerdemaeker, J. and Nicolaï, B. (2000). Light penetration properties of NIR radiation in fruit with respect to non-destructive quality assessment. *Postharvest Biology and Technology* 18(2): 121–132. DOI: 10.1016/S0925-5214(99)00071-X.

Le Deun, E., Van der Werf, R., Le Bail, G., Le Quéré, J. M. and Guyot, S. (2015). HPLC-DAD-MS profiling of polyphenols responsible for the yellow-orange color in apple juices of different French cider apple varieties. *Journal of Agricultural and Food Chemistry* 63(35): 7675–7684. DOI: 10.1021/acs.jafc.5b00988.

Lea, A. and Drilleau, J.-F. (2003). Cider-making. In: Lea, A. (Ed.) *Fermented Beverage Production*. Springer, Boston, pp. 59–87. DOI: https://doi.org/10.1007/978-1-4615-0187-9_4

Lee, T. H., Lee, C. H., Wong, S., Ong, P. Y., Hamdan, N. and Azmi, N. A. (2021). UPLC-Orbitrap-MS/MS based characterization of phytochemical compounds from Malaysia purple corn (*Zea mays*). *Biocatalysis and Agricultural Biotechnology* 32: 101922. DOI: 10.1016/j.bcab.2021.101922.

Lianfu, Z. and Zelong, L. (2008). Optimization and comparison of ultrasound/microwave assisted extraction (UMAE) and ultrasonic assisted extraction (UAE) of lycopene from tomatoes. *Ultrasonics Sonochemistry* 15(5): 731–737. DOI: 10.1016/j.ultsonch.2007.12.001.

Luca, S. V., Minceva, M., Gertsch, J. and Skalicka-Woźniak, K. (2021). LC-HRMS/MS-based phytochemical profiling of Piper spices: global association of piperamides with endocannabinoid system modulation. *Food Research International* 141: 110123. DOI: 10.1016/j.foodres.2021.110123.

Ma, Q., Cai, S., Liu, X., Shi, J., and Yi, J. (2021). Characterization of phytochemical components and identification of main antioxidants in *Crateva unilocalaris* Buch.

shoots by UHPLC-Q-Orbitrap-MS[2] analysis. *Food Research International* 143: 963-969. DOI: 10.1016/j.foodres.2021.110264.

Machado, A. P. D. F., Pasquel-Reátegui, J. L., Barbero, G. F. and Martínez, J. (2015). Pressurized liquid extraction of bioactive compounds from blackberry (*Rubus fruticosus* L.) residues: a comparison with conventional methods. *Food Research International* 77: 675–683. DOI: 10.1016/j.foodres.2014.12.042.

Machado, N., Domínguez-Perles, R., Ramos, A., Rosa, E. A. and Barros, A. I. (2017). Spectrophotometric versus NIR-MIR assessments of cowpea pods for discriminating the impact of freezing. *Journal of the Science of Food and Agriculture* 97(13): 4285-4294. DOI: 10.1002/JSFA.8251.

Maulidiani, M., Abdul-Hamid, N. A., Abas, F., Park, Y. S., Park, Y., Kim, Y. M. and Gorinstein, S. (2019). Detection of bioactive compounds in persimmon (*Diospyros kaki*) using UPLC-ESI-Orbitrap-MS/MS and fluorescence analyses. *Microchemical Journal* 149: 103978. DOI: 10.1016/j.microc.2019.103978.

McGoverin, C. M., Weeranantanaphan, J., Downey, G. and Manley, M. (2010). Review: the application of near infrared spectroscopy to the measurement of bioactive compounds in food commodities. *Journal of Near Infrared Spectroscopy* 18(2): 87-111. DOI: 10.1255/jnirs.874.

Meireles, M. A. A. (2008). *Extracting Bioactive Compounds for Food Products*. CRC Press. DOI: 10.1201/9781420062397.

Mirzadeh, M., Arianejad, M. R. and Khedmat, L. (2020). Antioxidant, antiradical, and antimicrobial activities of polysaccharides obtained by microwave-assisted extraction method: a review. *Carbohydrate Polymers* 229: 115421. DOI: 10.1016/J.CARBPOL.2019.115421.

Mittal, R. D. (2015). Tandem mass spectroscopy in diagnosis and clinical research. *Indian Journal of Clinical Biochemistry* 30(2): 121–123. DOI: 10.1007/s12291-015-0498-9.

Mohd Hazli, U. H. A., Abdul-Aziz, A., Mat-Junit, S., Chee, C. F. and Kong, K. W. (2019). Solid-liquid extraction of bioactive compounds with antioxidant potential from *Alternanthera sesillis* (red) and identification of the polyphenols using UHPLC-QqQ-MS/MS. *Food Research International* 115: 241-250. DOI: 10.1016/J.FOODRES.2018.08.094.

Musio, B., Ragone, R., Todisco, S., Rizzuti, A., Latronico, M., Mastrorilli, P., Pontrelli, S., Intini, N., Scapicchio, P., Triggiani, M., Di Noia, T., Acquotti, D., Airoldi, C., Assfalg, M., Barge, A., Bateman, L., Benevelli, F., Bertelli, D., Bertocchi, F., Bieliauskas, A., Borioni, A., Caligiani, A., Callone, E., Čamra, A., Cesare Marincola, F., Chalasani, D., Consonni, R., Dambruoso, P., Davalli, S., David, T., Diehl, B., Donarski, J., Gil, A. M., Gobetto, R., Goldoni, L., Hamon, E., Harwood, J. S., Kobrlová, A., Longobardi, F., Luisi, R., Mallamace, D., Mammi, S., Martin-Biran, M., Mazzei, P., Mele, A., Milone, S., Molero Vilchez, D., Mulder, R. J., Napoli, C., Ragno, D., Randazzo, A., Rossi, M. C., Rotondo, A., Šačkus, A., Sáez Barajas, E., Schievano, E., Sitaram, B., Stevanato, L., Takis, P. G., Teipel, J., Thomas, F., Torregiani, E., Valensin, D., Veronesi, M., Warren, J., Wist, J., Zailer-Hafer, E., Zuccaccia, C. and Gallo, V. (2020). A community-built calibration system: the case study of quantification of metabolites in grape juice by qNMR spectroscopy. *Talanta* 214: 120855. DOI: 10.1016/j.talanta.2020.120855.

Nadeem, M., Ubaid, N., Qureshi, T. M., Munir, M. and Mehmood, A. (2018). Effect of ultrasound and chemical treatment on total phenol, flavonoids and antioxidant properties on carrot-grape juice blend during storage. *Ultrasonics Sonochemistry* 45: 1–6. DOI: 10.1016/J.ULTSONCH.2018.02.034.

Naveen, M. and Kavitha, R. (2021). HPLC, GC-MS and FT-IR analysis of bioactive phytochemicals present in aqueous extract of leaf of *Clerodendrum phlomidis*: a further evidence for its medicinal diversity. *Materials Today: Proceedings*. Epub ahead of print 12 August 2021. DOI: 10.1016/j.matpr.2021.05.414.

Nowicka, P., Wojdyło, A. and Laskowski, P. (2019). Principal component analysis (PCA) of physicochemical compounds' content in different cultivars of peach fruits, including qualification and quantification of sugars and organic acids by HPLC. *European Food Research and Technology* 245(4): 929–938. DOI: 10.1007/S00217-019-03233-Z.

Parveen, S., Saleem, H., Sarfraz, M., Khurshid, U., Habib, M., Nazir, M., Akhtar, N., Nasim, F., Rashid, U. and Chotana, G. A. (2021). Phytochemical profiling, in vitro antioxidant and identification of urease inhibitory metabolites from *Erythrina suberosa* flowers by GC-MS analysis and docking studies. *South African Journal of Botany* 143: 422–427. DOI: 10.1016/j.sajb.2021.05.020.

Pascale, R., Bianco, G., Cataldi, T. R. I., Kopplin, P. S., Bosco, F., Vignola, L., Uhl, J., Lucio, M. and Milella, L. (2018). Mass spectrometry-based phytochemical screening for hypoglycemic activity of Fagioli di Sarconi beans (*Phaseolus vulgaris* L.). *Food Chemistry* 242: 497–504 DOI: 10.1016/j.foodchem.2017.09.091.

Pinela, J., Prieto, M. A., Pereira, E., Jabeur, I., Barreiro, M. F., Barros, L. and Ferreira, I. C. F. R. (2019). Optimization of heat- and ultrasound-assisted extraction of anthocyanins from *Hibiscus sabdariffa* calyces for natural food colorants. *Food Chemistry* 275: 309–321. DOI: 10.1016/J.FOODCHEM.2018.09.118.

Poojary, M. M. and Passamonti, P. (2015). Extraction of lycopene from tomato processing waste: kinetics and modelling. *Food Chemistry* 173: 943–950. DOI: 10.1016/J.FOODCHEM.2014.10.127.

Praveen, A., Prasad, D., Mishra, S., Nagarajan, S. and Chaudhari, S. R. (2021). Facile NMR approach for profiling curcuminoids present in turmeric. *Food Chemistry* 341(2): 128646. DOI: 10.1016/j.foodchem.2020.128646.

Radošević, K., Ćurko, N., Gaurina Srček, V., Cvjetko Bubalo, M., Tomašević, M., Kovačević Ganić, K. and Radojčić Redovniković, I. (2016). Natural deep eutectic solvents as beneficial extractants for enhancement of plant extracts bioactivity. *LWT* 73: 45–51. DOI: 10.1016/j.lwt.2016.05.037.

Renard, C. M. G. C. (2018). Extraction of bioactives from fruit and vegetables: state of the art and perspectives. *LWT*. Academic Press 93: 390–395. DOI: 10.1016/j.lwt.2018.03.063.

Richter, B. E., Jones, B. A., Ezzell, J. L., Porter, N. L., Avdalovic, N. and Pohl, C. (1996). Accelerated solvent extraction: a technique for sample preparation. *Analytical Chemistry* 68(6): 1033–1039. DOI: 10.1021/ac9508199.

Roohinejad, S., Koubaa, M., Barba, F. J., Greiner, R., Orlien, V. and Lebovka, N. I. (2016). Negative pressure cavitation extraction: a novel method for extraction of food bioactive compounds from plant materials. *Trends in Food Science and Technology* 52: 98–108. DOI: 10.1016/J.TIFS.2016.04.005.

Rubert-Nason, K. F., Holeski, L. M., Couture, J. J., Gusse, A., Undersander, D. J. and Lindroth, R. L. (2013). Rapid phytochemical analysis of birch (Betula) and poplar (Populus) foliage by near-infrared reflectance spectroscopy. *Analytical and Bioanalytical Chemistry* 405(4): 1333–1344. DOI: 10.1007/s00216-012-6513-6.

Sahu, A., Balhara, A., Singh, D. K., Kataria, Y., and Singh, S. (2019). Nuclear magnetic resonance spectroscopy | LC-NMR and LC-NMR-MS. In: *Encyclopedia of Analytical Science*. Academic Press, pp. 220–247. DOI: 10.1016/B978-0-12-409547-2.14074-0.

Santos, A. D. D. C., Fonseca, F. A., Dutra, L. M., Santos, M. F. C., Menezes, L. R. A., Campos, F. R., Nagata, N., Ayub, R. and Barison, A. (2018). 1H HR-MAS NMR-based metabolomics study of different persimmon cultivars (*Diospyros kaki*) during fruit development. *Food Chemistry* 239: 511–519. DOI: 10.1016/J.FOODCHEM.2017.06.133.

Santos, A. D. C., Fonseca, F. A., Lião, L. M., Alcantara, G. B. and Barison, A. (2015). High-resolution magic angle spinning nuclear magnetic resonance in foodstuff analysis. *TrAC Trends in Analytical Chemistry* 73: 10–18. DOI: 10.1016/j.trac.2015.05.003.

Santos, W. N. Ld dos, da Silva Sauthier, M. C., dos Santos, A. M. P., de Andrade Santana, D., Almeida Azevedo, R. S. and da Cruz Caldas, J. (2017). Simultaneous determination of 13 phenolic bioactive compounds in guava (*Psidium guajava* L.) by HPLC-PAD with evaluation using PCA and Neural Network Analysis (NNA). *Microchemical Journal* 133: 583–592. DOI: 10.1016/j.microc.2017.04.029.

Santos, D. T., Veggi, P. C. and Meireles, M. A. A. (2012). Optimization and economic evaluation of pressurized liquid extraction of phenolic compounds from jabuticaba skins. *Journal of Food Engineering* 108(3): 444–452. DOI: 10.1016/j.jfoodeng.2011.08.022.

Sato, T., Eguchi, K., Hatano, T. and Nishiba, Y. (2008). Use of near-infrared reflectance spectroscopy for the estimation of the isoflavone contents of soybean seeds. *Plant Production Science* 11(4): 481–486. DOI: 10.1626/pps.11.481.

Setford, P. C., Jeffery, D. W., Grbin, P. R. and Muhlack, R. A. (2017). Factors affecting extraction and evolution of phenolic compounds during red wine maceration and the role of process modelling. *Trends in Food Science and Technology* 69: 106–117. DOI: 10.1016/J.TIFS.2017.09.005.

Shettar, A. K., Sateesh, M. K., Kaliwal, B. B. and Vedamurthy, A. B. (2017). In vitro antidiabetic activities and GC-MS phytochemical analysis of Ximenia americana extracts. *South African Journal of Botany* 111: 202–211. DOI: 10.1016/j.sajb.2017.03.014.

Souza, L. S., Puziol, L. C., Tosta, C. L., Bittencourt, M. L. F., Ardisson, J. S., Kitagawa, R. R., Filgueiras, P. R. and Kuster, R. M. (2019). Analytical methods to access the chemical composition of an *Euphorbia tirucalli* anticancer latex from traditional Brazilian medicine. *Journal of Ethnopharmacology* 237: 255–265. DOI: 10.1016/j.jep.2019.03.041.

Sukor, N. F., Selvam, V. P., Jusoh, R., Kamarudin, N. S. and Rahim, S. A. (2021). Intensified DES mediated ultrasound extraction of tannic acid from onion peel. *Journal of Food Engineering* 296: 110437. DOI: 10.1016/J.JFOODENG.2020.110437.

Tiwari, B. K. (2015). Ultrasound: a clean, green extraction technology. *TrAC Trends in Analytical Chemistry* 71: 100–109. DOI: 10.1016/j.trac.2015.04.013.

Tsiokanos, E., Tsafantakis, N., Termentzi, A., Aligiannis, N., Skaltsounis, L. A. and Fokialakis, N. (2021). Phytochemical characteristics of bergamot oranges from the Ionian Islands of Greece: a multi-analytical approach with emphasis in the distribution of neohesperidose flavanones. *Food Chemistry* 343: 128400. DOI: 10.1016/j.foodchem.2020.128400.

Turner, C. and Waldebäck, M. (2013). Principles of pressurized fluid extraction and environmental, food and agricultural applications. In: Rizvi, S. S. H. (Ed.) *Separation, Extraction and Concentration Processes in the Food, Beverage and Nutraceutical Industries*. Woodhead Publishing, pp. 39–70. DOI: 10.1533/9780857090751.1.67.

Vadivel, V., Ravichandran, N., Rajalakshmi, P., Brindha, P., Gopal, A. and Kumaravelu, C. (2018). Microscopic, phytochemical, HPTLC, GC-MS and NIRS methods to differentiate herbal adulterants: pepper and papaya seeds. *Journal of Herbal Medicine* 11: 36–45. DOI: 10.1016/j.hermed.2018.01.004.

van Agthoven, M. A., Lam, Y. P. Y., O'Connor, P. B., Rolando, C. and Delsuc, M. A. (2019). Two-dimensional mass spectrometry: new perspectives for tandem mass spectrometry. *European Biophysics Journal* 48(3): 213–229. DOI: 10.1007/S00249-019-01348-5.

Wang, J. and Sporns, P. (2000). MALDI-TOF MS analysis of food flavonol glycosides. *Journal of Agricultural and Food Chemistry* 48(5): 1657–1662.

Wang, Y. S., Jin, Y. P., Gao, W., Xiao, S. Y., Zhang, Y. W., Zheng, P. H., Wang, J., Liu, J. X., Sun, C. H. and Wang, Y. P. (2016). Complete 1H-NMR and13C-NMR spectral assignment of five malonyl ginsenosides from the fresh flower buds of *panax ginseng*. *Journal of Ginseng Research* 40(3): 245–250. DOI: 10.1016/j.jgr.2015.08.003.

Wu, H., Guo, J., Chen, S., Liu, X., Zhou, Y., Zhang, X. and Xu, X. (2013). Recent developments in qualitative and quantitative analysis of phytochemical constituents and their metabolites using liquid chromatography-mass spectrometry. *Journal of Pharmaceutical and Biomedical Analysis* 72: 267–291. DOI: 10.1016/j.jpba.2012.09.004.

Yuenyong, J., Pokkanta, P., Phuangsaijai, N., Kittiwachana, S., Mahatheeranont, S. and Sookwong, P. (2021). GC-MS and HPLC-DAD analysis of fatty acid profile and functional phytochemicals in fifty cold-pressed plant oils in Thailand. *Heliyon* 7(2): e06304. DOI: 10.1016/J.HELIYON.2021.E06304.

Zhang, G., Li, P., Zhang, W. and Zhao, J. (2017). Analysis of multiple soybean phytonutrients by near-infrared reflectance spectroscopy. *Analytical and Bioanalytical Chemistry* 409(14): 3515–3525. DOI: 10.1007/s00216-017-0288-8.

Zhang, Q. W., Lin, L. G. and Ye, W. C. (2018). Techniques for extraction and isolation of natural products: a comprehensive review. *Chinese Medicine (United Kingdom)* 13(1): 20. DOI: 10.1186/s13020-018-0177-x.

Zheng, G., Yang, X., Chen, B., Chao, Y., Hu, P., Cai, Y., Wu, B. and Wei, M. (2020). Identification and determination of chemical constituents of *Citrus reticulata* semen through ultra high performance liquid chromatography combined with Q Exactive Orbitrap tandem mass spectrometry. *Journal of Separation Science* 43(2): 438–451. DOI: 10.1002/JSSC.201900641.

Zielinski, A. A. F., Haminiuk, C. W. I. and Beta, T. (2016). Multi-response optimization of phenolic antioxidants from white tea (*Camellia sinensis* L. Kuntze) and their identification by LC-DAD-Q-TOF-MS/MS. *LWT - Food Science and Technology* 65: 897–907. DOI: 10.1016/j.lwt.2015.09.020.

Zubarev, R. A. and Makarov, A. (2013). Orbitrap mass spectrometry. *Analytical Chemistry* 85(11): 5288–5296. DOI: 10.1021/AC4001223.

Chapter 11

Agronomic factors affecting phytochemical compounds in fruits and vegetables

Jiangtao Hu, Li Zhang, Zheng Wang, Jie Peng, Xiao Yang and Qichang Yang, Institute of Urban Agriculture, Chinese Academy of Agricultural Sciences, Chengdu National Agricultural Science and Technology Center, China

1 Introduction

Phytochemical compounds, one of the most critical indicators of food quality, can be strongly affected by environmental conditions and agronomic practices during agricultural production. An increasing number of studies suggest that regular consumption of vegetables and fruits is associated with a lower risk of several chronic diseases due to their high content of phytochemicals. The increasing individual requirements of consumers and the highly competitive nature of the fresh crop market create unprecedented demand for large quantities of high-quality vegetables and fruits. Therefore, novel strategies for agricultural practice have been developed and adopted to meet these requirements. The technological revolution in agriculture made it possible to control environmental conditions and improve the efficiency of agronomic practices, and therefore significantly enhance food quality.

This chapter aims to provide a comprehensive understanding of how widely reported environmental factors (e.g. lighting, temperature, relative humidity, and carbon dioxide) and agronomic practices (e.g. cultivar selection, fertilizer, phytohormones, abiotic stress, and harvest time) affect the accumulation of

http://dx.doi.org/10.19103/AS.2022.0101.13

phytochemicals in vegetables and fruits grown in controlled environments. Moreover, we describe a case study carried out by members of our group, in which we performed experiments to identify and quantify phytochemicals in lettuce and investigate the effects of different forms and concentrations of nitrogen on the accumulation of health-promoting compounds in lettuce.

2 Phytochemicals in fruits and vegetables

Fruits and vegetables are rich sources of health-promoting compounds, including phenolic compounds, carotenoids, alkaloids, nitrogen-containing compounds, organosulfur compounds, and phytosterols (Liu, 2013; Kyriacou and Rouphael, 2018). Some fruits and vegetables also contain specific phytochemicals that have been reported to protect against various chronic diseases, such as lycopene in tomato (Saini et al., 2020) and glucosinolates in *Brassica* species (Traka, 2016). Due to their potential beneficial effects on human health, the World Health Organization has recommended a daily fruit and vegetable intake of at least 400 g, preferably in five 80-g servings (Nishida

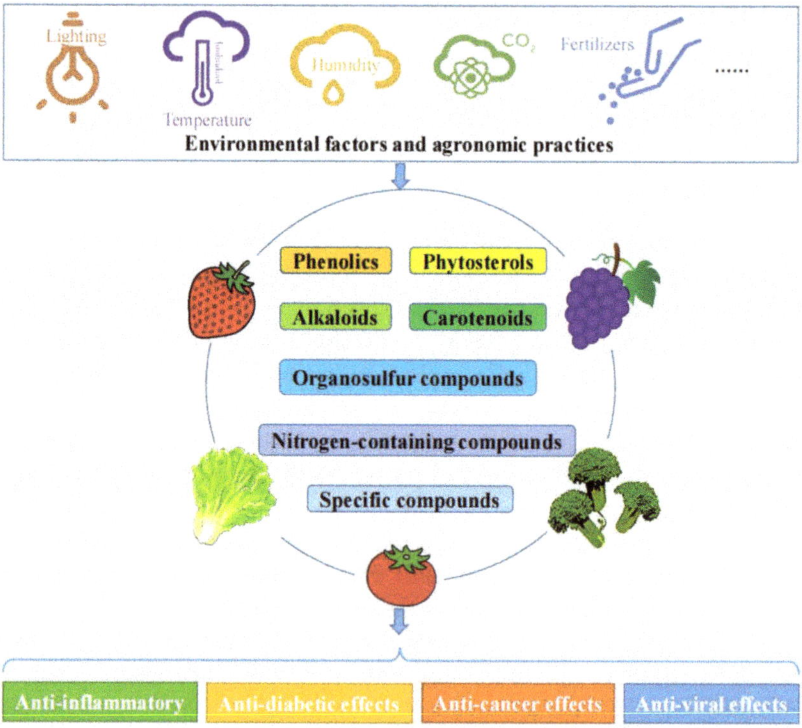

Figure 1 Phytochemicals in fruits and vegetables and their potential benefits on human health.

et al., 2004). A large number of strategies have been developed and adopted to improve the quality of fruits and vegetables in response to consumers' ever-increasing demands for high-quality food. Therefore, it is expected that precise control of environmental factors and manipulation of agronomic management could enhance the accumulation of health-promoting compounds in fruits and vegetables. Consequently, such developments hold promise for improving the health status of individuals (Fig. 1).

3 Environmental factors that affect phytochemicals in fruits and vegetables

3.1 Lighting

Lighting is vital for photosynthesis as well as the accumulation of phytochemicals in plants. Numerous review papers have described how light, including light quality, light intensity, and light duration, can be used to strategically manipulate the accumulation of health-promoting compounds in microgreens, lettuce, and postharvest vegetables and fruits (Ruangrak and Khummueng, 2019; Alrifai et al., 2019; Nassarawa et al., 2021; Zhang et al., 2020). Light-emitting diodes (LEDs), high-pressure sodium lamps, and fluorescent lamps have been widely used as light sources for vegetables and fruits produced in the field, greenhouse, and plant factory (Fig. 2).

Light quality is a crucial factor that can alter the nutritional value of vegetables and fruits grown in a controlled environment. Both photosynthetically efficient and less-efficient light qualities contribute to phytochemical changes in fruits and vegetables (Table 1). Red light generally increases the levels of phytochemical compounds in various food crops, though a small number of studies found that blue and far-red light negatively affect the accumulation of specific compounds. Evidence indicates that light signal transduction pathways, including the corresponding photoreceptors, transcriptional factors,

Figure 2 Supplemental lighting strategies for production of vegetables and fruits.

Table 1 Effects of light quality on the accumulation of phytochemicals in fruits and vegetables

Light quality	Species	Application method	Phytochemical change	Reference
Red	Bilberry	7.8 μmol m^{-2} s^{-1}, 48 h after harvest	Increased anthocyanins	Zoratti et al. (2014)
	Buckwheat sprouts	30 μmol m^{-2} s^{-1}, 16 h/day	Increased flavonoids	Nam et al. (2018)
	Buckwheat sprouts	50 μmol m^{-2} s^{-1}, 16 h/day	Increased catechins	Thwe et al. (2014)
	Chinese cabbage	16 h/day	Increased ferulic acid	Kim et al. (2015)
	Grape	80 μmol m^{-2} s^{-1}, 48 h after harvest	Increased stilbenic compounds	Ahn et al. (2015)
	Grape	15 μmol m^{-2} s^{-1}, during the natural photoperiod	Increased phenolic compounds	Gonzalez et al. (2015)
	Kale sprouts	80 μmol m^{-2} s^{-1}, 24 h before harvest	Increased phenolic compounds	Deng et al. (2017)
	Lettuce	130 μmol·m^{-2}·s^{-1}, 16 hours/day	Increased phenolic compounds	Li and Kubota (2009)
	Lettuce	210 μmol m^{-2} s^{-1}, 16 h/day	Increased ascorbic acid and glucose	Samuolienė et al. (2012)
	Orange	150 μmol m^{-2} s^{-1}, 10 days after harvest	Increased flavone and flavanone glycosides	Liu et al. (2019)
	Pepper	16 μmol m^{-2} s^{-1}, 16 h/day	Increased phenolic compounds	Azad et al. (2011)
	Rocket	150 μmol m^{-2} s^{-1}, 14 h/day	Increased total glucosinolates	Signore et al. (2020)
	Strawberry	100 μmol m^{-2} s^{-1}, 16 h/day	Increased proanthocyanidins	Zhang et al. (2018b)
	Tomato	113 μmol m^{-2} s^{-1}, 10–15 days after harvest	Increased lycopene, β-carotene, total flavonoid and phenolics	Panjai et al. (2017)
	Tomato	15.5 W m^{-2}, 4 days after harvest	Increased lycopene	Liu et al. (2009)
	Wheat sprouts	16 h/day	Increased ferulic acid and *p*-coumaric acid	Cuong et al. (2019)
Blue	Arugula	300 μmol m^{-2} s^{-1}, 24 h/day	Increased flavonoids	Taulavuori et al. (2018)
	Basil	300 μmol m^{-2} s^{-1}, 24 h/day	Increased phenolic acids	Taulavuori et al. (2018)
	Bilberry	8.1 μmol m^{-2} s^{-1}, 48 h after harvest	Increased anthocyanins	Zoratti et al. (2014)
	Broccoli sprouts	41 μmol m^{-2} s^{-1}, 5 days before harvest	Increased β-carotene, violaxanthin, total xanthophyll cycle pigments, glucoraphanin, epiprogoitrin, and aliphatic glucosinolates	Kopsell and Sams (2013)

Buckwheat sprouts	30 $\mu mol\ m^{-2}\ s^{-1}$, 16 h/day	Increased C-glycosyl flavonols	Nam et al. (2018)
Buckwheat sprouts	177 $\mu mol\ m^{-2}\ s^{-1}$, 96 h before harvest	Increased anthocyanins	Seo et al. (2015)
Buckwheat sprouts	50 $\mu mol\ m^{-2}\ s^{-1}$, 16 h/day	Increased rutin and cyanidin 3-*O*-rutinoside	Thwe et al. (2014)
Canola	50 $\mu mol\ m^{-2}\ s^{-1}$, 16 h/day	Increased glucosinolates and phenolic compounds	Park et al. (2019)
Chili	50 $\mu mol\ m^{-2}\ s^{-1}$, 3 days after harvest	Increased total phenolics and vitamin C	Pola et al. (2020)
Chinese cabbage	16 h/day	Increased phenolic compounds	Kim et al. (2015)
Grape	80 $\mu mol\ m^{-2}\ s^{-1}$, 48 h after harvest	Increased stilbenic compounds	Ahn et al. (2015)
Grape	15 $\mu mol\ m^{-2}\ s^{-1}$, during the natural photoperiod	Increased phenolic compounds	Gonzalez et al. (2015)
Grape	12.4 $\mu mol\ m^{-2}\ s^{-1}$, at night	Increased anthocyanins	Kondo et al. (2014)
Ice plant	150 $\mu mol\ m^{-2}\ s^{-1}$, 14 h/day	Increased myo-inositol and pinitol	Kim et al. (2018)
Ice plant	120 $\mu mol\ m^{-2}\ s^{-1}$, 14 h/day	Increased phenolic compounds	Kim et al. (2018)
Kale	40 W m^{-2}, 12 h/day	Increased sinigrin and glucoraphanin	Li et al. (2020b)
Kale sprouts	30 $\mu mol\ m^{-2}\ s^{-1}$, 15 h/day	Decreased gluconapin; increased phenolic compounds, and anthocyanins	Qian et al. (2016)
Lettuce	130 $\mu mol\ m^{-2}\ s^{-1}$, 16 h/day	Increased carotenoids and anthocyanins	Li and Kubota (2009)
Lettuce	300 $\mu mol\ m^{-2}\ s^{-1}$, 14 h/day	Increased vitamin C	Amoozgar et al. (2017)
Mustard	90 $\mu mol\ m^{-2}\ s^{-1}$, 16 h/day	Increased phenolic compounds	Park et al. (2020)
Pepper	49 $\mu mol\ m^{-2}\ s^{-1}$, 16 h/day	Increased anthocyanins	Azad et al. (2011)
Strawberry	100 $\mu mol\ m^{-2}\ s^{-1}$, 16 h/day	Increased anthocyanins	Zhang et al. (2018b)
Strawberry	40 $\mu mol\ m^{-2}\ s^{-1}$, 12 days after harvest	Increased anthocyanins	Xu et al. (2014)
Strawberry	95 $\mu mol\ m^{-2}\ s^{-1}$, 12 hours/day	Increased anthocyanins	Kadomura-Ishikawa et al. (2013)

(*Continued*)

Table 1 (*Continued*)

Light quality	Species	Application method	Phytochemical change	Reference
	Wheat sprouts	16 h/day	Increased gallic acid and quercetin; decreased *p*-coumaric acid and epicatechin	Cuong et al. (2019)
Far-red	Bilberry	7.6 μmol m^{-2} s^{-1}, 48 h after harvest	Increased anthocyanins	Zoratti et al. (2014)
	Brassica napus	36 μmol m^{-2} s^{-1}, 16 h/day	Decreased flavonoids and quercetin glycosides	Gerhardt et al. (2008)
	Brassica napus	10 μmol m^{-2} s^{-1}, 35 days	Increased glucosinolate and progoitrin	Molmann et al. (2021)
	Lettuce	18 μmol m^{-2} s^{-1}, 16 h/day	Decreased anthocyanins and carotenoids	Li and Kubota (2009)
	Pepper	5 μmol m^{-2} s^{-1}, 16 h/day	Decreased ascorbic acid	Azad et al. (2011)
UV-A	Broccoli sprouts	9.47 W m^{-2}, 2 h	Increased phenolic compounds	Moreira-Rodríguez et al. (2017)
	Kale	40 W m^{-2}, 12 h/day	Increased sinigrin and glucoraphanin	Li et al. (2020b)
	Lettuce	18 μmol m^{-2} s^{-1}, 16 h/day	Increased anthocyanins	Li and Kubota (2009)
	Pak choi	40 W m^{-2}, 12 h/day	Increased glucoraphanin, sinigrin and glucobrassicin	Li et al. (2020b)
UV-B	Apple	0.16–0.2 W m^{-2}, 72 h	Increased anthocyanin, glycosides and procyanidins	Lancaster et al. (2000)
	Basil	2 kJ m^{-2} day^{-1}, 1 h/day	Increased phenolic compounds, anthocyanin and ascorbic acid	Sakalauskaite et al. (2012)
	Broccoli sprouts	7.16 W m^{-2}, 2 h	Increased aliphatic and indole glucosinolates	Moreira-Rodríguez et al. (2017)
	Brassica napus	7 kJ m^{-2} day^{-1}, 16 h/day	Increased quercetin glycosides	Gerhardt et al. (2008)
	Pak choi microgreens	12.4 μmol m^{-2} s^{-1}, 16 h/day	Increased phenolic compounds, anthocyanins and ascorbic acid	Brazaityte et al. (2015)
	Tomato	6.132 kJ m^{-2} day^{-1}, 12 h/day	Increased flavonol quercetin-3-*O*-rutinoside	Dzakovich et al. (2016)
UV-C	Orange	100 μmol m^{-2} s^{-1}, 10 days after harvest	Increased flavone and flavanone glycosides	Liu et al. (2019)

	Peanut sprouts	1.8 W m^{-2}, 2 days	Increased phenolic compounds, flavonoids and stilbene	Zhu et al. (2020)
	Tomato	22.8 W m^{-2}, 4 days after harvest	Increased lycopene	Liu et al. (2009)
	Tomato	4 or 8 kJ m^{-2}, 35 days	Increased phenolic compounds	Liu et al. (2012)
	Tomato	0.282 W m^{-2}, 12 h after harvest	Increased lycopene, chlorogenic acid and ferulic acid; decreased β-carotene, naringenin and rutin	Bravo et al. (2013)
Green	Basil	22 μmol m^{-2} s^{-1}, 16 h/day	Increased plant yield without changing phytochemical compounds	Dou et al. (2019)
Mixed spectrum	Arugula, kale and mustard microgreens	300 μmol m^{-2} s^{-1}, 16 h/day, R:B = 8:2	Increased total ascorbate	Ying et al. (2021)
	Arugula, kale and red cabbage microgreens	300 μmol m^{-2} s^{-1}, 16 h/day, R:B = 7:3	Increased anthocyanins	Ying et al. (2021)
	Broccoli	210 μmol m^{-2} s^{-1}, 2 days, R:B:FR = 4:1:2	Increased glucosinolates	Jo and Lee (2018)
	Buckwheat sprout	50 μmol m^{-2} s^{-1}, 16 h/day, white LEDs	Increased carotenoids	Tuan et al. (2013)
	Buckwheat sprouts	137 μmol m^{-2} s^{-1}, 96 h before harvest, R:B = 7:3	Increased rutin and chlorogenic acid	Seo et al. (2015)
	Chinese cabbage sprouts	350 μmol m^{-2} s^{-1}, 2 days, R:B:FR = 12:3:20	Increased phenolic compounds; decreased progoitrin	Jo and Lee (2018)
	Gynura bicolor	38.1 μmol m^{-2} s^{-1} 16 h/day, R:B = 7:3	Increased antioxidants	Ren et al. (2014)
	Ice plant	150 μmol m^{-2} s^{-1}, 14 h/day, R:B:W = 8:1:1	Increased phenolic compounds and flavonoids	Kim et al. (2016)
	Kohlrabi microgreens	315 μmol m^{-2} s^{-1}, 16 h/day, R:B:FR = 84:9:7 or R:B = 87:13	Increased anthocyanins	Craver et al. (2017)

(Continued)

Table 1 (*Continued*)

Light quality	Species	Application method	Phytochemical change	Reference
	Lettuce	300 μmol m^{-2} s^{-1}, 14 h/day, R:B = 7:3	Increased carotenoids	Amoozgar et al. (2017)
	Lettuce	171 μmol m^{-2} s^{-1}, 12 h/day, R:B = 65:35, R:B = 53:47, or R:B = 41:59	Increased phenolic compounds and flavonoids	Son and Oh (2013)
	Lettuce	200 μmol m^{-2} s^{-1}, 18 h/day, R:B = 8:2, R:B = 1:1, or R:B = 1:4	Increased phenolic compounds	Spalholz et al. (2020)
	Lettuce	200 μmol m^{-2} s^{-1}, 24 h, R:B = 4:1	Increased phenolic compounds	Bian et al. (2016)
	Lettuce	151 μmol m^{-2} s^{-1}, 12 h/day, R:B = 9:1	Increased phenolic compounds	Son et al. (2017)
	Mustard, beet and parsley microgreens	302.5 μmol m^{-2} s^{-1}, 16 h/day, R:B:FR = 100:20:1	Increased carotenoids, α- and β-carotenes, lutein, violaxanthin and zeaxanthin	Samuoliene et al. (2017)
	Mustard, beet and parsley microgreens	302.5 μmol m^{-2} s^{-1}, 16 h/day, R:B:FR = 80:40:1	Increased tocopherols	Samuoliene et al. (2017)
	Strawberry	200 μmol m^{-2} s^{-1}, 16 h/day, R:B = 7:3	Increased anthocyanins	Choi et al. (2013)

and structural genes, are involved in these phytochemical changes (Tao et al., 2018; Landi et al., 2020). In addition, ultraviolet induces stress conditions that promote the accumulation of phytochemical compounds in plants. The effects of laser and pulsed lights have also been studied; laser light was suggested to increase vitamins, phenolic compounds, phenolic acids, flavonoids, and glutathione in buckwheat sprouts (Almuhayawi et al., 2021). Moreover, pulsed light improved the quality of microgreens (Vaštakaitė et al., 2017), spinach (Agüero et al., 2016), mango fruits (Charles et al., 2013; Lopes et al., 2016), and fruit juice (Kwaw et al., 2018; Vollmer et al., 2020).

An adequate light intensity is crucial for plants to receive sufficient photosynthetic active radiation. Generally, sprouts and microgreens accumulate more phytochemical compounds under lower light intensities ranging from 70 $\mu mol\ m^{-2}\ s^{-1}$ (12 h/day) to 100 $\mu mol\ m^{-2}\ s^{-1}$ (18 h/day) than at higher intensities (Gao et al., 2021), while mature plants such as hydroponically grown lettuce produce more phytochemicals at 250 $\mu mol\ m^{-2}\ s^{-1}$ (16 h/day) (Zhang et al., 2018a). Anthocyanins and carotenoids play essential roles in plant adaption to high light intensity; thus, under specific conditions, a relatively high light intensity can be used to produce more anthocyanins and carotenoids (Wang et al., 2009; Craver et al., 2017; Stetsenko et al., 2020). Shimomura et al. (2020) reported that high light intensity (200 $\mu mol\ m^{-2}\ s^{-1}$) increased the chlorogenic acid content of lettuce compared to 100 $\mu mol\ m^{-2}\ s^{-1}$. Fu et al. (2017) reported that increasing the light intensity (from 60 to 140 $\mu mol\ m^{-2}\ s^{-1}$) increased the concentration of vitamin C in 'Youmaicai' lettuce, though higher light intensities (140–220 $\mu mol\ m^{-2}s^{-1}$) decreased the vitamin C content. In another study, exposure of lettuce to high light conditions (800 $\mu mol\ m^{-2}\ s^{-1}$) for a short period was reported to increase the concentration of phytochemicals with antioxidant properties. High light intensities were suggested to increase the levels of carotenoids, glutathione, total phenols, and anthocyanins as a response to oxidative stress (Oh et al., 2009).

In addition to light quality and light intensity, photoperiodism is another factor that affects active photosynthetic radiation. Ali et al. (2009) and Song et al. (2020b) found that 12/12-h light/dark cycles led to the highest levels of phytochemical compounds in selected leafy vegetables and lettuce, respectively. Moreover, continuous LED illumination before harvest was reported as a powerful strategy to improve the accumulation of phenolic compounds and carotenoids in lettuce (Bian et al., 2016). Therefore, future studies are required to optimize the duration of light for specific crops due to the crucial role of light in the nutritional quality and energy use efficiency of vegetables.

3.2 Temperature

Temperature is one of the most relevant environmental factors that affect the accumulation of secondary metabolites in vegetables and fruits. For example, broccoli harvested in fall contains higher levels of sulforaphane, while total polyphenols are higher in spring (Pék et al., 2012). Moreover, cold acclimation increases the total flavonoid and ascorbic acid contents of sweet spinach (Watanabe and Ayugase, 2015).

In general, short-term low or high temperatures induce moderate stress within plants and increase the accumulation of phytochemical compounds. Specifically, short-term high temperatures increased glucosinolates in pak choi seedlings (40°C for 8 h; Rao et al., 2021), broccoli sprouts (33.1°C for 6 days; Pereira et al., 2002), and rocket sprouts (30°C for 5 days; Ragusa et al., 2017), while short-term low temperatures enhanced antioxidants in kale (4°C for 3 days; Lee and Oh, 2015) and tomato (10°C for 1 day; Khan et al., 2015). Low temperature also increased the levels of cyanidin-3-(6″-malonyl)-glucoside in lettuce leaves (Becker et al., 2014). In long-term cultivation studies, red chicory and garland chrysanthemum contained higher total phenols and flavonoids when grown at 25°C than at 20°C or 30°C (Lee et al., 2013). Swede roots grown at 15°C or 21°C contained more glucosinolates than plants grown at 9°C (Molmann et al., 2021; Johansen et al., 2016). Beet and lettuce grown at 20°C or 25°C had higher total polyphenol and flavonoid contents than plants grown at 30°C but accumulated more ascorbic acid at 30°C (Lee et al., 2015). However, Chinese broccoli cultivated at a root temperature of 10°C contained more glucosinolates than plants grown at 15°C or 20°C (He et al., 2020). *Dracocephalum palmatum*, an arctic plant, accumulated higher levels of antioxidants, free sugars, and polysaccharides at 1°C than at 20°C (Olennikov et al., 2017).

Fresh fruits and vegetables are generally susceptible to deterioration between harvest and consumption. Therefore, appropriate storage and transportation conditions are critical, and temperature is one of the most important conditions. Preserving the levels of health-promoting compounds throughout the shelf life of the produce requires proper cold storage throughout the entire chain. The levels of phytochemical compounds generally increase initially and then decrease during storage, though this process can be delayed by low-temperature storage. For instance, low temperatures (4°C or 5°C) delay the degradation of glucosinolates in pak choi (Yang et al., 2010), ascorbic acid and β-carotene in broccoli florets (Nath et al., 2011), and phenolic compounds in apples (Ma et al., 2019). Lycopene can be maintained in tomatoes by storage at 12°C for 6 days and then at 5°C for 7 days (Kubota et al., 2006). However, phenolic compounds and carotenoids were better preserved at higher temperatures than lower temperatures in several fruits, such as mango (20°C > 5°C; Talcott et al., 2005), grapefruit (23°C > 9°C; Chebrolu et al., 2012), papaya (25°C > 1°C; Rivera-Pastrana et al., 2010), and pitaya (15°C > 10°C > 5°C; Li et al., 2017). Notably, the stage of maturity should be considered when choosing the proper storage temperature for specific fruit.

For instance, strawberry harvested at the red tip stage maintained higher levels of anthocyanins and ascorbic acid at 10°C than at 5°C or 20°C (Jin et al., 2011; Shin et al., 2007), while 3°C was the optimal temperature for fruit harvested at the white tip stage (Shin et al., 2008a). The optimal storage temperature also has the potential to improve the levels of phytochemical compounds in fruits and vegetables. Phenolic compounds increase at low temperatures (0°C or 4°C) in peach (Manganaris et al., 2017), nectarine (Manganaris et al., 2017), sweet potato (Padda and Picha, 2008), and potato (Galani et al., 2017). Glucosinolates increase when broccoli is stored at 0°C for 4 days followed by 10°C for 3 days (Rybarczyk-Plonska et al., 2016). High temperatures (20°C and 23°C) also increase anthocyanins and phenolic compounds in sweet corn (Hong et al., 2021) and wounded carrots (Han et al., 2017).

3.3 Relative humidity

To date, very few studies have investigated the potential role of relative humidity on the accumulation of phytochemicals in vegetables and fruits. Lower relative humidity is thought to produce higher-quality fruit and vegetable crops. For instance, higher levels of phytochemicals were present in lettuce cultivated at 41% than 77% relative humidity (Amitrano et al., 2021). Peppers grown at lower relative humidity (38%) contained a higher concentration of carotenoids than fruit cultivated at a relative humidity of 43% (Lee et al., 2005). Moreover, lower relative humidity (60% compared with 90%) combined with light treatment contributed to the production of antioxidant-rich mung bean sprouts (Amitrano et al., 2020).

Relative humidity is also an important environmental factor that contributes to the preservation of health-promoting compounds during the shelf life of vegetables and fruits, though the influence of relative humidity appears to be species-specific. For example, higher relative humidity (50–75%) improved chlorogenic acid, ferulic acid, epicatechin, quercetin-3-rutinoside, and cyanidin-3-galactoside in apple peels (Zhang et al., 2016), while lower relative humidity (10–12%) maintained the quality of green lentils (Bragança et al., 2020). However, relative humidity has little effect on the fruit quality of strawberries during storage (Shin et al., 2007, 2008a).

3.4 CO_2

The effects of the CO_2 concentration on phytochemical compounds in fruits and vegetables remain controversial. Several studies reported that elevated CO_2 (600–820 ppm) beneficially improved health-promoting compounds and the morphological features of fruits and vegetables. Elevated CO_2 can alleviate abiotic stress in plants by increasing antioxidants (Xu et al., 2015; Li et al., 2020a) and reducing the induction of high levels of reactive oxygen species by oxidative stress (Romero et al., 2008; Lin et al., 2017), and enhance the consumption of electrons in photosynthetic carbon fixation (Salazar-Parra et al., 2012). Additionally, elevated CO_2 increases the levels of phytochemical compounds such as glucosinolates (Schonhof et al., 2007; Himanen et al., 2008; Klaiber et al., 2013; Almuhayawi et al., 2020), phenolic compounds (Reddy et al., 2004; Pérez-López et al., 2018), and flavonoids (Dong et al., 2018) in vegetables. However, in some species - such as strawberries - elevated CO_2 was reported to decrease the accumulation of phytochemicals, including phenolic compounds, anthocyanins, and flavonoids (Shin et al., 2008b; Bodelón et al., 2010). Elevated CO_2 also impaired the beneficial effects of arbuscular mycorrhizal fungi on phytochemical compounds in lettuce (Baslam et al., 2012). Additionally, Loladze et al. (2019) reviewed how elevated CO_2 decreased the carotenoid content of plants, though individual studies reported varied responses, especially in abiotically stressed plants. Hartley et al. (2000) suggested that the effects of CO_2 on the levels of phytochemical compounds in plants are species- and generation-specific. Therefore, further investigation is required to investigate the varied effects of elevated CO_2 on phytochemical compounds in various plant species.

In summary, numerous studies have investigated how a single environmental factor leads to phytochemical changes in vegetables and fruits. However, as plant metabolites are influenced by multiple environmental factors, it will be interesting to study the coupled effects of multiple environmental factors on the phytochemical contents of fruits and vegetables.

4 Agronomic practices

Agronomic practices have long been used to improve the yield and quality of vegetables and fruits. Practitioners' valuable experience in cultivar selection, applying fertilizers and phytohormones, inducing phytochemicals by abiotic stress, and optimization of harvest time have contributed to increased production and quality improvements in agriculture. This section discusses these valuable practices and their effects on the quality of horticultural crops.

4.1 Selection of varieties

Different vegetable and fruit species contain distinct primary health-promoting compounds, such as lycopene in tomato (Li et al., 2021), glucosinolates in *Brassicaceae* vegetables (Neugart et al., 2018), and chicoric acid and lactucopicrin in *Compositae* vegetables (Prohens-Tomás and Nuez, 2007). However, the levels of these distinct compounds can vary between different cultivars of the same species, such as peach (Manganaris et al., 2017), strawberry (Xie et al., 2014), pepper (Lee et al., 2005; Tripodi et al., 2019), broccoli (Pérez-Balibrea et al., 2011), soybean (Alghamdi et al., 2018), and potato (Samaniego et al., 2020b). Moreover, different environmental conditions can lead to varied concentrations and constitutions of phytochemicals in the same cultivar. For example, pitaya grown in Malaysia contained a high content of flavonoids, while those grown in Australia contained high levels of phenolic compounds (Nurul and Asmah, 2014).

Innovative breeding technologies provide a powerful approach to select cultivars in order to maintain the postharvest quality of vegetables and fruits, including improved phytochemical status (Damerum et al., 2020). In addition to the breeding strategies to improve resistance in crops proposed by Deng et al. (2020), the increasing availability of genome sequences and improved understanding of the underlying molecular mechanisms may help to breed food crops with high contents of beneficial compounds. Quantitative trait loci mapping, an effective novel breeding technology that has been used to understand the genetic control of secondary metabolites and identify potential molecular mechanisms, has dramatically improved the breeding efficiency of vegetables and fruits to enhance quality. Candidate genes co-located with quantitative trait loci in *Brassica oleracea* (Sotelo et al., 2014), sweet cherry (Calle et al., 2021), apple (Chagné et al., 2012), and cranberry (Diaz-Garcia et al., 2018) were found to be involved in the biosynthesis of glucosinolates, phenolic compounds, polyphenolic compositions, and anthocyanins, respectively. As expected, multiple candidate genes, such as *CYP79B2*, *PAL*, and *C3H*, are involved in these biosynthetic pathways. Thus, the identified quantitative trait loci and candidate genes may serve as molecular markers for breeding crops with high contents of phytochemical compounds in the future.

4.2 Fertilization

Fertilizers influence plant growth and development and sustain crop yield and quality. Recently, interest in comparing the effects of organic and inorganic forms of fertilizer on antioxidant accumulation has increased. Most investigations support the notion that the application of organic fertilizers such as vermicompost, extruded shrimp shells, and livestock droppings positively influences the biosynthesis of phytochemical compounds in vegetables (Nur et al., 2013; Øvsthus et al., 2015; Obidola et al., 2019). Moreover, relatively low concentrations of nutrients in solution were found to be beneficial for the accumulation of phytochemicals, as reported for onion (Ortega-García et al., 2015), lettuce (Song et al., 2020a,b), kale (Kopsell et al., 2017), pumpkin (Oloyede et al., 2012), and chicory (Sinkovič et al., 2015). For instance, the nitrogen supply negatively correlates with the accumulation of phytochemical compounds in different vegetables, such as rocket cultivated at a nitrogen concentration above 5 mM (Chun et al., 2017), peach at a nitrogen concentration above 45 kg ha^{-1} (Vashisth et al., 2017), spinach at a nitrogen concentration above 75 kg ha^{-1} (Machado et al., 2020), turnip at a nitrogen concentration above 80 kg ha^{-1} (Li et al., 2007), curly kale at a nitrogen concentration above 90 kg ha^{-1} (Groenbaek et al., 2014), and lettuce grown in nutrient solution with a nitrogen concentration above 26 ppm (Qadir et al., 2017). In comparison, nitrogen deficiency is beneficial for the biosynthesis of phenolic acids, flavonols, anthocyanins, and ascorbic acid in lettuce. Zhou et al. (2020) reported that short-term nitrogen limitation prior to harvest increased the levels of vitamin C and glutathione and the total phenolic content. Similar to nitrogen, application of a phosphorus concentration above 2 mM reduced the content of glucosinolates in rocket (Chun et al., 2017). Phosphorus deficiency also increased glucosinolates in pak choi (Yang et al., 2009). In contrast, higher potassium and sulfur concentrations increased glucosinolates in *Brassica* species (Li et al., 2007; Pék et al., 2012; Groenbaek et al., 2014; Chun et al., 2017). Notably, beneficial mineral elements also have the potential to increase phenolic compounds in fruits and vegetables: silicon, selenium, and calcium increased Se-methyl selenocysteine, caffeic acid, and chlorogenic acid in grape (Gomes et al., 2020), broccoli (Sepúlveda et al., 2013), and potato (Ngadze et al., 2014), respectively. Therefore, appropriate ratios of macroelements and mineral elements should be considered to obtain high-quality fruits and vegetables.

4.3 Phytohormones

Phytohormones play essential roles in the regulation of the biosynthesis of phytochemical compounds. For example, Van Meulebroek et al. (2015) reported that cis-12-oxo-phytodienoic acid, cucurbic acid, 2-oxindole-3-acetic acid, 1-acetylindole-3-carboxaldehyde, and cis-zeatin-*O*-glucoside were involved in carotenoid metabolism in tomato. Thus, it is possible to increase the levels of phytochemical compounds through the exogenous application of phytohormones. Researchers have investigated multiple phytohormones, and methyl jasmonate has been extensively studied. Applying 0.025 and 0.25 mM methyl jasmonate improved the content of glucosinolates in broccoli sprouts (Moreira-Rodríguez et al., 2017) and plants from the *Brassica* species (Yi et al., 2016), respectively. Application of 0.1 methyl jasmonate to sweet potato (Ghasemzadeh et al., 2016) and 0.5 mM methyl jasmonate to lettuce (Kim et al., 2007) improved phenolic compounds, 1 mM methyl jasmonate improved flavonoids in kiwi fruit (Öztürk and Yücedağ, 2021), and 10 mM methyl jasmonate increased anthocyanins in apple (Rudell et al., 2002). Moreover, these studies indicate that lower concentrations of methyl jasmonate should be applied to vegetables, whereas higher concentrations are beneficial for fruits. Abscisic acid and salicylic acid also contribute to the accumulation of phytochemical compounds. Application of abscisic acid at a concentration of 0.1 mM was reported to increase phenolic compounds in sweet potato (Ghasemzadeh et al., 2016) and Chinese cabbage (Thiruvengadam et al., 2015), while 1.1 mM abscisic acid increased phenolic compounds and anthocyanins in lettuce (Li et al., 2010). Applying salicylic acid at a concentration of 0.1 mM also increases phenolic compounds in sweet potato (Ghasemzadeh et al., 2016) and Chinese cabbage (Thiruvengadam et al., 2015). Other phytohormones also contribute to the production of phytochemicals. For instance, 50 $\mu L\ L^{-1}$ ethylene improved the contents of β-carotene and β-cryptoxanthin in citrus (Ma et al., 2015). Kinetin at a concentration of 0.9 mM improved the levels of rutin, epicatechin, and gallic acid in lentils (Giannakoula et al., 2012), and 1 μM 24-epibrassinolide increased phenolics and total anthocyanins in strawberries (Sun et al., 2019).

Phytohormones can also be applied to stressed plants to enhance stress resistance and the production of phytochemical compounds. For instance, 1000 $\mu L\ L^{-1}$ ethylene and 1.1 mM methyl jasmonate improved the content of neoglucobrassicin in broccoli subjected to wounding stress (Villarreal-García et al., 2016) and 5 μM salicylic acid improved flavonoids in red amaranth exposed to salt stress (Villarreal-García et al., 2016).

4.4 Phytochemicals induced by abiotic stress

Plants tend to accumulate more secondary metabolites under stress conditions. Therefore, short-term abiotic stress treatment represents a strategy to improve the levels of phytochemical compounds in plants. However, extreme environmental factors and high doses of phytohormones can also induce abiotic stress in plants. Thus, the abiotic stress-induced improvements in phytochemicals are dose-dependent, as reviewed by Duarte-Sierra et al. (2020). Therefore, such practices should be conducted carefully to only confer mild stress to the plants. For example, sodium chloride at concentrations of 50 and 100 mM improved the levels of phytochemical compounds in radish sprouts (Yuan et al., 2010), buckwheat sprouts (Lim et al., 2012), *Brassica* species (Šamec et al., 2021), lettuce (Garrido et al., 2014), and amaranth (Sarker et al., 2018); the periods of treatment were 5, 7, 1, 3, and 30 days, respectively. The sprouts were foliar sprayed, and the vegetable plants were drenched with sodium chloride. Amaranth is a salt-tolerant plant; thus, the period of treatment was relatively long.

In addition, heat (40°C for 10 min), chilling (4°C for 1 day), or high-light (800 $\mu mol\ m^{-2}\ s^{-1}$ for 1 day) stress can also be used to increase phenolic compounds in lettuce (Oh et al., 2009). Water stress improved anthocyanins in peach (Tavarini et al., 2011). Hot air treatment (37°C for 6 h) improved phenolic compounds and flavonoids in plum (Chang et al., 2019). Several studies have explored the mechanisms underlying these stress-induced phytochemical changes. Low oxygen increases the levels of anthocyanins, phenolic compounds, and carbohydrates in lettuce, partially by promoting ethylene production (Rajapakse et al., 2009). Water deficiency improves the quality of plants by inducing the production of abscisic acid, followed by altered stomatal closure, CO_2 absorption, and reactive oxygen species accumulation (González-Chavira et al., 2018). As abiotic stresses can be applied to stimulate the production of phytochemicals, and plants in specific areas suffer stress conditions, additional studies should be conducted to explore the underlying mechanisms and provide further information for breeding stress-resistant plants.

4.5 Harvest time and stage of maturation

Studies have investigated the phytochemical changes in fruits harvested in different seasons or stages of maturation. June-bearing, every-bearing, and day-neutral strawberry varieties harvested at the same stage of maturity have higher contents of ascorbic acid in October, but higher phenolic compounds in June (Pincemail et al., 2012). White-cultivar pears have a higher content of polyphenols, while yellow and red cultivars have higher betacyanin and betaxanthin contents at the end of August than those harvested at the end of October (Palmeri et al., 2020). Purple perilla has a higher content of phenolic compounds in mid-September than in mid-August (Kang and Lee, 2011). The ratio of rebaudioside A to stevioside in stevia increases between July and September (Tavarini and Angelini, 2013). The factors underlying these seasonal phytochemical changes may be due to the coupling of various environmental factors. Vegetables such as lettuce (El-Nakhel et al., 2020) and spinach (Barkat et al., 2018) have higher contents of total phenolic compounds at the juvenile stage, but higher pigment contents at the mature stage.

Phytochemical changes during different stages of maturity have been identified in cranberry (Oszmiański et al., 2018), blackberry (Samaniego et al., 2020a), wild grape (Srihanam, 2016), pomegranate (Galindo et al., 2014), and citrus (Wang et al., 2016). These studies suggest that color-related phytochemicals such as anthocyanins increase during the ripening of fruits, while phenolic compounds and flavonoids decrease. However, there are some exceptions; for example, phenolic compounds increase 2 weeks before commercial harvest in apples (Drogoudi and Pantelidis, 2011). In addition, tree-ripened apricot has a higher content of carotenoids and vitamin A than commercially matured fruit (Campbell et al., 2013). Moreover, although some beneficial phytochemicals are enriched in immature fruits, their phytochemical profiles may provide vital information for the food industry and enable the processing of fruits to create health-promoting products.

Overall, both breeding and cultivation techniques provide important strategies to improve the quality of fruits and vegetables; while breeding serves as the foundation, optimal cultivation techniques guarantee the quality of the produce. Thus, breeding and cultivation techniques should complement each other.

5 Case study

Lettuce, one of the most widely consumed leafy vegetables, is commercially available throughout all seasons as it can be cultivated in the open field, greenhouse, and plant factory. Lettuce is recognized as a rich source of health-promoting phytochemicals, such as flavonoids, carotenoids, and lactucins. Previously, our group standardized a nontargeted metabolomics methodology and established a metabolite library for lettuce to more precisely characterize the metabolic variations between different horticultural lettuce cultivars (Yang et al., 2018b; García et al., 2020).

Nitrogen is an essential fertilizer widely used in agricultural practice that sustains the growth and development of vegetables. Organic nitrogen represents 96–99% of total nitrogen in the soil; inorganic forms, such as nitrate, represent less than 4% of total nitrogen in the soil. Most studies suggest that a sufficient nitrate supply decreases the accumulation of phytochemicals, including phenolic acid and flavonoids, in lettuce (Becker et al., 2015). In contrast, several lines of evidence indicate that organic nitrogen significantly influences the growth, development, and nutritional quality of vegetables. Therefore, we performed a comparative study to explore how inorganic nitrogen (as nitrate) and organic nitrogen (as glycine) influence the accumulation of health-promoting compounds in lettuce (Fig. 3).

Lettuce was hydroponically cultivated in 9 mM glycine or 9 mM nitrate for 4 weeks. GC/MS and LC/MS analyses of primary and secondary metabolites

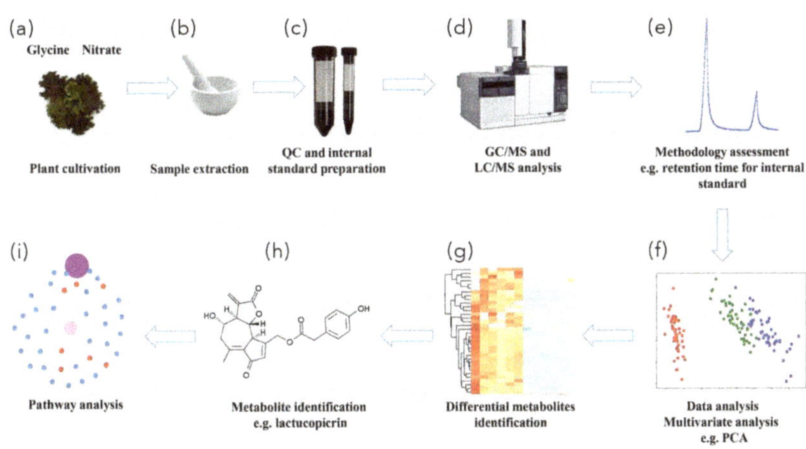

Figure 3 Workflow of our nontargeted metabolomics study.

suggested that 29 metabolites involved in phenylpropanoid metabolism, the flavonoid pathway, and the tricarboxylic acid cycle were significantly altered between the nitrate and glycine treatments. Glycine supply promoted the accumulation of quercetin and luteolin glycosides, ascorbic acid, and most amino acids, but reduced phenolic acids and tricarboxylic acid intermediates compared to the plants in the nitrate treatment (Yang et al., 2018a). Next, we selected two cultivars, Shenxuan 1 and Lollo Rossa, to analyze the effects of different glycine concentrations on the levels of health-promoting compounds and the antioxidant activity of the lettuce. The appropriate concentrations of exogenous glycine (18 mM for Shenxuan 1; 9 mM for Lollo Rossa) induced most glycosylated quercetin derivatives and luteolin derivatives, α-tocopherol, and antioxidant activity in lettuce. These results suggest that supplying exogenous glycine during lettuce cultivation could be a potential strategy to promote high levels of nutritional metabolites (including flavonoid glycosides, ascorbic acid, and amino acids) and increase the antioxidant activity, which could be beneficial for individuals' health (Yang et al., 2017).

6 Summary and future trends

Greenhouse cultivation significantly alleviates the influence of weather and greatly improves the quantity and quality of vegetable and fruit crops. This chapter provides a comprehensive understanding of the environmental factors and agricultural practices that affect phytochemical compounds in fruits and vegetables. Future research is required to improve the quality of vegetables and fruits through breeding programs and optimizing cultivation strategies, developing precise control approaches to manipulate the contents of health-promoting compounds, and explore the coupled effects of multiple agronomic factors on the accumulation of phytochemicals.

6.1 Breeding and cultivation techniques to improve the quality and flavor of vegetables and fruits

Breeding programs that aim to achieve high quality and flavor and improved cultivation techniques are useful approaches to enhance the quality and flavor of vegetables and fruits. Novel approaches such as high-throughput sequencing and big data analysis will help accelerate breeding processes. Additionally, the application of advanced environmental and fertigation equipment in modern horticultural practice will help to obtain vegetables and fruits with optimal quality and flavor.

6.2 Plant factory with artificial lighting as a promising technology for precise control of quality and flavor in vegetables and fruits

Plant factory with artificial lighting was introduced in recent years as precisely controlled environments that are independent of the geographical location and climatic factors (Kozai et al., 2019; Yang, 2019; SharathKumar et al., 2020). Plant factory has been rapidly developed and applied to meet the market requirements for safe and high-quality foods. These systems enable efficient agricultural production as the environmental conditions (e.g. lighting, humidity, CO_2 and temperature) and fertigation can be fully controlled; therefore, year-round production and precise control of the phytochemical compounds in the fruits and vegetables can be achieved. The agronomic practices introduced in this chapter can also be applied in plant factory. Based on the findings described in this chapter, plant factory represents a powerful technology to produce safe, high-quality foods. Further studies are required to optimize the specific environmental factors and fertilizers required to promote the accumulation of target metabolites using a combination of agricultural practices and artificial intelligence in such fully controlled environments.

6.3 Underlying coupled effects of multiple agronomic factors on the quality of fruits and vegetables

Laboratory experiments typically explore the effects of a single agronomic factor or two factors on phytochemical changes in vegetables and fruits. However, the accumulation of phytochemicals by vegetables and fruits in the field is influenced by coupling between numerous environmental factors and nutrients. Therefore, more research is necessary to reveal the mechanisms underlying these processes. In particular, identification of the metabolic networks that couple agronomic factors with signal transduction, transcriptional factors, structural genes, and metabolites could pave the way for precise control of phytochemical compounds in fruits and vegetables.

7 Where to look for further information

The following articles provide a good overview of the subject:

- Alrifai, O., Hao, X., Marcone, M. F. and Tsao, R. (2019), Current review of the modulatory effects of LED lights on photosynthesis of secondary metabolites and future perspectives of microgreen vegetables. *J. Agr. Food Chem.* 67(22): 6075–6090.

- Bantis, F., Smirnakou, S., Ouzounis, T., Koukounaras, A., Ntagkas, N. and Radoglou, K. (2018). Current status and recent achievements in the field of horticulture with the use of light-emitting diodes (LEDs). *Sci. Hortic.* 235: 437–451.
- Duarte-Sierra, A., Tiznado-Hernández, M. E., Jha, D. K., Janmeja, N. and Arul, J. (2020). Abiotic stress hormesis: An approach to maintain quality, extend storability, and enhance phytochemicals on fresh produce during postharvest. *Comp. Rev. Food Sci. F. Saf.* 19(6): 3659–3682.
- Leong, R. and Urano, D. (2018). Molecular breeding for plant factory: strategies and technology. In *Smart Plant Factory*, Springer, Singapore, pp. 301–323.
- Morand, C., De Roos, B., Garcia-Conesa, M. T., Gibney, E. R., Landberg, R., Manach, C., Milenkovic, D., Rodriguez-Mateos, A., Van de Wiele, T. and Tomas-Barberan, F. (2020). Why interindividual variation in response to consumption of plant food bioactives matters for future personalised nutrition. *P. Nutr. Soc.* 79(2): 225–235.
- Rouphael, Y., Kyriacou, M. C., Petropoulos, S. A., De Pascale, S. and Colla, G. (2018). Improving vegetable quality in controlled environments. *Sci. Hortic.* 234: 275–289.
- Thoma, F., Somborn-Schulz, A., Schlehuber, D., Keuter, V. and Deerberg, G. (2020). Effects of light on secondary metabolites in selected leafy greens: A review. *Front. Plant Sci.* 11: 497.
- Vázquez-Hernández, M. C., Parola-Contreras, I., Montoya-Gómez, L. M., Torres-Pacheco, I., Schwarz, D. and Guevara-González, R. G. (2019). Eustressors: Chemical and physical stress factors used to enhance vegetables production. *Sci. Hortic.* 250: 223–229.
- Yang, X., Gil, M. I., Yang, Q. and Tomás-Barberán, F. A. (2022). Bioactive compounds in lettuce: Highlighting the benefits to human health and impacts of preharvest and postharvest practices. *Comp. Rev. Food Sci. F. Saf.*, 21(1), 4-45.
- Zhang, X., Bian, Z., Yuan, X., Chen, X. and Lu, C. (2020). A review on the effects of light-emitting diode (LED) light on the nutrients of sprouts and microgreens. *Trends Food Sci. Technol.* 99: 203–216.

8 References

Agüero, M. V., Jagus, R. J., Martín-Belloso, O. and Soliva-Fortuny, R. (2016). Surface decontamination of spinach by intense pulsed light treatments: impact on quality attributes. *Postharvest Biol. Technol.* 121: 118–125.

Ahn, S. Y., Kim, S. A., Choi, S.-J. and Yun, H. K. (2015). Comparison of accumulation of stilbene compounds and stilbene related gene expression in two grape berries irradiated with different light sources. *Hortic. Environ. Biotechnol.* 56(1): 36–43.

Alghamdi, S. S., Khan, M. A., El-Harty, E. H., Ammar, M. H., Farooq, M. and Migdadi, H. M. (2018). Comparative phytochemical profiling of different soybean (*Glycine max* (L.) Merr) genotypes using GC-MS. *Saudi J. Biol. Sci.* 25(1): 15-21.

Ali, M. B., Khandaker, L. and Oba, S. (2009). Comparative study on functional components, antioxidant activity and color parameters of selected colored leafy vegetables as affected by photoperiods. *J. Food Agric. Environ.* 7: 392-398.

Almuhayawi, M. S., AbdElgawad, H., Al Jaouni, S. K., Selim, S., Hassan, A. H. A. and Khamis, G. (2020). Elevated CO_2 improves glucosinolate metabolism and stimulates anticancer and anti-inflammatory properties of broccoli sprouts. *Food Chem.* 328: 127102.

Almuhayawi, M. S., Hassan, A. H. A., Abdel-Mawgoud, M., Khamis, G., Selim, S., Al Jaouni, S. K. and AbdElgawad, H. (2021). Laser light as a promising approach to improve the nutritional value, antioxidant capacity and anti-inflammatory activity of flavonoid-rich buckwheat sprouts. *Food Chem.* 345: 128788.

Alrifai, O., Hao, X., Marcone, M. F. and Tsao, R. (2019). Current review of the modulatory effects of LED lights on photosynthesis of secondary metabolites and future perspectives of microgreen vegetables. *J. Agric. Food Chem.* 67(22): 6075-6090.

Amitrano, C., Arena, C., De Pascale, S. and De Micco, V. (2020). Light and low relative humidity increase antioxidants content in mung bean (*Vigna radiata* L.) sprouts. *Plants (Basel).* 9(9): 1093.

Amitrano, C., Rouphael, Y., De Pascale, S. and De Micco, V. (2021). Modulating vapor pressure deficit in the plant micro-environment may enhance the bioactive value of lettuce. *Horticulturae* 7(2): 32.

Amoozgar, A., Mohammadi, A. and Sabzalian, M. R. (2017). Impact of light-emitting diode irradiation on photosynthesis, phytochemical composition and mineral element content of lettuce cv. Grizzly. *Photosynthetica* 55(1): 85-95.

Azad, M. O. K., Chun, I. J., Jeong, J. H., Kwon, S. T. and Hwang, J. M. (2011). Response of the growth characteristics and phytochemical contents of pepper (*Capsicum annuum* L.) seedlings with supplemental LED light in glass house. *J. Bio-Environ. Control.* 20: 182-188.

Barkat, N., Singh, J., Jayaprakasha, G. K. and Patil, B. S. (2018). Effect of harvest time on the levels of phytochemicals, free radical-scavenging activity, α-amylase inhibition and bile acid-binding capacity of spinach (*Spinacia oleracea*). *J. Sci. Food Agric.* 98(9): 3468-3477.

Baslam, M., Garmendia, I. and Goicoechea, N. (2012). Elevated CO_2 may impair the beneficial effect of arbuscular mycorrhizal fungi on the mineral and phytochemical quality of lettuce. *Ann. Appl. Biol.* 161(2): 180-191.

Becker, C., Klaering, H. P., Kroh, L. W. and Krumbein, A. (2014). Cool-cultivated red leaf lettuce accumulates cyanidin-3-O-(6"-O-malonyl)-glucoside and caffeoylmalic acid. *Food Chem.* 146: 404-411.

Becker, C., Urlić, B., Jukić Špika, M., Kläring, H. P., Krumbein, A., Baldermann, S., Goreta Ban, S., Perica, S. and Schwarz, D. (2015). Nitrogen limited red and green leaf lettuce accumulate flavonoid glycosides, caffeic acid derivatives, and sucrose while losing chlorophylls, β-carotene and xanthophylls. *PLoS ONE* 10(11): e0142867.

Bian, Z. H., Cheng, R. F., Yang, Q. C., Wang, J. and Lu, C. (2016). Continuous light from red, blue, and green light-emitting diodes reduces nitrate content and enhances phytochemical concentrations and antioxidant capacity in lettuce. *J. Am. Soc. Hort. Sci.* 141(2): 186-195.

Bodelón, O. G., Blanch, M., Sanchez-Ballesta, M. T., Escribano, M. I. and Merodio, C. (2010). The effects of high CO_2 levels on anthocyanin composition, antioxidant activity and soluble sugar content of strawberries stored at low non-freezing temperature. *Food Chem.* 122(3): 673–678.

Bragança, G. C. M., Ziegler, V., Ávila, B. P., Monks, J. L. F., Peres, W. and Elias, M. C. (2020). Multivariate analysis of the conditions of temperature, moisture and storage time in the technological, chemical, nutritional parameters and phytochemical of green lentils. *J. Stored Prod. Res.* 87: 101617.

Bravo, S., García-Alonso, J., Martín-Pozuelo, G., Gómez, V., García-Valverde, V., Navarro-González, I. and Periago, M. J. (2013). Effects of postharvest UV-C treatment on carotenoids and phenolic compounds of vine-ripe tomatoes. *Int. J. Food Sci. Technol.* 48(8): 1744–1749.

Brazaityte, A., Virsile, A., Jankauskiene, J., Sakalauskiene, S., Samuoliene, G., Sirtautas, R., Novickovas, A., Dabasinskas, L., Miliauskiene, J., Vastakaite, V., Bagdonavičienė, A. and Duchovskis, P. (2015). Effect of supplemental UV-A irradiation in solid-state lighting on the growth and phytochemical content of microgreens. *Int. Agrophys.* 29(1): 13–22.

Calle, A., Serradilla, M. J. and Wünsch, A. (2021). QTL mapping of phenolic compounds and fruit colour in sweet cherry using a 6+ 9K SNP array genetic map. *Sci. Hortic.* 280: 109900.

Campbell, O. E., Merwin, I. A. and Padilla-Zakour, O. I. (2013). Characterization and the effect of maturity at harvest on the phenolic and carotenoid content of northeast USA apricot (*Prunus armeniaca*) varieties. *J. Agric. Food Chem.* 61(51): 12700–12710.

Chagné, D., Krieger, C., Rassam, M., Sullivan, M., Fraser, J., André, C., Pindo, M., Troggio, M., Gardiner, S. E., Henry, R. A., Allan, A. C., McGhie, T. K. and Laing, W. A. (2012). QTL and candidate gene mapping for polyphenolic composition in apple fruit. *BMC Plant Biol.* 12: 12.

Chang, X., Lu, Y., Li, Q., Lin, Z., Qiu, J., Peng, C., Brennan, C. S. and Guo, X. (2019). The combination of hot air and chitosan treatments on phytochemical changes during postharvest storage of 'Sanhua' plum fruits. *Foods* 8(8): 338.

Charles, F., Vidal, V., Olive, F., Filgueiras, H. and Sallanon, H. (2013). Pulsed light treatment as new method to maintain physical and nutritional quality of fresh-cut mangoes. *Innov. Food Sci. Emerg.* 18: 190–195.

Chebrolu, K. K., Jayaprakasha, G. K., Jifon, J. and Patil, B. S. (2012). Production system and storage temperature influence grapefruit vitamin C, limonoids, and carotenoids. *J. Agric. Food Chem.* 60(29): 7096–7103.

Choi, H. G., Kwon, J. K., Moon, B. Y., Kang, N. J., Park, K. S., Cho, M. W. and Kim, Y. C. (2013). Effect of different light emitting diode (LED) lights on the growth characteristics and the phytochemical production of strawberry fruits during cultivation. *Hortic. Sci. Technol.* 31: 56–64.

Chun, J. H., Kim, S., Arasu, M. V., Al-Dhabi, N. A., Chung, D. Y. and Kim, S. J. (2017). Combined effect of nitrogen, phosphorus and potassium fertilizers on the contents of glucosinolates in rocket salad (*Eruca sativa* Mill.). *Saudi J. Biol. Sci.* 24(2): 436–443.

Craver, J. K., Gerovac, J. R., Lopez, R. G. and Kopsell, D. A. (2017). Light intensity and light quality from sole-source light-emitting diodes impact phytochemical concentrations within *Brassica* microgreens. *J. Amer. Soc. Hort. Sci.* 142(1): 3–12.

Cuong, D. M., Ha, T. W., Park, C. H., Kim, N. S., Yeo, H. J., Chun, S. W., Kim, C. and Park, S. U. (2019). Effects of LED lights on expression of genes involved in phenylpropanoid biosynthesis and accumulation of phenylpropanoids in wheat sprout. *Agronomy* 9(6): 307.

Damerum, A., Chapman, M. A. and Taylor, G. (2020). Innovative breeding technologies in lettuce for improved post-harvest quality. *Postharvest Biol. Technol.* 168: 111266.

Deng, M., Qian, H., Chen, L., Sun, B., Chang, J., Miao, H., Cai, C. and Wang, Q. (2017). Influence of pre-harvest red light irradiation on main phytochemicals and antioxidant activity of Chinese kale sprouts. *Food Chem.* 222: 1-5.

Deng, Y., Ning, Y., Yang, D. L., Zhai, K., Wang, G. L. and He, Z. (2020). Molecular basis of disease resistance and perspectives on breeding strategies for resistance improvement in crops. *Mol. Plant* 13(10): 1402–1419.

Diaz-Garcia, L., Schlautman, B., Covarrubias-Pazaran, G., Maule, A., Johnson-Cicalese, J., Grygleski, E., Vorsa, N. and Zalapa, J. (2018). Massive phenotyping of multiple cranberry populations reveals novel QTLs for fruit anthocyanin content and other important chemical traits. *Mol. Genet. Genomics* 293(6): 1379–1392.

Dong, J., Gruda, N., Lam, S. K., Li, X. and Duan, Z. (2018). Effects of elevated CO_2 on nutritional quality of vegetables: A review. *Front. Plant Sci.* 9: 924.

Dou, H., Niu, G. and Gu, M. (2019). Photosynthesis, morphology, yield, and phytochemical accumulation in basil plants influenced by substituting green light for partial red and/or blue light. *Hortscience* 54(10): 1769–1776.

Drogoudi, P. D. and Pantelidis, G. (2011). Effects of position on canopy and harvest time on fruit physico-chemical and antioxidant properties in different apple cultivars. *Sci. Hortic.* 129(4): 752–760.

Duarte-Sierra, A., Tiznado-Hernández, M. E., Jha, D. K., Janmeja, N. and Arul, J. (2020). Abiotic stress hormesis: An approach to maintain quality, extend storability, and enhance phytochemicals on fresh produce during postharvest. *Compr. Rev. Food Sci. Food Saf.* 19(6): 3659–3682.

Dzakovich, M. P., Ferruzzi, M. G. and Mitchell, C. A. (2016). Manipulating sensory and phytochemical profiles of greenhouse tomatoes using environmentally relevant doses of ultraviolet radiation. *J. Agric. Food Chem.* 64(36): 6801–6808.

El-Nakhel, C., Pannico, A., Graziani, G., Kyriacou, M. C., Giordano, M., Ritieni, A., De Pascale, S. and Rouphael, Y. (2020). Variation in macronutrient content, phytochemical constitution and in vitro antioxidant capacity of green and red butterhead lettuce dictated by different developmental stages of harvest maturity. *Antioxidants (Basel)* 9(4): 300.

Fu, Y., Li, H., Yu, J., Liu, H., Cao, Z., Manukovsky, N. S. and Liu, H. (2017). Interaction effects of light intensity and nitrogen concentration on growth, photosynthetic characteristics and quality of lettuce (*Lactuca sativa* L. Var. youmaicai). *Sci. Hortic.* 214: 51–57.

Galani, J. H. Y., Mankad, P. M., Shah, A. K., Patel, N. J., Acharya, R. R. and Talati, J. G. (2017). Effect of storage temperature on vitamin C, total phenolics, UPLC phenolic acid profile and antioxidant capacity of eleven potato (*Solanum tuberosum*) varieties. *Hortic. Plant J.* 3(2): 73–89.

Galindo, A., Calín-Sánchez, A., Collado-González, J., Ondoño, S., Hernández, F., Torrecillas, A. and Carbonell-Barrachina, A. A. (2014). Phytochemical and quality attributes of pomegranate fruits for juice consumption as affected by ripening stage and deficit irrigation. *J. Sci. Food Agric.* 94(11): 2259–2265.

Gao, M., He, R., Shi, R., Zhang, Y., Song, S., Su, W. and Liu, H. (2021). Differential effects of low light intensity on broccoli microgreens growth and phytochemicals. *Agronomy* 11(3): 537.

García, C. J., Yang, X., Huang, D. and Tomás-Barberán, F. A. (2020). Can we trust biomarkers identified using different non-targeted metabolomics platforms? Multi-platform, inter-laboratory comparative metabolomics profiling of lettuce cultivars via UPLC-QTOF-MS. *Metabolomics* 16(8): 85.

Garrido, Y., Tudela, J. A., Marín, A., Mestre, T., Martínez, V. and Gil, M. I. (2014). Physiological, phytochemical and structural changes of multi-leaf lettuce caused by salt stress. *J. Sci. Food Agric.* 94(8): 1592–1599.

Gerhardt, K. E., Lampi, M. A. and Greenberg, B. M. (2008). The effects of far-red light on plant growth and flavonoid accumulation in *Brassica napus* in the presence of ultraviolet B radiation. *Photochem. Photobiol.* 84(6): 1445–1454.

Ghasemzadeh, A., Talei, D., Jaafar, H. Z., Juraimi, A. S., Mohamed, M. T. M., Puteh, A. and Halim, M. R. A. (2016). Plant-growth regulators alter phytochemical constituents and pharmaceutical quality in sweet potato (*Ipomoea batatas* L.). *BMC Complement. Altern. Med.* 16: 152.

Giannakoula, A. E., Ilias, I. F., Maksimović, J. J. D., Maksimović, V. M. and Živanović, B. D. (2012). The effects of plant growth regulators on growth, yield, and phenolic profile of lentil plants. *J. Food Compos. Anal.* 28(1): 46–53.

Gomes, T. M., Mazon, L. F., Panceri, C. P., Machado, B. D., Brighenti, A., Burin, V. M. and Bordignon-Luiz, M. T. (2020). Changes in vineyard productive attributes and phytochemical composition of Sauvignon blanc grape and wine induced by the application of silicon and calcium. *J. Sci. Food Agric.* 100(4): 1547–1557.

Gonzalez, C. V., Fanzone, M. L., Cortés, L. E., Bottini, R., Lijavetzky, D. C., Ballare, C. L. and Boccalandro, H. E. (2015). Fruit-localized photoreceptors increase phenolic compounds in berry skins of field-grown *Vitis vinifera* L. cv. Malbec. *Phytochemistry* 110: 46–57.

González-Chavira, M. M., Herrera-Hernández, M. G., Guzmán-Maldonado, H. and Pons-Hernández, J. L. (2018). Controlled water deficit as abiotic stress factor for enhancing the phytochemical content and adding-value of crops. *Sci. Hortic.* 234: 354–360.

Groenbaek, M., Jensen, S., Neugart, S., Schreiner, M., Kidmose, U. and Lakkenborg Kristensen, H. L. (2014). Influence of cultivar and fertilizer approach on curly kale (*Brassica oleracea* L. var. sabellica). 1. Genetic diversity reflected in agronomic characteristics and phytochemical concentration (*Brassica oleracea* L. var. sabellica). *J. Agric. Food Chem.* 62(47): 11393–11402.

Han, C., Li, J., Jin, P., Li, X., Wang, L. and Zheng, Y. (2017). The effect of temperature on phenolic content in wounded carrots. *Food Chem.* 215: 116–123.

Hartley, S. E., Jones, C. G., Couper, G. C. and Jones, T. H. (2000). Biosynthesis of plant phenolic compounds in elevated atmospheric CO_2. *Glob. Change Biol.* 6(5): 497–506.

He, F., Thiele, B., Santhiraraja-Abresch, S., Watt, M., Kraska, T., Ulbrich, A. and Kuhn, A. J. (2020). Effects of root temperature on the plant growth and food quality of Chinese broccoli (*Brassica oleracea* var. alboglabra Bailey). *Agronomy* 10(5): 702.

Himanen, S. J., Nissinen, A., Auriola, S., Poppy, G. M., Stewart, C. N., Holopainen, J. K. and Nerg, A. M. (2008). Constitutive and herbivore-inducible glucosinolate concentrations in oilseed rape (*Brassica napus*) leaves are not affected by Bt Cry1Ac insertion but change under elevated atmospheric CO_2 and O_3. *Planta* 227(2): 427–437.

Hong, H. T., Phan, A. D. T. and O'Hare, T. J. (2021). Temperature and maturity stages affect anthocyanin development and phenolic and sugar content of purple-pericarp supersweet sweetcorn during storage. *J. Agric. Food Chem.* 69(3): 922–931.

Jin, P., Wang, S. Y., Wang, C. Y. and Zheng, Y. (2011). Effect of cultural system and storage temperature on antioxidant capacity and phenolic compounds in strawberries. *Food Chem.* 124(1): 262-270.

Jo, J. S. and Lee, J. G. (2018). Evaluation of individual glucosinolates, phytochemical contents, and antioxidant activities under various red to far-red light ratios in three *Brassica* sprouts. *PHPF* 27(4): 415–423.

Johansen, T. J., Hagen, S. F., Bengtsson, G. B. and Mølmann, J. A. (2016). Growth temperature affects sensory quality and contents of glucosinolates, vitamin C and sugars in swede roots (*Brassica napus* L. ssp. rapifera Metzg.). *Food Chem.* 196: 228–235.

Kadomura-Ishikawa, Y., Miyawaki, K., Noji, S. and Takahashi, A. (2013). Phototropin 2 is involved in blue light-induced anthocyanin accumulation in *Fragaria × ananassa* fruits. *J. Plant Res.* 126(6): 847–857.

Kang, N. S. and Lee, J. H. (2011). Characterisation of phenolic phytochemicals and quality changes related to the harvest times from the leaves of Korean purple perilla (*Perilla frutescens*). *Food Chem.* 124(2): 556–562.

Khan, T. A., Fariduddin, Q. and Yusuf, M. (2015). Lycopersicon esculentum under low temperature stress: An approach toward enhanced antioxidants and yield. *Environ. Sci. Pollut. R. Int.* 22(18): 14178–14188.

Kim, H. J., Fonseca, J. M., Choi, J. H. and Kubota, C. (2007). Effect of methyl jasmonate on phenolic compounds and carotenoids of romaine lettuce (*Lactuca sativa* L.). *J. Agric. Food Chem.* 55(25): 10366–10372.

Kim, Y. J., Kim, H. M. and Hwang, S. J. (2016). Growth and phytochemical contents of ice plant as affected by light quality in a closed-type plant production system. *Hortic. Sci. Technol.* 34: 878–885.

Kim, Y. J., Kim, H. M., Kim, H. M., Jeong, B. R., Lee, H.-J., Kim, H.-J. and Hwang, S. J. (2018). Ice plant growth and phytochemical concentrations are affected by light quality and intensity of monochromatic light-emitting diodes. *Hortic. Environ. Biotechnol.* 59(4): 529–536.

Kim, Y. J., Kim, Y. B., Li, X., Choi, S. R., Park, S., Park, J. S., Lim, Y. P. and Park, S. U. (2015). Accumulation of phenylpropanoids by white, blue, and red light irradiation and their organ-specific distribution in Chinese cabbage (*Brassica rapa* ssp. pekinensis). *J. Agric. Food Chem.* 63(30): 6772-6778.

Klaiber, J., Dorn, S. and Najar-Rodriguez, A. J. (2013). Acclimation to elevated CO_2 increases constitutive glucosinolate levels of *Brassica* plants and affects the performance of specialized herbivores from contrasting feeding guilds. *J. Chem. Ecol.* 39(5): 653–665.

Kondo, S., Tomiyama, H., Rodyoung, A., Okawa, K., Ohara, H., Sugaya, S., Terahara, N. and Hirai, N. (2014). Abscisic acid metabolism and anthocyanin synthesis in grape skin are affected by light emitting diode (LED) irradiation at night. *J. Plant Physiol.* 171(10): 823–829.

Kopsell, D. A. and Sams, C. E. (2013). Increases in shoot tissue pigments, glucosinolates, and mineral elements in sprouting broccoli after exposure to short-duration blue light from light emitting diodes. *J. Amer. Soc. Hort. Sci.* 138(1): 31–37.

Kopsell, D. A., Sams, C. E. and Morrow, R. C. (2017). Interaction of light quality and fertility on biomass, shoot pigmentation and xanthophyll cycle flux in Chinese kale. *J. Sci. Food Agric.* 97(3): 911-917.

Kozai, T., Niu, G. and Takagaki, M. (2019). *Plant Factory: An Indoor Vertical Farming System for Efficient Quality Food Production*. London: Academic Press.

Kubota, C., Thomson, C. A., Wu, M. and Javanmardi, J. (2006). Controlled environments for production of value-added food crops with high phytochemical concentrations: Lycopene in tomato as an example. *Hortscience* 41(3): 522–525.

Kwaw, E., Ma, Y., Tchabo, W., Apaliya, M. T., Sackey, A. S., Wu, M. and Xiao, L. (2018). Effect of pulsed light treatment on the phytochemical, volatile, and sensorial attributes of lactic-acid-fermented mulberry juice. *Int. J. Food Prop.* 21(1): 213–228.

Kyriacou, M. C. and Rouphael, Y. (2018). Towards a new definition of quality for fresh fruits and vegetables. *Sci. Hortic.* 234: 463–469.

Lancaster, J. E., Reay, P. F., Norris, J. and Butler, R. C. (2000). Induction of flavonoids and phenolic acids in apple by UV-B and temperature. *J. Hortic. Sci. Biotechnol.* 75(2): 142–148.

Landi, M., Zivcak, M., Sytar, O., Brestic, M. and Allakhverdiev, S. I. (2020). Plasticity of photosynthetic processes and the accumulation of secondary metabolites in plants in response to monochromatic light environments: a review. *Biochim. Biophys. Acta Bioenerg.* 1861(2): 148131.

Lee, J. H. and Oh, M. M. (2015). Short-term low temperature increases phenolic antioxidant levels in kale. *Hortic. Environ. Biotechnol.* 56(5): 588–596.

Lee, J. J., Crosby, K. M., Pike, L. M., Yoo, K. S. and Leskovar, D. I. (2005). Impact of genetic and environmental variation on development of flavonoids and carotenoids in pepper (*Capsicum* spp.). *Sci. Hortic.* 106(3): 341–352.

Lee, S. G., Choi, C. S., Lee, H. J., Jang, Y. A. and Lee, J. G. (2015). Effect of air temperature on growth and phytochemical content of beet and ssamchoo. *Hortic. Sci. Technol.* 33(3): 303–308.

Lee, S. G., Choi, C. S., Lee, J. G., Jang, Y. A., Lee, H. J., Lee, H. J., Chae, W. B. and Um, Y. C. (2013). Influence of air temperature on yield and phytochemical content of red chicory and garland chrysanthemum grown in plant factory. *Hortic. Environ. Biotechnol.* 54(5): 399–404.

Li, B., Feng, Y., Zong, Y., Zhang, D., Hao, X. and Li, P. (2020a). Elevated CO_2-induced changes in photosynthesis, antioxidant enzymes and signal transduction enzyme of soybean under drought stress. *Plant Physiol. Biochem.* 154: 105–114.

Li, N., Wu, X., Zhuang, W., Xia, L., Chen, Y., Wu, C., Rao, Z., Du, L., Zhao, R., Yi, M., Wan, Q. and Zhou, Y. (2021). Tomato and lycopene and multiple health outcomes: umbrella review. *Food Chem.* 343: 128396.

Li, Q. and Kubota, C. (2009). Effects of supplemental light quality on growth and phytochemicals of baby leaf lettuce. *Environ. Exp. Bot.* 67(1): 59–64.

Li, S., Schonhof, I., Krumbein, A., Li, L., Stützel, H. and Schreiner, M. (2007). Glucosinolate concentration in turnip (*Brassica rapa* ssp. rapifera L.) roots as affected by nitrogen and sulfur supply. *J. Agric. Food Chem.* 55(21): 8452–8457.

Li, X., Li, M., Han, C., Jin, P. and Zheng, Y. (2017). Increased temperature elicits higher phenolic accumulation in fresh-cut pitaya fruit. *Postharvest Biol. Technol.* 129: 90–96.

Li, Y., Zheng, Y., Zheng, D., Zhang, Y., Song, S., Su, W. and Liu, H. (2020b). Effects of supplementary blue and UV-A LED Lights on morphology and phytochemicals of *Brassicaceae* baby-leaves. *Molecules* 25(23): 5678.

Li, Z., Zhao, X., Sandhu, A. K. and Gu, L. (2010). Effects of exogenous abscisic acid on yield, antioxidant capacities, and phytochemical contents of greenhouse grown lettuces. *J. Agric. Food Chem.* 58(10): 6503–6509.

Lim, J. H., Park, K. J., Kim, B. K., Jeong, J. W. and Kim, H. J. (2012). Effect of salinity stress on phenolic compounds and carotenoids in buckwheat (*Fagopyrum esculentum* M.) sprout. *Food Chem.* 135(3): 1065–1070.

Lin, Q., Lu, Y., Zhang, J., Liu, W., Guan, W. and Wang, Z. (2017). Effects of high CO_2 in-package treatment on flavor, quality and antioxidant activity of button mushroom (*Agaricus bisporus*) during postharvest storage. *Postharvest Biol. Technol.* 123: 112–118.

Liu, C. H., Cai, L. Y., Lu, X. Y., Han, X. X. and Ying, T. J. (2012). Effect of postharvest UV-C irradiation on phenolic compound content and antioxidant activity of tomato fruit during storage. *J. Integr. Agr.* 11(1): 159–165.

Liu, L. H., Zabaras, D., Bennett, L. E., Aguas, P. and Woonton, B. W. (2009). Effects of UV-C, red light and sun light on the carotenoid content and physical qualities of tomatoes during post-harvest storage. *Food Chem.* 115(2): 495–500.

Liu, R. H. (2013). Health-promoting components of fruits and vegetables in the diet. *Adv. Nutr.* 4(3): 384S–392S.

Liu, S., Hu, L., Jiang, D. and Xi, W. (2019). Effect of post-harvest LED and UV light irradiation on the accumulation of flavonoids and limonoids in the segments of Newhall navel oranges (*Citrus sinensis* Osbeck). *Molecules* 24(9): 1755.

Loladze, I., Nolan, J. M., Ziska, L. H. and Knobbe, A. R. (2019). Rising atmospheric CO_2 lowers concentrations of plant carotenoids essential to human health: A meta-analysis. *Mol. Nutr. Food Res.* 63(15): e1801047.

Lopes, M. M. A., Silva, E. O., Canuto, K. M., Silva, L. M. A., Gallão, M. I., Urban, L., Ayala-Zavala, J. F. and Miranda, M. R. A. (2016). Low fluence pulsed light enhanced phytochemical content and antioxidant potential of 'Tommy Atkins' mango peel and pulp. *Innov. Food Sci. Emerg.* 33: 216–224.

Ma, G., Zhang, L., Kato, M., Yamawaki, K., Kiriiwa, Y., Yahata, M., Ikoma, Y. and Matsumoto, H. (2015). Effect of the combination of ethylene and red LED light irradiation on carotenoid accumulation and carotenogenic gene expression in the flavedo of citrus fruit. *Postharvest Biol. Technol.* 99: 99–104.

Ma, Y., Ban, Q., Shi, J., Dong, T., Jiang, C. Z. and Wang, Q. (2019). 1-Methylcyclopropene (1-MCP), storage time, and shelf life and temperature affect phenolic compounds and antioxidant activity of 'Jonagold' apple. *Postharvest Biol. Technol.* 150: 71–79.

Machado, R. M. A., Alves-Pereira, I., Lourenço, D. and Ferreira, R. M. A. (2020). Effect of organic compost and inorganic nitrogen fertigation on spinach growth, phytochemical accumulation and antioxidant activity. *Heliyon* 6(9): e05085.

Manganaris, G. A., Drogoudi, P., Goulas, V., Tanou, G., Georgiadou, E. C., Pantelidis, G. E., Paschalidis, K. A., Fotopoulos, V. and Manganaris, A. (2017). Deciphering the interplay among genotype, maturity stage and low-temperature storage on phytochemical composition and transcript levels of enzymatic antioxidants in *Prunus persica* fruit. *Plant Physiol. Biochem.* 119: 189–199.

Molmann, J. A., Hansen, E. and Johansen, T. J. (2021). Effects of supplemental LED light quality and reduced growth temperature on swede (*Brassica napus* L. ssp. rapifera Metzg.) root vegetable development and contents of glucosinolates and sugars. *J. Sci. Food Agric.* 101(6): 2422–2427.

Moreira-Rodríguez, M., Nair, V., Benavides, J., Cisneros-Zevallos, L. and Jacobo-Velázquez, D. A. (2017). UVA, UVB light, and methyl jasmonate, alone or combined, redirect the biosynthesis of glucosinolates, phenolics, carotenoids, and chlorophylls in broccoli sprouts. *Int. J. Mol. Sci.* 18(11): 2330.

Nam, T. G., Lim, Y. J. and Eom, S. H. (2018). Flavonoid accumulation in common buckwheat (*Fagopyrum esculentum*) sprout tissues in response to light. *Hortic. Environ. Biotechnol.* 59(1): 19–27.

Nassarawa, S. S., Abdelshafy, A. M., Xu, Y., Li, L. and Luo, Z. (2021). Effect of light-emitting diodes (LEDs) on the quality of fruits and vegetables during postharvest period: A review. *Food Bioprocess Technol.* 14(3): 388–414

Nath, A., Bagchi, B., Misra, L. K. and Deka, B. C. (2011). Changes in post-harvest phytochemical qualities of broccoli florets during ambient and refrigerated storage. *Food Chem.* 127(4): 1510–1514.

Neugart, S., Baldermann, S., Hanschen, F. S., Klopsch, R., Wiesner-Reinhold, M. and Schreiner, M. (2018). The intrinsic quality of brassicaceous vegetables: How secondary plant metabolites are affected by genetic, environmental, and agronomic factors. *Sci. Hortic.* 233: 460–478.

Ngadze, E., Coutinho, T. A., Icishahayo, D. and Van der Waals, J. E. (2014). Effect of calcium soil amendments on phenolic compounds and soft rot resistance in potato tubers. *Crop Prot.* 62: 40–45.

Nishida, C., Uauy, R., Kumanyika, S. and Shetty, P. (2004). The joint WHO/FAO expert consultation on diet, nutrition and the prevention of chronic diseases: process, product and policy implications. Public Health Nutr. 7(1a), 245–250.

Nur, F. O., Siti, A. H. and Umi, K. Y. (2013). Comparative evaluation of organic and inorganic fertilizers on total phenolic, total flavonoid, antioxidant activity and cyanogenic glycosides in cassava (*Manihot esculenta*). *Afr. J. Biotechnol.* 12: 2414–2421.

Nurul, S. and Asmah, R. (2014). Variability in nutritional composition and phytochemical properties of red pitaya (*Hylocereus polyrhizus*) from Malaysia and Australia. *Int. Food Res. J.* 21 (4): 1689–1697.

Obidola, S. M., Ibrahim, I. I. and Agwom, R. Z. (2019). Influence of organic manure and inorganic fertilizer on the growth, yield and phytochemical constituents of cabbage (*Brassica oleracea*). *Asian J. Agric. Hortic. Res.* 4(1): 1–9.

Oh, M. M., Carey, E. E. and Rajashekar, C. B. (2009). Environmental stresses induce health-promoting phytochemicals in lettuce. *Plant Physiol. Biochem.* 47(7): 578–583.

Olennikov, D. N., Chirikova, N. K., Kashchenko, N. I., Gornostai, T. G., Selyutina, I. Y. and Zilfikarov, I. N. (2017). Effect of low temperature cultivation on the phytochemical profile and bioactivity of Arctic plants: A case of Dracocephalum palmatum. *Int. J. Mol. Sci.* 18(12): 2579.

Oloyede, F. M., Agbaje, G. O., Obuotor, E. M. and Obisesan, I. O. (2012). Nutritional and antioxidant profiles of pumpkin (*Cucurbita pepo* Linn.) immature and mature fruits as influenced by NPK fertilizer. *Food Chem.* 135(2): 460–463.

Ortega-García, J. G., Montes-Belmont, R., Rodríguez-Monroy, M., Ramírez-Trujillo, J. A., Suárez-Rodríguez, R. and Sepúlveda-Jiménez, G. (2015). Effect of *Trichoderma asperellum* applications and mineral fertilization on growth promotion and the content of phenolic compounds and flavonoids in onions. *Sci. Hortic.* 195: 8–16.

Oszmiański, J., Lachowicz, S., Gorzelany, J. and Matłok, N. (2018). The effect of different maturity stages on phytochemical composition and antioxidant capacity of cranberry cultivars. *Eur. Food Res. Technol.* 244(4): 705–719.

Øvsthus, I., Breland, T. A., Hagen, S. F., Brandt, K., Wold, A. B., Bengtsson, G. B. and Seljasen, R. (2015). Effects of organic and waste-derived fertilizers on yield, nitrogen and glucosinolate contents, and sensory quality of broccoli (*Brassica oleracea* L. var. italica). *J. Agric. Food Chem.* 63(50): 10757–10767.

Öztürk, B. and Yücedağ, F. (2021). Effects of methyl jasmonate on quality properties and phytochemical compounds of kiwifruit (*Actinidia deliciosa* cv. Hayward). *Turk. J. Agric. For.* 45(2): 154–164.

Padda, M. S. and Picha, D. H. (2008). Effect of low temperature storage on phenolic composition and antioxidant activity of sweetpotatoes. *Postharvest Biol. Technol.* 47(2): 176–180.

Palmeri, R., Parafati, L., Arena, E., Grassenio, E., Restuccia, C. and Fallico, B. (2020). Antioxidant and antimicrobial properties of semi-processed frozen prickly pear juice as affected by cultivar and harvest time. *Foods* 9(2): 235.

Panjai, L., Noga, G., Fiebig, A. and Hunsche, M. (2017). Effects of continuous red light and short daily UV exposure during postharvest on carotenoid concentration and antioxidant capacity in stored tomatoes. *Sci. Hortic.* 226: 97–103.

Park, C. H., Kim, N. S., Park, J. S., Lee, S. Y., Lee, J. W. and Park, S. U. (2019). Effects of light-emitting diodes on the accumulation of glucosinolates and phenolic compounds in sprouting canola (*Brassica napus* L.). *Foods* 8(2): 76.

Park, C. H., Park, Y. E., Yeo, H. J., Kim, J. K. and Park, S. U. (2020). Effects of light-emitting diodes on the accumulation of phenolic compounds and glucosinolates in *Brassica juncea* Sprouts. *Horticulturae* 6(4): 77.

Pék, Z., Daood, H., Nagyné, M. G., Berki, M., Tóthné, M. M., Neményi, A. and Helyes, L. (2012). Yield and phytochemical compounds of broccoli as affected by temperature, irrigation, and foliar sulfur supplementation. *Horts* 47(11): 1646–1652.

Pereira, F. M. V., Rosa, E., Fahey, J. W., Stephenson, K. K., Carvalho, R. and Aires, A. (2002). Influence of temperature and ontogeny on the levels of glucosinolates in broccoli (*Brassica oleracea* var. italica) sprouts and their effect on the induction of mammalian phase 2 enzymes. *J. Agric. Food Chem.* 50(21): 6239–6244.

Pérez-Balibrea, S., Moreno, D. A. and García-Viguera, C. (2011). Genotypic effects on the phytochemical quality of seeds and sprouts from commercial broccoli cultivars. *Food Chem.* 125(2): 348–354.

Pérez-López, U., Sgherri, C., Miranda-Apodaca, J., Micaelli, F., Lacuesta, M., Mena-Petite, A., Quartacci, M. F. and Muñoz-Rueda, A. (2018). Concentration of phenolic compounds is increased in lettuce grown under high light intensity and elevated CO_2. *Plant Physiol. Biochem.* 123: 233–241.

Pincemail, J., Kevers, C., Tabart, J., Defraigne, J. O. and Dommes, J. (2012). Cultivars, culture conditions, and harvest time influence phenolic and ascorbic acid contents and antioxidant capacity of strawberry (*Fragaria × ananassa*). *J. Food Sci.* 77(2): C205–C210.

Pola, W., Sugaya, S. and Photchanachai, S. (2020). Color development and phytochemical changes in mature green chili (*Capsicum annuum* L.) exposed to red and blue light-emitting diodes. *J. Agric. Food Chem.* 68(1): 59–66.

Prohens-Tomás, J. and Nuez, F. (Eds.) (2007). *Vegetables I: Asteraceae, Brassicaceae, Chenopodicaceae, and Cucurbitaceae* (Vol. 1). New York: Springer Science & Business Media.

Qadir, O., Siervo, M., Seal, C. J. and Brandt, K. (2017). Manipulation of contents of nitrate, phenolic acids, chlorophylls, and carotenoids in lettuce (*Lactuca sativa* L.) via contrasting responses to nitrogen fertilizer when grown in a controlled environment. *J. Agric. Food Chem.* 65(46): 10003–10010.

Qian, H., Liu, T., Deng, M., Miao, H., Cai, C., Shen, W. and Wang, Q. (2016). Effects of light quality on main health-promoting compounds and antioxidant capacity of Chinese kale sprouts. *Food Chem.* 196: 1232–1238.

Ragusa, L., Picchi, V., Tribulato, A., Cavallaro, C., Lo Scalzo, R. and Branca, F. (2017). The effect of the germination temperature on the phytochemical content of broccoli and rocket sprouts. *Int. J. Food Sci. Nutr.* 68(4): 411–420.

Rajapakse, N. C., He, C., Cisneros-Zevallos, L. and Davies, F. T., Jr. (2009). Hypobaria and hypoxia affects growth and phytochemical contents of lettuce. *Sci. Hortic.* 122(2): 171–178.

Rao, S. Q., Chen, X. Q., Wang, K. H., Zhu, Z. J., Yang, J. and Zhu, B. (2021). Effect of short-term high temperature on the accumulation of glucosinolates in Brassica rapa. *Plant Physiol. Biochem.* 161: 222–233.

Reddy, G. V., Tossavainen, P., Nerg, A. M. and Holopainen, J. K. (2004). Elevated atmospheric CO_2 affects the chemical quality of Brassica plants and the growth rate of the specialist, *Plutella xylostella*, but not the generalist, *Spodoptera littoralis*. *J. Agric. Food Chem.* 52(13): 4185–4191.

Ren, J., Guo, S., Xu, C., Yang, C., Ai, W., Tang, Y. and Qin, L. (2014). Effects of different carbon dioxide and LED lighting levels on the anti-oxidative capabilities of *Gynura bicolor* DC. *Adv. Space Res.* 53(2): 353–361.

Rivera-Pastrana, D. M., Yahia, E. M. and González-Aguilar, G. A. (2010). Phenolic and carotenoid profiles of papaya fruit (*Carica papaya* L.) and their contents under low temperature storage. *J. Sci. Food Agric.* 90(14): 2358–2365.

Romero, I., Teresa Sanchez-Ballesta, M., Maldonado, R., Isabel Escribano, M. and Merodio, C. (2008). Anthocyanin, antioxidant activity and stress-induced gene expression in high CO_2-treated table grapes stored at low temperature. *J. Plant Physiol.* 165(5): 522–530.

Ruangrak, E. and Khummueng, W. (2019). Effects of artificial light sources on accumulation of phytochemical contents in hydroponic lettuce. *J. Hortic. Sci. Biotechnol.* 94(3): 378–388.

Rudell, D. R., Mattheis, J. P., Fan, X. and Fellman, J. K. (2002). Methyl jasmonate enhances anthocyanin accumulation and modifies production of phenolics and pigments in 'Fuji' Apples. *J. Am. Soc. Hortic. Sci.* 127(3): 435–441.

Rybarczyk-Plonska, A., Hagen, S. F., Borge, G. I. A., Bengtsson, G. B., Hansen, M. K. and Wold, A.-B. (2016). Glucosinolates in broccoli (*Brassica oleracea* L. var. italica) as affected by postharvest temperature and radiation treatments. *Postharvest Biol. Technol.* 116: 16–25.

Saini, R. K., Rengasamy, K. R., Mahomoodally, F. M. and Keum, Y. S. (2020). Protective effects of lycopene in cancer, cardiovascular, and neurodegenerative diseases: An update on epidemiological and mechanistic perspectives. Pharmacol. Res. 155: 104730.

Sakalauskaite, J., Viškelis, P., Duchovskis, P., Dambrauskiene, E., Sakalauskiene, S., Samuoliene, G. and Brazaityte, A. (2012). Supplementary UV-B irradiation effects on basil (*Ocimum basilicum* L.) growth and phytochemical properties. *J. Food Agric. Environ.* 10: 342–346.

Salazar-Parra, C., Aguirreolea, J., Sánchez-Díaz, M., Irigoyen, J. J. and Morales, F. (2012). Climate change (elevated CO_2, elevated temperature and moderate drought) triggers the antioxidant enzymes' response of grapevine cv. Tempranillo, avoiding oxidative damage. *Physiol. Plant.* 144(2): 99–110.

Samaniego, I., Brito, B., Viera, W., Cabrera, A., Llerena, W., Kannangara, T., Vilcacundo, R., Angós, I. and Carrillo, W. (2020a). Influence of the maturity stage on the phytochemical composition and the antioxidant activity of four Andean blackberry Cultivars (*Rubus glaucus* Benth) from Ecuador. *Plants (Basel)* 9(8): 1027.

Samaniego, I., Espin, S., Cuesta, X., Arias, V., Rubio, A., Llerena, W., Angós, I. and Carrillo, W. (2020b). Analysis of environmental conditions effect in the phytochemical composition of potato (*Solanum tuberosum*) cultivars. *Plants (Basel)* 9(7): 815.

Šamec, D., Linić, I. and Salopek-Sondi, B. (2021). Salinity stress as an elicitor for phytochemicals and minerals accumulation in selected leafy vegetables of *Brassicaceae*. *Agronomy* 11(2): 361.

Samuolienė, G., Sirtautas, R., Brazaitytė, A., Viršilė, A. and Duchovskis, P. (2012). Supplementary red-LED lighting and the changes in phytochemical content of two baby leaf lettuce varieties during three seasons. *J. Food Agric. Environ.* 10: 701-706.

Samuoliene, G., Virsile, A., Brazaityte, A., Jankauskiene, J., Sakalauskiene, S., Vastakaite, V., Novickovas, A., Viskeliene, A., Sasnauskas, A. and Duchovskis, P. (2017). Blue light dosage affects carotenoids and tocopherols in microgreens. *Food Chem.* 228: 50-56.

Sarker, U., Islam, M. T. and Oba, S. (2018). Salinity stress accelerates nutrients, dietary fiber, minerals, phytochemicals and antioxidant activity in Amaranthus tricolor leaves. *PLoS ONE* 13(11): e0206388.

Schonhof, I., Kläring, H. P., Krumbein, A. and Schreiner, M. (2007). Interaction between atmospheric CO_2 and glucosinolates in broccoli. *J. Chem. Ecol.* 33(1): 105-114.

Seo, J. M., Arasu, M. V., Kim, Y. B., Park, S. U. and Kim, S. J. (2015). Phenylalanine and LED lights enhance phenolic compound production in Tartary buckwheat sprouts. *Food Chem.* 177: 204-213.

Sepúlveda, I., Barrientos, H., Mahn, A. and Moenne, A. (2013). Changes in SeMSC, glucosinolates and sulforaphane levels, and in proteome profile in broccoli (*Brassica oleracea* var. Italica) fertilized with sodium selenate. *Molecules* 18(5): 5221-5234.

SharathKumar, M., Heuvelink, E. and Marcelis, L. F. M. (2020). Vertical farming: Moving from genetic to environmental modification. *Trends Plant Sci.* 25(8): 724-727.

Shimomura, M., Yoshida, H., Fujiuchi, N., Ariizumi, T., Ezura, H. and Fukuda, N. (2020). Continuous blue lighting and elevated carbon dioxide concentration rapidly increase chlorogenic acid content in young lettuce plants. *Sci. Hortic.* 272: 109550.

Shin, Y., Liu, R. H., Nock, J. F., Holliday, D. and Watkins, C. B. (2007). Temperature and relative humidity effects on quality, total ascorbic acid, phenolics and flavonoid concentrations, and antioxidant activity of strawberry. *Postharvest Biol. Technol.* 45(3): 349-357.

Shin, Y., Ryu, J. A., Liu, R. H., Nock, J. F. and Watkins, C. B. (2008a). Harvest maturity, storage temperature and relative humidity affect fruit quality, antioxidant contents and activity, and inhibition of cell proliferation of strawberry fruit. *Postharvest Biol. Technol.* 49(2): 201-209.

Shin, Y., Ryu, J. A., Liu, R. H., Nock, J. F., Polar-Cabrera, K. and Watkins, C. B. (2008b). Fruit quality, antioxidant contents and activity, and antiproliferative activity of strawberry fruit stored in elevated CO_2 atmospheres. *J. Food Sci.* 73(6): S339-S344.

Signore, A., Bell, L., Santamaria, P., Wagstaff, C. and Van Labeke, M. C. (2020). Red light is effective in reducing nitrate concentration in rocket by increasing nitrate reductase activity, and contributes to increased total glucosinolates content. *Front. Plant Sci.* 11: 604.

Sinkovič, L., Demšar, L., Žnidarčič, D., Vidrih, R., Hribar, J. and Treutter, D. (2015). Phenolic profiles in leaves of chicory cultivars (*Cichorium intybus* L.) as influenced by organic and mineral fertilizers. *Food Chem.* 166: 507-513.

Son, K. H., Lee, J. H., Oh, Y., Kim, D., Oh, M. M. and In, B. C. (2017). Growth and bioactive compound synthesis in cultivated lettuce subject to light-quality changes. *Hortscience* 52(4): 584-591.

Son, K. H. and Oh, M. M. (2013). Leaf shape, growth, and antioxidant phenolic compounds of two lettuce cultivars grown under various combinations of blue and red light-emitting diodes. *Hortscience* 48(8): 988-995.

Song, J., Huang, H., Hao, Y., Song, S., Zhang, Y., Su, W. and Liu, H. (2020a). Nutritional quality, mineral and antioxidant content in lettuce affected by interaction of light intensity and nutrient solution concentration. *Sci. Rep.* 10(1): 2796.

Song, J., Huang, H., Song, S., Zhang, Y., Su, W. and Liu, H. (2020b). Effects of photoperiod interacted with nutrient solution concentration on nutritional quality and antioxidant and mineral content in lettuce. *Agronomy* 10(7): 920.

Sotelo, T., Soengas, P., Velasco, P., Rodríguez, V. M. and Cartea, M. E. (2014). Identification of metabolic QTLs and candidate genes for glucosinolate synthesis in *Brassica oleracea* leaves, seeds and flower buds. *PLoS ONE* 9(3): e91428.

Spalholz, H., Perkins-Veazie, P. and Hernández, R. (2020). Impact of sun-simulated white light and varied blue: Red spectrums on the growth, morphology, development, and phytochemical content of green-and red-leaf lettuce at different growth stages. *Sci. Hortic.* 264: 109195.

Srihanam, C. W. P. (2016). Phytochemical, antioxidant and antibacterial activities in different maturity stages of wild grape (*Ampelocissus martinii* Planch.) fruits. *Int. J. Appl. Chem.* 12: 337–346.

Stetsenko, L. A., Pashkovsky, P. P., Voloshin, R. A., Kreslavski, V. D., Kuznetsov, V. V. and Allakhverdiev, S. I. (2020). Role of anthocyanin and carotenoids in the adaptation of the photosynthetic apparatus of purple-and green-leaved cultivars of sweet basil (*Ocimum basilicum*) to high-intensity light. *Photosynthetica* 58(4): 890–901.

Sun, Y., Asghari, M. and Zahedipour-Sheshgelani, P. (2019). Foliar spray with 24-epibrassinolide enhanced strawberry fruit quality, phytochemical content, and postharvest life. *J. Plant Growth Regul.*: 1–10.

Talcott, S. T., Moore, J. P., Lounds-Singleton, A. J. and Percival, S. S. (2005). Ripening associated phytochemical changes in mangos (*Mangifera indica*) following thermal quarantine and low-temperature storage. *J. Food Sci.* 70(5): C337–C341.

Tao, R., Bai, S., Ni, J., Yang, Q., Zhao, Y. and Teng, Y. (2018). The blue light signal transduction pathway is involved in anthocyanin accumulation in 'Red Zaosu' pear. *Planta* 248(1): 37–48.

Taulavuori, K., Pyysalo, A., Taulavuori, E. and Julkunen-Tiitto, R. (2018). Responses of phenolic acid and flavonoid synthesis to blue and blue-violet light depends on plant species. *Environ. Exp. Bot.* 150: 183–187.

Tavarini, S. and Angelini, L. G. (2013). Stevia rebaudiana Bertoni as a source of bioactive compounds: The effect of harvest time, experimental site and crop age on steviol glycoside content and antioxidant properties. *J. Sci. Food Agric.* 93(9): 2121–2129.

Tavarini, S., Gil, M. I., Tomas-Barberan, F. A., Buendia, B., Remorini, D., Massai, R., Degl'Innocenti, E. and Guidi, L. (2011). Effects of water stress and rootstocks on fruit phenolic composition and physical/chemical quality in Suncrest peach. *Ann. Appl. Biol.* 158(2): 226-233.

Thiruvengadam, M., Kim, S. H. and Chung, I. M. (2015). Exogenous phytohormones increase the accumulation of health-promoting metabolites, and influence the expression patterns of biosynthesis related genes and biological activity in Chinese cabbage (*Brassica rapa* spp. pekinensis). *Sci. Hortic.* 193: 136–146.

Thwe, A. A., Kim, Y. B., Li, X., Seo, J. M., Kim, S. J., Suzuki, T., Chung, S. O. and Park, S. U. (2014). Effects of light-emitting diodes on expression of phenylpropanoid biosynthetic genes and accumulation of phenylpropanoids in Fagopyrum tataricum sprouts. *J. Aqric. Food Chem.* 62(21): 4839–4845.

Traka, M. H. (2016). Health benefits of glucosinolates. Adv. Bot. Res. 80: 247–279.

Tripodi, P., Ficcadenti, N., Rotino, G. L., Festa, G., Bertone, A., Pepe, A., Caramanico, R., Migliori, C. A., Spadafora, D., Schiavi, M., Cardi, T. and Lo Scalzo, R. (2019). Genotypic and environmental effects on the agronomic, health-related compounds and antioxidant properties of chilli peppers for diverse market destinations. *J. Sci. Food Agric.* 99(10): 4550–4560.

Tuan, P. A., Thwe, A. A., Kim, Y. B., Kim, J. K., Kim, S. J., Lee, S., Chung, S. O. and Park, S. U. (2013). Effects of white, blue, and red light-emitting diodes on carotenoid biosynthetic gene expression levels and carotenoid accumulation in sprouts of Tartary buckwheat (*Fagopyrum tataricum* Gaertn.). *J. Agric. Food Chem.* 61(50): 12356–12361.

Van Meulebroek, L., Bussche, J. V., De Clercq, N., Steppe, K. and Vanhaecke, L. (2015). A metabolomics approach to unravel the regulating role of phytohormones towards carotenoid metabolism in tomato fruit. *Metabolomics* 11(3): 667–683.

Vashisth, T., Olmstead, M. A., Olmstead, J. and Colquhoun, T. A. (2017). Effects of nitrogen fertilization on subtropical peach fruit quality: Organic acids, phytochemical content, and total antioxidant capacity. *J. Amer. Soc. Hort. Sci.* 142(5): 393–404.

Vaštakaitė, V., Viršilė, A., Brazaitytė, A. R., Samuolienė, G., Jankauskienė, J., Novičkovas, A. and Duchovskis, P. (2017). Pulsed light-emitting diodes for a higher phytochemical level in microgreens. *J. Agric. Food Chem.* 65(31): 6529–6534.

Villarreal-García, D., Nair, V., Cisneros-Zevallos, L. and Jacobo-Velázquez, D. A. (2016). Plants as biofactories: Postharvest stress-induced accumulation of phenolic compounds and glucosinolates in broccoli subjected to wounding stress and exogenous phytohormones. *Front. Plant Sci.* 7: 45.

Vollmer, K., Chakraborty, S., Bhalerao, P. P., Carle, R., Frank, J. and Steingass, C. B. (2020). Effect of pulsed light treatment on natural microbiota, enzyme activity, and phytochemical composition of pineapple (*Ananas comosus* [L.] Merr.) juice. *Food Bioprocess Technol.* 13(7): 1095–1109.

Wang, H., Chen, G., Guo, X., Abbasi, A. M. and Liu, R. H. (2016). Influence of the stage of ripeness on the phytochemical profiles, antioxidant and antiproliferative activities in different parts of *Citrus reticulata* Blanco cv. Chachiensis. *LWT Food Sci. Technol.* 69: 67–75.

Wang, S. Y., Chen, C. T. and Wang, C. Y. (2009). The influence of light and maturity on fruit quality and flavonoid content of red raspberries. *Food Chem.* 112(3): 676–684.

Watanabe, M. and Ayugase, J. (2015). Effect of low temperature on flavonoids, oxygen radical absorbance capacity values and major components of winter sweet spinach (*Spinacia oleracea* L.). *J. Sci. Food Agric.* 95(10): 2095–2104.

Xie, Z., Fan, J., Charlebois, D., Roussel, D., Dubé, C., Charles, M. T. and Khanizadeh, S. (2014). Agronomic characteristics and phytochemical profiles of advanced June-bearing strawberry lines for the northern Canadian climate. *Agric. Food Sci.* 23(1): 38–47.

Xu, F., Cao, S., Shi, L., Chen, W., Su, X. and Yang, Z. (2014). Blue light irradiation affects anthocyanin content and enzyme activities involved in postharvest strawberry fruit. *J. Agric. Food Chem.* 62(20): 4778–4783.

Xu, Z., Jiang, Y. and Zhou, G. (2015). Response and adaptation of photosynthesis, respiration, and antioxidant systems to elevated CO_2 with environmental stress in plants. *Front. Plant Sci.* 6: 701.

Yang, J., Zhu, Z. and Gerendás, J. (2009). Interactive effects of phosphorus supply and light intensity on glucosinolates in pakchoi (*Brassica campestris* L. ssp. chinensis var. communis). *Plant Soil* 323(1–2): 323–333.

Yang, J., Zhu, Z., Wang, Z. and Zhu, B. (2010). Effects of storage temperature on the contents of carotenoids and glucosinolates in Pakchoi (*Brassica rapa* L. ssp. chinensis var. communis). *J. Food Biochem.* 34(6): 1186–1204.

Yang, Q. (2019). *Plant Factory*. Beijing: Tsinghua University Press.

Yang, X., Cui, X., Zhao, L., Guo, D., Feng, L., Wei, S., Zhao, C. and Huang, D. (2017). Exogenous glycine nitrogen enhances accumulation of glycosylated flavonoids and antioxidant activity in lettuce (*Lactuca sativa* L.). *Front. Plant Sci.* 8: 2098.

Yang, X., Feng, L., Zhao, L., Liu, X., Hassani, D. and Huang, D. (2018a). Effect of glycine nitrogen on lettuce growth under soilless culture: A metabolomics approach to identify the main changes occurred in plant primary and secondary metabolism. *J. Sci. Food Agric.* 98(2): 467–477.

Yang, X., Wei, S., Liu, B., Guo, D., Zheng, B., Feng, L., Liu, Y., Tomás-Barberán, F. A., Luo, L. and Huang, D. (2018b). A novel integrated non-targeted metabolomic analysis reveals significant metabolite variations between different lettuce (Lactuca sativa. L) varieties. Hortic. Res. 5: 1–14.

Yi, G. E., Robin, A. H. K., Yang, K., Park, J. I., Hwang, B. H. and Nou, I. S. (2016). Exogenous methyl jasmonate and salicylic acid induce subspecies-specific patterns of glucosinolate accumulation and gene expression in Brassica oleracea L. *Molecules* 21(10): 1417.

Ying, Q., Jones-Baumgardt, C., Zheng, Y. and Bozzo, G. (2021). The proportion of blue light from light-emitting diodes alters microgreen phytochemical profiles in a species-specific manner. *Hortscience* 56(1): 13–20.

Yuan, G., Wang, X., Guo, R. and Wang, Q. (2010). Effect of salt stress on phenolic compounds, glucosinolates, myrosinase and antioxidant activity in radish sprouts. *Food Chem.* 121(4): 1014–1019.

Zhang, M., Zhang, G., You, Y., Yang, C., Li, P. and Ma, F. (2016). Effects of relative air humidity on the phenolic compounds contents and coloration. In: The 'Fuji' apple (*Malus domestica* Borkh.) peel. *Sci. Hortic.* 201: 18–23.

Zhang, X., Bian, Z., Yuan, X., Chen, X. and Lu, C. (2020). A review on the effects of light-emitting diode (LED) light on the nutrients of sprouts and microgreens. *Trends Food Sci. Technol.* 99: 203–216.

Zhang, X., He, D., Niu, G., Yan, Z. and Song, J. (2018a). Effects of environment lighting on the growth, photosynthesis, and quality of hydroponic lettuce in a plant factory. *Int. J. Agr. Biol. Eng.* 11(2): 33–40.

Zhang, Y., Jiang, L., Li, Y., Chen, Q., Ye, Y., Zhang, Y., Luo, Y., Sun, B., Wang, X. and Tang, H. (2018b). Effect of red and blue light on anthocyanin accumulation and differential gene expression in strawberry (*Fragaria × ananassa*). *Molecules* 23(4): 820.

Zhou, W., Liang, X., Zhang, Y., Dai, P., Liang, B., Li, J., Sun, C. and Lin, X. (2020). Role of sucrose in modulating the low-nitrogen-induced accumulation of phenolic compounds in lettuce (Lactuca sativa L.). J. Sci. Food Agr. 100(15): 5412–5421.

Zhu, T., Yang, J., Zhang, D., Cai, Q., Zhou, D., Tu, S., Liu, Q. and Tu, K. (2020). Effects of white LED light and UV-C radiation on stilbene biosynthesis and phytochemicals accumulation identified by UHPLC–MS/MS during peanut (*Arachis hypogaea* L.) germination. *J. Agric. Food Chem.* 68(21): 5900–5909.

Zoratti, L., Sarala, M., Carvalho, E., Karppinen, K., Martens, S., Giongo, L., Häggman, H. and Jaakola, L. (2014). Monochromatic light increases anthocyanin content during fruit development in bilberry. *BMC Plant Biol.* 14: 377.

Chapter 12

Understanding processing of phytochemical compounds in fruits and vegetables in the gut

Stan Kubow, Lucas Roldos, Kailee Wark and Michèle M. Iskandar, McGill University, Canada

1 Introduction

1.1 Nutrients and non-nutrients: an explanation

The coffee cherry is a beautiful, bright red fruit of the coffee plant (*Coffea* L.). This plant was originally grown and traded in Africa and the Arabian Peninsula in the 15th century, then carefully cultivated in Central and South America, Asia, and Oceania (Ukers, 2020; USDA, 2021). Two seeds found inside each harvested coffee cherry provide the aromatic coffee beans that may be roasted, ground, and brewed. In the 1600s, coffee houses in Europe began to bring people together to enjoy the bitter beverage. As a result of its stimulatory effects on digestion and the central nervous system, coffee has been historically recommended for medicinal purposes (Wagner et al., 2001). In 1819, caffeine was isolated from coffee and has since been shown to exert the psychostimulant effects that many have desired from coffee consumption (Bizzo et al., 2015). But caffeine is not the only interesting constituent of coffee.

Regular coffee consumers obtain a multitude of important bioactive substances with every sip, namely phytochemicals.

Phytochemicals are compounds synthesized in plants that result from secondary metabolism. The consumption of these non-nutrient compounds by humans has been associated with a wide spectrum of biological activities that confer various health benefits, particularly when the phytochemicals are consumed as a part of plant-rich diets. The physiological effects of phytochemicals include antioxidant, anticancer, antimicrobial, anticlotting,

http://dx.doi.org/10.19103/AS.2022.0101.14

anti-inflammatory actions, and many more (Rodriguez-Casado, 2016). Additionally, some phytochemicals exert adverse health effects. The health-modulating characteristics of phytochemicals are largely determined by their chemical structures but are also complicated by many intrinsic and extrinsic factors. In this regard, gastrointestinal (GI) digestive processes play a fundamental role in modulating the biological properties of phytochemicals. Once consumed (and liberated from their food matrix), these compounds travel through the digestive system to undergo processing at various stages in the GI tract. Although phytochemicals may be absorbed in their native state, in most circumstances, these compounds are broken down into metabolites either by digestive enzymes and pH changes in the upper GI tract or by microbiota in the colonic regions. These digestive processes release metabolites with a wide variety of bioactivities that can differ greatly from the original plant compounds. As health properties of phytochemicals are modulated greatly by their interactions at the gut level, a considerable amount of research has focused on the variety of factors that affect the GI processing of phytochemicals. For more information on this topic, readers are directed to the reviews in Del Rio et al. (2013), Bohn et al. (2017), Reboul (2019), Chacón-Ordóñez et al. (2019), and Dima et al. (2020).

Did you know?

Just as nutrients have established relationships important to human health, non-nutrients also have biological activities in the body. Nutrients are used by the body for adenosine triphosphate (ATP) energy production, building materials, or as cofactors in metabolism.

Compounds are considered essential or non-essential based on whether they can be synthesized in adequate amounts for bodily needs, independent of diet.

Nutrient recommendations for healthy people are provided by values in the Dietary Reference Intakes (DRI). More detailed information can be found in the documentation provided by the Institute of Medicine (2006). The DRI is an evidence-based set of values that can help guide nutritional decision-making to support a healthy diet.

Certain phytochemicals, such as provitamin A carotenoids (β-carotene, α-carotene, and β-cryptoxanthin), have conversion factor values listed in the DRI. These conversion factors can be used to calculate the Retinol Activity Equivalent of these three carotenoids to ensure the recommended level of vitamin A (retinol) is met (IOM, 2006). Apart from such carotenoids, phytochemicals are currently considered non-nutrients, meaning they do not yet have an established DRI. Based on their health-promoting characteristics, some phytochemicals could be considered for future inclusion in the DRIs, particularly with the development of more sensitive omics technologies, which, together, contribute to the assessments of their health impact with large population datasets in longitudinal studies.

Complicating this evaluation is the data showing that phytochemical compounds can have synergistic and antagonistic effects on each other. Given that there are thousands of phytochemical compounds in plant foods, with many yet to be identified, this nutritional analysis is a complex undertaking. Until DRI values are assigned to phytochemicals, dietary variety with plenty of plant foods is still the best recommendation.

1.2 Phytochemicals: classification, dietary sources, health effects and dietary intakes

Phytochemicals can be classified into four main groups: phenolics, terpenoids, alkaloids, and sulphur compounds. These main classes and their subclasses and some of their dietary sources are listed in Table 1. Total daily phytochemical intake has been estimated to reach a range of 1 to 2 g/day (Dingeo et al., 2020), with the most common and well-researched phytochemicals, polyphenols, and carotenoids, having demonstrated average intakes ranging from 900 mg/day (Del Bo et al., 2019) to 12 mg/day (Böhm et al., 2021), respectively (depending on the database used for phytochemical level estimation, dietary assessment, and study population evaluated). Detailed discussion on the health benefits of various phytochemicals can be found in the reviews in Barba et al. (2016), Bergman et al. (2019), Del Bo et al. (2019), Del Rio et al. (2013), Epriliati and Ginjom (2012), Leena et al. (2020), Możkus et al. (2020), Moses et al. (2014), Peng et al. (2019), Rodriguez-Casado (2016), Ruhee et al. (2020), and Yahfoufi et al. (2018).

1.2.1 Phenolics

Phenolics, often referred to as polyphenols, are the largest category of phytochemicals and are frequently studied in the context of inflammation due to their antioxidant capacity (Tsao, 2010). This group includes large categories such as flavonoids, often found in colourful and aromatic fruits and vegetables, among many other foods (Del Rio et al., 2013). Various berries as well as dark chocolate (depending on the type of processing) are high in polyphenol content, and some seasonings have been found to have over 15 000 mg/100 g of polyphenols (Pérez-Jiménez et al., 2010). The total daily intake of polyphenols can exceed 1 g/day, with a third obtained from phenolic acids and the other two-thirds from flavonoids (flavanols, proanthocyanidins, and anthocyanins)

Table 1 Phytochemicals: classifications, examples, and dietary sources

Classes	Subclasses	Further subclasses	Examples	Dietary sources	References
Phenolics	Phenolic acids	Hydroxycinnamic acids, hydroxybenzoic acids	Chlorogenic acid, *p*-hydroxybenzoic acid, ferulic acid, caftaric acid, coutaric acid, fertaric acid; vanillic acid, *p*-coumaric acid, caffeic acid, cinnamic acid, gallic acid, ellagic acid	Apricots, blueberries, carrots, cereals, pears, cherries, citrus fruits, oilseeds, peaches, plums, spinach, tomatoes, eggplants, apples, grapes, blueberries, cereals, cranberries, oilseeds	Del Rio et al. (2013) Gu et al. (2004) Naczk and Shahidi (2006) Rochfort and Panozzo (2007) Scalbert and Williamson (2000) Tsao (2010)
	Flavonoids	Anthocyanidins (aglycone) anthocyanins (glycoside)	Pelargonidin, cyanidin, delphinidin, peonidin, petunidin, malvidin	Bilberries, black and red currants, blueberries, cherries, chokecherries, grapes, strawberries, apples	
		Chalcones	Phlorizidin, phloretin	Apples, beers	
		Flavan-3-ols, Flavanols,	Catechin, epicatechin	Apples, blueberries	
		Catechins	Epifzelechin, epicatechin gallate, epigallocatechin, epigallocatechin, gallate, theaflavins, thearubigins	Grapes, onions, lettuce, tea	
		Flavanonols	Astilbin, engeletin	Grapes	
		Flavanones	Naringenin, hesperetin, neohesperidin, naringin	Citrus fruits	
		Flavonols	Flavonol glycosides: rutin, hyperin, isoquercitrin, reynoutrin, avicularin, quercitrin); quercetin, myricetin, kaempferol, isorhamnetin	Apples, beans, blueberries, buckwheat, cranberries, endive, leeks, lettuce, onions, olive, pepper, tomatoes	
		Flavones	Apigenin, luteolin, wogonin, baicalein, vitexin, isovitexin, orientin, iso-orientin, nobiletin, tangeretin	Citrus fruits, celery, parsley, spinach, some herbs, rooibos tea	
		Isoflavones (phytoestrogens)	Daidzein (daidzin), genistein (genistin), coumestrol, formononetin, biochanin A	Soybeans, chickpeas	

		Xanthones		Mango, mangosteen	
		Dihydrochalcones	Phloridzin, aspalathin, nothofagin	Apples, rooibos tea	
	Hydrolysable Tannins		Glucose esters of gallic acid and ellagic acid; gallotannins, punicalagin, punicalin	Pomegranate, raspberries	
	Condensed tannins, Proanthocyanidins		Made of flavan-3-ol units; Procyanidins ((epi)catechin), Propelargonidins ((epi)afzelechin), Prodelphinidins ((epi)gallocatechin)	Apples, grapes, peaches, plums, pears, cinnamon, blueberries, strawberries, sorghum, red kidney beans, hazelnuts, pecans, blackcurrants	
	Polyphenolic amides	Avenanthramides		Oats	
		Capsaisinoids	Capsaicin	Pepper, chilli pepper	
	Other	Arbutin		Pears	
	Phenolics	Alk(en)ylresorcinols		Cereals	
		Coumarins		Carrots, celery, citrus fruits, parsley, parsnips	
		Lignans, phytoestrogens	Secoisolariciresinol, matairesinol, pinoresinol, lariciresinol, syringaresinol, sesamin, sesaminol, medioresinol	Buckwheat, flaxseed, sesame seed, rye, wheat	
		Secoiridoids		Olives	
		Stillbenes (phytoalexins)	Resveratrol, piceids, astringins, resveratrolosides, pterostillbenes	Grapes	
Terpenoids (isoprenoids) and terpenes	Monoterpenes (C10)		*p*-Menthane monoterpenoids (menthol)	Peppermint	Bergman et al. (2019) Dingeo et al. (2020) Khoo et al. (2011) Moran et al. (2018) Moses et al. (2014) Rochfort and Panozzo (2007)
	Sesquiterpenes (C15)		G ngerol, B-caryphyllene, curcumin	Ginger, clove, cannabis, rosemary, turmeric	
	Diterpenes (C20)				
	Sterols (C27-29)				

(*Continiued*)

Table 1 (*Continiued*)

Classes	Subclasses	Further subclasses	Examples	Dietary sources	References
	Triterpenes (C30)		Triterpenoid glycosides (some saponins)	Legumes (soybeans, beans, peas), oats, licorice root	
	Tetraterpenes (C40)	Carotenoids (carotenes, xanthophylls)	β-carotene, α-carotene, β-cryptoxanthin, lycopene, lutein, zeaxanthin	Tomato, spinach, carrots, kale, oranges, eggs, avocado, mango, loquat, peach, broccoli, maize, pumpkin, papaya, persimmon, sweet potato	
Alkaloids	Ornithine-derived alkaloids	Pyrrolizidine alkaloids	Jacobine, senecionine, echimidine	Honey, contaminant in grain products	Dingeo et al. (2020) Hartmann (2007) Koleva et al. (2012)
		Tropane alkaloids	Cocaine, atropine, scopolamine, hyoscyamine	Trace in eggs	
	Lysine derivatives	Piperidine alkaloids	Coniine, piperine	Peppers	
		Quinolizidine alkaloids	Sparteine, lupanine, α-isolupanine, 13α-hydroxylupanine, anagyrine	Lupines	
	Tyrosine derivatives	Isoquinoline alkaloids	Papaverine, idrastine, morphine, codeine, thebaine	Opium poppy	
	Tryptophan derivatives	Indole (carboline), β-carboline alkaloids	β-carbolines (harman, norharman, harmine, harmol), dihidro-β-carbolines (harmaline, harmalol), tetrahydro-β- carbolines	Cooked fish, cooked meat, bread, breakfast cereals, soups, alcoholic beverages, vinegar, soy and tabasco sauce, coffee, fruit juices, and jams	
		Quinoline alkaloids	Quinine	Tonic water, bitter lemon, vermouth	
		Ergot alkaloids	Ergotamine, ergometrine	Contaminant of crops (rye)	
	Purine alkaloids	Methoxanthine alkaloids	Caffeine, theophylline, theobromine, 7-methylxanthine	Coffee, tea, cola, cocoa beans	
	From nicotinic acid		Nicotine	Tobacco, potatoes, black tea, tomatoes, cauliflower, eggplant, green peppers	

			Myosmine	Tobacco, peanuts, hazelnuts, rice, corn, wheat flour, millet, almonds, cocoa, cereals, popcorn, tomatoes, potatoes, carrot, pineapple, kiwi fruit, apples	
	Glycoalkaloids	Steroid alkaloids Steroidal glycoalkaloids Solanins *Solanum* alkaloids	Solanidine, tomatidine, solasodine, α-solanine, α-chaconine, tomatine, solasonine, solanidine, α- and β-solamargines, demissine, demissidine, leptines, liptidine, leptinines, leptinidine, leptimidine, tomatidenol, alkamine	Potatoes, tomatoes, eggplant, bell peppers (capsicum)	
Sulphur compounds	Glucosinolates	Aliphatic, aromatic, indolyl-glucosinolates	Glucoraphanin, glucoiberin, dehydroerucin, sinigrin, glucobrassicin	Cruciferous vegetables; Cabbage, Brussel sprouts, broccoli, cauliflower, radish	Dingeo et al. (2020) Steinbrecher and Linseisen (2009)
	Organosulphur compounds		Alliin (allicin with cell damage)	Garlic, onions	
Others	Phytosterols, phytostanols		Sitosterol, campesterol, sitostanol, campestanol, β-sitosterol, sigmasterol	Vegetable oils, nuts, seeds, cereals	Dingeo et al. (2020) Rochfort and Panozzo (2007)
	Cyanogenic glycosides		Amygdalin, linamarin, lotaustralin, neolinustatin	Bitter almonds, seeds of apricot, peach, plum, black cherry, flax, cassava	

(Scalbert and Williamson, 2000). In a prospective study involving an urban Polish population of over 10 400 (45-69 years old), average daily polyphenol intake was found to be about 1.7 g/day with 52% coming from flavonoids and 46% from phenolic acids. Coffee was consumed by 83% of the cohort and contributed 40% of the total polyphenol intake, followed by tea and chocolate, as well as fruits, vegetables and vegetable oils, legumes, and cereals.

Notable fruit and vegetable contributors were apples, oranges, berries, spinach, and potatoes (Grosso et al., 2014). Other cohorts have shown differing intakes of polyphenols with food sources that reflect cultural, regional, and socio-economic dietary differences. Major polyphenol contributors include coffee and tea (Northern and Central Europe), fruit alcoholic beverages (countries near the Mediterranean), apples, vegetables, green tea (varying parts of Asia), and tea, citrus, and legumes (United States) (Del Bo et al., 2019). Dietary intakes of phenolic subclasses have also been identified, such as flavanols (mean intake of 50 mg/day in the Netherlands, and intake range of 17.1–38.6 mg/day in the United States), condensed tannins (mean intake of 53.6 mg/day in the United States), and lignans (mean intake of 1 mg/day in Western countries) (Gu et al., 2004; Del Rio et al., 2013).

Polyphenols have been extensively studied both in vitro and in animal and human studies where they have shown to exert antioxidant (Li et al., 2014), immunomodulatory, anti-inflammatory (Yahfoufi et al., 2018), neuroprotective (Figueira et al., 2016), anticancer (Fantini et al., 2015), and antimicrobial (Álvarez-Martínez et al., 2020) activities among others. A myriad of molecular and cellular mechanisms has been proposed to describe their modes of action, such as their ability to inhibit specific inflammatory pathways and enzymes responsible for the generation of reactive oxygen species, or to increase the activities of endogenous antioxidant enzymes as well as direct free radical scavenging abilities (Yahfoufi et al., 2018; Li et al., 2014).

1.2.2 Terpenoids

Terpenoids (or isoprenoids) and terpenes are made of isoprene hydrocarbon units that contribute to the colouration, taste, and fragrance of plant foods. Terpenoids include more well-known phytochemicals such as saponins from legumes and carotenoids from colourful fruits and vegetables, with carotenoid intake ranging from 2 mg/day to 20 mg/day (Shi et al., 2004; Dingeo et al., 2020). Blood levels and adipose tissue concentrations of carotenoids show strong correlations with an intake of fruit and vegetables among populations, and carotenoid concentrations, significant in the skin, can be estimated by photographs of the skin (Böhm et al., 2021). Certain terpenoids have been found to exert bioactive systemic effects that include anticancer (Kudryavtseva et al., 2016, Rabi and Gupta, 2014), antimicrobial (Mahizan et al., 2019; Dwivedi et al., 2015; Sieniawska et al., 2017), anthelmintic (Mirza et al., 2020), and anti-inflammatory (Elsayed et al., 2014; De Cássia da Silveira e Sá et al., 2013) properties. In addition, recent evidence has emerged showing anti-ageing/geroprotective effects of terpenes (Kudryavtseva et al., 2016; Proshkina et al., 2020).

1.2.3 Alkaloids

Alkaloids are a numerous class of over 12 000 organic compounds containing nitrogen, produced by secondary metabolism involving amino acid precursors in plants undergoing exposure to stressors (Peng et al., 2019; Koleva et al., 2012). Alkaloids include phytochemicals with noticeable toxic and stimulatory effects, and are mostly known for their bitter taste and long history of use in medicine (Koleva et al., 2012). Common examples include quinine (antimalarial), morphine (analgesic), cocaine, and nicotine (stimulants), with comparably few bioactive alkaloids found in the diet (Koleva et al., 2012; Vuong, 2021). The methylxanthine caffeine (from coffee and tea) and lysine-derived piperine (from pepper) are two dietary alkaloids that exert stimulant and antihyperglycemic effects (Atal et al., 2012), respectively. The above two compounds are implicated in improving the absorption and/or bioavailability of certain phytochemicals (Fernández-Lázaro et al., 2020; Weiser and Weigmann, 2019; Yoovathaworn et al., 1986). Betaine is an alkaloid mostly found in cereals, as well as spinach and beets, which may enhance non-enzymatic antioxidant defences (Zhang et al., 2016). Some alkaloids can be more toxic, such as glycoalkaloids that are concentrated in younger parts of tubers (i.e. potato sprouts and leaves), although beneficial effects at lower doses have also been indicated (Friedman, 2006).

1.2.4 Sulphur-containing compounds

Sulphur compounds are pungent components commonly found in root vegetable staples such as garlic and onions. The two main subclasses of sulphur-containing compounds are organosulphur compounds and glucosinolates. Organosulphur compounds are primarily found in allium vegetables (i.e. onion and garlic), while glucosinolates are generally obtained from cruciferous vegetables (i.e. broccoli, cabbage, cauliflower) (Miękus et al., 2020). Both organosulphur and glucosinolate compounds appear to benefit human health by activating detoxification and antioxidant enzymes, scavenging free radicals, and inducing immune signalling (Ruhee et al., 2020). Glucosinolates are not biologically active as parent compounds, but the foods they are found in, along with some gut microbiota, contain enzymes that convert these sulphur-containing phytochemicals into more bioactive and bioavailable forms.

Glucosinolates also have antinutrient effects such as binding to iodine and preventing its absorption (Felker et al., 2016).

The Future of Agriculture and Nutrition: Phytochemical Farming

Plants have an array of phytochemicals, many of which provide the sensory joys of taste, colour, and aroma. Beyond their sensory properties, phytochemicals increase food storage shelf life and have an impressive impact on human health (Ochmian et al., 2020; Rodriguez-Casado, 2016). So how can agricultural practices help further enhance the nutritional health properties of plant foods? Plants that grow in more challenging conditions can produce more beneficial phytochemicals, as plant stress amplifies production of phytochemicals as the secondary metabolites. Plants use their secondary metabolites to defend against stressors and to adapt to their environment (Isah, 2019). As global agriculture efforts seek solutions to the growing energy and nutrient demands of populations, identification of agricultural practices that improve the phytochemical characteristics of plant produce may also be beneficial. A plant that produces more secondary metabolites could be more resistant to environmental stressors and benefit the consumer by providing additional bioactive compounds. Several agricultural approaches including organic farming and agronomic techniques are being explored to enhance plant food quality; manipulating cultivation conditions has the potential to optimize phytochemical profiles and quality of the produce (Ochmian et al., 2020, García-Mier et al., 2013).

---

2 The digestion process

GI digestion includes salivary, gastric, and intestinal digestive processes, each of which has specific pH conditions along with the release of proteases, lipases, and carbohydrate-digesting enzymes involved in the hydrolysis of dietary macromolecules. These enzymatic and non-enzymatic processes determine the release of phytochemicals from food matrices but can also degrade their native form depending upon the stability of the parent compound during digestion. The impact of digestive processes on the food matrix determines the rate and degree of release of phytochemicals from the food (i.e. bioaccessibility) and the fraction that is absorbed for use in terms of physiologic functions (i.e. bioavailability). Most major phytochemicals are located inside plant cells, within organelles. For example, carotenoids are found in chromoplasts or in the chloroplast membrane, while polyphenols can be found in vacuoles (Bohn et al., 2015). In addition, many phenolic compounds form complexes with cell wall polysaccharides. This renders most phytochemicals inaccessible to digestive enzymes within the human GI tract, as digestive enzymes are unable to break down most plant cell wall material.

The degree to which digestive enzymes can have access to components within the cell wall varies greatly across plants. For example, in vitro studies have shown that some plants such as potatoes, mango, and banana have cell walls that are relatively permeable to digestive enzymes, in comparison with wheat, kidney beans, peas, and chickpeas. In addition, food processing such as chopping, milling, grinding, heat treatment, and mastication in the mouth affects the plant tissue structure to varying degrees (Holland et al., 2020). The digestive stability and release from the food matrix of phytochemicals determine their bioaccessibility to allow for their intestinal absorption. The above multitude of complicating factors alters greatly the type of digestive end-products generated from phytochemical digestion. The variabilities in the structural characteristics of the digestive phytochemical end-products lead to major differences in their health-promoting properties. Consequently, a more comprehensive understanding is needed for modifying dietary factors involved in phytochemical digestion to better delineate the health benefits of phytochemicals.

2.1 Oral and gastric digestion

Upon ingestion, foods reach the stomach in the form of particles of different shapes and sizes with cell walls damaged to varying degrees, further increasing the variability with which phytochemicals are rendered accessible to human digestion (Holland et al., 2020). The extent of phytochemical bioaccessibility is impacted by various factors, including those intrinsic to the plant food itself, the degree of processing prior to ingestion, and the effect of food mixtures, but much is yet to be understood. Very little is known regarding the fate of phytochemicals in the oral cavity and stomach. Wine is rich in compounds that are not sequestered within plant cell walls, and their responses are easier to identify during digestion. For example, flavonols and proanthocyanidins interact with proteins in saliva. Albumin and mucins in saliva can also interact with tannins, leading to their precipitation. The presence of saliva facilitates the solubilization of polyphenols in plant beverages, thereby promoting their adherence to oral surfaces and improving the redox status of the oral cavity (Epriliati and Ginjom, 2012).

Different polyphenols (including anthocyanins) have different stabilities at different pH levels. For example, quercetin is stable at pH 2, whereas quinones are unstable at high pH where they are oxidized to diketones. Flavonols and proanthocyanidins are not significantly affected by gastric digestion, but they can be broken down when gastric pH is sufficiently low (Epriliati and Ginjom,

2012). Anthocyanins are stable in the stomach but degrade easily in the small intestine at higher pH (David et al., 2019). At gastric pH, flavonoid oligomers are broken down into smaller units.

The extent of enzymatically driven breakdown of phytochemicals in the mouth and stomach is also poorly understood. It has been shown that certain dietary flavonoids are hydrolysed from their glycoside form to aglycones in the oral cavity with high interindividual variability (Walle et al., 2005). However, many studies indicate that phenolic compounds, in general, are not significantly affected by digestion in the stomach (Wojtunik-Kulesza et al., 2020).

2.2 Small intestinal digestion

The enzymatic action and pH conditions in the small intestine can lead to structural modifications of more soluble phytochemical compounds such as phenolic acids (Kroll et al., 2003). The bioavailability of phytochemicals is based on the proportion of these compounds available for absorption following digestive processes.

Depending upon the structural characteristics of the phytochemicals or their metabolites, they can be available via absorption pathways in the small intestine and colon. The varying pH conditions of the different upper GI tract regions can lead to the degradation of phytochemicals such as phenolics and flavonoids or lead to polymerization reactions with enzymes via covalent or non-covalent bonding. For example, phenolics are unstable in an alkaline or neutral small intestinal environment. A variety of oxidation and polymerization reactions during intestinal processing can also lead to different types of phenolic moieties such as quinones and chalcones (Altunkaya et al., 2016; Gil-Izquierdo et al., 2002). Polymerization can diminish the bioaccessibility or solubility of phytochemicals but can also impede the activities of digestive tract enzymes, ultimately limiting protein, lipid, and glucose absorption. In this regard, phytochemicals can lead to antinutritional effects such as the intake of high concentrations of tannins leading to nonspecific binding and precipitation of dietary proteins (He et al., 2007). Conversely, anti-obesity and antidiabetic properties of dietary phytochemicals have been attributed to their inhibition of enzymes from the small intestine involved with lipid (Costamagna et al., 2016) and glucose (da Silva et al., 2014) absorption, respectively.

Once available for absorption, phytochemicals interact with the mucus layer and tight junctions, and can then undergo cellular uptake involving either active or passive transport (Dima et al., 2020; Lee et al., 2018). While some phytochemicals are actively transported, most are absorbed via passive transport together with varying levels of efflux. Apart from the food matrix effects, passive transport of phytochemicals depends primarily on their chemical properties which, as mentioned above, can be modified by various processes throughout upper intestinal digestion. The polar surface area of the phytochemical relates to its hydrogen bonding potential, which determines how well it can break from the aqueous environment, while its molecular mass and lipophilicity determine its ability to permeate through the intestinal membrane (Selby-Pham et al., 2017). Phenolic phytochemicals tend to be absorbed through passive diffusion and experience efflux, although some can also modulate efflux transporters, while some terpenoids can be actively transported (Bohn et al., 2017; Li et al., 2015). Lipophilic phytochemicals, such as carotenoids, must be emulsified into lipid droplets and micellarized along with bile salts and fatty acids to be absorbed by enterocytes (Mapelli-Brahm et al., 2017). Certain apical transport proteins present in the small and large intestines, including scavenger receptor class B Type 1 (SR-B1) and cluster of differentiation 36 (CD36), have been shown to facilitate the absorption of select lipophilic compounds, such as vitamins D and K and several carotenoids (Reboul, 2019). To overcome relatively poor GI stability and absorption of many phytochemicals, pharmacological drug delivery systems including nanoparticles, liposomes, and micelles are beginning to be applied to phytochemicals (Aqil et al., 2013).

2.3 Colonic digestion

The large intestine is composed of the cecum, the colon (ascending, transcending, descending, and sigmoid colon), and the rectum. After the small intestine, the watery chyme passes through the ileocecal valve and into the caecal segment of the large intestine. The chyme flows in carrying mostly indigestible materials such as dietary fibres, inorganic matter, and water. The large intestine is responsible for the remaining nutrient absorption, extracting water, compacting, and storing faecal matter, and housing trillions of microorganisms. These microorganisms comprise what is known as the gut microbiota and include the multitude of bacteria, viruses, bacteriophages, fungi, archaea, protozoa, and other microorganisms that colonize the human large intestine. The gut microbiota function as a community centre for the hosts' intestinal homeostasis and human health. This collection of commensal microbes forms a mutualistic relationship through their major roles in nutrient metabolism, hormone synthesis, energy harvesting, and immune system development. Over time, the gut microbiota have co-evolved with human metabolism, creating a multifaceted relationship where the hosted microorganisms generate bioactive metabolites important to the host's intestinal health, biochemistry, immune system, metabolism, mental health, and beyond (Nicholson et al., 2012). Comprehensive reviews on this subject can be found in Arumugam et al. (2011) and Catalkaya et al. (2020).

The gut microbiota function by acting on components that reach the large intestine through catabolism, fermentation, and biotransformation, which increases the bioavailability and bioactivity of nutrients and non-nutrients. The substrates available to the gut microbiota depend on overall diet, food processing, meal composition, food matrix, and transit time, with transit time itself being affected by diet, exercise, genetics, sex, drugs, and psychological status of the host (Oliphant and Allen-Vercoe, 2019; Degen and Phillips, 1996). From these substrates, the colonic microbiota synthesizes several vitamins, produces gases, and ferments dietary fibre to produce short-chain fatty acids (SCFAs). SCFAs (primarily butyrate, propionate, and acetate) are absorbed prior to excretion and used by the host and intestinal epithelial cells for energy. SCFAs also serve to promote tight junctions, increase local acidity and inhibit pathogen growth, and support anti-inflammatory immune system processes, thus exemplifying the role of gut microbiota in intestinal health maintenance (Oliphant and Allen-Vercoe, 2019; Nicholson et al., 2012).

Dietary phytochemicals are poorly absorbed from the small intestine and up to 95% reach the large intestine (Louis et al., 2014). The vast and diverse metabolic activity of the gut microbiota may extend the metabolic capacity of the host to biotransform unabsorbed phytochemicals into more bioactive and bioavailable forms and/or metabolites (Dey, 2019). Polyphenols are generally present in a glycosylated form (conjugated to a sugar) within foods, which inhibits their absorption until reaching the colon (Marín et al., 2015). Colonic bacteria can hydrolyse glycosylated phytochemicals, fermenting the sugar moiety for energy and SCFA production, and leaving the aglycone form that is more easily absorbed or further metabolized by microbiota via multi-enzymatic reactions into more simple phenolic compounds with altered bioactivity and bioavailability (Catalkaya et al., 2020; Dingeo et al., 2020).

Tetrahydrocurcumin is a bacterial metabolite of the polyphenol curcumin (found in turmeric), which exhibits greater antioxidant activity (Aggarwal et al., 2014).

Glucosinolates and organosulphur compounds are hydrolysed by the glucosidase enzymes myrosinase and allinase, respectively, which are released upon plant cell rupture (i.e. chopping and/or mastication) (Więkus et al., 2020). These enzymes can be inactivated depending on the type of processing the food may undergo. This prevents the absorption of the sulphur-containing phytochemicals until they reach the colon where they can undergo metabolism via gut bacterial enzymes. Glucosinolates can be biotransformed by gut microbes into isothiocyanates, which are highly reactive metabolites that can induce phase II detoxification enzymes, modulate the cell cycle and apoptosis, and exhibit fungicidal and bactericidal activity (Barba et al., 2016).

Microbial biotransformation of polyphenols and glucosinolates is relatively better understood than that of terpenoids and alkaloids. Beyond increased bioaccessibility of carotenoids due to their release from fibre or protein as shown via *in vitro* colonic fermentation studies (Goñi et al., 2006), or gut microbial production of carotenoids (Karlsson et al., 2012), little is known regarding the effect of the gut microbiome on carotenoid metabolism. Some evidence suggests that gut microbiota may catabolize alkaloids to use them as a source of nitrogen for nucleic acid metabolism (Farag et al., 2020), such as caffeine and theobromine from coffee and tea, while other microbiota may degrade toxic alkaloids, such as nicotine (Gunasekaran et al., 2020). Berberine, an alkaloid from the tart barberry, is biotransformed by the gut microbiota into dihydroberberine, which exhibits a fivefold greater intestinal absorption and is reverted (oxidized) back into berberine following absorption (Feng et al., 2015). On the other hand, caffeine is rapidly and almost completely absorbed (Blanchard and Sawers, 1983; Teekachunhatean et al., 2013) to be extensively metabolized by the liver, while metabolites are mostly found in urine and only 2–5% are excreted in the stool (Willson, 2018). Gut microbial biotransformation of phytochemicals is discussed in greater detail in the reviews in Aura (2008), Barba et al. (2016), Dey, (2019), Farag et al. (2020), Feng et al. (2015), and Rowland et al. (2018).

Phytochemicals have been shown to impact colonic health by modulating gut microbial composition, gut barrier integrity, and host immune function, along with the maintenance of redox homeostasis for normal cellular metabolism and function (Dingeo et al., 2020). Polyphenols, as well as carotenoids to a lesser extent, have been shown to exert prebiotic and antibacterial effects on the gut microbiome, illustrated by enhancements in favourable bacterial phyla (i.e. *Bifidobacterium*, *Lactobacillus*, and *Enterococcus*) and inhibition of pathogenic phyla (i.e. Clostridium), although differences in bacterial alterations can occur depending on the specific phytochemical (Yin et al., 2019). The alkaloid berberine has been shown to inhibit bile acid transforming microbiota, which appears to be how it exerts its antidiabetic influence (Zhang et al., 2020). Carotenoids may induce improvements in the synthesis of intestinal immunoglobulin A (IgA), a critical regulator of mucosal microbiota balance that supports the gut immune system and prevents gut microbial dysbiosis (Lyu et al., 2018). Certain polyphenols have been shown to promote the growth of specific mucin-degrading bacteria that stimulate mucin production (i.e. *Akkermansia*), while they seem to inhibit mucin-degrading bacteria (Catalkaya et al., 2020). This positive impact of phytochemicals on colonic health seems to serve as a form of feedback that may improve phytochemical metabolism and enhance the mutualistic relationship between symbiotic gut bacteria and the host. The role of phytochemicals in modifying gut microbial composition is further discussed in the reviews in Dingeo et al. (2020) and Yin et al. (2019).

2.4 Case study: a dynamic multistage GI model for the study of phytochemical biotransformation by gut microbiota

In vitro digestion models are typically used for the assessment of phytochemical products of digestion, as human trials are technically difficult and impractical due to their high cost and complexity. Animal models are also used to achieve these objectives, but they present distinct disadvantages. Although the mouse and human microbiomes share similarities, there is a degree of dissimilarity that necessitates caution when interpreting and translating findings from animal models to humans (Nguyen et al., 2015). In addition, animal models are relatively lengthier, costlier, and involve more complex ethical considerations than *in vitro* models. Among available in vitro GI model systems, many studies have used static models of digestion that do not take into consideration transit dynamics or the fact that different segments of the GI tract are characterized by different digestive conditions and microbial profiles. An innovative approach involves the use of the dynamic computer-controlled multistage GI digestion model using human faecal inoculum. Such models can be used to assess the effects of phytochemicals on gut microbiome profiles but also the biotransformation of phytochemicals by the gut microbiota, leading to the production of secondary metabolites, a notable proportion of which exert significant bioactivities.

The multi-compartmental dynamic GI model includes the regulation of pH, the dynamic flows of food, bicarbonate, bile, and digestive enzymes in the different GI compartments to simulate the complex gastric, small intestinal, and colonic environments (Fig. 1). It allows for consideration of the dynamics of digestion as well as the real-time sampling of gastric and intestinal digesta. This system has been validated via the monitoring of changes in SCFA profiles, gas production, and enzymatic activities (Molly et al., 1993). This model can be utilized to explore the digestion and biotransformation of any substrate, whether an individual phytochemical, food, or supplement, or combinations. By collecting faecal samples, faecal pellets, and faecal water, the metabolites produced by the biotransformation of phytochemicals in each segment of the GI tract can be assessed, as well as the effects of the substrates on the gut microbiome, inflammation and oxidant status, and more. In addition, as the system can be inoculated with individual bacteria or with human faecal matter, it is possible to evaluate the influence of specific diseases on the digestive processes via inoculation of the system with faecal matter from donors with disease conditions (Fig. 1). Reviews on *in vitro* gut models to study plant metabolite digestion and bioaccessibility are provided in Alminger et al. (2014), Dima et al. (2020), and Wojtunik-Kulesza et al. (2020).

Such a model has been used to examine the biotransformation of potato polyphenols and synthetic polyphenol mixtures throughout the three colonic segments: the ascending, transverse, and descending colonic regions

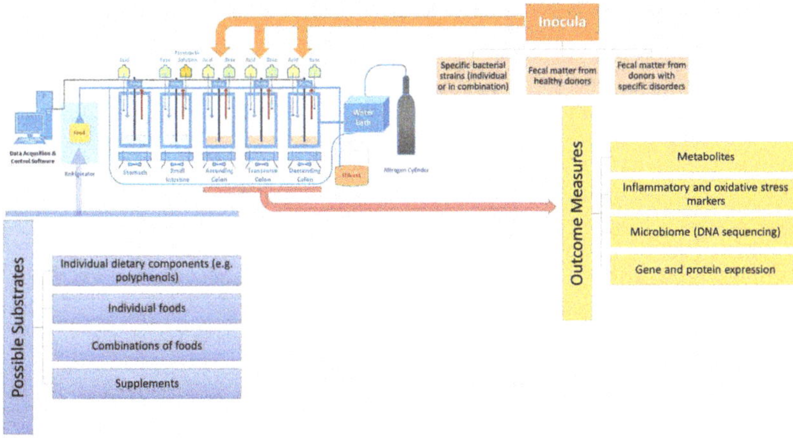

Figure 1 Computer-controlled *in vitro* dynamic gastrointestinal model. The model is compartmentalized allowing the study of the gastric and intestinal digestion of phytochemicals as well as their biotransformation by the gut microbiota. Different substrates can be assessed (whole foods, food components, food combinations, individual phytochemicals) and a panoply of outcomes can be studied to determine the effects of the substrate on the microbiota. In addition, the colonic vessels can be inoculated with individual bacterial strains, or with faecal matter from healthy humans or animals, or with faecal matter from humans or animals with particular diseases. The latter allows the assessment of the effects of specific diseases on the digestive processes, and conversely, the potential modulation of gut dysbiosis by substrates under study.

(Sadeghi Ekbatan et al., 2016; Khairallah et al., 2018). For example, Fig. 2 shows the relative disappearance of parent polyphenols and the appearance of metabolites derived from them, as digestion progresses from the stomach; small intestine; ascending, transverse, and descending colonic segments.

3 Factors affecting digestion

3.1 The food matrix

By altering the chemical structure of phytochemicals, GI digestion plays a central role in modulating the bioavailability and bioactivity of these compounds. In most circumstances, the absorbed bioactive molecules are not the native phytochemical compounds present in the food but rather are metabolites generated following digestive processes. The release of phytochemical molecules from plant foods is strongly affected by the food matrix that constitutes the physical form of the food. The food matrix refers to the composite of the physical structure and organization of the food components. These food components range from molecular constituents such as phytochemicals and micronutrients to the plant tissue structure itself.

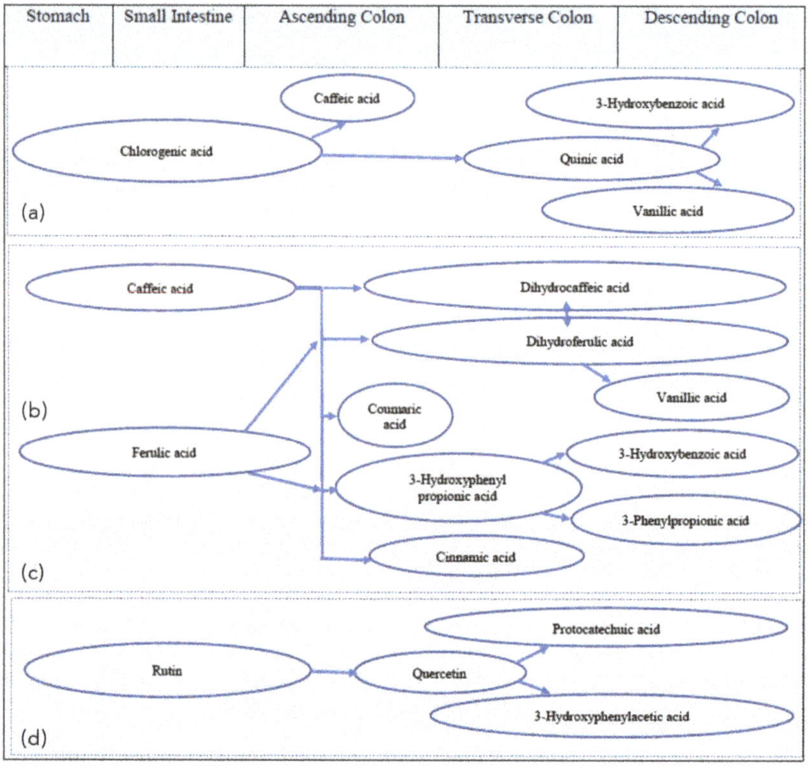

Figure 2 Bacterial biotransformation of chlorogenic acid as assessed by liquid chromatography-mass spectrometry. Cleavage of chlorogenic acid by esterase activity of colonic bacteria in the ascending and transverse colon generates caffeic acid and quinic acid. Caffeic acid undergoes a series of biotransformation processes to generate 3-phenylpropionic acid and other smaller phenolics. Further biotransformation of quinic acid generates vanillic and benzoic acids in the transverse and descending colon compartments (Sadeghi Ekbatan et al., 2016; Khairallah et al., 2018; Farah et al., 2008; Rechner et al., 2001; Aura, 2008).

The bioavailability, pharmacokinetics, and bioactivities of phytochemicals are modulated by the food matrix. Continual postprandial liberation of matrix-bound phytochemicals has been noted allowing for more prolonged systemic exposure in the bloodstream as compared to the phytochemicals provided as purified supplements (Rondini et al., 2004; Vitaglione et al., 2008; Cömert and Gökmen, 2017). Such differences can lead to misinterpretation of the biological properties of phytochemicals present within foods when compared to studies examining the isolated compound. For example, quercetin was ineffective in diminishing inflammation in a colitis mouse model while quercetin in the conjugated form of rutin (naturally found in foods) induced potent anti-inflammatory effects (Kwon et al., 2005). The food matrix also affects the capacity of colonic microbiota to biotransform phytochemicals into bioactive phytochemical moieties. Reviews on the role of the food matrix can be found in Holland et al. (2020), Marín et al. (2015), Oliphant and Allen-Vercoe (2019).

Phytochemicals are affected by their interactions with various food components during digestion such as proteins, fats, carbohydrates, fibres, and minerals (Alminger et al., 2014; Ortega et al., 2011). For instance, dietary fat enhances the absorption of β-carotene from vegetables (Jalal et al., 1998). The format of the food in terms of liquid versus solid food matrix also plays a role in terms of phytochemical bioavailability. More rapid absorption and greater peak blood concentrations of isoflavonoids have been noted with soy milk versus solid soy foods (Cassidy et al., 2006).

Different types of food processing (fermentation, mechanical and heat treatment) alter the complex matrices of foods, which can affect phytochemical bioaccessibility and bioavailability. Thermal and mechanical processing disrupts plant cell walls, which improves the bioavailability of carotenoids such as β-carotene and lycopene (Chacón-Ordóñez et al., 2019). Food fermentation involving lactic acid bacteria breaks down polyphenol structures producing lower molecular weight phenolic metabolites that are more bioavailable (Septembre-Malaterre et al., 2018). For instance, soy fermentation leads to deconjugation of the glucoside conjugates of soy isoflavones producing more bioavailable aglycone forms (Cassidy et al., 2006). On the other hand, processing can lead to adverse effects on phytochemical bioactivities. The capacity of garlic to retard the bioactivation of an experimental mammary carcinogen has been shown to be inhibited by heat treatment (Song and Milner, 1999). Much more research is needed to elucidate the impact of varying types of food processing on the bioactivity of phytochemicals.

A wide variety of single or multiple isolated phytochemicals are presently being marketed as supplements in various formats such as capsules, gels, pills, powders, or liquids. Phytochemical supplements are promoted for their biological properties that include antioxidant, anti-inflammatory, anticarcinogenic, hypotensive, vasodilatory, antiangiogenic, antiviral, antimicrobial, immune-enhancing, and antithrombotic characteristics, among many others (Zhang et al., 2016). The evidence for the health-promoting properties of such supplements is less compelling than what has been demonstrated with plant-rich diets containing a diverse mixture of phytochemical components (Zhang et al., 2016). A large body of evidence has accumulated from large-scale population studies showing that diets rich in whole grains, fruits, and vegetables provide significant protection against many non-communicable diseases. The discrepancies between supplements and whole foods are attributable to different biological effects exerted by isolated phytochemicals versus those seen from phytochemical interactions with other constituents within the plant food matrix.

The myriad of phytochemical by-products generated from the digestion of plant foods requires the applications of advanced omics and bioinformatics to identify the absorbed molecules with systemic bioactivities. In this regard, the science of foodomics uses a variety of large-scale omics platforms (i.e. metabolomics, transcriptomics, proteomics, lipidomics) to simultaneously analyse metabolites generated from the digestion of multiple components in foods such as proteins, lipids, and phytochemicals. In this regard, the future applications of systems biology can lead to a more detailed understanding of the interaction of phytochemical digestive by-products with their biological targets affecting human health.

3.2 Background diet

The protective effects of plant-derived foods are closely linked to the concept of food synergy. Food synergy refers to the combination of plant compounds producing greater bioactivities than the sum of their individual effects. This synergy involves both major phytochemical constituents in foods and minute amounts of unknown food components. Antioxidant synergistic interactions have been reported among phytochemicals such as flavonoids and tocopherols present naturally together within plant foods (Leena et al., 2020). Synergistic combinations of different phytochemicals from mixtures of fruit and vegetables have also demonstrated protective effects against chronic diseases. For example, studies have shown that: (a) tomato and broccoli combinations more effectively slow tumour growth in a prostate cancer rat model than either tomato or broccoli alone (Canene-Adams et al., 2007); and (b) synergistic effects for inhibition of α-glucosidase have been noted from the combination of anthocyanin-rich black currants and chlorogenic acid-rich rowanberry for the maintenance of normoglycemic levels in type 2 diabetes (Boath et al., 2012). A key feature underlying food synergy is the enhanced bioavailability that can occur for phytonutrients and phytochemicals from their intrinsic combinations within plant foods or different plant food combinations. Among such examples are: (a) the combination of lemon and green tea enhancing by tenfold the bioavailability of the major catechin (epigallocatechin gallate) in tea (Tewari et al., 2000); (b) combining black pepper with turmeric or turmeric-based foods increasing the bioavailability of curcumin 1000-fold (Shoba et al., 1998).

Fibre, as an indigestible plant material critical to a healthful diet, acts as a prebiotic. Gut bacteria feed on prebiotics, encouraging competitive selection for growth and activity of beneficial microbes, which in turn supports the host (Holscher, 2017). As colonic bacteria possess a myriad of carbohydrate-active enzymes that, beyond fermenting dietary fibre, can hydrolyse glycosylated molecules such as phytochemicals and proteins (i.e. mucin), diets lacking in fibre lead microbiota to resort to digesting mucin from the intestinal mucus barrier, increasing susceptibility to pathogens (Laville et al., 2019; Desai et al., 2016). Adequate fibre consumption therefore reduces detrimental mucin degradation and entraps cholesterol, bile salts, minerals, and phytochemicals within its matrix, bringing along these compounds to the large intestine (Anderson et al., 2009; Louis et al., 2014). Microbial metabolism of prebiotics in turn releases the entrapped minerals and phytochemicals, thereby increasing their overall bioaccessibility and potential for further biotransformation.

Gut processes in immunity and digestion may be further supported by probiotics that could provide a more supportive environment to enhance phytochemical bioavailabilities and bioactivities. The non-absorbable portion of phytochemicals can promote colonic health by acting as substrates for the microorganisms in the gut.

Prebiotics differ from probiotics as probiotics are live organisms administered with the goal of conferring health benefits to the host. Both prebiotics and probiotics ideally support a healthy intestinal microbiota profile (Xavier-Santos et al., 2020). If administered together, the combination of prebiotics and probiotics is called synbiotics. Prebiotic substrates may selectively support the role and metabolism of probiotics and desirable pre-established colonic microbes, therefore increasing the microbial stability, SCFA production, and promotion of more resilient niches.

3.3 Host/intrinsic factors affecting digestion

GI tract health depends on adequate nutrition, and diet is reciprocally an influential way to support microbial - and overall - health. Many factors impact the development and composition of the gut microbiota, such as birth delivery methods, xenobiotic use, lifestyle, disease, and diet (Fig. 3). Diets low in fruits

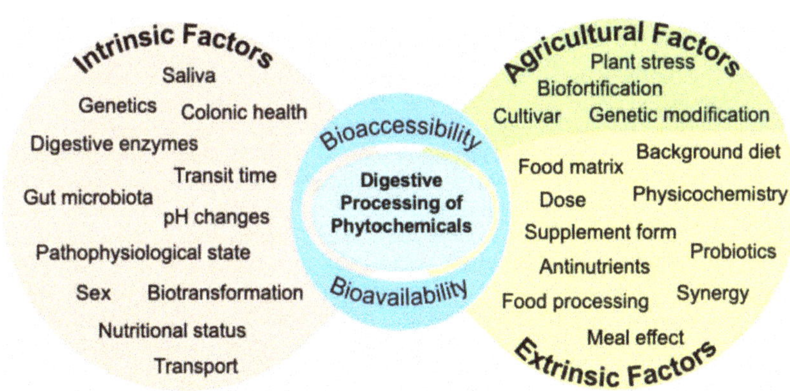

Figure 3 The impact of intrinsic and extrinsic factors on the digestive processing of phytochemicals.

and vegetables can lead to poor microbial diversity and reduced capacity to maintain host intestinal homeostasis, therefore increasing risk of infection, inflammation, and disease. Everyone has a unique microbial composition, and organisms may vary markedly between individuals, even of a 'healthy' status, and thus taxonomic data do not provide the whole picture (Huttenhower et al., 2012). Such intraindividual variabilities can lead to different capabilities towards microbial biotransformation of phytochemicals to their bioactive constituents in the colonic regions. Despite such variances among individuals, emerging research has indicated a certain degree of ubiquitous functional redundancy and stability of the gut microbiota throughout the population.

Gut microbiota develop into an ecosystem for each individual, which may be classified into enterotypes, with each enterotype comprising a cluster of metagenomes from signature genera. Enterotypes may respond differently to perturbations, such as drugs and short-term diet changes, and may be associated with long-term diet (Wu et al., 2011). Once established in an individual, the gut microbial community is difficult to modulate. This developed resilience can serve towards protection against infectious diseases; however, a gut microbial community may also develop into a steady, yet dysbiotic state. A state of dysbiosis is therefore implicated in the manifestation of chronic disease and resistance to treatment and can be detrimental to the host-microbial relationship (Sommer et al., 2017).

Depending on the gut microbial composition, individuals may or may not have the capacity to produce the more bioavailable and bioactive metabolites (postbiotics) of some phytochemicals, such as urolithins and equol from ellagitannins and isoflavones, respectively (see Equol Sidebar) (Cortés-Martín et al., 2020).

The equol metabotype

One of the most researched examples of phytochemical processing by gut microbiota is that of the isoflavones daidzin, genistin, and glycitin and their metabolite equol (Mayo et al., 2019). Isoflavones are phenolic compounds found in roots and seeds, where soybeans are the richest source. These phytochemicals resemble the hormone estrogen and exhibit estrogen-like activity; however, as with most polyphenols, they are primarily found in glycosylated forms that are not readily absorbed and show low levels of bioactivity. Only 1% of ingested isoflavones may be absorbed in the upper GI tract (Grace et al., 2004), leaving the rest to be biotransformed by certain colonic microbiota into isoflavone metabolites. *O*-desmethylangolensin is the more common metabolite, but equol garners most attention as it is more stable, easily absorbed, and bioactive than its parent compounds or its 'cousin' metabolite (Frankenfeld, 2011; Setchell and Clerici, 2010).

The trouble is that equol seemingly can only be produced by certain bacteria and under certain circumstances. Only a fraction of the human population appears to produce equol (Ideno et al., 2018; Liu et al., 2010), and isoflavone interventions may not influence the level of expression of equol-related genes (and thus bacteria) (Vázquez et al., 2017). Even in individuals that have microbiota with the metabolic capacity to metabolize isoflavones into equol, it appears that few individuals have this microbial capability. There are a number of modifying host factors that appear to determine the latter microbial capacity including blood pressure (Hong et al., 2012), lipid metabolism (Zheng et al., 2019), gut microbial composition, and dietary patterns (Yoshikata et al., 2019). Moreover, the process of equol production may require a concerted effort of different bacteria (Mayo et al., 2019; Wang et al., 2007). Such confounding variables demonstrate the complexity of understanding the role of gut microbiota in phytochemical processing and interactions that influence the subsequent health impacts of fruit and vegetable intake.

Understanding the specific microbiota responsible for the production of certain metabolites could allow for the establishment of enterotypes (specific microbial communities) and/or metabotypes (certain metabolic profiles) for the prediction of phytochemical bioavailability and/or bioactivity across individuals (Landberg et al., 2019). This will require large-scale studies using -omics approaches, including metagenomics, metatranscriptomics, metaproteomics, and metabolomics, to understand the genes that are present, their regulation and expression, the proteins that are produced, and the end-products of their activity (metabolites), respectively (Rowland et al., 2018). Such studies have been suggested to be critical for the future of personalized nutrition and may help explain observed interindividual differences in response to phytochemical intervention (Morand et al., 2020). More detailed discussion can be found in the reviews by Landberg et al. (2019), Moran et al. (2018), Wu et al. (2011), and Yoshikata et al. (2019).

4 Conclusion

Plant-based foods are the source of a wide spectrum of health-modulating phytochemicals with numerous biological activities that can promote human well-being. The *in vitro* activities of these food components cannot be directly related to their *in vivo* properties since phytochemical structures generally undergo substantial modifications via chemical, enzymatic, and microbial digestive processes within the GI tract. The physiological functions of phytochemicals are also modified by complicating intrinsic and extrinsic factors that affect their bioavailability and bioaccessibility. In that regard, there remain many digestion-related mechanisms as well as dietary, processing, and host-related factors that require more detailed investigation.

Further studies involving omics technologies can better delineate factors involved in the digestive breakdown of the food matrix that affects the release and biotransformation of phytochemicals. Due to the inherent expense and time-consuming nature of human trials, *in vitro* dynamic GI models can be used as a higher throughput approach to assess the varying conditions affecting the digestion of phytochemicals. The insights derived from the above studies are crucial to provide a better understanding of the role of phytochemicals in human health and to develop agronomic practices, dietary recommendations, and food processing strategies to harness their health-promoting properties.

5 References

Aggarwal, B. B., Deb, L. and Prasad, S. (2014). Curcumin differs from tetrahydrocurcumin for molecular targets, signaling pathways and cellular responses. *Molecules (Basel, Switzerland)* 20(1), 185–205. doi:10.3390/molecules20010185.

Alminger, M., Aura, A. M., Bohn, T., Dufour, C., El, S. N., Gomes, A., Karakaya, S., Martínez-Cuesta, M. C., McDougall, G. J., Requena, T. and Santos, C. N. (2014). In vitro models for studying secondary plant metabolite digestion and bioaccessibility. *Comprehensive Reviews in Food Science and Food Safety* 13(4), 413–436. doi:10.1111/1541-4337.12081.

Altunkaya, A., Gökmen, V. and Skibsted, L. H. (2016). pH dependent antioxidant activity of lettuce (L. sativa) and synergism with added phenolic antioxidants. *Food Chemistry* 190, 25–32. doi:10.1016/j.foodchem.2015.05.069.

Álvarez-Martínez, F. J., Barrajón-Catalán, E., Encinar, J. A., Rodríguez-Díaz, J. C. and Micol, V. (2020). Antimicrobial capacity of plant polyphenols against gram- positive bacteria: A comprehensive review. *Current Medicinal Chemistry* 27(15), 2576–2606. doi:10.2174/0929867325666181008115650.

Anderson, J. W., Baird, P., Davis Jr., R. H., Ferreri, S., Knudtson, M., Koraym, A., Waters, V. and Williams, C. L. (2009). Health benefits of dietary fiber. *Nutrition Reviews* 67(4), 188–205. doi:10.1111/j.1753-4887.2009.00189.x.

Aqil, F., Munagala, R., Jeyabalan, J. and Vadhanam, M. V. (2013). Bioavailability of phytochemicals and its enhancement by drug delivery systems. *Cancer Letters* 334(1), 133–141. doi:10.1016/j.canlet.2013.02.032.

Arumugam, M., Raes, J., Pelletier, E., Le Paslier, D., Yamada, T., Mende, D. R., Fernandes, G. R., Tap, J., Bruls, T., Batto, J. M., Bertalan, M., Borruel, N., Casellas, F., Fernandez, L., Gautier, L., Hansen, T., Hattori, M., Hayashi, T., Kleerebezem, M., Kurokawa, K., Leclerc, M., Levenez, F., Manichanh, C., Nielsen, H. B., Nielsen, T., Pons, N., Poulain, J., Qin, J., Sicheritz-Ponten, T., Tims, S., Torrents, D., Ugarte, E., Zoetendal, E. G., Wang, J., Guarner, F., Pedersen, O., de Vos, W. M., Brunak, S., Doré, J., Antolín, M., Artiguenave, F., Blottiere, H. M., Almeida, M., Brechot, C., Cara, C., Chervaux, C., Cultrone, A., Delorme, C., Denariaz, G., Dervyn, R., Foerstner, K. U., Friss, C., van de Guchte, M., Guedon, E., Haimet, F., Huber, W., van Hylckama-Vlieg, J., Jamet, A., Juste, C., Kaci, G., Knol, J., Lakhdari, O., Layec, S., Le Roux, K., Maguin, E., Mérieux,

A., Melo Minardi, R., M'rini, C., Muller, J., Oozeer, R., Parkhill, J., Renault, P., Rescigno, M., Sanchez, N., Sunagawa, S., Torrejon, A., Turner, K., Vandemeulebrouck, G., Varela, E., Winogradsky, Y., Zeller, G., Weissenbach, J., Ehrlich, S. D. and Bork, P. (2011). Enterotypes of the human gut microbiome. *Nature* 473(7346), 174–180. doi:10.1038/nature09944.

Atal, S., Agrawal, R. P., Vyas, S., Phadnis, P. and Rai, N. (2012). Evaluation of the effect of piperine per se on blood glucose level in alloxan-induced diabetic mice. *Acta Poloniae Pharmaceutica* 69(5), 965–969.

Aura, A.-M. (2008). Microbial metabolism of dietary phenolic compounds in the colon. *Phytochemistry Reviews* 7(3), 407–429. doi:10.1007/s11101-008-9095-3.

Barba, F. J., Nikmaram, N., Roohinejad, S., Khelfa, A., Zhu, Z. and Koubaa, M. (2016). Bioavailability of glucosinolates and their breakdown products: Impact of processing. *Frontiers in Nutrition* 3, 24. doi:10.3389/fnut.2016.00024.

Bergman, M. E., Davis, B. and Phillips, M. A. (2019). Medically useful plant terpenoids: Biosynthesis, occurrence, and mechanism of action. *Molecules (Basel, Switzerland)* 24(21), 3961. doi:10.3390/molecules24213961.

Bizzo, M. L. G., Farah, A., Kemp, J. A. and Scancetti, L. B. (2015). In *Coffee in Health and Disease Prevention* (pp. 11–17). Essay, Academic Press Is an Imprint of Elsevier.

Blanchard, J. and Sawers, S. J. A. (1983). The absolute bioavailability of caffeine in man. *European Journal of Clinical Pharmacology* 24(1), 93–98. doi:10.1007/BF00613933.

Boath, A. S., Stewart, D. and McDougall, G. J. (2012). Berry components inhibit α- glucosidase in vitro: Synergies between acarbose and polyphenols from black currant and rowanberry. *Food Chemistry* 135(3), 929–936. doi:10.1016/j.foodchem.2012.06.065.

Böhm, V., Lietz, G., Olmedilla-Alonso, B., Phelan, D., Reboul, E., Bánati, D., Borel, P., Corte-Real, J., de Lera, A. R., Desmarchelier, C., Dulinska-Litewka, J., Landrier, J. F., Milisav, I., Nolan, J., Porrini, M., Riso, P., Roob, J. M., Valanou, E., Wawrzyniak, A., Winklhofer-Roob, B. M., Rühl, R. and Bohn, T. (2021). From carotenoid intake to carotenoid blood and tissue concentrations – Implications for dietary intake recommendations. *Nutrition Reviews* 79(5), 544–573. doi:10.1093/nutrit/nuaa008.

Bohn, T., Desmarchelier, C., Dragsted, L. O., Nielsen, C. S., Stahl, W., Rühl, R., Keijer, J. and Borel, P. (2017). Host-related factors explaining interindividual variability of carotenoid bioavailability and tissue concentrations in humans. *Molecular Nutrition and Food Research* 61(6), 28101967. doi:10.1002/mnfr.201600685.

Bohn, T., McDougall, G. J., Alegría, A., Alminger, M., Arrigoni, E., Aura, A. M., Brito, C., Cilla, A., El, S. N., Karakaya, S., Martínez-Cuesta, M. C. and Santos, C. N. (2015). Mind the gap–Deficits in our knowledge of aspects impacting the bioavailability of phytochemicals and their metabolites–A position paper focusing on carotenoids and polyphenols. *Molecular Nutrition and Food Research* 59(7), 1307–1323. doi:10.1002/mnfr.201400745.

Canene-Adams, K., Lindshield, B. L., Wang, S., Jeffery, E. H., Clinton, S. K. and Erdman, J. W. Jr. (2007). Combinations of tomato and broccoli enhance antitumor activity in dunning r3327-h prostate adenocarcinomas. *Cancer Research* 67(2), 836–843. doi:10.1158/0008-5472.CAN-06-3462.

Cassidy, A., Brown, J. E., Hawdon, A., Faughnan, M. S., King, L. J., Millward, J., Zimmer-Nechemias, L., Wolfe, B. and Setchell, K. D. (2006). Factors affecting the bioavailability of soy isoflavones in humans after ingestion of physiologically relevant levels from different soy foods. *Journal of Nutrition* 136(1), 45-51. doi:10.1093/jn/136.1.45.

Catalkaya, G., Venema, K., Lucini, L., Rocchetti, G., Delmas, D., Daglia, M., Filippis, A. D., Xiao, H., Quiles, J. L., Xiao, J. and Capanoglu, E. (2020). Interaction of dietary polyphenols and gut microbiota: Microbial metabolism of polyphenols, influence on the gut microbiota, and implications on host health. *Food Frontiers* 1(2), 109-133. doi:10.1002/fft2.25.

Chacón-Ordóñez, T., Carle, R. and Schweiggert, R. (2019). Bioaccessibility of carotenoids from plant and animal foods. *Journal of the Science of Food and Agriculture* 99(7), 3220-3239. doi:10.1002/jsfa.9525.

Cömert, E. D. and Gökmen, V. (2017). Antioxidants bound to an insoluble food matrix: Their analysis, regeneration behavior, and physiological importance. *Comprehensive Reviews in Food Science and Food Safety* 16(3), 382-399. doi:10.1111/1541-4337.12263.

Cortés-Martín, A., Selma, M. V., Tomás-Barberán, F. A., González-Sarrías, A. and Espín, J. C. (2020). Where to look into the puzzle of polyphenols and health? The postbiotics and gut microbiota associated with human metabotypes. *Molecular Nutrition and Food Research* 64(9), e1900952. doi:10.1002/mnfr.201900952.

Costamagna, M. S., Zampini, I. C., Alberto, M. R., Cuello, S., Torres, S., Pérez, J., Quispe, C., Schmeda-Hirschmann, G. and Isla, M. I. (2016). Polyphenols rich fraction from Geoffroea decorticans fruits flour affects key enzymes involved in metabolic syndrome, oxidative stress and inflammatory process. *Food Chemistry* 190, 392-402. doi:10.1016/j.foodchem.2015.05.068.

da Silva, S. M., Koehnlein, E. A., Bracht, A., Castoldi, R., de Morais, G. R., Baesso, M. L., Peralta, R. A., de Souza, C. G. M., de Sá-Nakanishi, A. B. and Peralta, R. M. (2014). Inhibition of salivary and pancreatic α-amylases by a pinhão coat (Araucaria angustifolia) extract rich in condensed tannin. *Food Research International* 56, 1-8. doi:10.1016/j.foodres.2013.12.004.

David, L., Danciu, V., Moldovan, B. and Filip, A. (2019). Effects of in vitro gastrointestinal digestion on the antioxidant capacity and anthocyanin content of cornelian cherry fruit extract. *Antioxidants* 8(5), 114. doi:10.3390/antiox8050114.

de Cássia da Silveira e Sá, R., Andrade, L. N. and de Sousa, D. P. (2013) A review on anti-inflammatory activity of monoterpenes. *Molecules* 18(1), 1227-1254. doi:10.3390/molecules18011227.

Degen, L. P. and Phillips, S. F. (1996). Variability of gastrointestinal transit in healthy women and men. *Gut* 39(2), 299-305. doi:10.1136/gut.39.2.299.

Del Bo, C., Bernardi, S., Marino, M., Porrini, M., Tucci, M., Guglielmetti, S., Cherubini, A., Carrieri, B., Kirkup, B., Kroon, P., Zamora-Ros, R., Liberona, N. H., Andres-Lacueva, C. and Riso, P. (2019). Systematic review on polyphenol intake and health outcomes: Is there sufficient evidence to define a health-promoting polyphenol-rich dietary pattern? *Nutrients* 11(6). doi:10.3390/nu11061355.

Del Rio, D., Rodriguez-Mateos, A., Spencer, J. P. E., Tognolini, M., Borges, G. and Crozier, A. (2013). Dietary (poly)phenolics in human health: Structures, bioavailability, and evidence of protective effects against chronic diseases. *Antioxidants and Redox Signaling* 18(14), 1818-1892. doi:10.1089/ars.2012.4581.

Desai, M. S., Seekatz, A. M., Koropatkin, N. M., Kamada, N., Hickey, C. A., Wolter, M., Pudlo, N. A., Kitamoto, S., Terrapon, N., Muller, A., Young, V. B., Henrissat, B., Wilmes, P.,

Stappenbeck, T. S., Núñez, G. and Martens, E. C. (2016). A dietary fiber-deprived gut microbiota degrades the colonic mucus barrier and enhances pathogen susceptibility. *Cell* 167(5), 1339-1353.e1321. doi:10.1016/j.cell.2016.10.043.

Dey, P. (2019). Gut microbiota in phytopharmacology: A comprehensive overview of concepts, reciprocal interactions, biotransformations and mode of actions. *Pharmacological Research* 147, 104367. doi:10.1016/j.phrs.2019.104367.

Dima, C., Assadpour, E., Dima, S. and Jafari, S. M. (2020). Bioavailability and bioaccessibility of food bioactive compounds; overview and assessment by in vitro methods. *Comprehensive Reviews in Food Science and Food Safety* 19(6), 2862-2884. doi:10.1111/1541-4337.12623.

Dingeo, G., Brito, A., Samouda, H., Iddir, M., La Frano, M. R. and Bohn, T. (2020). Phytochemicals as modifiers of gut microbial communities. *Food and Function* 11(10), 8444-8471. doi:10.1039/D0FO01483D.

Dwivedi, U. N., Tiwari, S., Singh, P., Singh, S., Awasthi, M. and Pandey, V. P. (2015). Treponema pallidum putative novel drug target identification and validation - Rethinking syphilis therapeutics with plant-derived terpenoids. *Omics* 19(2), 104-114. doi:10.1089/omi.2014.0154.

Elsayed, E. A., El Enshasy, H., Wadaan, M. A. M. and Aziz, R. (2014). Mushrooms: A potential natural source of anti-inflammatory compounds for medical applications. *Mediators of Inflammation* 2014, 805841. doi:10.1155/2014/805841.

Epriliati, I. and Ginjom, I. R. (2012). *Bioavailability of Phytochemicals, Phytochemicals - A Global Perspective of their Role in Nutrition and Health*. In: Tech Dr. Rao, V. (Ed.). ISBN: 978-953-51-0296-0. http://www.intechopen.com/books/phytochemicals-a-global-perspective-of-their-role-in-nutrition-and-health/bioavailability-of-phytochemicals.

Fantini, M., Benvenuto, M., Masuelli, L., Frajese, G. V., Tresoldi, I., Modesti, A. and Bei, R. (2015). In vitro and in vivo antitumoral effects of combinations of polyphenols and anticancer drugs: Perspectives on cancer treatment. *International Journal of Molecular Sciences* 16(5), 9236-9282. doi:10.3390/ijms16059236.

Farag, M. A., Abdelwareth, A., Sallam, I. E., el Shorbagi, M., Jehmlich, N., Fritz- Wallace, K., Serena Schäpe, S., Rolle-Kampczyk, U., Ehrlich, A., Wessjohann, L. A. and von Bergen, M. (2020). Metabolomics reveals impact of seven functional foods on metabolic pathways in a gut microbiota model. *Journal of Advanced Research* 23, 47-59. doi:10.1016/j.jare.2020.01.001.

Farah, A., Monteiro, M., Donangelo, C. M. and Lafay, S. (2008). Chlorogenic acids from green coffee extract are highly bioavailable in humans. *Journal of Nutrition* 138(12), 2309-2315. doi:10.3945/jn.108.095554.

Felker, P., Bunch, R. and Leung, A. M. (2016). Concentrations of thiocyanate and goitrin in human plasma, their precursor concentrations in brassica vegetables, and associated potential risk for hypothyroidism. *Nutrition Reviews* 74(4), 248-258. doi:10.1093/nutrit/nuv110.

Feng, R., Shou, J. W., Zhao, Z. X., He, C. Y., Ma, C., Huang, M., Fu, J., Tan, X. S., Li, X. Y., Wen, B. Y., Chen, X., Yang, X. Y., Ren, G., Lin, Y., Chen, Y., You, X. F., Wang, Y. and Jiang, J. D. (2015). Transforming berberine into its intestine- absorbable form by the gut microbiota. *Scientific Reports* 5(1), 12155. doi:10.1038/srep12155.

Fernández-Lázaro, D., Mielgo-Ayuso, J., Córdova Martínez, A. and Seco-Calvo, J. (2020). Iron and physical activity: Bioavailability enhancers, properties of black pepper (Bioperine®) and potential applications. *Nutrients* 12(6). doi:10.3390/nu12061886.

Figueira, I., Menezes, R., Macedo, D., Costa, I. and Dos Santos, C. N. (2016). Polyphenols beyond barriers: A glimpse into the brain. *Current Neuropharmacology* 15(4), 562–594. doi:10.2174/1570159X14666161026151545.

Frankenfeld, C. L. (2011). O-desmethylangolensin: The importance of equol's lesser known cousin to human health. *Advances in Nutrition (Bethesda, Md.)* 2(4), 317–324. doi:10.3945/an.111.000539.

Friedman, M. (2006). Potato glycoalkaloids and metabolites: Roles in the plant and in the diet. *Journal of Agricultural and Food Chemistry* 54(23), 8655– 8681. doi:10.1021/jf061471t.

García-Mier, L., Guevara-González, R. G., Mondragón-Olguín, V. M., Del Rocío Verduzco-Cuellar, B. and Torres-Pacheco, I. (2013). Agriculture and bioactives: Achieving both crop yield and phytochemicals. *International Journal of Molecular Sciences* 14(2), 4203–4222. doi:10.3390/ijms14024203.

Gil-Izquierdo, A., Zafrilla, P. and Tomás-Barberán, F. A. (2002). An in vitro method to simulate phenolic compound release from the food matrix in the gastrointestinal tract. *European Food Research and Technology* 214(2), 155–159. doi:10.1007/s00217-001-0428-3.

Goñi, I., Serrano, J. and Saura-Calixto, F. (2006). Bioaccessibility of β-carotene, lutein, and lycopene from fruits and vegetables. *Journal of Agricultural and Food Chemistry* 54(15), 5382–5387. doi:10.1021/jf0609835.

Grace, P. B., Taylor, J. I., Low, Y. L., Luben, R. N., Mulligan, A. A., Botting, N. P., Dowsett, M., Welch, A. A., Khaw, K. T., Wareham, N. J., Day, N. E. and Bingham, S. A. (2004). Phytoestrogen concentrations in serum and spot urine as biomarkers for dietary phytoestrogen intake and their relation to breast cancer risk in European prospective investigation of cancer and nutrition-Norfolk. *Cancer Epidemiology, Biomarkers and Prevention* 13(5), 698–708.

Grosso, G., Stepaniak, U., Topor-Mądry, R., Szafraniec, K. and Pająk, A. (2014). Estimated dietary intake and major food sources of polyphenols in the Polish arm of the HAPIEE study. *Nutrition* 30(11–12), 1398–1403. doi:10.1016/j.nut.2014.04.012.

Gu, L., Kelm, M. A., Hammerstone, J. F., Beecher, G., Holden, J., Haytowitz, D., Gebhardt, S. and Prior, R. L. (2004). Concentrations of proanthocyanidins in common foods and estimations of normal consumption. *The Journal of Nutrition* 134(3), 613–617. doi:10.1093/jn/134.3.613.

Gunasekaran, M., Lalzar, M., Sharaby, Y., Izhaki, I. and Halpern, M. (2020). The effect of toxic pyridine-alkaloid secondary metabolites on the sunbird gut microbiome. *Npj Biofilms and Microbiomes* 6(1), 53. doi:10.1038/s41522-020-00161-9.

Hartmann, T. (2007). From waste products to ecochemicals: Fifty years research of plant secondary metabolism. *Phytochemistry* 68(22–24), 2831–2846. doi:10.1016/j.phytochem.2007.09.017.

He, Q., Lv., Y. and Yao, K. (2007). Effects of tea polyphenols on the activities of α- amylase, pepsin, trypsin and lipase. *Food Chemistry* 101(3), 1178–1182. doi:10.1016/j.foodchem.2006.03.020.

Holland, C., Ryden, P., Edwards, C. H. and Grundy, M. M.-L. (2020). Plant cell walls: Impact on nutrient bioaccessibility and digestibility. *Foods* 9(2), 201. doi:10.3390/foods9020201.

Holscher, H. D. (2017). Dietary fiber and prebiotics and the gastrointestinal microbiota. *Gut Microbes* 8(2), 172–184. doi:10.1080/19490976.2017.1290756.

Hong, K. W., Ko, K. P., Ahn, Y., Kim, C. S., Park, S. J., Park, J. K., Kim, S. S. and Kim, Y. (2012). Epidemiological profiles between equol producers and nonproducers: A genomewide association study of the equol-producing phenotype. *Genes and Nutrition* 7(4), 567–574. doi:10.1007/s12263-012-0292-8.

Huttenhower, C., Gevers, D., Knight, R., Abubucker, S., Badger, J. H., Chinwalla, T., Creasy, H., Earl, A. M., FitzGerald, M. G., Fulton, R. S., Giglio, M. G. Hallsworth-Pepin, K., Lobos, E. A., Madupu, R., Magrini, V, Martin, J. C., Mitreva, M., Muzny, D. M., Sodergen, E. J., ... and White, O. The Human Microbiome Project Consortium. (2012). Structure, function and diversity of the healthy human microbiome. *Nature* 486(7402), 207–214. doi:10.1038/nature11234.

Ideno, Y., Hayashi, K., Nakajima-Shimada, J., Onizuka, Y., Kishi, M., Ueno, T. and Uchiyama, S. (2018). Optimal cut-off value for equol-producing status in women: The Japan Nurses' Health Study urinary isoflavone concentration survey. *PLoS ONE* 13(7), e0201318. doi:10.1371/journal.pone.0201318.

Institute of Medicine (2006). *Dietary Reference Intakes: The Essential Guide to Nutrient Requirements*. Washington, DC: The National Academies Press.

Isah, T. (2019). Stress and defense responses in plant secondary metabolites production. *Biological Research* 52(1), 39. doi:10.1186/s40659-019-0246-3.

Jalal, F., Nesheim, M. C., Agus, Z., Sanjur, D. and Habicht, J. P. (1998). Serum retinol concentrations in children are affected by food sources of beta-carotene, fat intake, and anthelmintic drug treatment. *American Journal of Clinical Nutrition* 68(3), 623–629. doi:10.1093/ajcn/68.3.623.

Karlsson, F. H., Fåk, F., Nookaew, I., Tremaroli, V., Fagerberg, B., Petranovic, D., Bäckhed, F. and Nielsen, J. (2012). Symptomatic atherosclerosis is associated with an altered gut metagenome. *Nature Communications* 3(1), 1245. doi:10.1038/ncomms2266.

Khairallah, J., Sadeghi Ekbatan, S., Sabally, K., Iskandar, M. M., Hussain, R., Nassar, A., Sleno, L., Rodes, L., Prakash, S., Donnelly, D. J. and Kubow, S. (2018). Microbial biotransformation of a polyphenol-rich potato extract affects antioxidant capacity in a simulated gastrointestinal model. *Antioxidants (Basel)* 7(3), 43. doi:10.3390/antiox7030043.

Khoo, H. E., Prasad, K. N., Kong, K. W., Jiang, Y. and Ismail, A. (2011). Carotenoids and their isomers: Color pigments in fruits and vegetables. *Molecules (Basel, Switzerland)* 16(2), 1710–1738. doi:10.3390/molecules16021710.

Koleva, I. I., van Beek, T. A., Soffers, A. E. M. F., Dusemund, B. and Rietjens, I. M. C. M. (2012). Alkaloids in the human food chain – Natural occurrence and possible adverse effects. *Molecular Nutrition and Food Research* 56(1), 30–52. doi:10.1002/mnfr.201100165.

Kroll, J., Rawel, H. M. and Rohn, S. (2003). Reactions of plant phenolics with food proteins and enzymes under special consideration of covalent bonds. *Food Science and Technology Research* 9(3), 205–218. doi:10.3136/fstr.9.205.

Kudryavtseva, A., Krasnov, G., Lipatova, A., Alekseev, B., Maganova, F., Shaposhnikov, M., Fedorova, M., Snezhkina, A. and Moskalev, A. (2016). Effects of *Abies sibirica* terpenes on cancer- and aging-associated pathways in human cells. *Oncotarget* 7(50), 83744–83754. doi:10.18632/oncotarget.13467.

Kwon, J. H., Keates, A. C., Anton, P. M., Botero, M., Goldsmith, J. D. and Kelly, C. P. (2005). Topical antisense oligonucleotide therapy against LIX, an enterocyte- expressed CXC chemokine, reduces murine colitis. *American Journal of Physiology. Gastrointestinal and Liver Physiology* 289(6), G1075–G1083. doi:10.1152/ajpgi.00073.2005.

Landberg, R., Manach, C., Kerckhof, F. M., Minihane, A. M., Saleh, R. N. M., De Roos, B., Tomas-Barberan, F., Morand, C. and Van de Wiele, T. (2019). Future prospects for dissecting inter-individual variability in the absorption, distribution and elimination of plant bioactives of relevance for cardiometabolic endpoints. *European Journal of Nutrition* 58 (Suppl. 2), 21–36. doi:10.1007/s00394-019-02095-1.

Laville, E., Perrier, J., Bejar, N., Maresca, M., Esque, J., Tauzin, A. S., Bouhajja, E., Leclerc, M., Drula, E., Henrissat, B., Berdah, S., Di Pasquale, E., Robe, P. and Potocki-Veronese, G. (2019). Investigating host microbiota relationships through functional metagenomics. *Frontiers in Microbiology* 10(1286), 1286. doi:10.3389/fmicb.2019.01286.

Lee, B., Moon, K. M. and Kim, C. Y. (2018). Tight junction in the intestinal epithelium: Its association with diseases and regulation by phytochemicals. *Journal of Immunology Research* 2018, 2645465. doi:10.1155/2018/2645465.

Leena, M. M., Silvia, M. G., Vinitha, K., Moses, J. A. and Anandharamakrishnan, C. (2020). Synergistic potential of nutraceuticals: Mechanisms and prospects for futuristic medicine. *Food and Function* 11(11), 9317–9337. doi:10.1039/d0fo02041a.

Li, A. N., Li, S., Zhang, Y. J., Xu, X. R., Chen, Y. M. and Li, H. B. (2014). Resources and biological activities of natural polyphenols. *Nutrients* 6(12), 6020–6047. doi:10.3390/nu6126020.

Li, Z., Jiang, H., Xu, C. and Gu, L. (2015). A review: Using nanoparticles to enhance absorption and bioavailability of phenolic phytochemicals. *Food Hydrocolloids* 43, 153–164. doi:10.1016/j.foodhyd.2014.05.010.

Liu, B., Qin, L., Liu, A., Uchiyama, S., Ueno, T., Li, X. and Wang, P. (2010). Prevalence of the equol-producer phenotype and its relationship with dietary isoflavone and serum lipids in healthy Chinese adults. *Journal of Epidemiology* 20(5), 377–384. doi:10.2188/jea.JE20090185.

Louis, P., Hold, G. L. and Flint, H. J. (2014). The gut microbiota, bacterial metabolites and colorectal cancer. *Nature Reviews. Microbiology* 12(10), 661–672. doi:10.1038/nrmicro3344.

Lyu, Y., Wu, L., Wang, F., Shen, X. and Lin, D. (2018). Carotenoid supplementation and retinoic acid in immunoglobulin A regulation of the gut microbiota dysbiosis. *Experimental Biology and Medicine* 243(7), 613– 620. doi:10.1177/1535370218763760.

Mahizan, N. A., Yang, S. K., Moo, C. L., Song, A. A.-L., Chong, C. M., Chong, C. W., Abushelaibi, A., Lim, S.-H. E. and Lai, K. S. (2019). Terpene derivatives as a potential agent against antimicrobial resistance (AMR) pathogens. *Molecules* 24(14), 2631. doi:10.3390/molecules24142631.

Mapelli-Brahm, P., Corte-Real, J., Meléndez-Martínez, A. J. and Bohn, T. (2017). Bioaccessibility of phytoene and phytofluene is superior to other carotenoids from selected fruit and vegetable juices. *Food Chemistry* 229, 304–311. doi:10.1016/j.foodchem.2017.02.074.

Marín, L., Miguélez, E. M., Villar, C. J. and Lombó, F. (2015). Bioavailability of dietary polyphenols and gut microbiota metabolism: Antimicrobial properties. *BioMed Research International* 2015, 905215. doi:10.1155/2015/905215.

Mayo, B., Vázquez, L. and Flórez, A. B. (2019). Equol: A bacterial metabolite from the daidzein isoflavone and its presumed beneficial health effects. *Nutrients* 11(9), 2231. doi:10.3390/nu11092231.

Miękus, N., Marszałek, K., Podlacha, M., Iqbal, A., Puchalski, C. and Świergiel, A. H. (2020). Health benefits of plant-derived sulfur compounds, glucosinolates, and organosulfur compounds. *Molecules* 25(17), 3804. doi:10.3390/molecules25173804.

Mirza, Z., Soto, E. R., Hu, Y., Nguyen, T. T., Koch, D., Aroian, R. V. and Ostroff, G. R. (2020). Antihelmintic activity of yeast particle-encapsulated terpenes. *Molecules* 25(13), 2958. doi:10.3390/molecules25132958.

Molly, K., Vande Woestyne, M. and Verstraete, W. (1993). Development of a 5- step multi-chamber reactor as a simulation of the human intestinal microbial ecosystem. *Applied Microbiology and Biotechnology* 39(2), 254–258. doi:10.1007/BF00228615.

Moran, N. E., Mohn, E. S., Hason, N., Erdman, J. W. and Johnson, E. J. (2018). Intrinsic and extrinsic factors impacting absorption, metabolism, and health effects of dietary carotenoids. *Advances in Nutrition* 9(4), 465–492. doi:10.1093/advances/nmy025.

Morand, C., De Roos, B., Garcia-Conesa, M. T., Gibney, E. R., Landberg, R., Manach, C., Milenkovic, D., Rodriguez-Mateos, A., Van de Wiele, T. and Tomas- Barberan, F. (2020). Why interindividual variation in response to consumption of plant food bioactives matters for future personalised nutrition. *Proceedings of the Nutrition Society* 79(2), 225–235. doi:10.1017/S0029665120000014.

Moses, T., Papadopoulou, K. K. and Osbourn, A. (2014). Metabolic and functional diversity of saponins, biosynthetic intermediates and semi-synthetic derivatives. *Critical Reviews in Biochemistry and Molecular Biology* 49(6), 439–462. doi:10.3109/10409238.2014.953628.

Naczk, M. and Shahidi, F. (2006). Phenolics in cereals, fruits and vegetables: Occurrence, extraction and analysis. *Journal of Pharmaceutical and Biomedical Analysis* 41(5), 1523–1542. doi:10.1016/j.jpba.2006.04.002.

Nguyen, T. L. A., Vieira-Silva, S., Liston, A. and Raes, J. (2015). How informative is the mouse for human gut microbiota research? *Disease Models and Mechanisms* 8(1), 1–16. doi:10.1242/dmm.017400.

Nicholson, J. K., Holmes, E., Kinross, J., Burcelin, R., Gibson, G., Jia, W. and Pettersson, S. (2012). Host-gut microbiota metabolic interactions. *Science* 336(6086), 1262–1267. doi:10.1126/science.1223813.

Ochmian, I., Błaszak, M., Lachowicz, S. and Piwowarczyk, R. (2020). The impact of cultivation systems on the nutritional and phytochemical content, and microbiological contamination of highbush blueberry. *Scientific Reports* 10(1), 16696. doi:10.1038/s41598-020-73947-8.

Oliphant, K. and Allen-Vercoe, E. (2019). Macronutrient metabolism by the human gut microbiome: Major fermentation by-products and their impact on host health. *Microbiome* 7(1), 91. doi:10.1186/s40168-019-0704-8.

Ortega, N., Macià, A., Romero, M.-P., Reguant, J. and Motilva, M.-J. (2011). Matrix composition effect on the digestibility of carob flour phenols by an in-vitro digestion model. *Food Chemistry* 124(1), 65–71. doi:10.1016/j.foodchem.2010.05.105.

Peng, J., Zheng, T. T., Li, X., Liang, Y., Wang, L. J., Huang, Y. C. and Xiao, H. T. (2019). Plant-derived alkaloids: The promising disease-modifying agents for inflammatory bowel disease. *Frontiers in Pharmacology* 10(351), 351. doi:10.3389/fphar.2019.00351.

Pérez-Jiménez, J., Neveu, V., Vos, F. and Scalbert, A. (2010). Identification of the 100 richest dietary sources of polyphenols: An application of the phenol-Explorer database. *European Journal of Clinical Nutrition* 64(3), S112–S120. doi:10.1038/ejcn.2010.221.

Proshkina, E., Plyusnin, S., Babak, T., Lashmanova, E., Maganova, F., Koval, L., Platonova, E., Shaposhnikov, M. and Moskalev, A. (2020). Terpenoids as potential geroprotectors. *Antioxidants (Basel)* 9(6), 529. doi:10.3390/antiox9060529.

Rabi, T. and Gupta, S. (2014). Dietary terpenoids and prostate cancer chemoprevention. *Frontiers in Bioscience: A Journal and Virtual Library* 13, 3457–3469.

Reboul, E. (2019). Mechanisms of carotenoid intestinal absorption: Where do we stand? *Nutrients* 11(4): 838. doi:10.3390/nu11040838.

Rechner, A. R., Spencer, J. P. E., Kuhnle, G., Hahn, U. and Rice-Evans, C. A. (2001). Novel biomarkers of the metabolism of caffeic acid derivatives in vivo. *Free Radical Biology and Medicine* 30(11), 1213–1222. doi:10.1016/S0891-5849(01)00506-8.

Rochfort, S. and Panozzo, J. (2007). Phytochemicals for health, the role of pulses. *Journal of Agricultural and Food Chemistry* 55(20), 7981–7994. doi:10.1021/jf071704w.

Rodriguez-Casado, A. (2016). The health potential of fruits and vegetables phytochemicals: Notable examples. *Critical Reviews in Food Science and Nutrition* 56(7), 1097–1107. doi:10.1080/10408398.2012.755149.

Rondini, L., Peyrat-Maillard, M. N., Marsset-Baglieri, A., Fromentin, G., Durand, P., Tome, D., Prost, M. and Berset, C. (2004). Bound ferulic acid from bran is more bioavailable than the free compound in rat. *Journal of Agricultural and Food Chemistry* 52(13), 4338–4343. doi:10.1021/jf0348323.

Rowland, I., Gibson, G., Heinken, A., Scott, K., Swann, J., Thiele, I. and Tuohy, K. (2018). Gut microbiota functions: Metabolism of nutrients and other food components. *European Journal of Nutrition* 57(1), 1–24. doi:10.1007/s00394-017-1445-8.

Ruhee, R. T., Roberts, L. A., Ma, S. and Suzuki, K. (2020). Organosulfur compounds: A review of their anti-inflammatory effects in human health. *Frontiers in Nutrition* 7, 64. doi:10.3389/fnut.2020.00064.

Sadeghi Ekbatan, S., Iskandar, M. M., Sleno, L., Sabally, K., Khairallah, J., Prakash, S. and Kubow, S. (2016). Biotransformation of polyphenols in a dynamic multistage gastrointestinal model. *Foods* 7(1), 8. doi:10.3390/foods7010008.

Scalbert, A. and Williamson, G. (2000). Dietary intake and bioavailability of polyphenols. *The Journal of Nutrition* 130(8), 2073S–2085S. doi:10.1093/jn/130.8.2073S.

Selby-Pham, S. N. B., Miller, R. B., Howell, K., Dunshea, F. and Bennett, L. E. (2017). Physicochemical properties of dietary phytochemicals can predict their passive absorption in the human small intestine. *Scientific Reports* 7(1), 1931. doi:10.1038/s41598-017-01888-w.

Septembre-Malaterre, A., Remize, F. and Poucheret, P. (2018) Fruits and vegetables, as a source of nutritional compounds and phytochemicals: Changes in bioactive compounds during lactic fermentation. *Food Research International* 104, 86–99. doi:10.1016/j.foodres.2017.09.031.

Setchell, K. D. R. and Clerici, C. (2010). Equol: Pharmacokinetics and biological actions. *The Journal of Nutrition* 140(7), 1363S–1368S. doi:10.3945/jn.109.119784.

Shi, J., Arunasalam, K., Yeung, D., Kakuda, Y., Mittal, G. and Jiang, Y. (2004). Saponins from edible legumes: Chemistry, processing, and health benefits. *Journal of Medicinal Food* 7(1), 67–78. doi:10.1089/109662004322984734.

Shoba, G., Joy, D., Joseph, T., Majeed, M., Rajendran, R. and Srinivas, P. S. (1998). Influence of piperine on the pharmacokinetics of curcumin in animals and human volunteers. *Planta Medica* 64(4), 353–356:/doi:10.1055/s-2006-957450.

Sieniawska, E., Swatko-Ossor, M., Sawicki, R., Skalicka-Woźniak, K. and Ginalska, G. (2017). Natural terpenes influence the activity of antibiotics against isolated *Mycobacterium tuberculosis. Medical Principles and Practice* 26(2), 108–112. doi:10.1159/000454680.

Sommer, F., Anderson, J. M., Bharti, R., Raes, J. and Rosenstiel, P. (2017). The resilience of the intestinal microbiota influences health and disease. *Nature Reviews. Microbiology* 15(10), 630–638. doi:10.1038/nrmicro.2017.58.

Song, K. and Milner, J. A. (1999). Heating garlic inhibits its ability to suppress 7, 12-dimethylbenz(a)anthracene-induced DNA adduct formation in rat mammary tissue. *Journal of Nutrition* 129(3), 657–661. doi:10.1093/jn/129.3.657.

Steinbrecher, A. and Linseisen, J. (2009). Dietary intake of individual glucosinolates in participants of the EPIC-Heidelberg Cohort Study. *Annals of Nutrition and Metabolism* 54(2), 87–96. doi:10.1159/000209266.

Teekachunhatean, S., Tosri, N., Rojanasthien, N., Srichairatanakool, S. and Sangdee, C. (2013). Pharmacokinetics of caffeine following a single administration of coffee enema versus oral coffee consumption in healthy male subjects. *ISRN Pharmacology* 2013, 147238. doi:10.1155/2013/147238.

Tewari, S., Gupta, V. and Bhattacharya, S. (2000). Comparative study of antioxidant potential of tea with and without additives. *Indian Journal of Physiology and Pharmacology* 44(2), 215–219.

Tsao, R. (2010). Chemistry and biochemistry of dietary polyphenols. *Nutrients* 2(12), 1231–1246. doi:10.3390/nu2121231.

Ukers, W. (2020). *All About Coffee*. Oxfordshire: Benediction Classics.

USDA, N. R. C. S. (2021). *The PLANTS Database* (http://plants.usda.gov, 28 April 2021). Greensboro, NC: National Plant Data Team, pp. 27401–24901.

Vázquez, L., Guadamuro, L., Giganto, F., Mayo, B. and Flórez, A. B. (2017). Development and use of a real-time quantitative PCR method for detecting and quantifying equol-producing bacteria in human faecal samples and slurry cultures. *Frontiers in Microbiology* 8, 1155. doi:10.3389/fmicb.2017.01155.

Vitaglione, P., Napolitano, A. and Fogliano, V. (2008). Cereal dietary fiber: A natural functional ingredient to deliver phenolic compounds into the gut. *Trends in Food Science and Technology* 19(9), 451–463. doi:10.1016/j.tifs.2008.02.005.

Vuong, Q. V. (2021). *Utilisation of Bioactive Compounds from Agricultural and Food Waste*. Boca Raton: CRC Press.

Wagner, R., Hempstead, W. H., Villegas, B. and Rothkirch, C. V. (2001). *The History of Coffee in Guatemala*. Bogotá, D.C.: Villegas Editores.

Walle, T., Browning, A. M., Steed, L. L., Reed, S. G. and Walle, U. K. (2005). Flavonoid glucosides are hydrolyzed and thus activated in the oral cavity in humans. *Journal of Nutrition* 135(1), 48–52. doi:10.1093/jn/135.1.48.

Wang, X. L., Kim, H. J., Kang, S. I., Kim, S. I. and Hur, H. G. (2007). Production of phytoestrogen S-equol from daidzein in mixed culture of two anaerobic bacteria. *Archives of Microbiology* 187(2), 155–160. doi:10.1007/s00203-006-0183-8.

Weiser, T. and Weigmann, H. (2019). Effect of caffeine on the bioavailability and pharmacokinetics of an acetylsalicylic acid-paracetamol combination: Results of a phase I study. *Advances in Therapy* 36(3), 597–607. doi:10.1007/s12325-019-0891-5.

Willson, C. (2018). The clinical toxicology of caffeine: A review and case study. *Toxicology Reports* 5, 1140–1152. doi:10.1016/j.toxrep.2018.11.002.

Wojtunik-Kulesza, K., Oniszczuk, A., Oniszczuk, T., Combrzyński, M., Nowakowska, D. and Matwijczuk, A. (2020). Influence of in vitro digestion on composition, bioaccessibility and antioxidant activity of food polyphenols - A non- systematic review. *Nutrients* 12(5), 1401. doi:10.3390/nu12051401.

Wu, G. D., Chen, J., Hoffmann, C., Bittinger, K., Chen, Y. Y., Keilbaugh, S. A., Bewtra, M., Knights, D., Walters, W. A., Knight, R., Sinha, R., Gilroy, E., Gupta, K., Baldassano, R., Nessel, L., Li, H., Bushman, F. D. and Lewis, J. D. (2011) Linking long-term dietary patterns with gut microbial enterotypes. *Science (New York, N.Y.)* 334(6052), 105-108. doi:10.1126/science.1208344.

Xavier-Santos, D., Bedani, R., Lima, E. D. and Saad, S. M. I. (2020). Impact of probiotics and prebiotics targeting metabolic syndrome. *Journal of Functional Foods* 64. doi:10.1016/j.jff.2019.103666, 103666.

Yahfoufi, N., Alsadi, N., Jambi, M. and Matar, C. (2018). The immunomodulatory and anti-inflammatory role of polyphenols. *Nutrients* 10(11), 1618. doi:10.3390/nu10111618.

Yin, R., Kuo, H. C., Hudlikar, R., Sargsyan, D., Li, S., Wang, L., Wu, R. and Kong, A. N. (2019). Gut microbiota, dietary phytochemicals, and benefits to human health. *Current Pharmacology Reports* 5(5), 332–344. doi:10.1007/s40495-019-00196-3.

Yoovathaworn, K. C., Sriwatanakul, K. and Thithapandha, A. (1986). Influence of caffeine on aspirin pharmacokinetics. *European Journal of Drug Metabolism and Pharmacokinetics* 11(1), 71–76. doi:10.1007/BF03189777.

Yoshikata, R., Myint, K. Z., Ohta, H. and Ishigaki, Y. (2019). Inter-relationship between diet, lifestyle habits, gut microflora, and the equol-producer phenotype: Baseline findings from a placebo-controlled intervention trial. *Menopause* 26(3), 273–285. doi:10.1097/GME.0000000000001202.

Zhang, M., Zhang, H., Li, H., Lai, F., Li, X., Tang, Y., Min, T. and Wu, H. (2016). Antioxidant mechanism of betaine without free radical scavenging ability. *Journal of Agricultural and Food Chemistry* 64(42), 7921–7930. doi:10.1021/acs.jafc.6b03592.

Zhang, Y., Gu, Y., Ren, H., Wang, S., Zhong, H., Zhao, X., Ma, J., Gu, X., Xue, Y., Huang, S., Yang, J., Chen, L., Chen, G., Qu, S., Liang, J., Qin, L., Huang, Q., Peng, Y., Li, Q., Wang, X., Kong, P., Hou, G., Gao, M., Shi, Z., Li, X., Qiu, Y., Zou, Y., Yang, H., Wang, J., Xu, G., Lai, S., Li, J., Ning, G. and Wang, W. (2020). Gut microbiome-related effects of berberine and probiotics on type 2 diabetes (the PREMOTE study). *Nature Communications* 11(1), 5015. doi:10.1038/s41467-020-18414-8.

Zheng, W., Ma, Y., Zhao, A., He, T., Lyu, N., Pan, Z., Mao, G., Liu, Y., Li, J., Wang, P., Wang, J., Zhu, B. and Zhang, Y. (2019). Compositional and functional differences in human gut microbiome with respect to equol production and its association with blood lipid level: A cross-sectional study. *Gut Pathogens* 11, 20. doi:10.1186/s13099-019-0297-6.

Part 5

Case studies

Chapter 13

Advances in understanding and improving the nutraceutical properties of cranberries

Oliver Chen, Biofortis Research, Merieux NutriSciences and Tufts University, USA; and Eunice Mah, Biofortis Research, Merieux NutriSciences, USA

1 Introduction

American cranberry (*Vaccinium macrocarpon*) is commonly consumed as juice, dried fruit, or dietary supplement. It is mainly cultivated in the United States and Canada (Neto and Vinson, 2011). Native Americans used them as food, dye for rugs and blankets, and healing agent for wounds. On average, Americans eat 2.8 cranberries per day, equivalent to 0.25 mL of juice, making them an under-consumed fruit in the United States (Neto and Vinson, 2011). Cranberries are known for their tart taste and are generally consumed after being sweetened, for example, in the form of cranberry juice cocktails containing a variety of sweet juices, such as apples and grapes. Cranberry juice is perhaps best known for the prevention and treatment of urinary tract infections (UTIs). This health benefit has received a US FDA qualified health claim in 2020 (FDA, 2020). Nevertheless, the health benefits of cranberry products extend beyond the protection from infection in the urogenital tract to bacteria-related health conditions, such as dental and gastrointestinal health, and others, such as cardiometabolic and skin health, cognition, and cancer.

In this chapter, we first review clinical evidence supporting the health benefits of consumption of cranberry products, followed by a discussion on future directions in the elucidation of the mechanism of actions for health

http://dx.doi.org/10.19103/AS.2022.0101.16

benefits, as well as approaches to maximizing bioefficacy of cranberry-related foods/products. In this chapter, all discussed information is pertinent only to the American cranberry and related products but not European cranberry (*V. oxycoccos*).

2 Nutrient composition

Nutrition information on raw cranberries is presented in Table 1. Raw cranberries (1 cup, 100 g) have 46 kcal (mainly from carbohydrates), 87.3 g of water, and 3.6 g of fiber (Central, 2020). Cranberries are also a rich source of phytochemicals, including flavonoids (particularly flavan-3-ols), A-type proanthocyanidins (PACs), anthocyanins (cyanidin-3-arabinoside, cyanidin-3-galactoside, and cyanidin-3-glucoside), benzoic acid (quinic, citric, ellagic, and malic acids), ursolic acid, and triterpenoids. Due to the distinctive tart taste and astringency of phenolic compounds ascribed to their capability of binding saliva proteins, sweeteners are generally added to cranberry products, such as juices and dry fruit, to enhance palatability. The contents of anthocyanins, flavonols, and phenolic acids in raw cranberries can be found in the Phenol-Explorer (http://phenol-explorer.eu/) (Phenol-Explorer) and the USDA Database for the Flavonoid Content of Selected Foods (https://data.nal.usda.gov/dataset/usda-database-flavonoid-content-selected-foods-release-32-november-2015) (Bhagwat and Haytowitz, 2016). As reported in the Phenol-Explorer, in each subfamily, peonidin 3-O-galactoside, quercetin 3-O-galactoside, and benzoic acid are the most abundant with a mean content of 22.02, 10.81, and 48.10 mg/100 fresh weight, respectively. Total catechins in raw cranberry average 17 mg/100 g, with epicatechin being the most abundant (Harnly et al., 2006). PACs are the products of the polymerization of flavan-3-ols. A-type PACs (C2→O→C7 linkage between epicatechin units) are present in cranberries instead of the B-type PACs commonly found in most plant foods (Gu et al., 2003). As reported in the USDA Database for the Proanthocyanidin Content of Selected Foods (https://data.nal.usda.gov/dataset/usda-database-proanthocyanidin-content-selected-foods-release-2-2015) (Bhagwat and Haytowitz, 2015), the most abundant PACs in raw cranberries are polymers (>10 flavan-3-ols unit) at 217.6 mg per 100 g, followed by 4–6 mers at 56.84 mg/100 g. The average concentration of PACs is 231 mg/L in cranberry juice (Gu et al., 2004). PACs with a degree of polymerization larger than four are not absorbable in the small intestine because their large molecular size prohibits them from passing the gut barrier (Ou et al., 2012).

The variability in bioactive compounds among cranberry products partially explains the non-reproducibility of clinical evidence between studies on cranberries and UTI and other health indications. Among constituents in cranberries, A-type PACs have been recognized for UTI benefits and thus

Table 1 Nutrient composition in raw cranberries (100 g)[a]

Kcal	46
Water (g)	87.3
Carbohydrate (g)	12
Fat (g)	0.13
Protein (g)	0.46
Sugar (g)	4.27
Fiber (g)	3.6
Vitamin C (mg)	14
alpha-Tocopherol (mg)	1.32
beta-Carotene (μg)	38
Lutein + zeaxanthin (μg)	91
Peonidin 3-O-galactoside (mg)	22.02
Peonidin-3-O-arabinoside (mg)	9.61
Peonidin 3-O-glucoside (mg)	4.16
Cyanidin 3-O-galactoside (mg)	8.89
Cyanidin 3-O-arabinoside (mg)	4.47
Cyanidin 3-O-glucoside (mg)	0.74
Quercetin 3-O-galactoside (mg)	10.81
Quercetin 3-O-arabinoside (mg)	4.94
Quercetin 3-O-rhamnoside (mg)	6.17
Myricetin 3-O-arabinoside (mg)	5.30
Kaemperfol 3-O-glucoside (mg)	0.87
Benzoic acid (mg)	48.1
2,4-Didydroxybenzoic acid (mg)	0.8
3-Hydroxybenzoic acid (mg)	0.41
4-Hydroxybenzoic acid (mg)	0.42
Vanillic acid (mg)	0.69
Caffeic acid (mg)	0.38
Cinnamic acid (mg)	0.16
Ferulic acid (mg)	0.81
PACs (mg)	217.6
Phloridzin (mg)	12
Ursolic acid (mg)	60–100

[a] Values are obtained from https://fdc.nal.usda.gov/fdc-app.html#/food-details/171722/nutrients, http://phenol-explorer.eu/contents/food/74, Gu et al. (2004), Turner et al. (2007), Neto and Vinson (2011).

their contents have been employed for the standardization of cranberry products. PACs are generally measured using HPLC and 4-(dimethylamino) cinnamaldehyde (DMAC) colorimetric assays (Birmingham et al., 2021,

Gao et al., 2018). While HPLC can quantify individual PACs with different degrees of polymerization, DMAC colorimetric assays measure the overall PAC content and are more commonly used to standardize cranberry products, particularly nutritional supplements. Standardizing cranberry products with a particular cranberry constituent such as PACs is crucial because the quality of cranberry products can be tainted by adulterations with other ingredients, such as anthocyanin-rich ingredients, with a similar color.

Cranberries also contain phloridzin, a dihydrochalcone (12 mg/100 g), and ursolic acid, a triterpenoid (60–100 mg/100 g), and insignificant amounts of lutein (Turner et al., 2007, Neto and Vinson, 2011). Soluble xyloglucan and pectic oligosaccharides, which are non-digestible carbohydrates, are the most recently recognized bioactives in cranberries, particularly in cranberry pomace, a by-waste of juice production (Coleman and Ferreira, 2020). These oligosaccharides are present at ~20% *w/w* in many cranberry materials, especially dehydrated cranberry pomace powder, and display anti-adhesion activity against bacteria (Sun et al., 2019).

3 Health benefits

3.1 Urinary tract infections

Urinary tract infections (UTIs) affect 150 million people each year worldwide. In the United States alone, approximately 10.5 million office visits (Schappert and Rechtsteiner, 2011) and 1.5 million emergency department visits (Niska et al., 2010) occur every year due to UTI symptoms. Clinically, UTIs are categorized as uncomplicated or complicated, with the latter defined as those associated with factors that compromise the urinary tract or host defense, including urinary obstruction, urinary retention caused by neurological disease, immunosuppression, renal failure, renal transplantation, and the presence of foreign bodies such as calculi, indwelling catheters, or other drainage devices, as well as UTIs that occur in women during pregnancy and men (Gupta et al., 2011, Nicolle and AMMI Canada Guidelines Committee, 2005, Dason et al., 2011). Uncomplicated UTIs typically affect individuals who are otherwise healthy women without structural or neurological urinary tract abnormalities. The self-reported annual incidence of uncomplicated UTI in women is 12%, and by the age of 32 years, half of all women report having had at least one UTI (Foxman and Brown, 2003). Additionally, the rate of recurrence in the 12 months following an initial uncomplicated UTI in women is high, ranging from 25% to 44% (Mabeck, 1972, Foxman, 1990, Ikäheimo et al., 1996). The high incidence and recurrence rate of uncomplicated UTI among women, along with the rapid rise of multi-drug resistant uropathogens, provide support for the timely assessment of cranberries for decreased risk of uncomplicated recurrent UTI among women.

In 2020, the US FDA approved a qualified health claim that consuming one serving (8 oz) each day of a cranberry juice beverage or 500 mg each day of cranberry dietary supplement may help reduce the risk of recurrent UTI in healthy women (FDA, 2020). In approving the qualified health claim, the FDA relied on the evidence provided by seven publications reporting on eight intervention studies (five intervention studies on cranberry juice beverages, i.e. Barbosa-Cesnik et al., 2011, Maki et al., 2016, Stapleton et al., 2012, Stothers, 2002, Takahashi et al., 2013; and three intervention studies on cranberry dietary supplement, i.e. Stothers, 2002, Vostalova et al., 2015, Walker et al., 1997). Of the five studies on cranberry juice, two (Maki et al., 2016, Stothers, 2002) demonstrated a statistically significant benefit and one demonstrated mixed results whereby no statistically significant beneficial effect was observed when subjects with uncomplicated UTI (n = 170) consumed a daily dose of 125 mL of a cranberry juice beverage (65% of cranberry juice), but a statistically significant beneficial effect was observed in a sub-analysis of women aged 50 years and older (n = 118). The remaining two studies (Barbosa-Cesnik et al., 2011, Stapleton et al., 2012) showed no effect of consuming a cranberry juice beverage on risk reduction of recurrent UTI. With regard to cranberry supplementation, the three identified studies (Stothers, 2002, Vostalova et al., 2015, Walker et al., 1997) all demonstrated a statistically significant benefit of a cranberry dietary supplement consumption on the risk of recurrent UTI in healthy women with a history of UTI. Because of the limited number of randomized controlled studies (i.e. three on cranberry supplementation and five on cranberry juice with high and moderate methodological quality but with inconsistent results), FDA has concluded that the current evidence provides only qualified support for the scientific validity of the relationship between cranberry juice/supplements and recurrent UTIs.

The specificity of the qualified health in recurrent UTIs is consistent with evidence suggesting a difference in the efficacy of cranberries against uncomplicated vs. complicated UTIs. According to a meta-analysis by Wang et al. (2012), cranberry-containing products were effective in women with recurrent UTIs, but not in populations with potentially complicated UTIs such as those with neuropathic bladder, elderly patients, and pregnant patients. A Cochrane meta-analysis published in the same year as the Wang et al. (2012) meta-analysis concluded that cranberries are not effective against UTIs (Jepson et al., 2012). A review by Liska et al. (2016) suggested that one of the reasons for the difference between the Cochrane and Wang et al. (2012) meta-analysis conclusions is that the overall conclusion made by Jepson et al. (2012) was more heavily influenced by results from studies in populations with complicated UTIs whereas the Wang et al. (2012) meta-analysis weighted the evidence relatively equally across the populations. A more recent meta-analysis by Fu et al. (2017) focused only on uncomplicated UTIs (recurrent

UTIs in healthy women) and concluded that cranberry reduced the risk of UTIs. Additionally, another recent meta-analysis (Raguzzini et al., 2020) showed that cranberries are ineffective against UTIs in patients with spinal cord injury when compared with control. Although the mechanism by which cranberries affect UTI is still under investigation, the differential effects of cranberry products on uncomplicated and complicated UTI may lie in their pathogenesis and the anti-adhesion activities of cranberry constituents. For example, uropathogenic *E. coli* is the most common causative pathogen of uncomplicated UTI, accounting for ~80% of infections but only ~21% for complicated UTI (Stamm and Hooton, 1993, Hidron et al., 2008).

3.2 H. pylori-*induced gastric infection*

Helicobacter pylori (*H. pylori*), a gram-negative bacterium pathogen, was estimated to affect more than half of the world's population and is associated with chronic gastritis, peptic ulcer disease, mucosa-associated lymphoid tissue (MALT) lymphoma, and gastric cancer (Guevara and Cogdill, 2020). *H. pylori* infection begins with the adhesion of the bacteria to the gastric epithelium, followed by inflammation and ulcer (Wang et al., 2014). Standard regimens of *H. pylori* eradication include an array of antibiotic treatments, but their substantially reduced effectiveness due to antibiotic resistance has raised a need for alternative treatments (Savoldi et al., 2018). Since the adhesion of *H. pylori* to the gastric epithelium is the initial step of the infection, cranberry constituents are anticipated to be capable of diminishing *H. pylori* infection via their anti-bacterial adhesion activity. The potential benefit of cranberries in *H. pylori* infection was first reported in a prospective, randomized, double-blind, placebo-controlled trial with 189 adults with *H. pylori* infection in Shandong, China (Zhang et al., 2005). This trial showed that daily consumption of 500 mL cranberry juice for 90 days resulted in *H. pylori* eradication, assessed using a 13C-urea breath test, in 14% of participants as compared to 5% consuming placebo ($P<0.05$). While this result is significant, it appears the positive response rate is low. Additionally, such effect was only observed in women, but not in men. The comparable eradication efficacy of cranberry was also reported in children (Gotteland et al., 2008).

Because there is no need for absorption and transportation of specific active constituents to the target tissue, theoretically, any cranberry constituents can play a role in the eradication of *H. pylori*. A recent PAC dose-response trial with 522 *H. pylori*-positive adults in Shandong, China, was designed to examine whether PACs would be the main contributor (Li et al., 2020). The results of this double blind, randomized, placebo-controlled trial showed that consumption of cranberry juice providing 88 mg for 8 weeks led to a 20% *H. pylori*-negative rate as compared to 7.6, 4.5, and 7.3% following 44, 23, and 0 mg/d (placebo)

PAC, respectively. Since other non-PAC polyphenols were not controlled, whether PACs are the primary effector remains obscure. Interestingly, the null effect was observed when cranberry polyphenols were consumed in an encapsulated powder form (Li et al., 2020), suggesting the effectiveness of cranberry constituents against *H. pylori* depends on formulation forms.

As antibiotic treatments are standard for *H. pylori* infection, cranberry constituents may work with antibiotics in a synergistic/additive manner. This concept was suggested in a meta-analysis showing that the addition of antioxidants (vitamin C or E, N-acetylcysteine, curcumin, cranberry) to amoxicillin-clarithromycin-based therapy enhanced the eradication rate from 68.6% to 81.3% (Yang-Ou et al., 2018). Shmuely et al. (Shmuely et al., 2007) also noted that a combination of triple therapy (omeprazole, amoxicillin, and clarithromycin) with daily consumption of 500 mL cranberry juice significantly increased the *H. pylori* eradication rate in Israeli female patients from 86% with triple therapy alone to 95%. Similarly, an improved eradication rate (74% vs. 89%) was reported in another trial with Iranian patients with peptic ulcer disease receiving lansoprazole, clarithromycin, and amoxicillin and 500 mg/d cranberry capsules (Seyyedmajidi et al., 2016). Cranberry alone or together with standard antibiotic therapies at adequate PAC or polyphenol contents has the potential for suppressing *H. pylori* infection in some responders. Future studies are warranted to characterize factors for the responder effect and identify the actual compounds responsible for the *H. pylori* eradication.

3.3 Dental health

Oral health is essential to supporting nutrition needs and thus is an integral element of overall health and well-being (Peres et al., 2019). Oral diseases are chronic and progressive in nature and include dental caries (tooth decay), periodontal (gum) disease, and oral cancers. Dental caries is the most prevalent disease in the world and is caused by acid by-products produced from bacterial fermentation of free sugars in the dental plaque biofilm composing a complex microbial community (Peres et al., 2019). Thus, bacterial populations residing in the human mouth have a marked impact on oral health as colonization and proliferation of pathogenic ones can increase the risk of dental caries and periodontal disease (Marsh, 2018, Seneviratne et al., 2011). Given that cranberry constituents, particularly polyphenols, display anti-inflammatory, anti-microbial, and anti-bacteria adhesion, cranberry products are anticipated to help maintain/improve oral health (Philip and Walsh, 2019).

Most evidence on the beneficial effects of cranberry constituents on dental caries and periodontal disease were gathered from *in vitro* experiments (Greene et al., 2020, Feghali et al., 2012, Sánchez et al., 2020, de Medeiros et al., 2016). For example, Philip et al. (2019) demonstrated in an *in vitro* experiment that cranberry extracts reduced saliva-derived polymicrobial biofilm biomass, acidogenicity, exopolysaccharide (EPS)/microbial biovolumes, bacterial counts, and the relative abundance of specific caries- (*Streptococcus sobrinus* and *Provotella denticola*) and health-associated bacteria (*Streptococcus sanguinis*).

Clinical evidence supporting the potential of cranberry products for oral health remains scarce. A recent human trial tested the effect of adding cranberry polyphenol-rich extract to a casein phosphopeptide-amorphous calcium phosphate (CPP-ACP) dentifrice changed the ecology of oral microbiome in a parallel, 3-group, double-blind, randomized, and controlled trial with 90 participants with at least 4 fully erupted permanent maxillary teeth (Philip et al., 2020). The results showed that cranberry polyphenols were more effective in decreasing the bacterial loads of two caries-associated bacterial species (*Streptococcus mutans* and *Veillonella parvula*) and increasing two health-associated bacterial species (*Neisseria flavescens* and *Streptococcus sanguinis*) as compared to CPP-ACP and fluoride control dentifrices, suggesting the possibility of formulating polyphenol-rich cranberry extracts in oral care products to support healthy oral microbial communities. An earlier parallel randomized trial with 50 patients with gingivitis showed that as compared to water, daily consumption of 750 mL cranberry functional beverage consisting of cranberry juice (20% by volume), apple juice (80%), and ground cinnamon (0.25 g/100 mL) inhibited dental plaque deposition and gingival inflammation but did not affect the count of *Streptococcus mutans* in dental plaques (Woźniewicz et al., 2018).

3.4 Gut health via microbiota

Thousands of microbes are residing in/on the human body, particularly in the colon where they interact intimately with foods and human host. Although a healthy microbiome profile has not been established, the composition and metabolic capacity of the gut microbiota appear critical to human health locally and systematically, such as the case for obesity and insulin resistance, type 2 diabetes mellitus, gastrointestinal disorders, and recurrent infection (Fassarella et al., 2021). Microbial diversity, metabolic flexibility, microbe-microbe and host-microbe interactions are subject to perturbations by fecal microbiota transplantation, supplementation with probiotics or non-digestible carbohydrates, and dietary modifications (Fassarella et al., 2021). Given that cranberry constituents decrease UTIs via their bactericidal and anti-bacteria adhesion activities, their impact on the gut microbiome is anticipated (Zhao et al., 2020).

Several human trials have been conducted to evaluate how cranberry modulates the gut microbiome. A very recent trial showed that daily consumption of a cranberry beverage for 24 weeks did not affect the overall microbiome composition, microbial diversity, functional pathways, and relative abundances of most bacterial taxa in women with UTI but decreased one unnamed *Flavonifractor* species (OTU41) (Straub et al., 2021). This bacteria is associated with negative human health and relates to the transport and metabolism of tryptophan and cobalamin. The effect of cranberry constituents on the gut microbiome was also observed in a small-scale randomized, crossover, controlled-feeding trial with 11 healthy adults (Rodríguez-Morató et al., 2018). As compared to the animal-based control diet, consumption of cranberry diet consisting of 30 g/day freeze-dried whole cranberry powder for 5 days modified 9 taxonomic clades, including a decrease in the abundance of *Firmicutes* and increase in *Bacteroidetes*, *Lachnospira,* and *Anaerostipes* and attenuated the increase in carcinogenesis-related secondary bile acids and decrease in acetic and butyric acids caused by the animal-based control diet. Similarly, another small-scale trial showed that 2-week consumption of sweetened dried cranberries (42 g/d) with lunch tended to change the relative abundance of several bacterial taxa, particularly with an increase in *Akkermansia* bacteria and a decrease in the ratio of *Firmicutes* to *Bacteroidetes* (Bekiares et al., 2018). The main difference in study products (juice vs. whole cranberry) and dose may contribute to the discrepancy between the two latter trials (Rodríguez-Morató et al., 2018, Bekiares et al., 2018) and that of the first more robust study (Straub et al., 2021). Ursolic acid, undetectable in commercial cranberry juice, may also contribute to the gut microbiome modulating effect of whole cranberry products used in the two latter trials (Rodríguez-Morató et al., 2018, Bekiares et al., 2018) as suggested by the decrease in the ratio of *Firmicutes* to *Bacteroidetes* and increase in the growth of short-chain fatty acid (SCFA)-producing bacteria in preclinical studies (Neto, 2007, Hao et al., 2020). Additionally, non-digestible complex cranberry carbohydrates, such as xyloglucans, can be utilized by microbes in the large intestine to generate products that can influence the gut microbiome via a cross-feeding interaction between microbes (Özcan et al., 2017).

3.5 Cardiometabolic diseases

Cardiometabolic diseases are a cluster of diseases such as diabetes, hypertensive and ischemic heart diseases, and metabolic dysfunction-associated fatty liver disease that share a number of common risk factors, including obesity, prehypertension, pre-diabetics, hyperlipidemia, and insulin resistance. Drug therapies are frequently and effectively applied, although their undesirable side effects support and promote lifestyle modifications, such as diet and exercise, for the management of these risk factors. Given that cranberry products contain antioxidant phytochemicals, their potential for the amelioration of these risk factors has been examined in preclinical and clinical studies. A recent meta-analysis of 12 randomized trials comprising 496 participants showed that cranberry products significantly improved systolic blood pressure (−3.63 mmHg; 95% CI: −6.27, −0.98) and body mass index (−0.30 kg/m^2, 95% CI: −0.57 to −0.02) but did not affect lipid profile, blood glucose and insulin, diastolic blood pressure, waist circumference, C-reactive protein, and intercellular adhesion molecule (Pourmasoumi et al., 2020). Interestingly, the efficacy of cranberry products is age-dependent such that systolic pressure improved in people > 50 years and HDL improved in people < 50 years. This divergence leads to a notion that personalized regimens for the prevention or management of cardiometabolic diseases must be made based on individual's metabolic, genetic, and physiological conditions. Additionally, the impact of product type (juice vs. supplement), bioactive composition of intervention products, subject demographics, dose, and intervention on the efficacy of cranberry products on risk factors of cardiometabolic diseases must be elucidated. For example, Rodriguez-Mateos et al. (2016) identified 12 plasma polyphenol metabolites that significantly correlated with enhanced endothelial function, as measured using a flow-mediated dilation test, including ferulic and caffeic acid sulfates, quercetin-3-O-ß-D-glucuronide and a γ-valerolactone sulfate, suggesting bioactive compounds may not be present in cranberry products but derived from bacterial metabolism of cranberry polyphenols.

Metabolic dysfunction-associated fatty liver disease shares many risk factors with other cardiometabolic diseases, such as obesity and insulin resistance (Eslam et al., 2020). Polyphenol-rich fruits, spices, teas, coffee, and other plant foods are helpful for fatty liver disease by ameliorating hepatic steatosis, oxidative stress, inflammation, and apoptosis (Li et al., 2021). A recent double-blind placebo-controlled randomized trial with 41 patients with the non-alcoholic fatty liver disease showed a larger improvement in alanine aminotransferase (index of liver damage and inflammation) and insulin resistance following 12-week supplementation of a cranberry supplement (288 mg extract/day, equivalent to 26 g dried cranberry) compared to placebo (Hormoznejad et al., 2020).

3.6 Cognition

Dietary supplementations with fruit or vegetable extracts ameliorate age-related declines in learning, memory, motor performance, and neuronal signal transduction in a rat model (Shukitt-Hale et al., 2005). Additionally, the potential cognitive benefits of polyphenols consumed in the form of an extract or whole food have been supported by preclinical and observational studies and some clinical studies (Fraga et al., 2019, Gildawie et al., 2018). Increased intake of flavonoids and anthocyanins, particularly from berries, were strongly associated with a slower rate of cognitive decline in 16010 older women (Devore et al.,

2012). Another observational study also showed intake of anthocyanins, flavones, flavonols, and phenolic acids was associated with enhanced performance in verbal and episodic memory tasks 13 years later but was negatively associated with executive functioning in 2574 middle-aged adults (Kesse-Guyot et al., 2012). Several studies have reported positive clinical evidence with selected flavonoid subfamilies and foods, such as cocoa flavanols and anthocyanin-rich foods (Mastroiacovo et al., 2015, Sloan et al., 2021, Whyte et al., 2018, Travica et al., 2020, Kent et al., 2017). A recent study showed that 16-week supplementation of 160 mg/day of purified anthocyanins derived from bilberry and blackcurrant improved memory and executive function in older adults with mild cognitive impairment (Bergland et al., 2019). Natural PACs are reported to attenuate pathological features of Alzheimer's disease, including extracellular amyloid deposits and neurofibrillary tangles (Zhao et al., 2019). The exact mechanisms by which polyphenols modulate cognitive functions remain to be elucidated, and emerging mechanisms include increasing cerebral blood flow (CBF), accelerating brain oxygenation, improving insulin sensitivity, and modulating brain activity (Fraga et al., 2019, Gratton et al., 2020).

As cranberry products contain flavanol, anthocyanins, and other polyphenols, their consumption is anticipated to be beneficial to cognition, but preclinical and clinical evidence is scarce. In an older trial with 50 cognitively intact older adults, as compared to placebo, consumption of 32 ounces/day of a beverage containing 27% cranberry juice for 6 weeks did not affect thinking processes and mood even though the ability to remember tended to improve (Crews et al., 2005). A positive effect of dried water extract of frozen cranberries (2% of diet) was noted in an aged rat experiment (Shukitt-Hale et al., 2005). Additionally, water-soluble compounds in cranberries, including some polyphenols, enhanced neuronal signal transduction as measured by striatal dopamine release and ameliorated deficits in motor performance (muscle tone and strength and balance) and hippocampal HSP70 neuroprotection. Thus, the effect of dietary polyphenols, including those in cranberries, on cognitive functions is promising, but more clinical evidence is needed to substantiate the benefits via diet, supplementation, or both.

3.7 Skin health

Human skin is not only a physical barrier against biological, chemical, and physical insults but also contributes to water and electrolyte balance, thermoregulation, and immunity (Michalak et al., 2021, Feng et al., 2021). Skin health is subject to the influence of natural aging and photoaging, with the latter mainly caused by free radicals derived from environmental factors, including UV radiation, pollution, and cigarette smoke (Michalak et al., 2021). Among them, the harm of UV radiation is well appreciated as it decreases the synthesis of collagen fibrin and elastic fibrin, augments abnormal elastic fibrin, enables disappearance of the extracellular matrix, and causes protein and DNA damages (Feng et al., 2021). Thus, nutrients that support or enhance antioxidant defense are anticipated to promote skin health. For example, carotenoids (e.g. lutein) and polyphenols (e.g. apigenin, quercetin, curcumin, silymarin, and PACs), and antioxidant vitamins and minerals are appreciated for their potential benefits in protection against photoaging (Dini and Laneri, 2019, Hernandez et al., 2020). Additionally, a balanced diet with an adequate intake of proteins, carbohydrates, and fats is essential to skin health because the skin is a tissue with a high turnover rate and requires ample nutrients for cellular generation (Cao et al., 2020).

While clinical evidence of cranberry products on skin health remains lacking, their potential is partially supported by positive results of other polyphenol-rich foods or nutraceuticals (Pérez-Sánchez et al., 2018). Several human trials had shown that Pycnogenol®, a standardized French maritime pine bark extract rich in catechins and B-type procyanidins and phenolic acids (caffeic, ferulic, and p-hydroxybenzoic acids), was protective against UV light-induced skin damage and promoted de novo skin collagen (Saliou et al., 2001, Marini et al., 2012, Furumura et al., 2012). A supplement containing citrus and rosemary extracts, NutroxSun®, enriched in naringenin, phenolics, and diterpenes (carnosic acid) was also noted to increase UV radiation-induced minimal erythema dose (a measure of skin photosensitivity) by 56% and decrease wrinkle depth and increase skin elasticity (Pérez-Sánchez et al., 2014, Nobile et al., 2016). Whether cranberry polyphenols and other constituents possess properties toward skin health is uncertain at this time, a clinical trial showed that oral intake of cranberry proanthocyanidin and antioxidative vitamins A, C, and E significantly decreased the degree of pigmentation (epidermal melisma) in the left malar and right malar regions of 60 middle-aged Filipino (Handog et al., 2009), suggesting a potential of cranberry products for treating natural aging and photoaging-related skin pigmentation.

3.8 Cancer

Cranberry and its components, particularly flavonols, anthocyanins, and PCAs, display chemopreventive properties by reducing oxidative stress and inflammation, inhibiting cell proliferation and angiogenesis, inducing cell apoptosis, and attenuating metastasis (Mantzorou et al., 2019, Zhao et al., 2020). The putative benefits of cranberry are demonstrated in preclinical studies. Zhao et al. (2020) recently summarized that cranberry juice extract and PCA and flavonoid fractions diminished the cell viability of 41 cell lines, which included 16 cancer types. However, the utility of the *in vitro* data is questionable, such as in lung cancer (Kresty et al., 2011), because most cranberry PCAs are not absorbable (Walsh et al., 2016). Nevertheless, the anti-cancer potential of cranberry constituents for gastrointestinal cancers warrants further examination as bioactives in cranberry can exert actions without the initial absorption process. A preclinical study showed that triterpenes, sterols, and polyphenol-rich fractions were effective in suppressing tumor metrics in mice with colitis-associated colon cancer (Wu et al., 2020). Additionally, the efficacy of cranberry polyphenols against colon cancer may be amplified by combining with cell wall biomass of Lactobacillus probiotics (Desrouillères et al., 2020), supporting the development of products containing both probiotics and cranberry polyphenols. Moreover, the potential additive/synergistic effects of different cranberry constituents warrant exploration. For example, Xiao et al. (2015) noted that dried cranberries were more protective than a polyphenol-rich cranberry extract against dextran sodium sulfate-induced colitis symptoms in mice. While preclinical evidence supports the promising anti-cancer potentials of cranberry and its components, clinical and observational data remain scarce. Student et al. (2016) reported in a randomized, parallel trial that 30 days of 1500 mg cranberry fruit powder supplementation decreased serum prostate-specific antigen in patients with prostate cancer but did not affect cancer staging scores evaluated after prostatectomy, as compared to placebo. In an earlier study, the same group also observed a reduction in serum prostate-specific antigen (PSA) in men at risk of prostate disease and with elevated PSA and clinically confirmed chronic non-bacterial prostatitis (Vidlar et al., 2010).

4 Future directions

Cranberry products are protective against recurrent UTIs, but the exact mechanism(s) of action has not been elucidated (González de Llano et al., 2020, Scharf et al., 2020). The observed benefits have been postulated to be due to A-type PACs in cranberry via their anti-adhesive activity against uropathogenic *E. coli* by their interaction with [D-Gal-(1-4)-β-D-Gal]-binding P-fimbriae (Scharf et al., 2020). However, PACs are poorly bioavailable and do not accumulate to amounts with detectable anti-adhesive efficacy *in vitro* (de Llano et al., 2015, Walsh et al., 2016), suggesting that other mechanisms may be responsible. Additionally, other constituents in whole cranberries may work additively or synergistically with polyphenols and others to support health. The result of studies on the anti-adhesive activity of PAC-free cranberry extracts supports the presence of other bioactive constituents, such as soluble xyloglucan and pectic oligosaccharides (Sun et al., 2019, Rafsanjany et al., 2015, González de Llano et al., 2020, Coleman et al., 2019). Xyloglucan and pectic oligosaccharides are present at ~20% *w/w* in many cranberry materials, but their effects on UTI and other health conditions remained to be examined in human trials. Furthermore, the potential contribution of cranberry-derived changes in composition and metabolic capacity of the gut microbiota in the prevention of UTI recurrence needs to be tested because the intestine is a possible source of uropathogenic bacteria. Concrete evidence on the exact mechanisms of action is expected to inform the development of cranberry products formulated with the most efficacious components against UTI.

Bioefficacy of cranberry constituents may be enhanced when they are formulated with other ingredients that can generate synergistic or additive interactions. Combining cranberry extracts and *Lactobacillus* probiotic decreased the number of women who experienced recurrent UTIs than placebo in a 26-week trial with pre-menopausal adult women (Koradia et al., 2019). Along the same lines, the *H. pylori* eradication rate of cranberry products can be improved by combining with other alternative treatments, such as omega-3 fatty acids (Zare Javid et al., 2020). However, these two studies were not designed to examine whether there was a synergistic or additive interaction between cranberry and other functional ingredients, although it should be noted that a synergy between *Lactobacillus acidophilus* and cranberry c-PAC in reduction in invasiveness of pathogenic *E. coli* was noted *in vitro* (Polewski et al., 2016). Regardless, the development of cranberry products containing other functional ingredients, such as probiotics and others, for all health conditions should be pursued further (González de Llano et al., 2020).

More research is warranted to establish the causal evidence of cranberry products in cardiometabolic diseases, cognitive functions, skin health, and chemoprevention (Grammatikopoulou et al., 2020, Zhao et al., 2020). Additionally, cranberry constituents, particularly polyphenols, display enormous potential for supporting oral health, but there is a need for more clinical evidence, as well as the development of innovative oral health products containing cranberry constituents such as cranberry polyphenols with probiotics. Moreover, the benefits of novel products, such as cranberry seed oil containing flavonoids, salicylic acid, omega-3 fatty acids, for cardiometabolic health must be substantiated in adequately designed human trials. Clinical investigation for the development of cranberry products to alleviate metabolic dysfunction-associated fatty liver disease is also encouraged because of the increasing prevalence of the disease in developed and developing countries.

In the personalized nutrition era, the effectiveness of cranberry products in health is individual-dependent, likely due to variability in the absorption and metabolism of active constituents in the small intestine, microbial metabolism in the large intestine, background diet, host genetics, and other physiological factors. For example, phenyl-γ-valerolactones, proven *in vitro* to have anti-adhesive activity against uropathogenic *E. coli* in bladder epithelial cells (Mena et al., 2017), may contribute to UTI protection but their production can vary between individuals due to variation in the gut microbiota composition (Feliciano et al., 2017). Understanding the magnitude of the individual variability will help inform advancement in cranberry product developments for certain health conditions and specific populations.

The bioefficacy of cranberry products depends mainly on the bioavailability of bioactive constituents and/or their metabolites in target tissues. For example, the protection against uncomplicated UTI is anticipated to be maximized if cranberry A-type PACs can reach the urinary tract at higher concentrations by using novel formulation technologies, such as encapsulation delivery systems (Delfanian and Sahari, 2020). Thus, research on technologies that enhance the bioavailability of bioactive constituents in target tissues is warranted.

Products derived from foods are generally considered safe but are not totally without risk. Untoward effects of some specific phytochemicals consumed at a high dose have been appreciated, and the adversities can be escalated particularly when there are interactions with drugs (Ronis et al., 2018). Cranberry products, including supplements, are safe for moderate consumption by most people. However, there are some mixed reports in the literature showing an interaction between cranberry and warfarin, in which significant bleeding can occur (Tan and Lee, 2021, Srinivas, 2013), possibly through the modulation of cytochrome P450 isozymes responsible for warfarin metabolism (Suvarna et al., 2003). Interestingly, warfarin pharmacokinetics appear not to be affected by consumption of cranberry juice but altered by cranberry supplements, suggesting a need for caution when using concentrated cranberry products. When consuming concentrated cranberry products, the potential drug interaction may also be dependent on the length of consumption, with deleterious effects observed >3 weeks (Srinivas, 2013). While more clinical evidence is warranted to substantiate the cranberry-warfarin interaction in susceptible populations, it is prudent for warfarin users not to substantially increase cranberry intake, particularly products containing cranberry phytochemical extracts.

5 Conclusion

American cranberries are a rich source of phytochemicals, particularly A-type PAC and anthocyanins. Cranberries and cranberry-derived products are commonly consumed in the American diet, and their consumption is best known for protection against the recurrence of UTIs. More significantly, this health benefit has received a US FDA qualified health claim in 2020. The health benefits of cranberries can also be extended to bacteria-related health conditions, such as dental and gastrointestinal health, and others, such as cardiometabolic and skin health, cognition, and cancer (Table 2). Among these health benefits, reduction in blood pressure and body weight is supported by the results of a meta-analysis (Pourmasoumi et al., 2020). Other health benefits require more clinical evidence to realize the potential of cranberry

Table 2 Summary of health benefits of cranberries

Health benefits	Mechanism of actions	Evidence[a]	Level of evidence[b]
Uncomplicated UTI	• Anti-*E. coli* adhesion activities	• Clinic • Positive	Convincing
Complicated UTI		• Clinic • Null	Convincing
***H. pylori* infection**	• Anti-bacterial adhesion activity	• Clinic • Positive	Probable
Dental health	• Anti-inflammatory activity • Anti-microbial activity • Anti-bacterial adhesion activity	• Clinic • *In vitro* • Positive	Possible
Gut health	• Bactericidal activity • Anti-bacterial adhesion activity	• Clinic • Positive	Possible
Cardiometabolic diseases	• Antioxidant activity • Modulating insulin sensitivity	• Clinic • Positive (blood pressure and body weight)	Convincing
Cognition	• Increasing cerebral blood flow • Accelerating brain oxygenation • Improving insulin sensitivity • Modulating brain activity	• Pre-clinic • Positive	Insufficient
Skin health	• Antioxidant activity	• Clinic • Positive	Possible
Cancer	• Antioxidant activity • Anti-inflammatory activity • Inhibiting cell proliferation • Inhibiting angiogenesis • Inducing cell apoptosis • Attenuating metastasis	• Pre-clinic • Positive	Insufficient

Abbreviation: UTI, urinary tract infection
[a] Clinic: human data; pre-clinic: animal and/or cell culture data.
[b] Convincing evidence is designated with support of systematic review and meta-analysis; probably evidence support of several human studies; possible support of a few human studies; insufficient support of preclinical studies.

products. Furthermore, more research needs to be conducted to elucidate the mechanisms of actions by which cranberry constituents support health benefits, for example, whether other cranberry constituents, such as xyloglucans, besides A-type PAC contribute to the protection of UTI recurrence and modulation of gut microbiota. This information is anticipated to inform the development of cranberry products formulated with the most efficacious components. Efforts on establishing additive/synergistic interaction of cranberry components with other ingredients are encouraged, such as antibiotics and probiotics for bacteria-related health issues. Finally, the development of novel technologies that enhance the bioavailability of bioactive constituents of cranberries in target tissues is warranted to maximize bioefficacy.

6 Where to look for further information

New information about the nutraceutical properties of cranberries is anticipated to be added to the literature and can be found in the PubMed database (https://pubmed.ncbi.nlm.nih.gov/), using cranberry and nutraceuticals/functional foods as keywords for the search.

7 References

Barbosa-Cesnik, C., Brown, M. B., Buxton, M., Zhang, L., Debusscher, J. and Foxman, B. 2011. Cranberry juice fails to prevent recurrent urinary tract infection: Results from a randomized placebo-controlled trial. *Clin. Infect. Dis.* 52(1), 23–30.

Bekiares, N., Krueger, C. G., Meudt, J. J., Shanmuganayagam, D. and Reed, J. D. 2018. Effect of sweetened dried cranberry consumption on urinary proteome and fecal microbiome in healthy human subjects. *Omics* 22(2), 145–153.

Bergland, A. K., Soennesyn, H., Dalen, I., Rodriguez-Mateos, A., Berge, R. K., Giil, L. M., Rajendran, L., Siow, R., Tassotti, M., Larsen, A. I. and Aarsland, D. 2019. Effects of anthocyanin supplementation on serum lipids, glucose, markers of inflammation and cognition in adults with increased risk of dementia – A pilot study. *Front. Genet.* 10, 536.

Bhagwat, S. and Haytowitz, D. B. 2015. USDA database for the proanthocyanidin content of selected foods, release. Available at: https://data.nal.usda.gov/dataset/usda-database-proanthocyanidin-content-selected-foods-release-2-2015[Accessed April 6, 2021]. Nutrient Data Laboratory, Beltsville Human Nutrition Research Center, Anaesthetics Research Society, United States Department of Agriculture (vol. 2).

Bhagwat, S. and Haytowitz, D. B. 2016. USDA database for the flavonoid content of selected foods. Release 3.2 (November 2015). Available at: https://data.nal.usda.gov/dataset/usda-database-flavonoid-content-selected-foods-release-32-november-2015 [Accessed April 6, 2021]. Nutrient Data Laboratory, Beltsville Human Nutrition Research Center, Anaesthetics Research Society, United States Department of Agriculture.

Birmingham, A. D., Esquivel-Alvarado, D., Maranan, M., Krueger, C. G. and Reed, J. D. 2021. Inter-laboratory validation of 4-(dimethylamino) cinnamaldehyde (DMAC) assay using cranberry proanthocyanidin standard for quantification of soluble proanthocyanidins in cranberry foods and dietary supplements, first action official MethodSM: 2019.06. *J. AOAC Int.* 104(1), 216–222.

Cao, C., Xiao, Z., Wu, Y. and Ge, C. 2020. Diet and skin aging - From the perspective of food nutrition. *Nutrients* 12(3), 870.

Central, F. 2020. Cranberries, raw. Available at: https://fdc.nal.usda.gov/fdc-app.html#/food-details/1102706/attributes [Accessed April 6, 2021]. Agricultural Research Service, United States Department of Agriculture.

Coleman, C. M., Auker, K. M., Killday, K. B., Azadi, P., Black, I. and Ferreira, D. 2019. Arabinoxyloglucan oligosaccharides may contribute to the antiadhesive properties of porcine urine after cranberry consumption. *J. Nat. Prod.* 82(3), 589–605.

Coleman, C. M. and Ferreira, D. 2020. Oligosaccharides and complex carbohydrates: A new paradigm for cranberry bioactivity. *Molecules* 25(4), 881.

Crews, W. D., JR, Harrison, D. W., Griffin, M. L., Addison, K., Yount, A. M., Giovenco, M. A. and Hazell, J. 2005. A double-blinded, placebo-controlled, randomized trial of the neuropsychologic efficacy of cranberry juice in a sample of cognitively intact older adults: Pilot study findings. *J. Altern. Complement. Med.* 11(2), 305–309.

Dason, S., Dason, J. T. and Kapoor, A. 2011. Guidelines for the diagnosis and management of recurrent urinary tract infection in women. *Can. Urol. Assoc. J.* 5(5), 316–322.

de Llano, D. G., Esteban-Fernández, A., Sánchez-Patán, F., Martínlvarez, P. J., Moreno-Arribas, M. V. and Bartolomé, B. 2015. Anti-adhesive activity of cranberry phenolic compounds and their microbial-derived metabolites against uropathogenic *Escherichia coli* in bladder epithelial cell cultures. *Int. J. Mol. Sci.* 16(6), 12119–12130.

de Medeiros, A. K. B., De Melo, L. A., Alves, R. A. H., Barbosa, G. A. S., De Lima, K. C. and Porto Carreiro, A. D. F. 2016. Inhibitory effect of cranberry extract on periodontopathogenic biofilm: an integrative review. *J. Indian Soc. Periodontol.* 20(5), 503–508.

Delfanian, M. and Sahari, M. A. 2020. Improving functionality, bioavailability, nutraceutical and sensory attributes of fortified foods using phenolics-loaded nanocarriers as natural ingredients. *Food Res. Int.* 137, 109555.

Desrouillères, K., Millette, M., Bagheri, L., Maherani, B., Jamshidian, M. and Lacroix, M. 2020. The synergistic effect of cell wall extracted from probiotic biomass containing Lactobacillus acidophilus CL1285, L. casei LBC80R, and L. rhamnosus CLR2 on the anticancer activity of cranberry juice-HPLC fractions. *J. Food Biochem.* 44(5), e13195.

Devore, E. E., Kang, J. H., Breteler, M. M. and Grodstein, F. 2012. Dietary intakes of berries and flavonoids in relation to cognitive decline. *Ann. Neurol.* 72(1), 135–143.

Dini, I. and Laneri, S. 2019. Nutricosmetics: A brief overview. *Phytother. Res.* 33(12), 3054–3063.

Eslam, M., Newsome, P. N., Sarin, S. K., Anstee, Q. M., Targher, G., Romero-Gomez, M., Zelber-Sagi, S., Wai-Sun Wong, V., Dufour, J. F., Schattenberg, J. M., Kawaguchi, T., Arrese, M., Valenti, L., Shiha, G., Tiribelli, C., Yki-Järvinen, H., Fan, J. G., Grønbæk, H., Yilmaz, Y., Cortez-Pinto, H., Oliveira, C. P., Bedossa, P., Adams, L. A., Zheng, M. H., Fouad, Y., Chan, W. K., Mendez-Sanchez, N., Ahn, S. H., Castera, L., Bugianesi, E., Ratziu, V. and George, J. 2020. A new definition for metabolic dysfunction-associated fatty liver disease: an international expert consensus statement. *J. Hepatol.* 73(1), 202–209.

Fassarella, M., Blaak, E. E., Penders, J., Nauta, A., Smidt, H. and Zoetendal, E. G. 2021. Gut microbiome stability and resilience: Elucidating the response to perturbations in order to modulate gut health. *Gut* 70(3), 595-605.

FDA. 2020. *Qualified Health Claim for Certain Cranberry Products and Urinary Tract Infections*. FDA. Available from: https://www.fda.gov/food/cfsan-constituent-updates/fda-announces-qualified-health-claim-certain-cranberry-products-and-urinary-tract-infections#:~:text=%E2%80%9CConsuming%20one%20serving%20(8%20oz,claim%20is%20limited%20and%20inconsistent.%E2%80%9D.

Feghali, K., Feldman, M., La, V. D., Santos, J. and Grenier, D. 2012. Cranberry proanthocyanidins: Natural weapons against periodontal diseases. *J. Agric. Food Chem.* 60(23), 5728-5735.

Feliciano, R. P., Mills, C. E., Istas, G., Heiss, C. and Rodriguez-Mateos, A. 2017. Absorption, metabolism and excretion of cranberry (poly)phenols in humans: A dose response study and assessment of inter-individual variability. *Nutrients* 9(3), 268. doi: 10.3390/nu9030268.

Feng, M., Zheng, X., Wan, J., Pan, W., Xie, X., Hu, B., Wang, Y., Wen, H. and Cai, S. 2021. Research progress on the potential delaying skin aging effect and mechanism of tea for oral and external use. *Food Funct.* 12(7), 2814-2828.

Foxman, B. 1990. Recurring urinary tract infection: Incidence and risk factors. *Am. J. Public Health* 80(3), 331-333.

Foxman, B. and Brown, P. 2003. Epidemiology of urinary tract infections: Transmission and risk factors, incidence, and costs. *Infect. Dis. Clin. North Am.* 17(2), 227-241.

Fraga, C. G., Croft, K. D., Kennedy, D. O. and Tomás-Barberán, F. A. 2019. The effects of polyphenols and other bioactives on human health. *Food Funct.* 10(2), 514-528.

Fu, Z., Liska, D., Talan, D. and Chung, M. 2017. Cranberry reduces the risk of urinary tract infection recurrence in otherwise healthy women: A systematic review and meta-analysis. *J. Nutr.* 147(12), 2282-2288.

Furumura, M., Sato, N., Kusaba, N., Takagaki, K. and Nakayama, J. 2012. Oral administration of French maritime pine bark extract (Flavangenol®) improves clinical symptoms in photoaged facial skin. *Clin. Interv. Aging* 7, 275-286.

Gao, C., Cunningham, D. G., Liu, H., Khoo, C. and Gu, L. 2018. Development of a Thiolysis HPLC method for the analysis of procyanidins in cranberry products. *J. Agric. Food Chem.* 66(9), 2159-2167.

Gildawie, K. R., Galli, R. L., Shukitt-Hale, B. and Carey, A. N. 2018. Protective effects of foods containing flavonoids on age-related cognitive decline. *Curr. Nutr. Rep.* 7(2), 39-48.

González de Llano, D., Moreno-Arribas, M. V. and Bartolomé, B. 2020. Cranberry polyphenols and prevention against urinary tract infections: relevant considerations. *Molecules* 25(15), 3523.

Gotteland, M., Andrews, M., Toledo, M., Muñoz, L., Caceres, P., Anziani, A., Wittig, E., Speisky, H. and Salazar, G. 2008. Modulation of *Helicobacter pylori* colonization with cranberry juice and Lactobacillus johnsonii La1 in children. *Nutrition* 24(5), 421-426.

Grammatikopoulou, M. G., Gkiouras, K., Papageorgiou, S. T, Myrogiannis, I., Mykoniatis, I., Papamitsou, T., Bogdanos, D. P. and Goulis, D. G. 2020. Dietary factors and supplements influencing prostate specific-antigen (PSA) concentrations in men with prostate cancer and increased cancer risk: An evidence analysis review based on randomized controlled trials. *Nutrients* 12(10).

Gratton, G., Weaver, S. R., Burley, C. V., Low, K. A., Maclin, E. L., Johns, P. W., Pham, Q. S., Lucas, S. J. E., Fabiani, M. and Rendeiro, C. 2020. Dietary flavanols improve cerebral cortical oxygenation and cognition in healthy adults. *Sci. Rep.* 10(1), 19409.

Greene, A. C., Acharya, A. P., Lee, S. B., Gottardi, R., Zaleski, E. and Little, S. R. 2020. Cranberry extract-based formulations for preventing bacterial biofilms. *Drug Deliv. Transl. Res.* 11(3), 1144–1155.

Gu, L., Kelm, M. A., Hammerstone, J. F., Beecher, G., Holden, J., Haytowitz, D., Gebhardt, S. and Prior, R. L. 2004. Concentrations of proanthocyanidins in common foods and estimations of normal consumption. *J. Nutr.* 134(3), 613–617.

Gu, L., Kelm, M. A., Hammerstone, J. F., Beecher, G., Holden, J., Haytowitz, D. and Prior, R. L. 2003. Screening of foods containing proanthocyanidins and their structural characterization using LC-MS/MS and thiolytic degradation. *J. Agric. Food Chem.* 51(25), 7513–7521.

Guevara, B. and Cogdill, A. G. 2020. Helicobacter pylori: A review of current diagnostic and management strategies. *Dig. Dis. Sci.* 65(7), 1917–1931.

Gupta, K., Hooton, T. M., Naber, K. G., Wullt, B., Colgan, R., Miller, L. G., Moran, G. J., Nicolle, L. E., Raz, R., Schaeffer, A. J., Soper, D. E., Infectious Diseases Society of America and European Society for Microbiology and Infectious Diseases. 2011. International clinical practice guidelines for the treatment of acute uncomplicated cystitis and pyelonephritis in women: A 2010 update by the Infectious Diseases Society of America and the European Society for Microbiology and Infectious Diseases. *Clin. Infect. Dis.* 52(5), e103–e120.

Handog, E. B., Galang, D. A., De Leon-Godinez, M. A. and Chan, G. P. 2009. A randomized, double-blind, placebo-controlled trial of oral procyanidin with vitamins A, C, E for melasma among Filipino women. *Int. J. Dermatol.* 48(8), 896–901.

Hao, W., Kwek, E., He, Z., Zhu, H., Liu, J., Zhao, Y., MA, Ma, K. Y., He, W. S. and Chen, Z. Y. 2020. Ursolic acid alleviates hypercholesterolemia and modulates the gut microbiota in hamsters. *Food Funct.* 11(7), 6091–6103.

Harnly, J. M., Doherty, R. F., Beecher, G. R., Holden, J. M., Haytowitz, D. B., Bhagwat, S. and Gebhardt, S. 2006. Flavonoid content of U.S. fruits, vegetables, and nuts. *J. Agric. Food Chem.* 54(26), 9966–9977.

Hernandez, D. F., Cervantes, E. L., Luna-Vital, D. A. and Mojica, L. 2020. Food-derived bioactive compounds with anti-aging potential for nutricosmetic and cosmeceutical products. *Crit. Rev. Food Sci. Nutr.*, 1–16.

Hidron, A. I., Edwards, J. R., Patel, J., Horan, T. C., Sievert, D. M., Pollock, D. A., Fridkin, S. K., National Healthcare Safety Network Team and Participating National Healthcare Safety Network Facilities 2008. NHSN annual update: Antimicrobial-resistant pathogens associated with healthcare-associated infections: Annual summary of data reported to the National Healthcare Safety Network at the Centers for Disease Control and Prevention, 2006–2007. *Infect. Control Hosp. Epidemiol.* 29(11), 996–1011.

Hormoznejad, R., Mohammad Shahi, M., Rahim, F., Helli, B., Alavinejad, P. and Sharhani, A. 2020. Combined cranberry supplementation and weight loss diet in non-alcoholic fatty liver disease: A double-blind placebo-controlled randomized clinical trial. *Int. J. Food Sci. Nutr.* 71(8), 991–1000.

Ikäheimo, R., Siitonen, A., Heiskanen, T., Kärkkäinen, U., Kuosmanen, P., Lipponen, P. and Mäkelä, P. H. 1996. Recurrence of urinary tract infection in a primary care setting: Analysis of a 1-year follow-up of 179 women. *Clin. Infect. Dis.* 22(1), 91–99.

Jepson, R. G., Williams, G. and Craig, J. C. 2012. Cranberries for preventing urinary tract infections. *Cochrane Database Syst. Rev.* 10, CD001321.

Kent, K., Charlton, K. E., Netzel, M. and Fanning, K. 2017. Food-based anthocyanin intake and cognitive outcomes in human intervention trials: A systematic review. *J. Hum. Nutr. Diet.* 30(3), 260–274.

Kesse-Guyot, E., Fezeu, L., Andreeva, V. A., Touvier, M., Scalbert, A., Hercberg, S. and Galan, P. 2012. Total and specific polyphenol intakes in midlife are associated with cognitive function measured 13 years later. *J. Nutr.* 142(1), 76–83.

Koradia, P., Kapadia, S., Trivedi, Y., Chanchu, G. and Harper, A. 2019 Probiotic and cranberry supplementation for preventing recurrent uncomplicated urinary tract infections in premenopausal women: A controlled pilot study. *Expert Rev. Anti Infect. Ther.* 17(9), 733–740.

Kresty, L. A., Howell, A. B. and Baird, M. 2011. Cranberry proanthocyanidins mediate growth arrest of lung cancer cells through modulation of gene expression and rapid induction of apoptosis. *Molecules* 16(3), 2375–2390.

Li, H. Y., Gan, R. Y., Shang, A., Mao, Q. Q., Sun, Q. C., Wu, D. T., Geng, F., He, X. Q. and Li, H. B. 2021. Plant-based foods and their bioactive compounds on fatty liver disease: Effects, mechanisms, and clinical application. *Oxid. Med. Cell. Longev.* 2021, 6621644.

Li, Z. X., Ma, J. L., Guo, Y., Liu, W. D., Li, M., Zhang, L. F., Zhang, Y., Zhou, T., Zhang, J. Y., Gao, H. E., Guo, X. Y., Ye, D. M., Li, W. Q., You, W. C. and Pan, K. F. 2020. Suppression of *Helicobacter pylori* infection by daily cranberry intake: A double-blind, randomized, placebo-controlled trial. *J. Gastroenterol. Hepatol.* 36(4), 927–935.

Liska, D. J., Kern, H. J. and Maki, K. C. 2016. Cranberries and urinary tract infections: How can the same evidence lead to conflicting advice? *Adv. Nutr.* 7(3), 498–506.

Mabeck, C. E. 1972. Treatment of uncomplicated urinary tract infection in non-pregnant women. *Postgrad. Med. J.* 48(556), 69–75.

Maki, K. C., Kaspar, K. L., Khoo, C., Derrig, L. H., Schild, A. L. and Gupta, K. 2016. Consumption of a cranberry juice beverage lowered the number of clinical urinary tract infection episodes in women with a recent history of urinary tract infection. *Am. J. Clin. Nutr.* 103(6), 1434–1442.

Mantzorou, M., Zarros, A., Vasios, G., Theocharis, S., Pavlidou, E. and Giaginis, C. 2019. Cranberry: A promising natural source of potential nutraceuticals with anticancer activity. *Anti Cancer Agents Med. Chem.* 19(14), 1672–1686.

Marini, A., Grether-Beck, S., Jaenicke, T., Weber, M., Burki, C., Formann, P., Brenden, H., Schönlau, F. and Krutmann, J. 2012. Pycnogenol® effects on skin elasticity and hydration coincide with increased gene expressions of collagen type I and hyaluronic acid synthase in women. *Skin Pharmacol. Physiol.* 25(2), 86–92.

Marsh, P. D. 2018. In Sickness and in health-what does the oral microbiome mean to us? An ecological perspective. *Adv. Dent. Res.* 29(1), 60–65.

Mastroiacovo, D., Kwik-Uribe, C., Grassi, D., Necozione, S., Raffaele, A., Pistacchio, L., Righetti, R., Bocale, R., Lechiara, M. C., Marini, C., Ferri, C. and Desideri, G. 2015. Cocoa flavanol consumption improves cognitive function, blood pressure control, and metabolic profile in elderly subjects: The Cocoa, Cognition, and Aging (CoCoA) Study–A randomized controlled trial. *Am. J. Clin. Nutr.* 101(3), 538–548.

Mena, P., González De Llano, D., Brindani, N., Esteban-Fernández, A., Curti, C., Moreno-Arribas, M. V., Del Rio, D. and Bartolomé, B. 2017. 5-(3′,4′-dihydroxyphenyl)-γ-va lerolactone and its sulphate conjugates, representative circulating metabolites of flavan-3-ols, exhibit anti-adhesive activity against uropathogenic *Escherichia coli* in bladder epithelial cells. *J. Funct. Foods* 29, 275–280.

Michalak, M., Pierzak, M., Kręcisz, B. and Suliga, E. 2021. Bioactive compounds for skin health: A review. *Nutrients* 13(1), 203.

Neto, C. C. 2007. Cranberry and its phytochemicals: A review of in vitro anticancer studies. *J. Nutr.* 137(1), 186S–193S.

Neto, C. C. and Vinson, J. A. 2011. Cranberry. In: Benzie Iff, W.-G. S. (Ed.) *Herbal Medicine: Biomolecular and Clinical Aspects* (2nd edn.). Boca Raton, FL: CRC Press Press/ Taylor & Francis.

Nicolle, L. E. and AMMI Canada Guidelines Committee. 2005. Complicated urinary tract infection in adults. *Can. J. Infect. Dis. Med. Microbiol.* 16(6), 349–360.

Niska, R., Bhuiya, F. and Xu, J. 2010. National hospital ambulatory medical care survey: 2007 emergency department summary. *Natl Health Stat. Rep.* 26, 1–31.

Nobile, V., Michelotti, A., Cestone, E., Caturla, N., Castillo, J., Benavente-García, O., Pérez-Sánchez, A. and Micol, V. 2016. Skin photoprotective and antiageing effects of a combination of rosemary (*Rosmarinus officinalis*) and grapefruit (*Citrus paradisi*) polyphenols. *Food Nutr. Res.* 60, 31871.

Ou, K., Percival, S. S., Zou, T., Khoo, C. and Gu, L. 2012. Transport of cranberry A-type procyanidin dimers, trimers, and tetramers across monolayers of human intestinal epithelial Caco-2 cells. *J. Agric. Food Chem.* 60(6), 1390–1396.

Özcan, E., Sun, J., Rowley, D. C. and Sela, D. A. 2017. A human gut commensal ferments cranberry carbohydrates to produce formate. *Appl. Environ. Microbiol.* 83(17), 3637–3644.

Peres, M. A., Macpherson, L. M. D., Weyant, R. J., Daly, B., Venturelli, R., Mathur, M. R., Listl, S., Celeste, R. K., Guarnizo-Herreño, C. C., Kearns, C., Benzian, H., Allison, P. and Watt, R. G. 2019. Oral diseases: A global public health challenge. *Lancet* 394(10194), 249–260.

Pérez-Sánchez, A., Barrajón-Catalán, E., Caturla, N., Castillo, J., Benavente-García, O., Alcaraz, M. and Micol, V. 2014. Protective effects of citrus and rosemary extracts on UV-induced damage in skin cell model and human volunteers. *J. Photochem. Photobiol. B* 136, 12–18.

Pérez-Sánchez, A., Barrajón-Catalán, E., Herranz-López, M. and Micol, V. 2018. Nutraceuticals for skin care: A comprehensive review of human clinical studies. *Nutrients* 10(4).

Phenol-Explorer. American cranberry. Available at: http://phenol-explorer.eu/contents/food/74 [Accessed April 6, 2021].

Philip, N., Bandara, H. M. H. N., Leishman, S. J. and Walsh, L. J. 2019. Effect of polyphenol-rich cranberry extracts on cariogenic biofilm properties and microbial composition of polymicrobial biofilms. *Arch. Oral Biol.* 102, 1–6.

Philip, N., Leishman, S. J., Bandara, H. M. H. N., Healey, D. L. and Walsh, L. J. 2020. Randomized controlled study to evaluate microbial ecological effects of CPP-ACP and cranberry on dental plaque. *JDR Clin. Trans. Res.* 5(2), 118–126.

Philip, N. and Walsh, L. J. 2019. Cranberry polyphenols: Natural weapons against dental caries. *Dent. J. (Basel)* 7(1), 20.

Polewski, M. A., Krueger, C. G., Reed, J. D. and Leyer, G. 2016. Ability of cranberry proanthocyanidins in combination with a probiotic formulation to inhibit in vitro invasion of gut epithelial cells by extra-intestinal pathogenic *E. coli*. *J. Funct. Foods* 25, 123–134.

Pourmasoumi, M., Hadi, A., Najafgholizadeh, A., Joukar, F. and Mansour-Ghanaei, F. 2020. The effects of cranberry on cardiovascular metabolic risk factors: A systematic review and meta-analysis. *Clin. Nutr.* 39(3), 774–788.
Rafsanjany, N., Senker, J., Brandt, S., Dobrindt, U. and Hensel, A. 2015. In vivo consumption of cranberry exerts ex vivo antiadhesive activity against FimH-dominated uropathogenic *Escherichia coli*: A combined in vivo, ex vivo, and in vitro Study of an extract from Vaccinium macrocarpon. *J. Agric. Food Chem.* 63(40), 8804–8818.
Raguzzini, A., Toti, E., Sciarra, T., Fedullo, A. L. and Peluso, I. 2020. Cranberry for bacteriuria in individuals with spinal cord injury: A systematic review and meta-analysis. *Oxid. Med. Cell. Longev.* 2020, 9869851.
Rodriguez-Mateos, A., Feliciano, R. P., Boeres, A., Weber, T., Dos Santos, C. N., Ventura, M. R. and Heiss, C. 2016. Cranberry (poly)phenol metabolites correlate with improvements in vascular function: A double-blind, randomized, controlled, dose-response, crossover study. *Mol. Nutr. Food Res.* 60(10), 2130–2140.
Rodríguez-Morató, J., Matthan, N. R., Liu, J., De La Torre, R. and Chen, C. O. 2018. Cranberries attenuate animal-based diet-induced changes in microbiota composition and functionality: A randomized crossover controlled feeding trial. *J. Nutr. Biochem.* 62, 76–86.
Ronis, M. J. J., Pedersen, K. B. and Watt, J. 2018. Adverse effects of nutraceuticals and dietary supplements. *Annu. Rev. Pharmacol. Toxicol.* 58, 583–601.
Saliou, C., Rimbach, G., Moini, H., McLaughlin, L., Hosseini, S., Lee, J., Watson, R. R. and Packer, L. 2001. Solar ultraviolet-induced erythema in human skin and nuclear factor-kappa-B-dependent gene expression in keratinocytes are modulated by a French maritime pine bark extract. *Free Radic. Biol. Med.* 30(2), 154–160.
Sánchez, M. C., Ribeiro-Vidal, H., Bartolomé, B., Figuero, E., Moreno-Arribas, M. V., Sanz, M. and Herrera, D. 2020. New evidences of antibacterial effects of cranberry against periodontal pathogens. *Foods* 9(2), 246.
Savoldi, A., Carrara, E., Graham, D. Y., Conti, M. and Tacconelli, E. 2018. Prevalence of antibiotic resistance in Helicobacter pylori: A systematic review and meta-analysis in World Health Organization regions. *Gastroenterology* 155(5), 1372–1382.e17.
Schappert, S. M. and Rechtsteiner, E. A. 2011. Ambulatory medical care utilization estimates for 2007. *Vital Health Stat.* 13(169), 1–38.
Scharf, B., Schmidt, T. J., Rabbani, S., Stork, C., Dobrindt, U., Sendker, J., Ernst, B. and Hensel, A. 2020. Antiadhesive natural products against uropathogenic *E. coli*: What can we learn from cranberry extract? *J. Ethnopharmacol.* 257, 112889.
Seneviratne, C. J., Zhang, C. F. and Samaranayake, L. P. 2011. Dental plaque biofilm in oral health and disease. *Chin. J. Dent. Res.* 14(2), 87–94.
Seyyedmajidi, M., Ahmadi, A., Hajiebrahimi, S., Seyedmajidi, S., Rajabikashani, M., Firoozabadi, M. and Vafaeimanesh, J. 2016. Addition of cranberry to proton pump inhibitor-based triple therapy for *Helicobacter pylori* eradication. *J. Res. Pharm. Pract.* 5(4), 248–251.
Shmuely, H., Yahav, J., Samra, Z., Chodick, G., Koren, R., Niv, Y. and Ofek, I. 2007. Effect of cranberry juice on eradication of *Helicobacter pylori* in patients treated with antibiotics and a proton pump inhibitor. *Mol. Nutr. Food Res.* 51(6), 746–751.
Shukitt-Hale, B., Galli, R. L., Meterko, V., Carey, A., Bielinski, D. F., McGhie, T. and Joseph, J. A. 2005. Dietary supplementation with fruit Polyphenolics ameliorates age-related deficits in behavior and neuronal markers of inflammation and oxidative stress. *Age (Dordr)* 27(1), 49–57.

Sloan, R. P., Wall, M., Yeung, L. K., Feng, T., Feng, X., Provenzano, F., Schroeter, H., Lauriola, V., Brickman, A. M. and Small, S. A. 2021. Insights into the role of diet and dietary flavanols in cognitive aging: Results of a randomized controlled trial. *Sci. Rep.* 11(1), 3837.

Srinivas, N. R. 2013. Cranberry juice ingestion and clinical drug-drug interaction potentials; review of case studies and perspectives. *J. Pharm. Pharm. Sci.* 16(2), 289–303.

Stamm, W. E. and Hooton, T. M. 1993. Management of urinary tract infections in adults. *N. Engl. J. Med.* 329(18), 1328–1334.

Stapleton, A. E., Dziura, J., Hooton, T. M., Cox, M. E., Yarova-Yarovaya, Y., Chen, S. and Gupta, K. 2012. Recurrent urinary tract infection and urinary *Escherichia coli* in women ingesting cranberry juice daily: A randomized controlled trial. *Mayo Clin. Proc.* 87(2), 143–150.

Stothers, L. 2002. A randomized trial to evaluate effectiveness and cost effectiveness of naturopathic cranberry products as prophylaxis against urinary tract infection in women. *Can. J. Urol.* 9(3), 1558–1562.

Straub, T. J., Chou, W. C., Manson, A. L., Schreiber, H. L. T., Walker, B. J., Desjardins, C. A., Chapman, S. B., Kaspar, K. L., Kahsai, O. J., Traylor, E., Dodson, K. W., Hullar, M. A. J., Hultgren, S. J., Khoo, C. and Earl, A. M. 2021. Limited effects of long-term daily cranberry consumption on the gut microbiome in a placebo-controlled study of women with recurrent urinary tract infections. *BMC Microbiol.* 21(1), 53.

Student, V., Vidlar, A., Bouchal, J., Vrbkova, J., Kolar, Z., Kral, M., Kosina, P. and Vostalova, J. 2016. Cranberry intervention in patients with prostate cancer prior to radical prostatectomy: Clinical, pathological and laboratory findings. *Biomed. Pap. Med. Fac. Univ. Palacky Olomouc Czech Repub.* 160(4), 559-565.

Sun, J., Deering, R. W., Peng, Z., Najia, L., Khoo, C., Cohen, P. S., Seeram, N. P. and Rowley, D. C. 2019. Pectic oligosaccharides from cranberry prevent quiescence and persistence in the uropathogenic *Escherichia coli* CFT073. *Sci. Rep.* 9(1), 19590.

Suvarna, R., Pirmohamed, M. and Henderson, L. 2003. Possible interaction between warfarin and cranberry juice. *BMJ* 327(7429), 1454.

Takahashi, S., Hamasuna, R., Yasuda, M., Arakawa, S., Tanaka, K., Ishikawa, K., Kiyota, H., Hayami, H., Yamamoto, S., Kubo, T. and Matsumoto, T. 2013. A randomized clinical trial to evaluate the preventive effect of cranberry juice (UR65) for patients with recurrent urinary tract infection. *J. Infect. Chemother.* 19(1), 112–117.

Tan, C. S. S. and Lee, S. W. H. 2021. Warfarin and food, herbal or dietary supplement interactions: A systematic review. *Br. J. Clin. Pharmacol.* 87(2), 352–374.

Travica, N., D'cunha, N. M., Naumovski, N., Kent, K., Mellor, D. D., Firth, J., Georgousopoulou, E. N., Dean, O. M., Loughman, A., Jacka, F. and Marx, W. 2020. The effect of blueberry interventions on cognitive performance and mood: A systematic review of randomized controlled trials. *Brain Behav. Immun.* 85, 96–105.

Turner, A., Chen, S. N., Nikolic, D., Van Breemen, R., Farnsworth, N. R. and Pauli, G. F. 2007. Coumaroyl iridoids and a depside from cranberry (Vaccinium macrocarpon). *J. Nat. Prod.* 70(2), 253–258.

Vidlar, A., Vostalova, J., Ulrichova, J., Student, V., Stejskal, D., Reichenbach, R., Vrbkova, J., Ruzicka, F. and Simanek, V. 2010. The effectiveness of dried cranberries (Vaccinium macrocarpon) in men with lower urinary tract symptoms. *Br. J. Nutr.* 104(8), 1181–1189.

Vostalova, J., Vidlar, A., Simanek, V., Galandakova, A., Kosina, P., Vacek, J., Vrbkova, J., Zimmermann, B. F., Ulrichova, J. and Student, V. 2015. Are high proanthocyanidins key to cranberry efficacy in the prevention of recurrent urinary tract infection? *Phytother. Res.* 29(10), 1559–1567.

Walker, E. B., Barney, D. P., Mickelsen, J. N., Walton, R. J. and Mickelsen, R. A., Jr. 1997. Cranberry concentrate: UTI prophylaxis. *J. Fam. Pract.* 45(2), 167–168.

Walsh, J. M., Ren, X., Zampariello, C., Polasky, D. A., Mckay, D. L., Blumberg, J. B. and Chen, C. Y. 2016. Liquid chromatography with tandem mass spectrometry quantification of urinary proanthocyanin A2 dimer and its potential use as a biomarker of cranberry intake. *J. Sep. Sci.* 39(2), 342–349.

Wang, C. H., Fang, C. C., Chen, N. C., Liu, S. S., Yu, P. H., Wu, T. Y., Chen, W. T., Lee, C. C. and Chen, S. C. 2012. Cranberry-containing products for prevention of urinary tract infections in susceptible populations: A systematic review and meta-analysis of randomized controlled trials. *Arch. Intern. Med.* 172(13), 988–996.

Wang, F., Meng, W., Wang, B. and Qiao, L. 2014. Helicobacter pylori-induced gastric inflammation and gastric cancer. *Cancer Lett.* 345(2), 196–202.

Whyte, A. R., Cheng, N., Fromentin, E. and Williams, C. M. 2018. A randomized, double-blinded, placebo-controlled study to compare the safety and efficacy of low dose enhanced wild blueberry powder and wild blueberry extract (ThinkBlue™) in maintenance of episodic and working memory in older adults. *Nutrients* 10(6), 660. doi: 10.3390/nu10060660.

Woźniewicz, M., Nowaczyk, P. M., Kurhańska-Flisykowska, A., Wyganowska-Świątkowska, M., Lasik-Kurdyś, M., Walkowiak, J. and Bajerska, J. 2018. Consumption of cranberry functional beverage reduces gingival index and plaque index in patients with gingivitis. *Nutr. Res.* 58, 36–45.

Wu, X., Xue, L., Tata, A., Song, M., Neto, C. C. and Xiao, H. 2020. Bioactive components of polyphenol-rich and non-polyphenol-rich cranberry fruit extracts and their chemopreventive effects on colitis-associated colon cancer. *J. Agric. Food Chem.* 68(25), 6845–6853.

Xiao, X., Kim, J., Sun, Q., Kim, D., Park, C. S., Lu, T. S. and Park, Y. 2015. Preventive effects of cranberry products on experimental colitis induced by dextran sulphate sodium in mice. *Food Chem.* 167, 438–446.

Yang-Ou, Y. B., Hu, Y., Zhu, Y. and Lu, N. H. 2018. The effect of antioxidants on Helicobacter pylori eradication: A systematic review with meta-analysis. *Helicobacter* 23(6), e12535.

Zare Javid, A., Maghsoumi-Norouzabad, L., Bazyar, H., Aghamohammadi, V. and Alavinejad, P. 2020. Effects of concurrent Omega-3 and cranberry juice consumption Along with standard antibiotic therapy on the eradication of *Helicobacter pylori*, gastrointestinal symptoms, some serum inflammatory and oxidative stress markers in adults with *Helicobacter pylori* infection: A study protocol for a randomized controlled trial. *Infect. Drug Resist.* 13, 3179–3185.

Zhang, L., MA, Ma, J., Pan, K., Go, V. L., Chen, J. and You, W. C. 2005. Efficacy of cranberry juice on Helicobacter pylori infection: A double-blind, randomized placebo-controlled trial. *Helicobacter* 10(2), 139–145.

Zhao, S., Liu, H. and Gu, L. 2020. American cranberries and health benefits – An evolving story of 25 years. *J. Sci. Food Agric.* 100(14), 5111–5116.

Zhao, S., Zhang, L., Yang, C., Li, Z. and Rong, S. 2019. Procyanidins and Alzheimer's disease. *Mol. Neurobiol.* 56(8), 5556–5567.

Chapter 14

Advances in understanding and improving the nutraceutical properties of apples

Lia Noemi Gerschenson, Eliana Noemi Fissore and Carolina Bélen Gómez Vargas, Universidad de Buenos Aires and CONICET, Argentina

1 Introduction

Apple trees belong to the Rosaceae family. The scientific nomenclature for apples has changed since Linnaeus denominated it as *Pyrus malus*. Other denominations in the past have included *Malus communis*, *Malus sylvestris*, *Malus pumila*, and *M. domestica*.

The apple tree is a hybrid originating from a combination of wild species with *Malus sieversii* the main contributor. The domesticated apple is the result of an interspecific hybridization and is named *Malus x domestica* Borkh. This tree adapts well to different climates, meaning apple trees can be cultivated from northern Europe down to the tropics where two crops can be obtained at high altitudes. As a consequence, it has been introduced in South America, South Africa, New Zealand, and Australia (Oshima et al., 2019; Pereira Lorenzo et al., 2009).

http://dx.doi.org/10.19103/AS.2022.0101.17

According to the Food and Agriculture Organization, global apple production was around 87 million tonnes in 2019 (FAOSTAT, 2019).

According to the USDA database (US Department of Agriculture, Agricultural Research Service, 2019), apples (raw and with the skin) of the varieties Red delicious, Golden delicious, Gala, Granny Smith, and Fuji have the following composition per 100 g of product: 85.6 g water, 0.26 g protein, 0.17 g lipid, 0.19 g ash, 13.8 g carbohydrate including 2.4 g total dietary fibre, 10.4 g sugars comprising 2.07 g sucrose, 2.43 g glucose (dextrose), 5.9 g fructose, and 0.05 g starch. With respect to minerals, they contain 6 mg calcium, 0.12 mg iron, 5 mg magnesium, 11 mg phosphorus, 107 mg potassium, 1 mg sodium, 0.04 mg zinc, 0.027 mg copper, 0.035 mg manganese, and 3.3 μg fluoride. The vitamins present are vitamin C (total ascorbic acid) 4.6 mg, thiamin (vitamin B1) 0.017 mg, riboflavin (B2) 0.026 mg, niacin 0.091 mg, panthotenic acid 0.061 mg, vitamin B6 0.041 mg, folate 3 μg, vitamin A 54 IU, beta carotene 27 μg, beta criptoxanthin 11 μg, lutein + zeaxanthin 29 μg, vitamin E (alpha tocopherol) 0.18 mg, and vitamin K (phylloquinone) 2.2 μg. Phenolic compounds in apples (raw, with skin) of the same varieties showed the following content of flavonoids per 100 g of edible portion: cyanidin (anthocyanidin) 2.17 mg; between the flavan-3-ols: (−)-epicatechin 7.53 mg, (−)-epigallocatechin 0.26 mg, (−)-epigallocatechin-3-gallate 0.19 mg, (+)-catechin 1.30 mg; and between the flavonols: quercetin 4.01 mg (Bhagwat and Haytowitz, 2015).

Smanalieva et al. (2020) studied the nutritional value of fully ripened apples (*Malus sieversii* var. *kirgizorum*) produced during 2018 and 2019 in two different locations in the forests of southern Kyrgyzstan. They observed that the fruits had a vitamin C content between 11.48 and 13.60 mg 100 g^{-1}, total crude fibre between 1.99 and 1.77 g 100 g^{-1}, ashes between 0.31 and 0.42 g 100 g^{-1}, total phenolic content between 439 and 476 mg, gallic acid equivalent 100 g^{-1}, and antioxidant activity towards 2,2-diphenyl-1-picrylhydrazyl (half maximal inhibitory concentration, IC50) between 11.2 and 10.0 μg mL^{-1}. The minerals presented values per 100 g of fresh tissue of 16.87–15.16 mg for K, 12.43–1.98 mg for Ca, and 2.29–2.32 mg for iron. These results, in general, differ from those reported by the USDA database, likely due to the incidence of fruit variety, climate, geographical location, year of production, and other factors.

Macit et al. (2021) studied the nutritional value of Anatolian fruits for 3 years between 2017 and 2019. The authors studied eight local apple varieties with different characteristics, reporting a vitamin C content of 2.31-7.66 mg g^{-1}, a total phenolic content of 2.0–8.0 mg, gallic acid equivalent 100 g^{-1}, antioxidant activity of 2.2–90.9%, and an iron content of 0.6–1.4 mg 100 g^{-1}.

Saveleva et al. (2021) studied the variability in the biochemical composition of apple varieties of *Malus x domestica* Borkh fruits (ascorbic acid, soluble solids, and titratable acids) through long-term studies, observing intervarietal differences in biochemical composition and variation over the years.

Kumar et al. (2018) studied 22 apple cultivars grown in Himachal Pradesh, India, which were harvested at commercial maturity. Different attributes (ascorbic acid, antioxidant activity, total carotenoids, sugars, organic acids, phenolic compounds, and minerals) were studied. They reported that ascorbic acid content ranged between 19.38 mg 100 g^{-1} and 32.08 mg 100 g^{-1} while the antioxidant activity varied between 2.64 μmol trolox equivalent g^{-1} and 13.20 μmol trolox equivalent g^{-1}. The highest total carotenoid content was 14.7 mg 100 g^{-1} and the lowest was 2.9 mg 100 g^{-1}. Fructose (average 50.79 g L^{-1}) was the most abundant sugar. Malic acid (average 6.03 mg L^{-1}) predominated among organic acids. Potassium (average 795.14 mg 100 g^{-1}) and iron (average 2.04 μg g^{-1}) were the predominant macro and micro elements, respectively. Chlorogenic acid was the major constituent among phenolic compounds (average 28.42 mg L^{-1}). The differences observed between varieties shows the effect of genotype on the results obtained. The higher phenolic content was associated with antioxidant and anti-inflammatory activity.

Piagentini and Pirovani (2017) studied the total phenolic content and antioxidant capacity of five cultivars (Granny Smith, Red delicious, Caricia, Eva, and Princesa) of apples harvested in Argentina. The study was performed on the peel and flesh of the fruit. These attributes were significantly different among cultivars, with Red delicious having the highest phenolic content and antioxidant capacity. Phenolic content (2–5 times) and antioxidant capacity (2–4 times) were higher in peel than in flesh for all varieties, highlighting the incidence not only of variety but also of the tissue.

Almeida et al. (2017) studied the composition (total phenolics, flavonoids, and anthocyanin contents) and antioxidant activity in the flesh and skin of fruits at the ripe stage of *M. domestica*. The trees pertained to the *Maçã de Alcobaça* of the protected geographic indication of Portugal. For this study, they used eight cultivars - Casa nova, Fuji, Golden delicious, Granny Smith, Jonagored, Reinette Grise, Royal Gala, and Starking - and concluded that the highest concentration of phytochemicals was located in the skin and not in the flesh of the fruits.

Navarro et al. (2018) studied the total polyphenolic compounds and antioxidant activity of *M. domestica* cultivars from Costa Rica. They also performed a qualitative analysis of phenolic-enriched extracts of the commercial cultivar of *M. domestica*, known as Anna cultivar. They observed higher values in apple peels that presented a total phenolic content of 619.6 mg gallic acid equivalents g^{-1} extract. The antioxidant capacity was evaluated by the DPPH (2,2-diphenyl-1-picrylhydrazyl) and ORAC (oxygen radical absorbance capacity) technique, and a value of IC50 of 4.54 g mL^{-1} was evaluated through the DPPH method and a value of 16.8 mmol trolox equivalent g^{-1} via the ORAC method. With respect to the qualitative analysis of apple peels, they detected, among other polyphenolics, the presence of procyanidin tetramers and pentamers that could be responsible for the higher antioxidant activity observed.

Veberic et al. (2005) studied 22 apple cultivars, 11 of which were organically grown and 11 were of integrated production from Austria and Slovenia. The fruits were studied through high-performance liquid chromatography (HPLC) with reference to phenolic compounds in peel and pulp. In peel, chlorogenic acid, *p*-coumaric acid, procyanidin B3, protocatechuic acid, (−)-epicatechin, phloridzin, rutin, and quercetin-3-rhamnoside were identified, and (+)-catechin was also identified in the pulp of the fruit for all cultivars. In the apple peel, phenolic content was similar for both types of cultivars. Organically grown apples exhibited a higher content of phenolic substances in the apple pulp, which might be ascribed to plant response to stress.

Manzoor et al. (2012) studied the phenolic compounds and antioxidant activity of *M. domestica x* Borkh using five cultivars from Pakistan (Red delicious, Golden delicious, Kashmiri amri, Kala kulu, and Sky spur). An 80:20 methanol-water (v/v) extract showed a yield of antioxidant compounds of 22.1 g 100 g^{-1} for peel and 14.2 g 100 g^{-1} for pulp on a dry weight basis. The amounts of total phenolics and total flavonoids in peel and pulp of different cultivars of apple ranged from 1907 mg to 2587 mg gallic acid equivalent 100 g^{-1} dry matter (DM), from 1214 mg to 1816 mg catechin equivalent 100 g^{-1} DM, from 1185 to 1475 mg gallic acid equivalent 100 g^{-1} DM, and from 712 mg to 999 mg catechin equivalent 100 g^{-1} DM, respectively. The inhibition of linoleic acid peroxidation and DPPH scavenging activity of the extracts varied from 72% to 85% and 67% to 81% in peel, and from 44% to 53% and 43% to 51% in pulp, respectively.

The information reported in the bibliography regarding the composition and antioxidant activity of the apple is highly variable since it is affected by different dimensions such as geographical location, climate, maturation degree, cultivar, and tissue.

2 Health effects of apple

According to Hyson (2011), there is a link between fruit and vegetable intake and improved health in humans.

Elhakem et al. (2021) studied various bioactive compounds of five Saudi Arabia apple varieties (Black, Apricot, Jester, Big Ariane, and Medium Ariane). The Apricot variety showed the highest antioxidant activity, which was evaluated through DPPH-2,2-diphenyl-1-picrylhydrazyl (93.67 μmol trolox equivalent g^{-1}, fresh weight), ABTS-2,2′-azinobis-(3-ethylbenzothiazoline-6-sulfonic acid) (7.71 μmol trolox equivalent g^{-1}, fresh weight), and FRAP-ferric reducing antioxidant power (85.98 μmol trolox equivalent g^{-1}, fresh weight) tests. The apples showed a total flavonoid content of 36.86–55.52 mg quercetin g^{-1}. The extracts of the apples showed anti-inflammatory activity values, with the Apricot variety having the highest bioactivity due to its high antioxidant and anti-inflammatory activity.

Zielinska et al. (2019) studied the effect of apple consumption on intestinal inflammation symptoms and linked the positive effect to the presence of phloretin (phenolic compound) and its glycoside, phloridzin. It was concluded that both compounds present in apples are associated with the benefits of apple consumption since phloretin (10–50 M) decreases the synthesis of pro-inflammatory molecules in interleukin-1 (IL-1) treated myofibroblasts in the colon cell line CCD-18Co and has an inhibitory potential for the formation of advanced glycation end products. Ulaszewska et al. (2018) reported that urinary phloretin seems applicable as the biomarker of apple intake because, for this fruit tissue only does it appear in urine.

In the case of cancer, studies demonstrated that apple products reduce malignant cell proliferation, increase apoptotic mechanisms, and modulate signal transduction pathways. It is difficult to obtain general conclusions from existing studies due to the variation in the composition of the extracts used in the experiments, such as different geographical origin, variety, and the degree of maturity of the apples studied (Mourtzinos and Goula, 2019; Calinoiu et al., 2017; Francini and Sebastiani, 2013; Boyer and Liu, 2004), as well as the procedures followed to obtain the extract and the apple tissues. It has been verified that the extracts obtained from apple contain different bioactives and that they are many times more effective than isolated compounds or controlled mixtures of them.

According to Boyer and Liu (2004), apples have strong antioxidant activity, inhibit cancer cell proliferation, decrease lipid oxidation, and lower cholesterol. These authors also reported that apples contain a variety of phytochemicals, including quercetin, catechin, phloridzin, and chlorogenic acid, all of which are strong antioxidants.

Harvard T. H. Chan School of Public Health (2021) reported that the fruit of *M. domestica* is a source of soluble and insoluble fibre, and polyphenolics such as quercetin, catechin, chlorogenic acid, and anthocyanins. Quercetin is a flavonoid that has antioxidant and anti-inflammatory effects. Pectin is a soluble fibre that can help reduce low-density lipoprotein cholesterol, being also fermented by colon bacteria that produce short-chain fatty acids (SCFAs) that can prevent gastrointestinal disorders and cancer.

Koutsos et al. (2015) reported that apple intake might help to prevent or relieve cardiovascular disease, type 2 diabetes, and cancer and assist with weight management.

To reach more definitive conclusions regarding the effect of different bioactives present in apples in terms of their effect on health, it is necessary to carry out further clinical studies as well as to clearly determine the chemical composition of the extracts used in each study.

3 Apple processing and pomace generation

The industrialization of the apple renders leftovers. After pressing apples to obtain juice, the solid that remains consists mainly of the skin, pulp, seeds, and the stalk of the fruit. It is the major by-product of the apple juice industry, representing 25% of the product, as approximately 75% of the fruit weight is extracted as juice (Sudha, 2011). The bagasse obtained from the different industrialization processes that use apple as its raw material is known as apple pomace. The implementation of sustainable strategies for the addition of value to apple pomace is of great importance because it contributes to the goals of the circular economy (Hamam et al., 2021). It is important to state that the stabilization of apple pomace involves its drying and that pectin degradation to values of 3.91% occurs for drum-drying due to thermal effects and to 6.09% for freeze-drying (Kosseva, 2013).

Perussello et al. (2017) analysed the possibility of transforming apple pomace into industrial commodities. They reported its use as a raw material for fermentation to obtain citric acid, lactic acid, enzymes, aroma compounds, and bioethanol (biofuels). It is also proposed to be used for the recovery of polyphenols, dietary fibres in general, and the soluble fibre pectin, which are bioactives that can contribute to the maintenance of human health.

According to Vukušić et al. (2021), the widely available apple pomace can be used for the recovery of pectin, xyloglucan, and polyphenols, as well as biofuel pellets with a net calorific value of 20.3 MJ/kg. Concentrated apple pomace extract can be used as a partial substitute for molasses in industrial bioprocesses. Apple pomace contains 11.9% of pectin and, following pectin extraction, free apple pomace extract can be used as a carbon source for the growth of baker's yeast. All these products can contribute to a zero discharge biorefinery process that converts waste into valuable products, while generating no additional wastes.

Guardia et al. (2019) reported that apple waste obtained from the juice and cider industries can be used as a precursor of porous carbons of different physicochemical properties. They proposed the stabilization of the residue by means of drying followed by hydrothermal carbonization at 200°C, obtaining a liquid fraction that contained phenolics, mainly constituted of catechol (1,2-dihydroxybenzene) and 5-hydroxymethyl-2-furfural with antioxidant activity. The remaining solid fraction comprised porous carbons that can be used for energy production.

An alkaline treatment of apple pomace resulted in a product containing 26% of cellulose, a uronide fraction of 10–18%, 19.2% water-soluble hemicelluloses, and 15.3% lignin. This product had a higher dietary fibre and phenolic content than the original pomace and a high antioxidant property that may allow its use as a natural substitute for synthetic antioxidants (Sudha, 2011).

Fernández et al. (2019) characterized apple pomace obtained from the processing of a mixture of apples, mainly comprised of the Royal gala variety. The industrial apple pomace obtained had an 81% water content. After drying, it was characterized by a protein content of 5 g 100 g^{-1} of its dry weight, while carbohydrates were the major components (72 g 100 g^{-1} dry weight basis). These included 18 g 100 g^{-1} of free sugars, mainly fructose (77 mol%), and 53 g 100 g^{-1} of polysaccharides, having as its main sugars, glucose (41 mol%), galacturonic acid (19 mol%), arabinose (12 mol%), xylose (10 mol%), and galactose (9 mol%). This composition reflects the presence of pectic polysaccharides as soluble dietary fibre and hemicelluloses and cellulose as insoluble dietary fibre.

Lobo and Dorta (2019) characterized apple pomace and reported that it accounted for about 25% of the original fruit mass and contained 66.4% to 78.2% moisture and 26.4% DM. From DM, 4.0 g 100 g^{-1} are proteins and 9.5-22 g 100 g^{-1} are carbohydrates. The carbohydrates comprised sugars, cellulose, and pectin (10–15 g 100 g^{-1} dry matter) and the latter can be recovered by acid extraction and precipitation with ethanol. Phenolic compounds contribute significantly to the antioxidant activity of apple pomace. Flavonols (quercetin glycosides), cinnamic acids (chlorogenic and caffeic acids), flavanols (catechin and epicatechin), procyanidins, dihydrochalcones (phloridzin), and anthocyanins (cyanidin glycosides) were all present. Phenolic components have been reported as antioxidants, inhibitors of colon cancer *in vitro,* and as an inhibitor of the proliferation of CaCo-2 cells.

Vidović et al. (2020) reported that the total dietary fibre content in industrial apple pomace was 74% and that apple pomace contained approximately 10–15% pectin on a dry weight basis. They reported that 16 phenolic acids were identified in apple pomace samples of the Gala and Fuji cultivars and the content of free phenolic acids was 29 mg g^{-1} and 16 mg g^{-1}, respectively.

4 Nutraceutical compounds in apples: phenolic compounds

Stephen DeFelice (1995) defined nutraceuticals as 'food or part of a food that provides medical or health benefits, including the prevention and/or treatment of a disease'. Nutraceuticals are known as bioactive substances that can be delivered in the form of dietary supplements or functional foods, supplying beneficial effects in addition to nutritionally essential components (Nasri et al., 2014). They comprise a wide range of bioactive derivatives accumulated in edible sources including antioxidants, phytochemicals, and probiotics. With either established or potential effects, nutraceuticals are well known for their role in disease treatment and prevention, their anti-ageing properties, and malignancy prevention (AlAli et al., 2021; Yang, 2021).

Among the emerging food forms, 'functional foods' first appeared in Japan in the 1980s. One important function of food is that of nourishment, but these foods are called functional because they also have physiological benefits and/or reduce the risk of chronic diseases (Tur and Bibiloni, 2016). Although both the definition and the current regulatory framework for functional foods have variations according to criteria and countries, these products are a growing commercial reality (Alderete, 2006).

From all the above, it can be concluded that, in the case of apple pomace, phenolic compounds, dietary fibre, and pectin are present, and their isolation techniques and health effects are evaluated below.

Phenolics are compounds that have in their chemical structure an aromatic ring with one or more hydroxyl groups. Polyphenols are compounds that have more than one phenolic hydroxyl group attached to one or more aromatic rings (Yu and Ahmedna, 2013). In this chapter, the term 'phenolic compounds' is used to refer to both types of compounds.

Natural phenolic compounds are present in plant tissues. They can be classified according to different criteria with diverse results (Tsimogiannis and Oreopoulou, 2019; Shahidi and Ambigaipalan, 2015; Abbas et al., 2017; Yu and Ahmedna, 2013). According to the chemical structure, the natural phenolic compounds found in foods can be classified as phenolic acids, flavonoids, lignans, and/or stilbenes (Yu and Ahmedna, 2013). Different subgroups constitute the flavonoids group: flavonols, flavononols, flavones, flavanols (catechin), flavanones, anthocyanidins, and isoflavonoids (Shahidi and Ambigaipalan, 2015).

4.1 Phenolic compounds extraction

Apple by-products contain phenolic compounds with valuable antioxidant properties. These bioactives are present in the cell wall as free or bound phenolic forms.

Phenolic acids such as hydroxycinnamic acid form ether linkage with lignins. Procyanidins are bound to polysaccharides of the cell wall. For this reason, it is highly likely to find phenolic compounds in dietary fibre isolated fractions. Lignolytic enzymes degrade lignocellulose, facilitating phenolic compound extraction. The use of water for the extraction of phenolic compounds is currently proposed as an alternative to methanol and acetone. The objective is to decrease process costs and to increase process safety. Another proposed green solvent is the subcritical carbon dioxide (CO_2). However, the addition of a co-solvent such as ethanol or methanol is required to increase polarity of subcritical CO_2. Alternative technologies such as ultrasound-assisted extraction and microwave-assisted extraction have proven to save energy and to speed up the extraction process (Rabetafika et al., 2014).

Zheng et al. (2008) studied the effect of enzymatic treatment with pectinase on the yield of phenolic compounds, observing that it improved the extraction at pH 3.6, with an enzyme to pomace ratio of 12%, a temperature of 37°C, and a reaction time of 11 h. The yield of total phenolic compounds increased by about 28% in comparison to the control without enzymes.

Schieber et al. (2003) reported a process for the simultaneous recovery of pectin and polyphenols from apple pomace which involved the extraction of apple pomace at pH 2.8, the heating of the extract at 60°C, and its separation with an AmberliteXAD16 HP resin, producing an effluent containing the pectin, which was precipitated by means of ethanol, obtaining a light-coloured pectin with a high degree of esterification. The phenolic compounds adsorbed in the resin were eluted with methanol, the solvent was removed under vacuum, and the residues were freeze-dried. The main phenolic compounds detected were phloridzin, chlorogenic acid, and quercetin-3-*O*-glycosides, with appreciable amounts of catechin and procyanidin.

4.2 Phenolic compounds and health effects

Phenolic compounds contribute significantly to the antioxidant activity of apple pomace. They have been reported not only as antioxidants but also as inhibitors of colon cancer *in vitro* and as an inhibitor of the proliferation of CaCo-2 cells. As a consequence, these compounds are of considerable interest due to their potential beneficial health effects.

In the last few years, research concerning phenolic compounds has been abundant. Lu and Foo (2000) studied the antioxidant properties of apple pomace phenolic compounds using DPPH assay. These compounds exhibited better DPPH-scavenging properties than vitamin C or vitamin E. Superoxide anion radical scavenging activities were evaluated by a celular xanthine/xanthine oxidase system and apple pomace phenolic compounds were found to be effective superoxide scavengers, in comparison to vitamins C and E.

Eberhardt et al. (2000) reported that polyphenols from apple extracts inhibit tumour cell proliferation *in vitro*. Djukic (2016) evaluated the effect of the aglycone (quercetin) of the phenolic compound quercetin-3-glucoside, which is abundant in apple pomace, on the proliferation and apoptosis of human epithelial colorectal adenocarcinoma cells (HT-29), observing that, at the highest concentration tested (400 μM in cell culture), the aglycone inhibited 82% of carcinoma cell growth.

As the researchers who performed clinical trials using fractions obtained from edible plant tissues, in general, did not report the exact composition and the chemical structure of those fractions in relation to phenolic compounds, it is not possible to arrive at definitive conclusions regarding the chemical composition–chemical structure–health effect relationship. However, there is increasing evidence that consumption of a variety of phenolic compounds present in foods can prevent the risk of health disorders, or, at least, decrease this risk. This effect has been mainly attributed to their antioxidant activity. Apples are an important source of flavonoids, especially procyanidins, which have an antiproliferative function that involves the induction of cell death through the apoptotic pathway rather than through pro-oxidative stress. Moreover, various factors, including bioavailability, structural factors, and cytokine signal transduction, are involved in *in vivo* studies of the effect of phenolic compounds in tumour cell death, a fact that makes the unambigous determination of the mechanisms involved more complex (Shoji and Miura, 2014).

5 Nutraceutical compounds in apples: dietary fibre

The European Commission (2019), on the basis of the European Union (EU) regulation 1169/2011, defined fibre as 'carbohydrate polymers with three or more monomeric units, which are neither digested nor absorbed in the human small intestine and belong to the following categories:

- edible carbohydrate polymers naturally occurring in the food as consumed,
- edible carbohydrate polymers which have been obtained from food raw material by physical, enzymatic or chemical means and which have a beneficial physiological effect demonstrated by generally accepted scientific evidence,
- edible synthetic carbohydrate polymers which have a beneficial physiological effect demonstrated by generally accepted scientific evidence.'

However, the Codex Alimentarius (FAO, 2009) makes a different consideration in relation to the EU definition regarding the number of monomeric units in the definition of fibre and states a number of ten or more units, leaving it to national authorities to decide whether to define carbohydrates with three to nine monomers as fibre.

5.1 Fibre extraction

In order to obtain fibre concentrates from plant tissues, treatment of apple residues with water or ethanolic solutions have been assayed for the purpose of decreasing the amount of free monosaccharides present. The aqueous treatment produces pectin-rich fibre with a significant amount of bioactive phenolic compounds. The alcoholic process is less efficient in relation to fibre concentration. In both cases, protein is present in the concentrates obtained.

A bleaching process with alkaline hydrogen peroxide at pH 12 is applied to the brownish apple pomace for the purpose of generating light-coloured fibre concentrates. The final product is enriched in cellulose due to the bleaching process modifying the fibre proportion because of extraction of lignin and a major part of the pectins. This also determines a low yield for the process. Other bleaching processes using ultrasound and ozone have also been proposed (Rabetafika et al., 2014).

Kołodziejczyk et al. (2007) reported that Poland produces up to 200 000 tons of clear apple juice concentrates annually. Approximately 12–20% of processed raw material remains as a waste product after juicing, containing carbohydrates, acids, and proteins. Pomace contains polyphenolics and fibre with beneficial effects for human health. The authors proposed to air dry the pomace from Red apple cultivars at 70°C. The dried pomace contained 61 g 100 g^{-1} of total dietary fibre, of which 8 g 100g^{-1} was soluble dietary fibre and the content of polyphenols in the examined pomace was 300 mg 100 g^{-1} dry matter. Dietary fibre fraction was obtained by means of dried apple pomace, which was grinded and extracted with ethanol. The remaining solid was hot air dried at 70°C. The dietary fibre fraction post-extraction constituted of 72 g 100g^{-1} of total dietary fibre (TDF) and 10 g 100g^{-1} of soluble dietary fibre (SDF), observing an SDF/TDF of 14.4 g 100 $g^{-1}$and 9 g 100 g^{-1} moisture content. The elimination of solvent from ethanol extract produces a solid polyphenol concentrate, which showed a total polyphenol content of 690 mg 100 g^{-1} DM. The analysis by HPLC of flavan-3-ols (procyanidins, epicatechin, quercetin glycosides, quercetin, and phloridzin) showed a concentration of 255 mg 100 g^{-1} DM, and the higher concentration was shown by quercentin glycosides.

Figuerola et al. (2005) used residues from juice extraction of Royal gala (Granny Smith and Liberty cultivars) apples as the fibre source. Apple pomace was washed with warm water (30°C), dried at 60°C for 30 min in an air tunnel drier, and ground to a particle size of 500–600 μm. The material was washed under mild conditions to avoid or minimize losses of soluble fibre components (such as pectins and pentosans) as well as bioactive components such as flavonoids, polyphenols, and carotenes. Washing allowed the reduction of free sugar and ash contents. The concentrates contained 61–90 g 100 g^{-1} dry matter of total dietary fibre. Lipid content (1.6–4.5 g) and moisture content (2 g 100 g^{-1} dry matter) were low. The fibre obtained showed values of water retention capacity of 1.6–1.8 g water per g of DM and a swelling capacity of 6.6–8.3 g water per g of DM. Fibre with strong hydration properties can increase stool weight, increase food viscosity, and diminish the rate of absorption from the intestine.

Procentese et al. (2018) reported the use of deep eutectic solvents (DESs) constituted by choline chloride:glycerol or choline chloride:ethylene glycol for the pre-treatment of apple residues from the European food and drink industries. DESs are eutectic mixtures with a eutectic point lower than that of the ideal liquid mixture. They are green solvents and possess the ability to pretreat selectively and dissolve the constituents of biomass, facilitating the production of value-added products (Kalhor and Ghandi, 2019). The operating conditions consisted of 3 h of treatment with a biomass to solvent ratio of 1:8 g g^{-1}, 1:16 g g^{-1}, and 1:32 g g^{-1}. The temperatures used were 60, 115, and 150°C. They observed a composition of apple residues (percentage calculated on biomass dry basis) of 21.2% of glucan, 12.2% of xylan, 2.5% of arabinan, and 18.5% of lignin. The use of both DESs produced a decrease of xylan, arabinan, and lignin content and an increase in glucan content in the products obtained with an increase in temperature or with an increase in biomass to solvent ratio at a certain temperature. The decrease in lignin content in the products obtained from DESs application might increase the enzymatic digestibility of the products obtained, affecting their functionality.

5.2 Fibre and health effects

American adults eat 10–15 g of total fibre per day, while the USDA recommends a daily amount of 25 g for women and 38 g for men, up to age 50. Women and men aged 50 and over should have a daily intake of 21 g and 30 g, respectively (McManus, 2019).

Dietary fibre reaches the large intestine and is fermented by colonic microflora with the production of SCFAs (acetate, propionate, and butyrate). Butyrate is considered to be the primary nutrient for the epithelial cells lining the colon, and SCFAs stimulate the proliferation of colonic epithelial cells.

Soluble fibre dissolves in water, forming a gel, helping to lower cholesterol levels, reducing the risk of heart disease, and regulating blood sugar levels. High insulin levels are linked to cardiovascular disease and diabetes. Insoluble fibre passes through the digestive system relatively intact, adding bulk to stools. This fibre prevents constipation and regulates bowel movements, removing waste from the body in a timely manner. Weight control is another benefit of high fibre diets as it helps the person feel full for longer after a meal or snack, and high fibre whole grains can help them eat less (Nishida and Martinez Nocito, 2007; Chew and Brownlee, 2018).

6 Nutraceutical compounds in apple: pectin

The primary cell wall of plant tissues is constituted mainly of the polysaccharides cellulose, hemicelluloses, and pectin.

Pectin is a soluble fibre and, from a chemical point of view, is a complex polysaccaride that consists mainly of homogalacturonan (HG), rhamnogalacturonan I (RG-I), and rhamnogalacturonan II (RG-II), which can be covalently linked to form a pectic network throughout the primary cell wall matrix and middle lamellae (Willats et al., 2001).

Pectin can be obtained from apple pomace and is used as a gelling agent, thickener, emulsifier, and stabilizer in the food industry and in the cosmetic and pharmaceutical industries.

6.1 Pectin extraction

Various processes have been used to extract pectins in apple by-products (Rabetafika et al., 2014). The conventional fractionation method of pectins from apple pomace consists of treating the raw materials with hot mineral acid such as hydrochloric, sulphuric, and nitric acids at a pH range of 1.5–3.0. Organic acids such as citric acid have also been assayed. The pH and temperature of extraction, as well as the solid to solvent ratio, influence the yield and quality of pectins. From the viscous extract obtained following filtration or centrifugation, the pectin is precipitated, usually by means of ethanol. In general, apple renders pectin with a high degree of esterification but the increase in acid concentration or the use of alkali or enzymes can produce pectin with a low degree of esterification.

Increasingly, environmentally friendly water-based processes are being developed to extract pectins from apple pomace to minimize the environmental impacts, and this water-based extraction can be combined with physical or enzymatic treatments to facilitate the extraction.POther non-conventional extractions can also be used, such as subcritical water extraction, ultrasound, microwave, and pulsed electric field-assisted extractions.

Canteri-Schemin et al. (2005) optimized, on a small scale, the extraction of pectin from apple pomace. The highest yields were obtained when pomace was dried and ground to obtain an apple flour (granulometry 106 μm) to be used as the raw material. Citric acid concentration was 6.2 g 100 mL^{-1}, and the time of extraction was 153 min. The apple variety in itself was not significant in pectin yield.

Wang et al. (2007) used a microwave-assisted extraction system at a frequency of 2450 MHz for extraction of pectin from dried apple pomace powder. The optimization of the process showed that extraction at pH 1.0 using HCl, a microwave power of 499.4 W, extraction time of 21 min, and a solid to liquid ratio of 0.069:1, produced a yield of 15.8 g 100 g^{-1} of dried apple pomace powder. They also concluded that microwave-assisted extraction reduced the pectin extraction time considerably.

Liang et al. (2018) studied a treatment with steam explosion prior to fibre obtention and analysed the effect of this treatment on the yield and functionality of dietary fibre. They applied an optimized procedure for this treatment, which comprised a steam pressure of 0.51 MPa, a residence time in the equipment of 168 s, and a sieving step with a mesh size of 60. After this treatment they undertook a fibre obtention procedure that comprised dispersion in petroleum ether, dispersion in ethanol, and filtration. Enzymatic treatment with cellulase (0.1%, w/w) produced a supernatant and sediment and the treatment of the former with 95% (v/v) ethanol followed by centrifugation produced a floculate that was dried at 70°C, which, after grinding, created a soluble dietary fibre powder. The direct procedure of soluble dietary fibre extraction produced a yield of 6.25%, which, after steam explosion, was 29.85%. The soluble fibre isolated after the pre-treatment exhibited higher porosity and roughness, showing a change in structure as a consequence of the pre-treatment.

6.2 Pectin and health effects

Mateos-Aparicio et al. (2020) studied the pectins contained in apple by-products. They reported that the side chains of the RG-I (arabinans, galactans, and arabinogalactans) can be used by colon microorganisms that produce acetate and butyrate. These SCFAs decreased the pH and enhanced intestinal mucosal cell proliferation. The bile acid binding effect led to a decrease in triglycerides and a rise in high-density lipoprotein cholesterol.

Zhang et al. (2015) reported that pectin prevents colon cancer. Pectin was modified to enhance its bioavailability and bioactivity through the generation of pectin fragments with low molecular mass and a low degree of esterification. This type of saccharide is reported to inhibit tumour growth, induce apoptosis, suppress metastasis, and modulate immunological responses. Anti-tumour activity of modified pectin fragments arises from intervention in ligand recognition by galectin-3. Pectin is also an adequate vehicle for anti-cancer drug delivery systems.

Delphi and Sepehri (2016) studied the effect of pectic acid in relation to breast cancer. They employed MTT (3-(4, 5-dimethylthiazol-2-yl)-2, 5-diphenyltetrazolium bromide) cell proliferation assays, double fluorescence staining (acridine orange/ethidium bromide), and cell cycle analysis to measure apoptosis *in vitro*. 4T1 cells were implanted into female BALB/c mice for *in vivo* studies. The results of *in vitro* studies showed that a 0.1% concentration of pectic acid could induce apoptosis, inhibiting cell growth. *In vivo* studies showed that pectic acid could inhibit the progression of tumours through over-expression of P53, increasing the number of apoptotic cells. The results demonstrated that this acid, a natural component of apple, can prevent metastasis in both cancer cell lines and primary tumours. This potential effect is mainly due to its ability to induce apoptosis.

Leclere et al. (2013) stated that modified pectin is one of the most promising anti-metastatic drugs, especially if used in combination with more conventional molecules. It exerts anti-tumour activity through a multiplicity of mechanisms that depend on the structure of pectin. Pectin is not digested in the upper gastrointestinal tract and can protect cells from mutagenic attacks. In the colon, pectin is fermented by bacteria, creating SCFAs, for example, butyrate, which inhibit colon inflammation and prevent carcinogenesis. HG-rich pectin also prevents inflammatory cell activation. Galactan-rich pectin (RG-I) is capable of interacting with galectin-3, inhibiting cancer cell metastasis, and some heat-modified pectin can trigger apoptosis in cancer cells in a galectin-3-independent manner.

The studies revealed that apple and apple pomace are important sources of phenolic compounds, total dietary fibre, and pectin, which have significant health effects. The isolation of these bioactives from apple pomace could contribute to the development of functional foods. Moreover, the use of apple pomace to obtain these compounds would add value to the raw material and create less environmental pollution.

7 Improving apple cultivation through research on food bioactives

The summer apples of the Coruh Valley, located in the northeastern part of Turkey, ripen between the end of July and the beginning of August. Their colours are varied, the peels ranging from yellow to purple, possessing an intense flavour. This valley is considered one of 34 hotspots for plant biodiversity by the International Union for Conservation of Nature. However, there has been no systematic study of the genotype and phenotypic characteristics of summer apples from this area.

Gecer et al. (2020) studied the horticultural characteristics of 22 genotypes of these apples, known as CVE1 to CVE22. Apples from 2018 were used in their state of commercial maturity, and their mechanical and biochemical characteristics were studied. The firmness of the genotypes used varied from values of 3.41 kg/cm^2 (CVE5) to 6.11 kg/cm^2 (CVE3). Significant differences ($P < .05$) were observed between genotypes for total phenolic content and for antioxidant capacity, observing that the genotype CVE16 presented the highest value (138.1 mg gallic acid equivalent 100 g^{-1}) and CVE1 showed the lowest value (72.0 mg gallic acid equivalent 100 g^{-1}). It is noteworthy that CVE16 was harvested on 8 August 2018 (mid-ripening) and CEV1 on 9 July 2018 (early ripening). The antioxidant capacity of the CVE16 genotype was higher (127.2 μmol trolox equivalents 100 g^{-1}), and the CVE1 genotype presented lower values (59.1 μmol trolox equivalents 100 g^{-}) than all the other genotypes.

The authors proposed that the results of this work could indicate that these genotypes have the potential to guide the development of new summer apple cultivars in breeding activities. Accordingly, the CVE3 genotype would obtain firmer apples, and the genotype CVE16 would allow the development of cultivars with a higher phenolic content and higher antioxidant activity, allowing farmers to choose genotype candidates for future breeding activities.

Similarly, Bilbrey et al. (2021) developed a platform based on information provided by a study of 124 apple varieties in terms of their chemical composition (i.e. sugars, acids, and antioxidant compounds) and their genetic characteristics. For this purpose, three sets of experimental progenies ('Honeycrisp' × 'Fuji', 'Goldrush' × 'Sweet 16', and 'Honeycrisp' × 'MSH 10-1'), 23 members of their pedigree-connected families, wild apples, and apples of reproductive interest were studied, and their genotype and phenotype were determined. Associated studies were carried out between the genotype and the phenotypic data acquired from LC-MS (liquid chromatography-mass spectrometry) techniques and 1H nuclear magnetic resonance, looking for candidate genes that could be responsible for the production of compounds with a positive effect on human health. The location in the genome of the *trait loci* linked to the production of these compounds was determined and validated by pedigree-based analysis.

The developed platform was tested (proof of concept) through the study, in particular, of the production of apples with a high load of chlorogenic acid, proving its usefulness. Consequently, this platform has proven to be adequate to allow the development of plant breeding based on the presence of genes responsible for the production of compounds with a positive effect on human health. Its usefulness could be extended to other selection factors of apple varieties, such as yield by hectare, tolerance to different climatic factors, and tolerance to diseases, among others. This platform could also help to shorten timescales for the development of new apple varieties, and shows the importance of basic research for improving apple cultivation.

8 Future trends

The information reported in the bibliography regarding the composition and antioxidant activity of the fruit of *M. domestica* is highly variable since it is affected by different dimensions such as geographical location, climate, maturation degree, cultivar, and tissue.

The studies reported show that apple and apple pomace are important sources of phenolic compounds, total dietary fibre, and pectin, which have significant health effects. The isolation of these bioactives from apple pomace would add value to the raw material and contribute to the reduction of environmental pollution and to the development of functional foods.

To improve knowledge on this subject, it is necessary to:

1 Increase the evaluation of bioactives present in apple varieties to assure the existence of reliable databases that compile systematic information.

2 Study alternative isolation techniques for these bioactives that produce minimal damage to the environment and contribute optimal health perfomance.
3 Carry out further clinical studies to reach definitive conclusions regarding the health effect of the different bioactives present in apples, their effectiveness, possible side effects, interactions, and dosage. These rigorous medical studies must be accompanied by knowledge of the chemical composition of the isolated fractions containing the bioactives used for these studies, to help state the link compound-health effect.
4 Increase the use of systematic information of bioactives present in different apple varieties to help the breeding process and to produce varieties of *M. domestica* with enhanced nutraceutical content. This will help obtain healthier fruits and a higher yield when these compounds are extracted from the agro-industrial residues of the fruit.

9 Where to find more information

More details on *M. domestica* production, agroindustrial residues and genotype-phenotype studies can be found at the following:

- http://www.fao.org/publications/card/en/c/e006e995-6921-4280-b19e-0cb42461b306/.
- https://www.cabi.org/isc/datasheet/31964.
- https://www.sciencedirect.com/topics/agricultural-and-biological-sciences/malus-domestica.
- http://www.fao.org/3/ca6030en/ca6030en.pdf.
- https://www.nature.com/articles/s41438-019-0190-y.

10 Acknowledgements

The authors acknowledge the financial support received from National Scientific and Technical Research Council - Argentina (CONICET), Buenos Aires University (UBA), and the National Agency of Technological and Scientific Research.

11 References

Abbas, M., Saeed, F., Anjum, F. M., Afzaal, M., Tufail, T., Bashir, M. S., Ishtiaq, A., Hussain, S. and Suleria, H. A. R. (2017). Natural polyphenols: An overview, *Int. J. Food Prop.* 20(8): 1689–1699.

AlAli, M., Alqubaisy, M., Aljaafari, M. N., AlAli, A. O., Baqais, L.,Molouki, A., Abushelaibi, A., Lai, K. S., Lim, S-H. E. (2021). Nutraceuticals: transformation of conventional foods into health promoters/disease preventers and safety considerations. *Molecules* 26(9): 2540. doi: 10.3390/molecules26092540.

Alderete, J. M. (2006). Alimentos funcionales. Consolidación de una tendencia, *Aliment. Argent.* 34: 15–17. In Spanish.

Almeida, D. P. F., Gião, M. S., Pintado, M. and Gomes, M. H. (2017). Bioactive phytochemicals in apple cultivars from the Portuguese protected geographical indication "Maçã de Alcobaça": Basis for market segmentation, *Int. J. Food Prop.* 20(10): 2206–2214.

Bhagwat, S. and Haytowitz, D. B. (2015). *USDA Database for the Flavonoid Content of Selected Foods Release 3.2*. United States Department of Agriculture, Agricultural Research Service. Nutrient Data Laboratory. http://www.ars.usda.gov/nutrientdata/flav. Accessed July 26, 2021.

Bilbrey, E. A., Williamson, K., Hatzakis, E., Miller, D. D., Fresnedo-Ramírez, J. and Cooperstone, J. L. (2021). Integrating genomics and multi-platform metabolomics enables metabolite QTLdetection in breeding-relevant apple germplasm, *New Phytol.* 232(5): 1944–1958 https://doi.org/10.1111/nph.17693.

Boyer, J. and Liu, R. H. (2004). Apple phytochemicals and their health benefits, *Nutr. J.* 3: 5. https://doi.org/10.1186/1475-2891-3-5.

Calinoiu, L. F., Mitrea, L., Precup, G., Binde, M., Rusus, B. (2017). Characterization of grape and apple peel wastes' bioactive compounds and their increased bioavailability after exposure to thermal process. *Bull. UASVM Food Sci. Technol.* 74(2). https://doi.org/ 10.15835/buasvmcn-fst: 0028.

Canteri-Schemin, M. H., Ramos Fertonani, H. C. R., Waszczynskyj, N. and Wosiacki, G. (2005). Extraction of pectin from apple pomace, *Braz. Arch. Biol. Technol.* 48(2): 259–266.

Chew, K. Y. and Brownlee, I. A. (2018). The impact of supplementation with dietary fibers on weight loss: A systematic review of randomised controlled trials, *Bioact. Carbohydr. Diet. Fibre* 14: 9–19.

DeFelice, S. L. (1995). The nutraceutical revolution: Its impact on food industry R&D, *Trends Food Sci. Technol.* 6(2): 59–61.

Delphi, L. and Sepehri, H. (2016). Apple pectin: A natural source for cancer suppression in 4T1 breast cancer cells *in vitro* and express p53 in mouse bearing 4T1 cancer tumors, *in vivo*, *Biomed. Pharmacother.* 84: 637–644.

Djukic, V. (2016). Apple pomace polyphenols and their effect on the proliferation of human epithelial colorectal adenocarcinoma cells (HT-29). A master thesis presented to The University of Guelph, Guelph, Ontario, Canada. https://atrium.lib.uoguelph.ca/xmlui/bitstream/handle/10214/10041/Djukic_Vanja_201609_Msc.pdf?sequence=3&isAllowed=y. Accessed July 25, 2021.

Eberhardt, M. V., Lee, C. Y. and Liu, R. H. (2000). Antioxidant activity of fresh apples, *Nature* 405(6789): 903–904.

Elhakem, A. H., Almatrafi, M. M., Benajiba, N., Koko, N. N. and Sami, R. (2021). Comparative analysis of bioactive compounds, antioxidant and anti-inflammatory activities of apple varieties, *Asian J. Plant Sci.* 20(1): 61–66.

European Commission (2019). Dietary fibre. https://ec.europa.eu/jrc/en/health-knowledge-gateway/promotion-prevention/nutrition/fibre.

FAO (2009). Codex Alimentarius Commission FAO/WHO distribution of the report of the 30th session of the Codex Committee on Nutrition and Foods for Special Dietary Uses (ALINORM 09/32/26).

FAOSTAT (2019). Food and Agriculture Organization of the United Nations. FAO Statistics Online Database. http://www.fao.org/faostat. Accessed July 25, 2021.

Fernández, P. A. R., Ferreira, S. S., Bastos, R., Ferreira, I., Cruz, M. T., Pinto, A., Coelho, E., Passos, C. P., Coimbra, M. A., Cardoso, S. M. and Wessel, D. F. (2019). Apple pomace extract as a sustainable food ingredient, *Antioxidants (Basel)* 8(6): 189-204.

Figuerola, F., Hurtado, M. L., Estevez, A. M., Chiffelle, I. and Asenjo, F. (2005). Fibre concentrates from apple pomace and citrus peel as potential fibre sources for food enrichment, *Food Chem.* 91(3): 395-401.

Francini, A. and Sebastiani, L. (2013). Phenolic compounds in apple (*Malus × domestica* Borkh.): Compounds characterization and stability during postharvest and after processing, *Antioxidants (Basel)* 2(3): 181-193.

Gecer, M. K., Ozkan, G., Sagbas, H. I., Ilhan, G., Gundogdu, M. and Ercisli, S. (2020). Some important horticultural properties of summer apple genotypes from Coruh Valley in Turkey, *Int. J. Fruit Sci.* 20(Suppl. 3): S1406-S1416.

Guardia, L., Suárez, L., Querejeta, N., Rodríguez Madrera, R., Suárez, B. and Centeno, T. A. (2019). Apple waste: A sustainable source of carbon materials and valuable compounds, *ACS Sustainable Chem. Eng.* 7(20): 17335-17343.

Hamam, M., Chinnici, G., Di Vita, G., Pappalardo, G., Pecorino, B., Maesano, G. and D'Amico, M. (2021). Circular economy models in agrofood systems: A review, *Sustainability* 13(6): 3453. https://doi.org/10.3390/su13063453.

Harvard, T. H. and Chan School of Public Health (2021). Apples. In: *The Nutrition Source*. https://www.hsph.harvard.edu/nutritionsource/food-features/apples/. Accessed July 15, 2021.

Hyson, D. A. (2011). A Comprehensive review of apples and apple components and their relationship to human health, *Adv. Nutr.* 2(5): 408-420.

Kalhor, P. and Ghandi, K. (2019). Deep eutectic solvents for pretreatment, extraction, and catalysis of biomass and food waste, *Molecules* 24(22): 4012-4048. https://doi.org/10.3390/molecules24224012.

Kołodziejczyk, K., Markowski, J., Kosmala, M., Król, B. and Płocharski, W. (2007). Apple pomace as a potential source of nutraceutical products, *Pol. J. Food Nutr. Sci.* 57(4): 291-295.

Kosseva, M. R. (2013). Recovery of commodities from food wastes using solid-state fermentation. In: Kosseva, M. R., Webb, C. and Chapter, V. (Eds). *Food Industry Wastes*. The Netherlands: Academic Press Imprint. Elsevier, pp. 77-102. https://doi.org/10.1016/b978-0-12-391921-2.00005-6.

Koutsos, A., Tuohy, K. M. and Lovegrove, J. A. (2015). Apples and cardiovascular health - Is the gut microbiota a core consideration?, *Nutrients* 7(6): 3959-3998.

Kumar, P., Sethi, S., Sharma, R. R., Singh, S., Saha, S., Sharma, V. K., Verma, M. K. and Sharma, S. K. (2018). Nutritional characterization of apple as a function of genotype, *J. Food Sci. Technol.* 55(7): 2729-2738.

Leclere, L., Cutsem, P. V. and Michiels, C. (2013). Anti-cancer activities of pH- or heat-modified pectin, *Front. Pharmacol.* 4(4): 128. http://doi.org/10.3389/fphar.2013.00128.

Liang, X., Ran, J., Sun, J., Wang, T., Jiao, Z., He, H. and Zhu, M. (2018). Steam-explosion-modified optimization of soluble dietary fiber extraction from apple pomace using response surface methodology, *CyTA J. Food* 16(1): 20-26.

Lobo, M. G. and Dorta, E. (2019). Chapter 19: Utilization and management of horticultural waste in postharvest technology of perishable horticultural commodities. In:

Yahia, E. M. (Ed.). *Postharvest Technology of Perishable Horticultural Commodities*. Amsterdam, The Netherlands: Woodhead Publishing Imprint. Elsevier, pp. 639–666.

Lu, Y. and Foo, L. Y. (2000). Antioxidant and radical scavenging activities of polyphenols from apple pomace, *Food Chem.* 68(1): 81–85.

Macit, İ, Aydın, E., Tas, A. and Gundogdu, M. (2021). Fruit quality properties of the local apple varieties of Anatolia, *Sustainability* 13(11): 6127–6136. https://doi.org/10.3390/su13116127.

Manzoor, M., Anwar, F., Saari, N. and Ashraf, M. (2012). Variations of antioxidant characteristics and mineral contents in pulp and peel of different apple (*Malus domestica* Borkh.) cultivars from Pakistan, *Molecules* 17(1): 390–407.

Mateos-Aparicio, I., De la Pena Armada, R., Perez-Cozar, M. L., Ruperez, P., Redondo-Cuenca, A. and Villanueva-Suarez, M. J. (2020). Apple by-product dietary fibre exhibits potential prebiotic and hypolipidemic effectsin high-fat fed Wistar rats, *Bioact. Carbohydr. Diet. Fibre* 23: 100219. https://doi.org/10.1016/j.bcdf.2020.100219.

McManus, K. D. (2019). Should I be eating more fiber?, *Harvard Health Publishing of the Harvard Medical School*. https://www.health.harvard.edu/blog/should-i-be-eating-more-fiber-2019022115927. Accessed July 25, 2021.

Mourtzinos, I. and Goula, A. (2019). Chapter 2: Polyphenols in agricultural by-products and food waste. In: Watson, R. R. (Ed.). *Polyphenols in Plants: Isolation, Purification and Extract Preparation*. Amsterdam, The Netherlands: Academic Press Imprint Elsevier, pp. 23–44.

Nasri, H., Baradaran, A., Shirzad, H. and Rafieian-Kopaei, M. (2014). New concepts in nutraceuticals as alternative for pharmaceuticals, *Int. J. Prev. Med.* 5(12): 1487–1499.

Navarro, M., Moreira, I., Arnaez, E., Quesada, S., Azofeifa, G., Vargas, F., Alvarado, D. and Chen, P. (2018). Polyphenolic characterization and antioxidant activity of *Malus domestica* and *Prunus domestica* cultivars from Costa Rica, *Foods* 7(2): 15–33. https://doi.org/10.3390/foods7020015.

Nishida, C. and Martinez Nocito, F. (2007). FAO/WHO scientific update on carbohydrates in human nutrition: Introduction, *Eur. J. Clin. Nutr.* 61 (Suppl. 1): S1–S4.

Oshima, R., Dagallier, B. and Kearns, P. W. E. (2019). *OECD Consensus Document on the Biology of Apple (Malus domestica Borkh), Series on Harmonisation of Regulatory Oversight in Biotechnology*.

Pereira-Lorenzo, S., Ramos-Cabrer, A. M. and Fischer, M. (2009). Breeding Apple (*Malus x domestica* Borkh). In: Mohan Jain, S. and Priyadarshan, P. M. (Eds). *Breeding Plantation Tree Crops: Temperate Species*. Cham: Springer Media, pp. 33–81.

Perussello, C. A., Zhang, Z., Marzocchella, A. and Tiwari, B. K. (2017). Valorization of apple pomace by extraction of valuable compounds, *Compr. Rev. Food Sci. Food Saf.* 16(5): 776–796.

Piagentini, A. M. and Pirovani, M. E. (2017). Total phenolics content, antioxidant capacity, physicochemical attributes, and browning susceptibility of different apple cultivars for minimal processing, *Int. J. Fruit Sci.* 17(1): 102–116.

Procentese, A., Raganati, F., Olivieri, G., Russo, M. E., Rehmann, L. and Marzocchella, A. (2018). Deep eutectic solvents pretreatment of agro-industrial food waste, *Biotechnol. Biofuels* 11: 37. https://doi.org/10.1186/s13068-018-1034-y.

Rabetafika, H. N., Bchir, Bchir, B., Blecker, C. and Richel, A. (2014). Fractionation of apple by-products as source of new ingredients: Current situation and perspectives, *Trends Food Sci. Technol.* 40(1): 99–114.

Saveleva, N., Borzykh, N., Chivilev, V., Yushkov, A., Zemisov, A. and Cherenkova, T. (2021). Biochemical composition of scab-immune apple fruits varieties (*Malus domestica* B.) as a valuable component of healthy dietary, *BIO Web Conf.* 30. https://doi.org/10.1051/bioconf/20213001018.

Schieber, A., Hilt, P., Streker, P., Endre, H. U., Rentschler, C. and Carle, R. (2003). A new process for the combined recovery of pectin and phenolic compounds from apple pomace, *Innov. Food Sci. Emer. Tech.* 4(1): 99–107.

Shahidi, F. and Ambigaipalan, P. (2015). Phenolics and polyphenolics in foods, beverages and spices: Antioxidant activity and health effects: A review, *J. Funct. Foods* 18: 820–897.

Shoji, T. and Miura, T. (2014). Chapter 104: Apple polyphenols in cancer prevention. In: Watson, R., Preedy, V. R. and Zibadi, S. (Eds). *Polyphenols in Human Health and Disease*. Amsterdam, The Netherlands: Academic Press Imprint Elsevier, pp. 1373–1383.

Smanalieva, J., Iskakova, J., Oskonbaeva, Z., Wichern, F. and Darr, D. (2020). Investigation of nutritional characteristics and free radical scavenging activity of wild apple, pear, rosehip, and barberry from the walnutfruit forests of Kyrgyzstan, *Eur. Food Res. Technol.* 246(5): 1095–1104.

Sudha, M. L. (2011). Chapter 36: Apple pomace (By-product of fruit juice industry) as a flour fortification strategy. In: Preedy, V., Watson, R. and Patel, V. (Eds). *Flour and Breads and Their Fortification in Health and Disease Prevention*. Amsterdam, The Netherlands: Academic Press Imprint, Elsevier, pp. 395–405.

Tsimogiannis, D. and Oreopoulou, V. (2019). Chapter 16: Classification of phenolic compounds in plants. In: Watson, R. R. (Ed.). *Polyphenols in Plants*. Amsterdam, The Netherlands: Academic Press Imprint Elsevier, pp. 263–284.

Tur, J. M. and Bibiloni, M. M. (2016). Functional foods. In: Caballero, B., Finglas, P. M. and Todrá, F. (Eds). *Encyclopedia of Food and Health (Reference Module in Food Science)*. The Netherlands: Academic Press Imprint Elsevier.

Ulaszewska, M., Vázquez-Manjarrez, N., Garcia-Aloy, M., Llorach, R., Mattivi, F., Dragsted, L. O., Praticò, G. and Manach, C. (2018). Food intake biomarkers for apple, pear, and stone fruit, *Genes Nutr.* 13: 29. https://doi.org/10.1186/s12263-018-0620-8.

U.S. Department of Agriculture and Agricultural Research Service (USDA) (2019). FoodData central. fdc.nal.usda.gov, *Database*. https://fdc.nal.usda.gov/fdc-app.html#/food-details/171688/nutrients. Accessed July 26, 2021.

Veberic, R., Trobec, M., Herbinger, K., Hofer, M., Grill, D. and Stampar, F. (2005). Phenolic compounds in some apple (*Malus domestica* Borkh) cultivars of organic and integrated production, *J. Sci. Food Agric.* 85(10): 1687–1694.

Vidović, S., Horecki, A. T., Vladić, J., Šumić, Z., Gavarić, A. and Vakula, A. (2020). Chapter 2: Apple. In: Galanakis, C. (Ed.). *Valorization of Fruit Processing By-Products*. Amsterdam, The Netherlands: Academic Press Imprint Elsevier, pp. 17–42.

Vukušić, J. L., Millenautzki, T., Cieplik, R., Obst, V., Abdechafik, S., Saaid, M., Clavijo, L., Zlatanovic, S., Hof, J., Mösche, M. and Barbe, S. (2021). Reshaping apple juice production into a zero discharge biorefinery process, *Waste Biomass Valoriz* 12: 3617–3627.

Wang, S., Chen, F., Wu, J., Wang, Z., Liao, X. and Hu, X. (2007). Optimization of pectin extraction assisted by microwave from apple pomace using response surface methodology, *J. Food Eng.* 78(2): 693–700.

Willats, W. G. T., McCartney, L., Mackie, W. and Knox, J. P. (2001). Pectin: Cell biology and prospects for functional analysis, *Plant Mol. Biol.* 47(1–2): 9–27.

Yang, M. (2021). Regulatory aspects of nutraceuticals: chinese perspective. Chapter 75, pp. 1281–1291. In: *Nutraceuticals: Efficacy, Safety and Toxicity (Second Edition)*, Gupta, R. C., Lall, R., Srivastava, A (Eds.). The Netherlands: Academic Press Imprint, Elsevier. https://doi.org/10.1016/B978-0-12-821038-3.00075-6.

Yu, J. and Ahmedna, M. (2013). Functional components of grape pomace: Their composition, biological properties and potential applications, *Int. J. Food Sci. Technol.* 48(2): 221–237.

Zhang, W., Xu, P. and Zhang, H. (2015). Pectic in cancer therapy: A review, *Trends Food Sci. Technol.* 44(2): 258–271.

Zheng, H. Z., Lee, H. R., Lee, S. H., Kim, S. and Chung, K. (2008). Pectinase assisted extraction of polyphenols from apple pomace, *Chin. J. Anal. Chem.* 36: 306–310.

Zielinska, D., Laparra-Llopis, J. M., Zielinski, H., Szawara-Nowak, D. and Giménez-Bastida, J. A. (2019). Role of apple phytochemicals, phloretin and phloridzin, in modulating processes related to intestinal inflammation, *Nutrients* 11(5): 1173. https://doi.org/10.3390/nu11051173.

Chapter 15

Advances in understanding and improving the nutraceutical properties of broccoli and other Brassicas

Elsa M. Gonçalves, Unidade de Tecnologia e Inovação, Instituto Nacional de Investigação Agrária e Veterinária and GeoBioTec – Geobiociências, Geoengenharias e Geotecnologias, Faculdade de Ciências e Tecnologia, Universidade Nova de Lisboa, Portugal; Carla Alegria, SFCOLAB – Associação Smart Farm COLAB Laboratório Colaborativo para a Inovação Digital na Agricultura, Rua Cândido dos Reis n°1, Espaço SFCOLAB and cE3c – Centre for Ecology, Evolution and Environmental Changes, Faculdade de Ciências, Universidade de Lisboa, Portugal; Ana Cristina Ramos, Unidade de Tecnologia e Inovação, Instituto Nacional de Investigação Agrária e Veterinária and GeoBioTec – Geobiociências, Geoengenharias e Geotecnologias, Faculdade de Ciências e Tecnologia, Universidade Nova de Lisboa, Portugal; and Marta Abreu, Unidade de Tecnologia e Inovação, Instituto Nacional de Investigação Agrária e Veterinária and LEAF, Linking Landscape, Environment, Agriculture and Food, School of Agriculture, Universidade de Lisboa, Portugal

1 Introduction

Biologically active molecules present in food are often known as nutraceuticals, since they have characteristics similar to both nutrients and pharmaceuticals. Nutraceuticals are natural bioactive chemical compounds, which have a nutritional role and provide health-promoting and disease curative or preventive properties (Sachdeva et al., 2020) and, consequently, delay the ageing process, increase life expectancy and support the structure or function

http://dx.doi.org/10.19103/AS.2022.0101.18

of the body. Numerous studies show that nutraceuticals are promising agents to prevent and/or treat specific diseases such as cardiovascular conditions, diabetes, atherosclerosis, cancer and neurological disorders (Aquila et al., 2019; McClements, 2019; Sarris et al., 2019; Ashwlayan and Nimesh, 2018; Sosnowska et al., 2017). Vegetables receive considerable attention regarding the human diet because they are safe to eat, low in calories and tasty. Other important reasons are their potential nutritional value and their therapeutic effects. Indeed, vegetables are considered nutraceutical foods because they are rich in health-promoting minerals, vitamins, fibre and phytochemical compounds. Vitamins and minerals are considered micronutrients and play essential roles in metabolism. Fibre is a class of nondigestible carbohydrates associated with numerous health benefits. Some types of fibre act as a prebiotic, a substance that alters the types and activities of bacteria or microbiota present in the human gut (Camara et al., 2017).

Phytochemical compounds also have many health-promoting effects, acting as antioxidants against many diseases or as antibacterial, antifungal, antiviral, antithrombotic, anti-inflammatory or cholesterol-lowering agents. Phytochemicals comprise an array of compounds such as polyphenols, alkaloids, saponins and terpenoids. These compounds are secondary metabolites with various identifiable structures, although a benzene ring with one or more hydroxyl groups is a common feature (Arya et al., 2019; Podsedek, 2007). These compounds are widely diffused in the vegetable kingdom, both in the edible parts of foods and in agro-industrial wastes. Their intake can be obtained by diet or recovered from the inedible food parts which can be used in the food industry to obtain functional foods or in the pharmaceutical industry to produce supplements.

The word 'Brassica' is a Latin derivation of 'cut off the head.' Brassica is the most important genus of the Brassicaceae family and consists of 37 different species. This genus comprises six interrelated species, three diploid (*B. nigra*, *B. oleracea* and *B. rapa*) and three amphidiploid (*B. carinata*, *B. juncea* and *B. napus*) species. *B. oleracea* and *B. rapa* incorporate most Brassica plants, including kale, cabbage, savoy cabbage, Brussels sprouts, cauliflower, Chinese kale and broccoli (Gupta, 2011).

Likely originating from the Mediterranean region, Brassicas are among the most commonly cultivated vegetables globally and some of the most consumed and nutritious vegetables that consumers can eat. According to the data provided by the Food and Agriculture Organization (FAO, 2019) and Food and Agriculture Organization statistics (FAOSTAT, 2019), production of Brassica vegetables in the world totals approximately 139 million tons - 105.7 million tons of cabbage and other Brassicas and 33.5 million tons of cauliflower and broccoli - and China is the biggest Brassica producer in the world.

In addition to the nutrients mentioned above, the Brassica vegetable family has phytonutrients known as glucosinolates (GLs), which are sulphur-containing chemicals. These unstable compounds degrade biologically active indoles and isothiocyanates under the influence of the myrosinase enzyme present in plant tissues. At different levels, the breakdown of GLs into biologically active compounds - promoted by myrosinase, often released when vegetable cells are damaged - is responsible for most of the characteristically strong aroma and bitter flavour of cruciferous vegetables (Kołodziejski et al., 2019). Glucosinolates have been shown to have beneficial effects in disease prevention, for example, by eliminating or neutralising carcinogenic and mutagenic factors.

The different parts of the Brassica plants (roots, stems, leaves, flowers and even seeds) lend themselves to many diverse ways of being consumed, from chopped raw and braised, to boiled in water or steamed. Studies have shown that the methods used for preparation, cooking, or storage of these vegetables can induce changes and interactions among the constituents, improving or reducing their nutritional value (Nugrahedi et al., 2016). Different preservation methods can also be used, such as refrigeration, drying or freezing following blanching, pickling and fermentation. Therefore, it is essential to know the effects of the different steps involved in processing Brassica products before they reach the consumer's table, identifying the critical phases and conditions that maximise their nutrient content and, consequently, their health benefits.

This review is based on the evaluation of electronically collated information published on the nutritional and health benefits of Brassica between 1987 and 2021, from major databases such as Chemical Abstracts, ScienceDirect, SciFinder, PubMed, Henriette's Herbal Homepage and Google Scholar databases. It includes the identification of the main phytochemical compounds related to promoting consumer health, with specific relevance to broccoli. It also discusses the role of different research studies on the compounds' bioavailability with respect to consumer health and the influence of processing methods on the healthful benefits of Brassica products, which may cast some light on consumer and industry methods of processing these vegetables and promoting the intake of health-protective compounds.

2 The main phytochemical and health-promoting compounds

The Brassicaceae family, commonly known as Cruciferae, plays an important role in human nutrition and health. These vegetables, which have unique tastes and aromas, act as a good source of macronutrients (e.g. fibre), micronutrients (e.g. vitamins and minerals) and phenols and GLs, which are associated with presumed health benefits and are not present in most other vegetables. For that reason, the Brassica spp. have been the subject of numerous scientific studies over the last few years, demonstrating a positive correlation between the direct consumption of these vegetables and protective effects against diseases (e.g. cancer, cardiovascular disease, neurodegenerative and other age-related diseases) and inflammatory conditions (e.g. arthritis) (Ramirez et al., 2020; Ilahy et al., 2020; Le et al., 2020). Among all the Brassicas, broccoli (*Brassica oleracea* L. var. *Italica*) is considered the best species regarding its chemical composition, richness in various bioactive compounds and multiple biological capacities (Hedges and Lister, 2007).

2.1 Phytochemical compounds and their effects on human health

One of the most striking features of the Brassicaceae family is the presence of health-promoting phytochemical compounds such as antioxidants, vitamins, essential minerals, carbohydrates, proteins, amino acids and fatty acids. Some phytonutrients (or phytochemicals) are bioactive compounds accumulated in plants, mainly through secondary metabolism. Phytonutrients are bioactive non-nutrient plant compounds and can be classified into several groups based on protective functions and the molecules' individual physical and chemical characteristics. These compounds contain essential properties such as antioxidant activity, antimicrobial effects, modulation of detoxification enzymes, immune system stimulation, decreased platelet aggregation, modulation of hormone metabolism, and anticancer properties. The following list includes the most relevant phytonutrients in the Brassica family, their appropriate medicinal values and other relevant characteristics (Table 1).

Although these vegetables are a significant source of healthy nutritive compounds, the Brassica group is also low in fat and high in fibre and essential minerals, which can help meet basic nutrient needs in humans. Nonetheless, each species of the family Brassicaccae has a distinct composition. The energy and nutrient content of 100 grams of some of the edible vegetables of the Brassica group are shown in Table 2.

In the consulted bibliography, there are few references to adverse effects associated with the consumption of Brassicas. An example is the breakdown compounds of GLs, namely thioglycosides metabolised to thiocyanates. These compounds could inhibit the transport of iodine and its incorporation into thyroglobulin, increasing the proliferation of TSH secretion and the risk of developing thyroid carcinomas. However, there is little epidemiological evidence that the goitrogenic effects of GL degradation products are a significant cause of thyroid carcinomas (Truong et al., 2010).

Table 1 Phytochemical/health-promoting compounds in the Brassicaceae family

Phytochemical compounds	Health-promoting effects and other functions	References
Phenolic compounds		
They are one of the most influential groups. Categorised into flavonoids (flavonols, flavones, flavan-3-ols, anthocyanidins, flavanones, isoflavones and others) and nonflavonoids (phenolic acids, hydroxycinnamates, stilbenes and others) Flavonoids (e.g. kaempferol, quercetin) and hydroxycinnamic acids (e.g. sinapic acid) are the most common and heterogeneous group of polyphenols in the Brassica species).	The essential action is their antioxidant capacity for multiple biological effects for human health, including anti-inflammatory, enzyme inhibition, antimicrobial, antiallergic, vascular and cytotoxic antitumour activity. Anthocyanins lower the risk of myocardial infarction. They are responsible for the colour of some vegetables and are involved in some flavour features (e.g. astringency), also serving as substrates for enzymatic deterioration.	Chu et al. (2000), de Pascual et al. (2010), Wei et al. (2011), Al-sane (2011), Cartea et al. (2011), Cassidy et al. (2013), Dam et al. (2013), Ms et al. (2019)
Glucosinolates (GLs)		
GLs are the main secondary metabolites found among the organosulphur compounds, nearly exclusively present in Brassicaceae. The GLs themselves are not beneficial but after hydrolysation, their breakdown products (isothiocyanates - ITCs) have been shown to exhibit promising effects on health promotion. Other metabolites such as oxozolidine-2-thiones, nitriles, epithionitriles and thiocyanates are also formed, mainly depending on the structure of the GLs side chain. Sulforaphane (SFN) is among the most potent bioactive component and is the leading GL found in broccoli.	They are involved in boosting immunity. Isothiocyanates, dithiolthiones and sulforaphane participate in blocking enzymes responsible for tumorous growth in liver, lung, breast and gastrointestinal tracts and immune-enhancing and cardiovascular protective properties. ITCs form when Brassica tissues are disrupted and are responsible for these vegetables' sharp and bitter-tasting flavours.	Vazquez and Miatello (2010), Björkman et al. (2011), Baskar et al. (2012), Kalpana et al. (2013), Branca et al. (2013), Sanlier and Guler (2018), Baenas and Wagner (2019), Ms et al. (2019), Argento et al. (2019), Ilahy et al. (2020)
Ascorbic Acid		
L-ascorbic acid (AA) is a water-soluble vitamin, known mainly as an essential nutrient. It is easily oxidised to form dehydroascorbic acid (DHAA).	Widely used as vitamins, regenerative and antiviral medication in treating various respiratory viral infections, including influenza, herpetic infections, viral hepatitis and other infectious diseases.	Davies et al. (1991), Gonçalves et al. (2013), Dominguez-Perles et al. (2014)
Vitamin E		
Vitamin E includes a group of potent, lipid-soluble chain-breaking antioxidants such as tocopherols.	Reduces the risk of cancer and can prevent the progression of precancerous lesions. It shows protective effects against coronary heart disease.	Schneider (2005), Pinheiro-Sant'ana et al. (2011)

(Continued)

Table 1 *(Continued)*

Phytochemical compounds	Health-promoting effects and other functions	References
Carotenoids		
The carotenoid family consists of carotenes and xanthophylls. These groups of bioactive compounds have not been deeply characterised for Brassicaceae species. Some reports indicate that the predominant ones are β-carotene and luteolin, but variable amounts of zeaxanthin, cryptoxanthin, neoxanthin and violaxanthin have also been detected.	High antioxidant activity, protecting tumours of the lung, colorectal, breast, uterine and prostate areas. Augment immune response and can protect skin cells against UV radiation.	Guzman et al. (2012), Yoon et al. (2012), Mahima et al. (2014)
Chlorophyll		
Two different types of chlorophyll (chlorophyll a and chlorophyll b) are found in plants, absorbing light at slightly different wavelengths.	Some research suggests that it may be important in protecting against various forms of cancer by binding to the mutant DNA to prevent proliferation. Chlorophyll is well known as the pigment that gives plants and algae their green colour, and it is the primary compound in photosynthesis.	Pareek et al. (2017), Neugart et al. (2018)
Indoles		
These are GLS breakdown products but structurally different from ITCs.	Bind with chemical carcinogens. It could be related to hormone-sensitive cancers such as prostate and breast due to its effect on oestrogen activity and metabolism.	Ms et al. (2019)
Terpenes		
Some terpenes such as tocotrienols, tocopherols and phytosterols can be found among some Brassicaceae. Phytosterols are present in green and yellow vegetables and their seeds.	They help alleviate the risk of cardiovascular diseases and block cancer development in various organs, especially colon, breast and prostate glands.	Awaisheh et al. (2013), Ms et al. (2019), Ramirez et al. (2020)
Alkaloids		
These are secondary metabolites synthesised from amino acids and reported as specific secondary cruciferous metabolites.	Exhibit a wide range of pharmacological properties such as antibacterial, analgesic, antidepressant and anticancer. Some have been proven to possess neuroprotective properties.	Kamarul-Zaman and Mohamad Azzeme (2018), Ramirez et al. (2020)

Table 2 Nutritive value of some Brassica vegetables (per 100 g of vegetable) (TCA, 2019)

	Broccoli	Brussels sprouts	Cabbage	Cauliflower	Galega cabbage	Portuguese cabbage
Water (g)	91.1	84.3	91.1	89.9	89.0	90.6
Energy (Kcal)	32.0	50.0	26.0	34.0	32.0	31.0
Protein (g)	3.4	3.5	2.4	3.7	2.4	2.2
Total fat (g)	0.8	1.4	0.2	0.2	0.4	0.4
Carbohydrates (g)	1.5	4	2.1	3.3	3.1	3.5
Fibre (g)	2.6	3.8	3.1	1.9	3.1	2.4
Sugar (g)	1.2	3.1	2.0	2.8	2.7	3.4
Minerals						
Calcium (mg)	67	26	51	21	290	76
Iron (mg)	1.3	0.7	0.5	0.5	1	1
Magnesium (mg)	22	8	12	22	18	28
Phosphorous (mg)	50	77	64	34	40	65
Potassium (mg)	370	450	250	380	180	270
Sodium (mg)	8	6	9	14	21	15
Zinc (mg)	0.6	0.5	0.3	0.7	0.5	0.4

2.2 Health benefits of broccoli (*Brassica oleracea* L. var. *Italica*)

Among the Brassicaceae family, broccoli (*Brassica oleracea* L.) has been the most exhaustively studied regarding its polyphenol composition, with much evidence supporting the health benefits of its consumption. Several studies have shown that this crop (leaves, florets and sprouts) contains a high antioxidant potential linked to a high level of phenolic compounds (caffeic, ferulic, p-coumaric and sinapic acids, quercetin, kaempferol, isorhamnetin, apigenin and luteolin). These phytochemicals are believed to help protect against chronic diseases such as heart disease and cancer and health problems associated with ageing. As mentioned above, the antioxidant phytochemicals are directly related to the health benefits (Table 1). Broccoli is an excellent source of health-promoting phytochemicals, including nitrogen-sulphur derivatives (GLs and isothiocyanates); polyphenols (flavonoids and chlorogenic and sinapic acid derivatives); minerals (selenium, zinc, iron, potassium and manganese) and vitamins (A, C, E, K and B_6) (Le et al., 2020). Among the various subspecies of *Brassica oleracea*, broccoli is the second richest species in terms of antioxidants such as carotene, tocopherol and ascorbate contents (Kurilich et al., 1999) and is described by many as a 'super-vegetable.' Recently, it has been reported that daily consumption of broccoli sprouts for two months (70 g a day of glucoraphanin) can effectively reduce the oxidative stress induced by *Helicobacter pylori* and may also be helpful in the prevention of gastritis in animals and humans (Manchali et al., 2012).

Several reviews focusing on the general health benefits of broccoli florets and related compounds have been published. For example, Le et al. (2020) published a review of the biologically functional properties of broccoli sprouts and microgreens, which might attract more attention regarding further applications in the food and nutraceutical industries or even in clinical studies on cancer as chemo-preventive agents. Table 3 gives an overview of the biological activities associated with broccoli intake.

In summary, the Brassica plant family features a vast phytochemical diversity (compound types and levels) associated with specific effects on human health. The nutritional and bioactive composition of cruciferous vegetables (e.g. phenolics, carotenoids and GLs) varies depending on the species or cultivar and the type of plant tissue. In addition to genetic factors, this composition is also influenced by environmental factors, including soil and climate conditions (e.g. temperature, UV light and water stress). Variations in nutrient concentrations might also be affected by the extraction and detection techniques used for its characterisation (Ramirez et al., 2020; Sanlier and Guler, 2018). These and other factors are discussed in the next section.

3 Research on bioavailability

Bioactive compounds are being studied intensively to evaluate their effects on health. The correlation between the beneficial health effects and high consumption of antioxidant-rich foods reported in epidemiological studies

Table 3 Biological activities of broccoli

Biological activities	Studies reference
Gastroprotective activity	Moon et al. (2010)
Antimicrobial activity	Benko-Iseppon et al. (2010), Caroling et al. (2013), Corrêa et al. (2014), Owis (2015)
Antioxidant activity	Riso et al. (2010), Bidchol et al. (2011), Jasmina et al. (2012), Bachiega et al. (2016), Chaudhary et al. (2018)
Anticancer activity	Dinkova-Kostova et al. (2006), Tang et al. (2006), Zhang et al. (2006), Munday et al. (2008), Ritz et al. (2007), Jasmina et al. (2012), Hashem et al. (2012), Wang et al. (2012), Hwang and Lim (2015), Chaudhary et al. (2018)
Hepatoprotective activity	Al-Howiriny (2008), Hashem et al. (2013)
Cardioprotective activity	Mukherjee et al. (2008), Vasanthi et al. (2012)
Anti-diabetic activity	Lee et al. (2009), Motawea et al. (2010)
Anti-inflammatory activity	Hwang and Lim (2014), Idrees et al. (2019)
Anti-obesity activity	Riso et al. (2010), Choi et al. (2014), Xu et al. (2018)
Immunomodulatory activity	Thejass and Kuttan (2007), Tilg (2012)

requires that bioactive compounds be bioavailable in food. For the expected impacts to occur, the relevant compounds must be effectively absorbed from the gut into the circulatory system, delivered at the target location, and used in normal physiological processes (Thakur et al., 2020). It is critical to understand how bioactive compounds act *in vivo*, change during digestion, interact with human cells and gut microbiota and then get absorbed and metabolised.

Bioavailability can be defined as the proportion of ingested food nutrients readily available for use in physiological functions (Thakur et al., 2020) or for storage in the body (Cardoso et al., 2015). Indeed, bioavailability encompasses compound release from the food matrix, absorption, metabolism, tissue distribution and bioactivity. Thus, two additional terms fall under the bioavailability concept: bioaccessibility and bioactivity. Bioaccessibility refers to the release of a compound from its food matrix, making it available for intestinal absorption and reaching the systemic circulation. In turn, bioactivity (i.e. biological activity) stands for a myriad of phenomena occurring after the compound reaches the systemic circulation, including transport to target tissues, interactions with biomolecules, metabolism and the physiological effects it promotes (Cardoso et al., 2015; Fernández-García et al., 2009).

Over the last few decades, Brassicaceae crops have been the focus of intense research due to their health benefits, including chemoprotective effects. Brassica vegetables are rich in bioactive compounds, particularly secondary metabolites such as phenolic compounds and GLs (Gonçalves et al., 2013). The bioactive compounds focused on in this section are well known and recognised for their preventive roles against certain types of cancer and cardiovascular diseases, depending on their bioavailability and metabolism.

3.1 Bioavailability of phenolic compounds

The beneficial health effects of Brassica vegetables are attributed to their complex mixture of phytochemicals with antioxidant activity. Among those, phenolic compounds are one of the essential groups due to their respective antioxidant capacity. With more than 8000 known compounds, phenolics are secondary metabolites widely distributed in the plant kingdom and are chemically characterised by having at least one aromatic ring with one or more hydroxyl groups. Phenolic compounds, ranging from low-molecular weight single aromatic-ringed compounds to complex tannins, are classified according to carbon number and arrangement into flavonoids (e.g. flavonols, flavones, flavan-3-ols, anthocyanidins, flavanones and isoflavones) and non-flavonoids (e.g. phenolic acids, hydroxycinnamates, stilbenes), frequently conjugated with sugars and organic acids.

In Brassica species, flavonoids - particularly flavonols, anthocyanins and hydroxycinnamic acids - are widespread (Fig. 1). The flavonols quercetin, kaempferol and isorhamnetin and their O-glycosides (mainly conjugated to glucose) or acylated forms with hydroxycinnamic acids are identified as the main Brassica crop flavonols (Le et al., 2020; Nawaz et al., 2018; Duchnowicz et al., 2012; Cartea et al., 2011). Likewise, anthocyanins are sugar-conjugated forms of anthocyanidins responsible for plants' red, blue and purple colouration. The structure of anthocyanins is responsible for compound stability, colour intensity and potential bioactivity (Mattioli et al., 2020; Castañeda-Ovando et al., 2009) and, in Brassica species, the most commonly found anthocyanins are cyanidin (the most common), delphinidin, malvidin, pelargonidin, peonidin and petunidin (Song et al., 2020; Ahmadiani et al., 2014). Regarding hydroxycinnamic acids (non-flavonoid phenolics), characterised by a C6-C3 structure, p-coumaric, ferulic and sinapic acids and their conjugated forms with sugars or other hydroxycinnamic acids are the most widespread in Brassica species (Nguyen et al., 2021; Shao et al., 2014; Cartea et al., 2011; Harbaum et al., 2007).

The phenolics mentioned above all have the potential to promote anti-inflammatory, enzyme inhibitory, antimicrobial, antiallergic, vascular and cytotoxic antitumor activities, but as a baseline, their antioxidant activity (AOx) stands out. The AOx of phenolic compounds derives from their chemical structure, responsible for the respective redox properties. Flavonoids and phenolic acid AOx are correlated to the number and position of hydroxyl groups of the molecule (increasing number of hydroxyl groups leading to higher AOx) and glycosylation degree (increasing number leading to reduced AOx) (Fukumoto and Mazza, 2000). Nevertheless, these compounds are able to

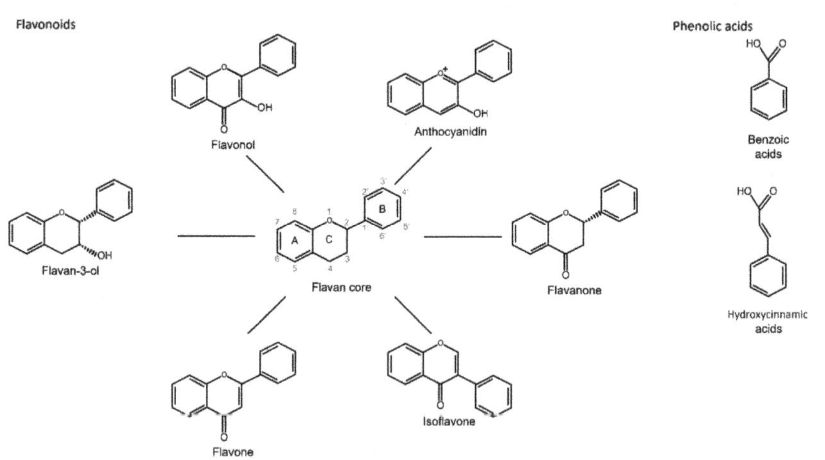

Figure 1 Phenolic compounds grouped as flavonoids and phenolic acids.

quench and neutralise reactive oxygen species (ROS) resulting from oxidation processes, leading to the production of damaging free oxygen radicals. The ROS promote oxidative modification of cellular membranes and intracellular molecules, resulting in lipid peroxidation and accumulation of lipid peroxides, which, in turn, can lead to ageing, atherosclerosis, cancer, inflammation and neurodegenerative diseases such as Parkinson's and Alzheimer's disease. Consequently, phenolic antioxidants are seen as protective agents that lower ROS oxidative damage in the human body and the oxidation of low-density lipoproteins, delaying the progress of chronic diseases. Moreover, the flavonoid composition of Brassica species is also linked to the modulation of crucial cellular pathways leading to the pathogenesis of those diseases (de Pascual-Teresa et al., 2010; Skandrani et al., 2010).

In the gastrointestinal (GI) tract, the metabolism of phenolic compounds follows a similar process. The physiological conditions of the mouth, oesophagus and stomach barely induce any changes in phenolic compounds, but in the small intestine and colon, some are metabolised or absorbed (Swallah et al., 2020; Hussain et al., 2019; Tarko et al., 2013; Cartea et al., 2011; Zhao and Moghadasian, 2010; Crozier et al., 2009; Tomás-Barberán et al., 2009). It is known that the structural properties of phenolic compounds influence the respective intestinal absorption and the occurrence of metabolites in plasma (Swallah et al., 2020; Hussain et al., 2019; Crozier et al., 2009). In fact, free phenolic compounds are mainly absorbed in the small intestine as aglycones (sugar-free flavonoids) or simple glucosides. In contrast, absorption of more complex phenolic forms (oligomers or polymers, rhamnosylglucosides, neohesperidosides, or esters with different compounds) occurs in the colon and are metabolised by gut microbiota to aglycones and simple phenolic acids (D'Archivio et al., 2010; Tomás-Barberán et al., 2009; Murota and Terao, 2003). The absorbed phenolic metabolites (either in the small or large intestine) are usually conjugated to produce glucuronide or methyl conjugates and then transported via the mesenteric vein up to the liver to be further conjugated by glucuronidation, sulphation or methylation (reactions led by phase II enzymes), taken up by peripheral tissues and then eliminated through urine or bile (Swallah et al., 2020; Hussain et al., 2019; Lafay and Gil-Izquierdo, 2008; Scalbert and Williamson, 2000).

Most phenolic compounds are poorly absorbed in the GI tract, particularly flavonoids - Brassica species' most representative phenolics - due to their high glycoside sugar moieties, increasing the molecules' hydrophilicity. In such cases, the gut microbiota plays a crucial role in flavonoid bioavailability. For instance, flavonols such as quercetin, rutin, phloroglucinol, myricetin and kaempferol are metabolised by gut microbiota (particularly *Clostridium*, *Butyrovibrio*, *Eubacterium* and *Peptostreptococcus* genera) by breaking the flavonoid C ring into simpler phenolic compounds derived from the respective

A and B rings (Kasikci and Bagdatlioglu, 2016; Guo and Bruno, 2015; Crozier et al., 2010; Rechner et al., 2004; Kim et al., 1998; Winter et al., 1991). In such cases, the colonic microbiota is responsible for breaking down the flavonoids' glycosidic bonds into aglycones that are easily absorbed in the colon due to their lipophilicity and further metabolised in the liver (Cartea et al., 2011; Tomás-Barberán et al., 2009; Silberberg et al., 2006; Murota and Terao, 2003). In effect, most studies on the bioavailability of flavonoids indicate conflicting results, as flavonoid absorption seems to be dependent on the type and position of attached sugar groups (Swallah et al., 2020; Cartea et al., 2011; Tomás-Barberán et al., 2009), with respective bioactivity being influenced by colonic microbiota metabolisation.

3.2 Bioavailability of glucosinolates

Despite being rich in many bioactive compounds, GLs are the primary contributors to the anticarcinogenic effects of Brassica vegetables. Chemoprevention can be described as the use of natural or synthesised agents as a preventive strategy to delay, inhibit or counteract carcinogenesis. Moreover, glusocinolates have also been reported to have fungicidal, fungistatic, nematicidal and bactericidal activities (Plaszkó et al., 2021; Barba et al., 2016; Sotelo et al., 2015; Ramos García et al., 2012).

Chemically, GLs consist of a thiohydroximate-O-sulphonate group linked to glucose and an alkyl, aralkyl or indolyl side chain, with reports of over 200 identified side groups (Franco et al., 2016; Deng et al., 2015; Agerbirk and Olsen, 2012; Fahey et al., 2001). Glucosinolates are reasonably stable in plant cells, with no related bioactivity (Tomás-Barberán et al., 2009). However, consequent to plant physical damage, these compounds can be hydrolysed to a range of bioactive compounds, such as isothiocyanates (ITC), by the plant-based physically segregated myrosinase (β-thioglucosidase) or by microbial myrosinase from gut bacteria.

Isothiocyanates (ITC) are the main breakdown products known for their antioxidant, anticarcinogenic and antimicrobial properties. However, a wide range of GL hydrolysis products resulting from myrosinase (β-thioglucosidase) activity have biological properties. The hydrolysis arises from the myrosinase secretion which is induced by physical vegetable damage, or microbial myrosinase from gut bacteria.

Glucosinolate breakdown products (Fig. 2) depend on the nature of the GL side chain, pH conditions, availability of ferrous ions and specific protein factors, namely epithiospecifier protein (ESP) (Fernández-León et al., 2017; Barba et al., 2016; Williams et al., 2009; Cartea and Velasco, 2008). Isothiocyanates, the formation of which is favoured at physiological pH, are highly reactive and considered inducers of phase II enzymes (e.g. sulphoraphane), inhibitors of

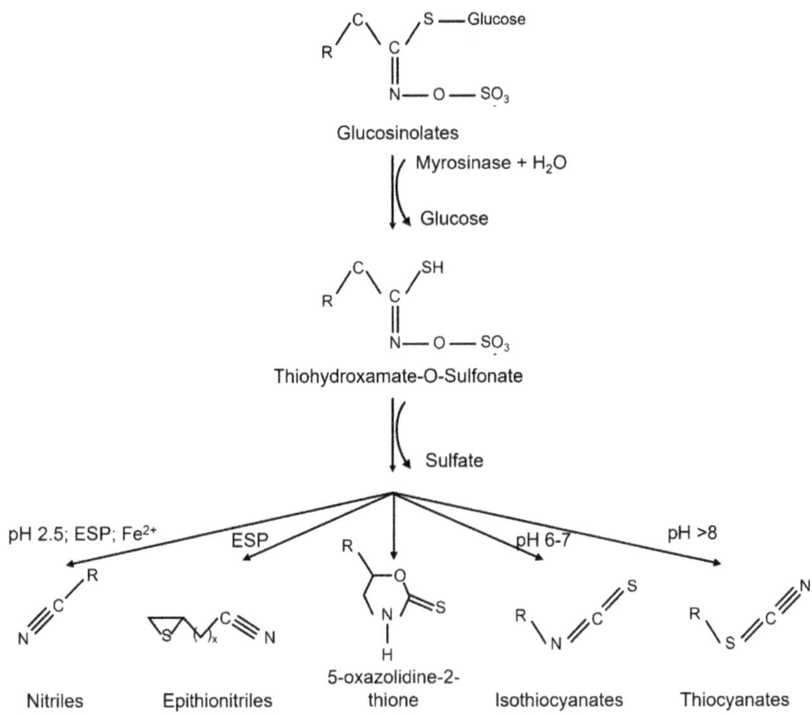

Figure 2 Glucosinolate hydrolysis by myrosinase and the resulting products. ESP, epithiospecifier protein.

mitosis (e.g. allyl isothiocyanates) and apoptosis promoters of tumour cells (e.g. benzyl isothiocyanate) (Zhu et al., 2017; Barba et al., 2016; Åsberg et al., 2015; Geng et al., 2011; Tomás-Barberán et al., 2009; Kwon et al., 2007; Pham et al., 2004; Smith et al., 2004).

Glucosinolates undergo gut bacterial metabolism (Prieto et al., 2019): intact molecules can be partially absorbed in the stomach while others transit through the GI tract, reaching the small intestine to be further metabolised by active plant myrosinase and the resulting ITCs absorbed (Angelino and Jeffery, 2014; Clarke et al., 2011; Song et al., 2005); non-hydrolysed GLs transit to the colon to be hydrolysed into ITCs by bacterial myrosinase and absorbed or excreted (Prieto et al., 2019; Barba et al., 2016; Dinkova-Kostova and Kostov, 2012; Cartea and Velasco, 2008). Gut microbiota *Bifidobacterium* strains (e.g. *B. longum*, *B. pseudocatenulatum* and *B. adolescentis*) are reported to be linked to the metabolisation of GLs to nitriles, which are eliminated through urine (Barba et al., 2016) or further metabolised to ammonia by he gut microbiota. Other bacterial strains related to GL degradation are identified as *Escherichia coli*, *Bacteroides thetaiotaomicron*, *Bacteroides vulgatus*, *Enterobacter cloacae* and *Lactobacillus agilis* (Tomás-Barberán et al., 2009).

Assimilation of ITCs occurs through passive diffusion from the small or large intestine into the plasma (Angelino and Jeffery, 2014; Bheemreddy and Jeffery, 2007). Absorbed ITCs are usually conjugated to glutathione (liver) and excreted as mercapturic acid, which is considered to be a biomarker since the respective concentration in urine can be traced to the amount of ingested ITCs (Prieto et al., 2019; Barba et al., 2016; Bheemreddy and Jeffery, 2007; Shapiro et al., 2001; Bollard et al., 1997; Mennicke et al., 1987). The assimilation/excretion of other GL breakdown products is thought to be similar to ITCs (Barba et al., 2016), but further studies are needed.

Regarding the bioactivity of ITCs attributed to the molecules' hydrophilicity, several studies have demonstrated a wide range of biological activities related to the glucosinolate-myrosinase in Brassica vegetables, as reviewed by Prieto et al. (2019). As mentioned, the main bioactivity related to Brassicas' GLs relates to the chemoprotective effects of ITCs, particularly sulforaphane (SFN). SFN, extensively found in broccoli, can up-regulate phase II detoxification enzymes (e.g. quinone reductase, glutathione-S-transferase and glucuronosyl transferases) and, simultaneously, inhibit the phase I CYP1A enzyme responsible for the bioactivation of pre-carcinogens (Gründemann and Huber, 2018; Manchali et al., 2012; Herr and Büchler, 2010; Wiktorska et al., 2009; Lai et al., 2008; Halkier and Gershenzon, 2006).

4 Methods for preserving Brassicas' bioactive composition through the supply chain

The supply chain of agro-food products involves many actors that play relevant roles, such as farmers, processors, retailers and consumers. The primary role of these intervenients is to maintain or create value from the raw vegetable throughout the entire food chain to provide high-quality, safe, nutritive, tasty and sensorily appealing products to consumers. In this part of this review, we detail the main actors and relevant steps in the food supply chain of Brassica vegetables and their influence on the nutraceutical properties of these vegetables, with particular attention to broccoli sprouts.

4.1 Effect of pre-harvest conditions

It is common knowledge that the biosynthesis and concentration of nutraceutical compounds, like phenolic compounds in plants, depend on genetic and environmental factors. The genetic influences on these groups of compounds, both within and among species, and even among crops of the same species, have been well-presented by other authors in different comprehensive and recent reviews (Jin et al., 2021; Ilahy et al., 2020; Cartea et al., 2011; Verkerk et al., 2009, McNaughton and Marks, 2003, Rosa, 1997). As a recent example, the biological activities of the primary metabolites and secondary metabolites of 69 green cabbage varieties were tested by Jin et al. (2021), who concluded that there are significant differences between the different amino acids and polyphenols of the 69 cabbage varieties, the most abundant being the amino acids and polyphenols glutamic acid and rutin, respectively. Both amino acids and polyphenols were found to have high genetic diversity. Based on 29 indices of amino acids and polyphenols, the authors could classify the cabbage varieties according to the level of amino acids and polyphenols, providing a theoretical basis for the genetic improvement of the nutritional quality of cabbage. These studies are essential tools for identifying plants with optimal health-promoting potential and could be very useful for producers and sellers to create final products with added value.

The tissue type (e.g. seeds, roots, leaves, shoots or flowers) and tissue developmental stages determine the type of phytochemical present and its respective content. Mature leaves generally have significantly higher levels of phenolic compounds than younger ones, and the roots are usually richer than shoots in content and diversity of GL compounds. For example, the concentration of SFN in broccoli is about ten times higher than for mature broccoli (1153 mg/100 g dw vs 44-171 mg/100 g dw, respectively). Even cultivars of the same species could show differences in their metabolite profile and in the concentrations of specific metabolites (Neugart et al., 2018).

In addition, the overall concentration of the health-related compounds is also affected by ecophysiological parameters such as temperature, water availability, light exposure and growing season. Light seems to be one of the most influential environmental stresses that cause significant changes in the bioactive composition of crops. Light (intensity and density) regulates plant growth and development and the biosynthesis of both primary and secondary metabolites. However, different plants have differentiated responses to light intensity changes (Borges et al., 2018). In broccoli, better productive development is achieved under average temperatures between 15.5°C and 18°C, with a maximum of 24°C (Maynard and Hochmuth, 2006). Prolonged periods at a temperature above 25°C delay the formation of inflorescences during the plant's vegetative growth, reducing the size and causing the development of leaves and bracts in the floral peduncles (Bjorkman and Pearson, 1998).

Fertilisation of soils and the level, type and value of mineral nutrient application directly influence the level of nutrients available in plants and indirectly influence the plant's physiology and the biosynthesis of secondary metabolites or phytonutrients (Borges et al., 2018; De Pascale et al., 2007). For example, selenium (Se) fertilisation led to decreased GL concentrations in broccoli florets, whereas GL levels in sprouts were not affected (Ávila et al., 2014). Moreover, Ramos et al. (2011) investigated 38 broccoli cultivars, of which

13 cultivars showed reduced total GL concentrations after Se fertilisation, while the remaining cultivars had total GL levels similar to that of control plants.

4.2 Influence of postharvest treatments

Although Brassica vegetables can be eaten raw, they are usually processed prior to consumption and undergo different preservation methods to guarantee their safety and quality. The most common methods of preparing/processing Brassica vegetables are steaming, roasting, boiling, frying, sautéing, sous vide, microwaving and pressure-cooking. These processes can both positively and negatively affect nutritional value, health benefits and attractiveness, depending on the duration, temperature, and degradative mechanisms involved.

At the beginning of the food chain, crops are typically stored in refrigerated conditions after harvesting. The effect of storage conditions on Brassica vegetables, namely broccoli, has been studied by many authors (Martínez et al., 2020; Aiamla-or et al., 2019; Vallejo et al., 2003; Bestard et al., 2001). It is known that lower temperature decreases metabolic rates and thereby slows down deterioration, being determinant to maintaining product quality and diminishing changes in bioactive compounds.

Minimally processed vegetables are obtained through operations such as trimming, peeling and cutting into an edible product that is subsequently washed (decontaminated) and packaged, usually using modified atmospheres (MAP). The minimally processed vegetable remains in a fresh state, characterised by living tissues; however, the mechanical wounding triggers complex reaction mechanisms and physical and physiological processes, leading to changes in the content of the bioactive compounds. In particular, Brassica vegetables are usually chopped before eating. Activation of myrosinase promotes the hydrolysis rate of GLs and the formation of active isothiocyanates (e.g. the anticarcinogenic SFN), thus enhancing the bioactive potential of fresh-cut broccoli as a functional food (Martinez et al., 2020). On the other hand, a complete breakdown of GLs by autolysis could occur in extreme cases. Following cutting and other preparatory operations, the fresh-cut product is washed in sanitising solutions with oxidising potential. At this stage, leaching of certain healthy compounds such as ascorbic acid, carotenoids, phenolics and even certain GL compounds is expected, resulting from leaching and oxidation phenomena (Alegria et al., 2009; Schreiner et al., 2007).

Preservation techniques, such as canning and blanching before freezing or drying, can extend vegetable shelf-life and simultaneously induce important changes in the composition and organoleptic quality of the product. Processing conditions involving heating the vegetable tissue (either in water or steam) before consumption could be responsible for the degradation of many chemical compounds. During thermal treatments, different mechanisms such as leaching, compound breakdown or enzymatic inactivation are promoted, depending on the treatment's temperature and time and on the physical and chemical properties of the vegetable (Gonçalves et al., 2009). Numerous studies have been conducted investigating the influence of processing and storage on Brassica vegetables (Oliviero et al., 2018; Duarte-Sierra et al., 2017; Song and Thornalley, 2007). In broccoli, these processes impact the bioaccessibility of bioactive compounds, particularly the content of GLs (notably SFN) and polyphenols, and therefore, interfere with associated antioxidant activity.

Enhanced antioxidant and anticarcinogenic activities associated with the enrichment of phenols, flavonoids and GLs were observed in steam-processed broccoli (Cartea et al., 2011). Mahn and Reyes (2012) stated that cooking or blanching cut broccoli at a temperature below 100°C favours anticarcinogenic properties, supporting processed broccoli as a functional food with improved health-promoting properties. Steam-processed broccoli showed a higher antioxidant capacity due to the significantly increased extractability of phenols and flavonoids, increasing the *in vivo* bioavailability of these compounds and thus improving their related health-promoting properties (Sun et al., 2021; Gliszczyńska-Świgło et al., 2006; Roy et al., 2009). In contrast, a significant decrease in polyphenols was observed in conventionally cooked and microwave-cooked broccoli compared to fresh broccoli (Zhang and Hamauzu, 2004). Blanching and cooking at a temperature ranging between 50°C and 90°C affects broccoli's GL content (Cieślik et al., 2007), mostly attributed to its loss from the vegetal tissues through leaching into the cooking water and being dependent on cooking conditions. Vegetable preservation by freezing involves the application of a low temperature (−18°C) to maintain the quality and nutritive value of the vegetables during long periods of storage. However, changes in bioactive compounds could occur due to damage to the vegetable tissue by the growth of water crystals and cellular disintegration. If temperature oscillation occurs during the frozen chain, frozen vegetables such as broccoli lose more of their ascorbic acid content, which decreases their shelf-life compared with the same products preserved at constant temperature (Gonçalves et al., 2011). Limited research regarding the influence of the freezing process on GL content is available at the moment.

The postharvest drying process and the methods adopted can result in changes in Brassicas' functional profile and physical properties. Hot air drying implies a thermal treatment, and therefore thermal degradation of polyphenols is expected. However, the decomposition of polyphenols is proven to depend on the food matrix and the processing conditions (Managa et al., 2020; Duarte et al., 2019; Yilmaz et al., 2019; Mrkìc et al., 2006). Moreover, this process enhances or depletes the antioxidant activity of vegetables, depending on the nature of the substrate. During drying, oxidation reactions could also take place and polyphenols with an intermediate oxidation state can exhibit a higher radical scavenging activity than non-oxidised polyphenols. High drying temperature could further cause the formation of Maillard reaction products that could act as pro- or antioxidants (Oliveira et al., 2015).

5 Future trends and conclusion

The Brassica genus includes popular horticultural vegetables with major global economic importance, particularly in the Mediterranean region. The consumption of Brassicaceae vegetables contributes in a relevant manner to human nutrition, as several epidemiological studies have indicated. These crops contain a broad array of health-promoting compounds, including rich sources of vitamin C, phenolic and GL compounds. Likewise, the importance of the food matrix on the absorption and metabolism of these healthy compounds is closely related to the range of health benefits attributed to its intake. However, less beneficial effects were observed when purified molecules were administered compared to the ingestion of horticultural products such as Brassicas.

Deepening our knowledge regarding Brassicas' bioactive compounds is crucial to fully understanding their respective bioavailability. Although studies can be found regarding the rate and metabolic changes of many phenolic compounds and GLs, the role of gut microbiota on the metabolism, assimilation, interaction with other molecules and delivery to target tissues of bioactive compounds needs to be further addressed. The generated knowledge could promote predictive models to study the bioavailability of Brassicas' main bioactive compounds. Moreover, pre- and post-harvest conditions (including processing) affect bioactive compounds' bioaccessibility, concentration, or even compromise compound chemical structure, which might have severe implications regarding its bioavailability. Further comprehensive studies are required and should be conducted to ascertain the *in vivo* prospects, to understand better the connections between food, nutrition and health.

The adequate management of Brassicas' supply chain conditions, including crop varieties, growth conditions, storage and processing and storage operations, could potentially increase or preserve their nutritive value. The health-promoting capacity of plant foods depends strictly on their processing history. Storage, processing and preparation conditions have been proven to have significant effects on the content of antioxidants, but the impact of processing on the antioxidant activity of vegetables is still a neglected area, and little information is available. The findings vary considerably for each bioactive component, depending on the treatments, conditions and matrices. Research studies leading to the optimisation of processing and storage conditions are therefore essential to improve the beneficial compounds of *Brassica* spp.

6 Where to look for further information

This next section has been developed to provide the reader with additional/complementary data concerning the subjects discussed in this chapter, namely improving the nutraceutical properties of broccoli and other Brassicas. Therefore, readers may wish to consider also:

6.1 International research projects

- New strategies for sustainable production of brassica crops in integrated and organic systems. (November 1999 to May 2000). Coordinated by UNITED VEGETABLES LTD, United Kingdom.
- DREAM. Design and development of REAlistic food Models with well-characterised micro- and macro-structure and composition. (May 2009 to October 2013). Coordinated by INSTITUT NATIONAL DE RECHERCHE POUR L'AGRICULTURE, L'ALIMENTATION ET L'ENVIRONNEMENT, France.
- ECOSEED. The impact of environmental stress on plant seed quality. (January 2013 to December 2016). Coordinated by UNIVERSITAET INNSBRUCK, Austria.
- EUROLEGUME. Enhancing of legumes growing in Europe through sustainable cropping for protein supply for food and feed. (January 2014 to December 2017). Coordinated by UNIVERSIDADE DE TRAS-OS-MONTES E ALTO DOURO, Portugal.
- HST-FOOD-TRAIN. Science based precision processing for future healthy, structured and tasteful fruit and vegetable based foods: an integrated research and training program. (January 2011 to December 2014). Coordinated by KATHOLIEKE UNIVERSITEIT LEUVEN, Belgium.
- Project BRESOV. Optimising Vegetable Crops for Organic Farming: Launch of 4-Year EU Research. (May 2018, duration 60 months). Coordinated by Università degli Studi di Catania, Italy.

6.2 Books and review articles

- Vegetable Brassicas and Related Crucifers. November. By: Geoffrey Dixon, University of Reading, UK, 2006.
- Nutritional attributes of Brassica vegetables. L J Hedges & C Lister. Report number: 1618, Affiliation: Crop & Food Research. 2006.
- Agricultural and Related Biotechnologies. Editor-in-Chief Murray Moo-Young. Comprehensive Biotechnology (Second Edition), 2011.
- Brassicaceae: Characterization, Functional Genomics and Health Benefits. Minglin Lang, PhD (Editor). Department of Biochemistry, Kansas State University, Manhattan, KS, USA. Series: Agriculture Issues and Policies, 2013.

- Brassicas: Cooking the World's Healthiest Vegetables: Kale, Cauliflower, Broccoli, Brussels Sprouts and More [A Cookbook] Hardcover – Illustrated, April 8, 2014 by Laura B. Russell (Author), Rebecca Katz (Foreword).
- Brassica: Characteristics and Properties. By J. W. Fahey. Encyclopedia of Food and Health, 2016.
- Nutraceuticals. Efficacy, Safety and Toxicity. Edited by Ramesh C. Gupta, 2016.
- Brassica Germplasm: Characterisation, Breeding and Utilisation. Edited by Mohamed Ahmed El-Esawi. IntechOpen, 2018.
- Pharmacoepigenetics. Edited by Ramón Cacabelos. Volume 10 in Translational Epigenetics, 2019.
- Soengas, P.; Velasco, P.; Fernández, J.C.; Cartea, M.E. New Vegetable Brassica Foods: A Promising Source of Bioactive Compounds. Foods 2021, 10, 2911. https://doi.org/10.3390/foods10122911

6.3 Research centres, professional organisations and other relevant websites

- International Legume Society:
  www.legumesociety.org.
- Nofima:
  https://nofima.com/.
- Yara:
  https://www.yara.co.uk/crop-nutrition/vegetable-brassicas/.
- HortInnovation:
  https://www.horticulture.com.au.
- FHIS:
  https://www.hutton.ac.uk.
- CGIAR ECPGR Brassica Working Group:
  https://www.ecpgr.cgiar.org/working-groups/brassica.
- ISHS, Workgroup Brassica:
  https://www.ishs.org/brassica.
- UK Brassica Research Community (UK-BRC):
  https://ukbrc.hosted.york.ac.uk/index.html.

6.4 Upcoming conferences, meetings and events

- ICGGBC 2022: XVI International Conference on Glucosinolates and Glucosinolates In Brassica Crops, 20-21 September, 2022. Lisbon, Portugal (https://waset.org/glucosinolates-and-glucosinolates-in-brassica-crops-conference-in-september-2022-in-lisbon).

- VIII International Symposium on Brassicas, ISHS Brassica Meetings, January 1, 2023. India (https://www.ishs.org/symposium/87).

7 References

Agerbirk, N. and Olsen, C. E. 2012. Glucosinolate structures in evolution. *Phytochemistry* 77: 16-45.

Ahmadiani, N., Robbind, R. J., Collins, T. M. and Giusti, M. M. 2014. Anthocyanins contents, profiles, and color characteristics of red cabbage extracts from different cultivars and maturity stages. *J. Agric. Food Chem.* 62(30): 7524-7531.

Aiamla-or, S., Shigyo, M. and Yamauchi, N. 2019. UV-B treatment controls chlorophyll degradation and related gene expression in broccoli (*Brassica oleracea* L. italica Group) florets during storage. *Sci. Hortic.* 243: 524-527.

Alegria, C., Pinheiro, J., Gonçalves, E. M., Fernandes, I., Moldão, M. and Abreu, M. 2009. Quality attributes of shredded carrot (*Daucus carota* L. cv. Nantes) as affected by alternative decontamination processes to chlorine. *Innov. Food Sci. Emerg. Technol.* 10(1): 61-69.

Al-Howiriny, T. 2008. Evaluation of hematoprotective activity of broccoli "*Brassica oleracea*" in rats. *Hung. Med. J.* 2(1): 145-156.

Al-sane, P. 2011 Anthocyanin tomato mutants: Overview and characterization of an anthocyanin-less somaclonal mutant. *Plant Biosyst.* 145(2): 436-444.

Angelino, D. and Jeffery, E. 2014. Glucosinolate hydrolysis and bioavailability of resulting isothiocyanates: Focus on glucoraphanin. *J. Funct. Foods* 7(1): 67-76.

Aquila, G., Marracino, L., Martino, V., Calabria, D., Campo, G., Caliceti, C. and Rizzo, P. 2019. The use of nutraceuticals to counteract atherosclerosis: The role of the notch pathway. *Oxid. Med. Cell. Longev.* 2019: 5470470.

Argento, S., Melilli, M. G. and Branca, F. 2019. Enhancing Greenhouse Tomato-Crop Productivity by using *Brassica macrocarpa* Guss. Leaves for controlling root-knot nematodes. *Agronomy* 9(12): 1-13.

Arya, M. S., Reshma, U. R., Syama, S., Anaswara, S. J. and Karishma, S. 2019. Nutraceuticals in vegetables: New breeding approaches for nutrition, food and health: A review. *Int. J. Pharmacogn. Phytochem.* 8(1): 677-882.

Åsberg, S. E., Bones, A. M. and Øverby, A. 2015. Allyl isothiocyanate affects the cell cycle of *Arabidopsis thaliana*. *Front. Plant Sci.* 6: 364.

Ashwlayan, V. D. and Nimesh, S. 2018. Nutraceuticals in the management of diabetes mellitus. *Int. J.* 6(2): 114-120.

Ávila, F. W., Yang, Y., Faquin, V., Ramos, S. J., Guilherme, L. R. G., Thannhauser, T. W. and Li, L. 2014. Impact of selenium supply on Se-methylselenocysteine and glucosinolate accumulation in selenium-biofortified Brassica sprouts. *Food Chem.* 165: 578-586.

Awaisheh, S. S., Khalifeh, M. S., Al-Ruwaili, M. A., Khalil, O. M., Al-Ameri, O. H. and Al-Groom, R. 2013. Effect of supplementation of probiotics and phytosterols alone or in combination on serum and hepatic lipid profiles and thyroid hormones of hypercholesterolemic rats. *J. Dairy Sci.* 96(1): 9-15.

Bachiega, P., Salgado, J. M., de Carvalho, J. E., Ruiz, A. L. T. G., Schwarz, K., Tezotto, T. and Morzelle, M. C. 2016. Antioxidant and antiproliferative activities in different maturation stages of broccoli (*Brassica oleracea italica*) biofortified with selenium. *Food Chem.* 190: 771-776.

Baenas, N. and Wagner, A. E. 2019. Pharmacoepigenetics of brassica-derived compounds. In: *Pharmacoepigenetics*, pp. 847–857.

Barba, F. J., Nikmaram, N., Roohinejad, S., Khelfa, A., Zhu, Z. and Koubaa, M. 2016. Bioavailability of glucosinolates and their breakdown products: Impact of processing. *Front. Nutr.* 3: 24.

Baskar, V., Gururani, M. A., Yu, J. W. and Park, S. W. 2012. Engineering glucosinolates in plants: Current knowledge and potential uses. *Appl. Biochem. Biotechnol.* 168(6): 1694–1717.

Benko-Iseppon, A. M., Galdino, S. L., Calsa, T., Kido, E. A., Tossi, A., Belarmino, L. C. and Crovella, S. 2010. Overview on plant antimicrobial peptides. *Curr. Protein Pept. Sci.* 11(3): 181–188.

Bestard, M. J., Sanjuán, N., Rosselló, C., Mulet, A. and Femenia, A. 2001. Effect of storage temperature on the cell wall components of broccoli (*Brassica oleracea* L. Var. *italica*) plant tissues during rehydration. *J. Food Eng.* 48(4): 317–323.

Bheemreddy, R. M. and Jeffery, E. H. 2007. The metabolic fate of purified glucoraphanin in F344 rats. *J. Agric. Food Chem.* 55(8): 2861–2866.

Bidchol, A. M., Wilfred, A., Abhijna, P. and Harish, R. 2011. Free radical scavenging activity of aqueous and ethanolic extract of *Brassica oleracea* L. var. *italica*. *Food Bioprocess Technol.* 4(7): 1137–1143.

Björkman, M., Klingen, I., Birch, A. N. E., Bones, A. M., Bruce, T. J. A., Johansen, T. J., Meadow, R., Mølmann, J., Seljåsen, R., Smart, L. E. and Stewart, D. 2011. Phytochemicals of Brassicaceae in plant protection and human health - Influences of climate, environment and agronomic practice. *Phytochemistry* 72(7): 538–556.

Bjorkman, T. and Pearson, K. J. 1998. High temperature arrest of inflorescence development in broccoli (*Brassica oleracea* var. *italica* L.). *J. Exp. Bot.* 49(318): 101–106.

Bollard, M., Stribbling, S., Mitchell, S. and Caldwell, J. 1997. The disposition of allyl isothiocyanate in the rat and mouse. *Food Chem. Toxicol.* 35(10–11): 933–943.

Borges, C. V., Junior, S. S., Ponce, F. S. and Lima, G. P. P. 2018. Agronomic factors influencing brassica productivity and phytochemical quality. In: *Brassica Germplasm - Characterisation, Breeding and Utilisation*. InTechOpen. https://www.intechopen.com/chapters/59888 (accessed 7 October 2021).

Branca, F., Lucia, R., Alessandro, T., Lo Scalzo, R., Picchi, V. and Argento, S. 2013. The glucosinolates and variation of antioxidant compounds in seeds and sprouts of broccoli (*Brassica oleracea* L. var. *italica*) and rocket (*Eruca sativa* L.) in relation to temperature and germinative stage. *Acta Hortic.* 1005: 271–278.

Camara, M., Fernandez-Ruiz, V., Morales, P. and Cortes Sanchez-Mata, M. C. 2017. Fiber compounds and human health. *Curr. Pharm. Des.* 23(19): 2835–2849.

Cardoso, C., Afonso, C., Lourenço, H., Costa, S. and Nunes, M. L. 2015. Bioaccessibility assessment methodologies and their consequences for the risk-benefit evaluation of food. *Trends Food Sci. Technol.* 41(1): 5–23.

Caroling, G., Tiwari, S. K., Ranjitham, M. and Suja, R. 2013. Biosynthesis of silver nanoparticles using aqueous broccoli extract-characterisation and study of antimicrobial, cytotoxic effects. *Asian J. Pharm. Clin. Res.* 6(4): 165–172.

Cartea, M. E., Francisco, M., Soengas, P. and Velasco, P. 2011. Phenolic compounds in Brassica vegetables. *Molecules* 16(1): 251–280.

Cartea, M. E. and Velasco, P. 2008. Glucosinolates in Brassica foods: Bioavailability in food and significance for human health. *Phytochem. Rev.* 7(2): 213–229.

Cassidy, A., Mukamal, K. J., Liu, L., Franz, M., Eliassen, A. H. and Rimm, E. B. 2013. Intakes of anthocyanins and flavones are associated with biomarkers of insulin resistance and inflammation in women. *J. Nutr.* 144(2): 202–208.

Castañeda-Ovando, A., Pacheco-Hernández, MdL., Páez-Hernández, M. E., Rodríguez, J. A. and Galán-Vidal, C. A. 2009. Chemical studies of anthocyanins: A review. *Food Chem.* 113(4): 859–871.

Chaudhary, A., Choudhary, S., Sharma, U., Vig, A. P., Singh, B. and Arora, S. 2018. Purple head broccoli (*Brassica oleracea* L. var. *italica* Plenck), a functional food crop for antioxidant and anticancer potential. *J. Food Sci. Technol.* 55(5): 1806–1815.

Choi, Y. J., Lee, W. S., Lee, E. G., Sung, M. S. and Yoo, W. H. 2014. Sulforaphane inhibits IL-1β-induced proliferation of rheumatoid arthritis synovial fibroblasts and the production of MMPs, COX-2, and PGE2. *Inflammation* 37(5): 1496–1503.

Chu, Y. H., Chang, C. L. and Hsu, H. F. 2000. Flavonoid content of several vegetables and their antioxidant activity. *J. Sci. Food Agric.* 80(5): 561–566.

Cieślik, E., Leszczyńska, T., Filipiak-Florkiewicz, A., Sikora, E. and Pisulewski, P. M. 2007. Effects of some technological processes on glucosinolate contents in cruciferous vegetables. *Food Chem.* 105(3): 976–981.

Clarke, J. D., Hsu, A., Riedl, K., Bella, D., Schwartz, S. J., Stevens, J. F. and Ho, E. 2011. Bioavailability and inter-conversion of sulforaphane and erucin in human subjects consuming broccoli sprouts or broccoli supplement in a cross-over study design. *Pharmacol. Res.* 64(5): 456–463.

Corrêa, C. B., Martin, J. G. P., Alencar, S. M. and Porto, E. 2014. Antilisterial activity of broccoli stems (*brassica oleracea*) by flow cytometry. *Int. Food Res. J.* 21(1): 395–399.

Crozier, A., Del Rio, D. and Clifford, M. N. 2010. Bioavailability of dietary flavonoids and phenolic compounds. *Mol. Aspects Med.* 31(6): 446–467.

Crozier, A., Jaganath, I. B. and Clifford, M. N. 2009. Dietary phenolics: Chemistry, bioavailability and effects on health. *Nat. Prod. Rep.* 26(8): 1001–1043.

Dam, R. M., Naidoo, N. and Landberg, R. 2013. Dietary flavanoids and the development of type 2 diabetes and cardiovascular diseases: Review of recent findings. *Curr. Opin. Lipidol.* 24(1): 25–33.

D'Archivio, M., Filesi, C., Vari, R., Scazzocchio, B. and Masella, R. 2010. Bioavailability of the polyphenols: Status and controversies. *Int. J. Mol. Sci.* 11(4): 1321–1342.

Davies, M. B., Partridge, D. A. and Austin, J. A. 1991. *Vitamin C: Its Chemistry and Biochemistry*. Royal Society of Chemistry, Cambridge.

Deng, Q., Zinoviadou, K. G., Galanakis, C. M., Orlien, V., Grimi, N., Vorobiev, E., Lebovka, N. and Barba, F. J. 2015. The effects of conventional and non-conventional processing on glucosinolates and its derived forms, isothiocyanates: Extraction, degradation, and applications. *Food Eng. Rev.* 7(3): 357–381.

De Pascale, S., Maggio, A., Pernice, R., Fogliano, V. and Barbieri, G. 2007. Sulphur fertilisation may improve the nutritional value of *Brassica rapa* L. subsp. sylvestris. *Eur. J. Agron.* 26(4): 418–424.

de Pascual-Teresa, S., Moreno, D. A. and Garcia-Viguera, C. 2010. Flavanols and anthocyanins in cardiovascular health: A review of current evidence. *Int. J. Mol. Sci.* 11(4): 1679–1703.

Dinkova-Kostova, A. T., Jenkins, S. N., Fahey, J. W., Ye, L., Wehage, S. L., Liby, K. T., Stephenson, K. K., Wade, K. L. and Talalay, P. 2006. Protection against UV-light-induced skin carcino-genesis in SKH-1 high-risk mice by sulforaphane-containing broccoli sprout extracts. *Cancer Lett.* 240(2): 243–252.

Dinkova-Kostova, A. T. and Kostov, R. V. 2012. Glucosinolates and isothiocyanates in health and disease. *Trends Mol. Med.* 18(6): 337–347.

Domínguez-Perles, R., Mena, P., García-Viguera, C. and Moreno, D. A. 2014. Brassica foods as a dietary source of vitamin C: A review. *Crit. Rev. Food Sci. Nutr.* 54(8): 1076–1091.

Duarte, C., Sousa, P., Rocha, S., Pinheiro, R. and Vaz Velho, M. 2019. The effect of different drying processes on physicochemical characteristics and antioxidant activity of Brassica spp. Cultivars from northern Atlantic Portugal. *Chem. Eng. Trans.* 75: 421–426.

Duarte-Sierra, A., Forney, C. F., Michaud, D., Angers, P. and Arul, J. 2017. Influence of hormetic heat treatment on quality and phytochemical compounds of broccoli florets during storage. *Postharvest Biol. Technol.* 128: 44–53.

Duchnowicz, P., Bors, M., Podsędek, A., Koter-Michalak, M. and Broncel, M. 2012. Effect of polyphenols extracts from Brassica vegetables on erythrocyte membranes (in vitro study). *Environ. Toxicol. Pharmacol.* 34(3): 783–790.

Fahey, J. W., Zalcmann, A. T. and Talalay, P. 2001. The chemical diversity and distribution of glucosinolates and isothiocyanates among plants. *Phytochemistry* 56(1): 5–51.

Fernández-García, E., Carvajal-Lérida, I. and Pérez-Gálvez, A. 2009. In vitro bioaccessibility assessment as a prediction tool of nutritional efficiency. *Nutr. Res.* 29(11): 751–760.

Fernández-León, A. M., Fernández-León, M. F., González-Gómez, D., Ayuso, M. C. and Bernalte, M. J. 2017. Quantification and bioaccessibility of intact glucosinolates in broccoli 'Parthenon' and savoy cabbage 'Dama.' *J. Food Compost. Anal.* 61: 40–46.

Food and Agriculture Organization 2019. *Statistics, FAOSTAT*. http://www.fao.org/faostat/en/#rankings/countries_by_commodity (accessed 7 October 2021).

Franco, P., Spinozzi, S., Pagnotta, E., Lazzeri, L., Ugolini, L., Camborata, C. and Roda, A. 2016. Development of a liquid chromatography – electrospray ionisation – tandem mass spectrometry method for the simultaneous analysis of intact glucosinolates and isothiocyanates in Brassicaceae seeds and functional foods. *J. Chromatogr. A* 1428: 154–161.

Fukumoto, L. R. and Mazza, G. 2000. Assessing antioxidant and prooxidant activities of phenolic compounds. *J. Agric. Food Chem.* 48(8): 3597–3604.

Geng, F., Tang, L., Li, Y., Yang, L., Choi, K. S., Kazim, A. L. and Zhang, Y. 2011. Allyl isothiocyanate arrests cancer cells in mitosis, and mitotic arrest in turn leads to apoptosis via Bcl-2 protein phosphorylation. *J. Biol. Chem.* 286(37): 32259–32267.

Gliszczyńska-Świgło, A., Ciska, E., Pawlak-Lemańska, K., Chmielewski, J., Borkowski, T. and Tyrakowska, B. 2006. Changes in the content of health-promoting compounds and antioxidant activity of broccoli after domestic processing. *Food Addit. Contam.* 23(11): 1088–1098.

Gonçalves, E. M., Abreu, M., Brandão, T. R. S. and Silva, C. L. M. 2011. Degradation kinetics of colour, vitamin C and drip loss in frozen broccoli (*Brassica oleracea* L. ssp. *Italica*) during storage at isothermal and non-isothermal conditions. *Int. J. Refrig.* 34(8): 2136–2144.

Gonçalves, E. M., Alegria, C. and Abreu, M. 2013. Benefits of Brassica nutraceutical compounds in human health. In: *Brassicaceae: Characterisation, Functional Genomics and Health Benefits*, Lang, M. (Ed.). Nova Press Publishers, Inc., New York, pp. 19–65. ISBN: 978-1-62808-869-4 (eBook).

Gonçalves, E. M., Pinheiro, J., Alegria, C., Abreu, M., Brandão, T. R. S. and Silva, C. L. M. 2009. Degradation kinetics of peroxidase enzyme, phenolic content, and physical and sensorial characteristics in broccoli (*Brassica oleracea* L. ssp. Italica) during blanching. *J. Agric. Food Chem.* 57(12): 5370-5375.

Gründemann, C. and Huber, R. 2018. Chemoprevention with isothiocyanates - From bench to bedside. *Cancer Lett.* 414: 26-33.

Guo, Y. and Bruno, R. S. 2015. Endogenous and exogenous mediators of quercetin bioavailability. *J. Nutr. Biochem.* 26(3): 201-210.

Gupta, U. S. 2011. *Brassica Vegetables, What's New About Crop Plants* (1st edn.). Science Publishers, Enfield, NH; Marketed and distributed by CRC Press, Boca Raton, FL, pp. 378-402.

Guzman, I., Yousef, G. G. and Brown, A. F. 2012. Simultaneous extraction and quantitation of carotenoids, chlorophylls, and tocopherols in brassica vegetables. *J. Agric. Food Chem.* 60(29): 7238-7244.

Halkier, B. A. and Gershenzon, J. 2006. Biology and biochemistry of glucosinolates. *Annu. Rev. Plant Biol.* 57(1): 303-333.

Harbaum, B., Hubbermann, E. M., Wolff, C., Herges, R., Zhu, Z. and Schwarz, K. 2007. Identification of flavonoids and hydroxycinnamic acids in Pak Choi varieties (*Brassica campestris* L. ssp. chinensis var. communis) by HPLC-ESI-MS n and NMR and their quantification by HPLC-DAD. *J. Agric. Food Chem.* 55(20): 8251-8260.

Hashem, F., Motawea, H., El-Shabrawy, A., El-Sherbini, S. M., Shaker, K. and Farrag, A. R. H. 2013. Hepatoprotective activity of *Brassica oleracea* L. var. *Italica* Egypt. *Pharm. J.* 12: 177-178.

Hashem, F. A., Motawea, H., El-Shabrawy, A. E., Shaker, K. and El-Sherbini, S. 2012. Myrosinase hydrolysates of *Brassica oleraceae* L. Var. *italica* Reduce the Risk of colon Cancer. *Phytother. Res.* 26(5): 743-747.

Hedges, L. J. and Lister, C. E. 2007. Nutritional attributes of herbs. *Crop and Food Research, 1891-89.* http://vegetables.designcom.co.nz/resources/1files/pdf/booklet_herbs_foodreport.pdf (07102021).

Herr, I. and Büchler, M. W. 2010. Dietary constituents of broccoli and other cruciferous vegetables: Implications for prevention and therapy of cancer. *Cancer Treat. Rev.* 36(5): 377-383.

Hussain, M. B., Hassan, S., Waheed, M., Javed, A., Farooq, M. A. and Tahir, A. 2019. Bioavailability and metabolic pathway of phenolic compounds. In: *Plant Physiological Aspects of Phenolic Compounds*, Soto-Hernández, M., García-Mateos, R. and Palma-Tenango, M. (Eds). IntechOpen.

Hwang, J. H. and Lim, S. B. 2014. Antioxidant and anti-inflammatory activities of broccoli florets in LPS-stimulated RAW 264.7 cells. *Prev. Nutr. Food Sci.* 19(2): 89-97.

Hwang, J. H. and Lim, S. B. 2015. Antioxidant and anticancer activities of broccoli by-products from different cultivars and maturity stages at harvest. *Prev. Nutr. Food Sci.* 20(1): 8-14.

Idrees, N., Tabassum, B., Sarah, R. and Hussain, M. K. 2019. Natural compound from genus brassica and their therapeutic activities. In: *Natural Bio-Active Compounds*, Akhtar, M., Swamy, M. and Sinniah, U. (Eds). Springer, Singapore.

Ilahy, R., Tlili, I., Pék, Z., Montefusco, A., Siddiqui, M. W., Homa, F., Hdider, C., R'Him, T., Lajos, H. and Lenucci, M. S. 2020. Pre- and post-harvest factors affecting glucosinolate content in broccoli. *Front. Nutr.* 7(9): 147.

Jasmina, Č., Adisa, P., Milka, M. and Kasim, B. 2012. Antioxidative and antitumor properties of in vitro-cultivated broccoli (*Brassica oleracea* var. *italica*). *Pharm. Biol.*, 50(2): 175-181.

Jin, N., Jin, L., Luo, S., Tang, Z., Liu, Z., Wei, S., Liu, F., Zhao, X., Yu, J. and Zhong, Y. 2021. Comprehensive evaluation of amino acids and polyphenols in 69 varieties of green cabbage (*Brassica oleracea* L. var. *capitata* L.) based on multivariate statistical analysis. *Molecules* 26(17): 5355.

Kalpana, D. P., Gayathri, R., Gunassekaran, G. R., Murugan, S. and Sakthisekaran, D. 2013. Apoptotic role of natural isothiocyanate from broccoli (*Brassica oleracea italica*) in experimental chemical lung carcinogenesis. *Pharm. Biol.* 51(5): 621-628..

Kamarul-Zaman, M. A. and Mohamad Azzeme, A. 2018. Plant toxins: Alkaloids and their toxicities. *GSC Biol. Pharm. Sci.* 6(2): 021-029.

Kasikci, M. B. and Bagdatlioglu, N. 2016. Bioavailability of quercetin. *Curr. Res. Nutr. Food Sci.* 4(SI2): 146-151.

Kim, D. H., Jung, E. A., Sohng, I. S., Han, J. A., Kim, T. H. and Han, M. J. 1998. Intestinal bacterial metabolism of flavonoids and its relation to some biological activities. *Arch. Pharm. Res.* 21(1): 17-23.

Kołodziejski, D., Piekarska, A., Hanschen, F. S., Pilipczuk, T., Tietz, F., Kusznierewicz, B. and Bartoszek, A. 2019. Relationship between conversion rate of glucosinolates to isothiocyanates/indoles and genotoxicity of individual parts of Brassica vegetables. *Eur. Food Res. Technol.* 245(2): 383-400.

Kurilich, A. C., Tsau, G. J., Brown, A., Howard, L., Klein, B. P., Jeffery, E. H., Kushad, M., Wallig, M. A. and Juvik, J. A. 1999. Carotene, tocopherol, and ascorbate contents in subspecies of *Brassica oleracea*. *J. Agric. Food Chem.* 47(4): 1576-1581.

Kwon, K. H., Barve, A., Yu, S., Huang, M. T. and Kong, A. N. 2007. Cancer chemoprevention by phytochemicals: Potential molecular targets, biomarkers and animal models. *Acta Pharmacol. Sin.* 28(9): 1409-1421.

Lafay, S. and Gil-Izquierdo, A. 2008. Bioavailability of phenolic acids. *Phytochem. Rev.* 7(2): 301-311.

Lai, R. H., Keck, A. S., Wallig, M. A., West, L. G. and Jeffery, E. H. 2008. Evaluation of the safety and bioactivity of purified and semi-purified glucoraphanin. *Food Chem. Toxicol.* 46(1): 195-202.

Le, T. N., Chiu, C. H. and Hsieh, P. C. 2020. Bioactive compounds and bioactivities of *Brassica oleracea* L. var. *Italica* Sprouts and microgreens: An Updated Overview from a nutraceutical Perspective. *Plants (Basel)* 9(8): 946.

Lee, J. J., Shin, H. D., Lee, Y. M., Kim, A. R. and Lee, M. Y. 2009. Effect of broccoli sprouts on cholesterol-lowering and anti-obesity effects in rats fed high fat diet. *Korean Soc. Food Sci. Nutr.* 38: 309-318.

Mahima, R. A., Verma, A. K., Amit, K., Tiwari, R. and Kapoor, S. 2014. Phytonutrients and nutraceuticals in vegetables and their multi-dimensional medicinal and health benefits for humans and their companion animals: A review. *J. Biol. Sci.* 14(1): 1-19.

Mahn, A. and Reyes, A. 2012. An overview of health-promoting compounds of broccoli (*Brassica oleracea* var. *italica*) and the effect of processing. *Food Sci. Technol. Int.* 18(6): 503-514.

Managa, M. G., Sultanbawa, Y. and Sivakumar, D. 2020. Effects of different drying methods on untargeted phenolic metabolites, and antioxidant activity in Chinese cabbage (*Brassica rapa* L. subsp. *chinensis*) and nightshade (*Solanum retroflexum* Dun.). *Molecules* 25(6): 1326.

Manchali, S., Murthy, K. N. C. and Patil, B. S. 2012. Crucial facts about health benefits of popular cruciferous vegetables. *J. Funct. Foods* 4(1): 94–106.

Martínez, S., Armesto, J., Gómez-Limia, L. and Carballo, J. 2020. Impact of processing and storage on the nutritional and sensory properties and bioactive components of Brassica spp. A review. *Food Chem.* 313: 126065.

Mattioli, R., Francioso, A., Mosca, L. and Silva, P. 2020. Anthocyanins: A comprehensive review of their chemical properties and health effects on cardiovascular and neurodegenerative diseases. *Molecules* 25(17): 3809.

Maynard, D. N. and Hochmuth, G. J. 2006. *Knott's Handbook for Vegetable Growers* (5th edn.). John Wiley & Sons, New York.

McClements, D. J. 2019. Nutraceuticals: Superfoods or superfads? In: *Future Foods*. Copernicus, Chem, pp. 167–201.

McNaughton, S. A. and Marks, G. C. 2003. Development of a food composition database for the estimation of dietary intakes of glucosinolates, the biologically active constituents of cruciferous vegetables. *Br. J. Nutr.* 90(3): 687–697.

Mennicke, W. H., Kral, T., Krumbiegel, G. and Rittmann, N. 1987. Determination of N-acetyl-S-(N-alkylthiocarbamoyl)-L-cysteine, a principal metabolite of alkyl isothiocyanates, in rat urine. *J. Chromatogr.* 414(1): 19–24.

Moon, J. K., Kim, J. R., Ahn, Y. J. and Shibamoto, T. 2010. Analysis and anti-Helicobacter activity of sulforaphane and related compounds present in broccoli (*Brassica oleracea* L.) sprouts. *J. Agric. Food Chem.* 58(11): 6672–6677.

Motawea, M. H., Hashem, F. A., El-Shabrawy, A. E. and El-Sherbini, S. 2010. *Brassica oleraceae* L var. *italica*: A nutritional supplement for weight loss. *Aust J Herb.* 22: 127–131.

Mrkìc, V., Cocci, E., Rosa, M. D. and Sacchetti, G. 2006. Effect of drying conditions on bioactive compounds and antioxidant activity of broccoli (*Brassica oleracea* L.). *J. Sci. Food Agric.* 86(10): 1559–1566.

Ms, A., Ur, R., Thampi, S. S., Sj, A. and Sebastian, K. 2019. Nutraceuticals in vegetables: New breeding approaches for nutrition, food and health : A review. *J. Pharmacogn. Phytochem.*, 8(1): 677–682.

Mukherjee, S., Gangopadhyay, H. and Das, D. K. 2008. Broccoli: A unique vegetable that protects mammalian hearts through the redox cycling of the thioredoxin superfamily. *J. Agric. Food Chem.* 56(2): 609–617.

Munday, R., Mhawech-Fauceglia, P., Munday, C. M., Paonessa, J. D., Tang, L., Munday, J. S., Lister, C., Wilson, P., Fahey, J. W., Davis, W. and Zhang, Y. 2008. Inhibition of urinary bladder carcinogenesis by broccoli sprouts. *Cancer Res.* 68(5): 1593–1600.

Murota, K. and Terao, J. 2003. Antioxidative flavonoid quercetin: Implication of its intestinal absorption and metabolism. *Arch. Biochem. Biophys.* 417(1): 12–17.

Nawaz, H., Shad, M. A. and Muzaffar, S. 2018. Phytochemical composition and antioxidant potential of brassica. In: *Brassica Germplasm - Characterisation, Breeding and Utilisation*, El-Esawi, M. A. (Ed.). IntechOpen, London. https://www.intechopen.com/chapters/61233 (07102021).

Neugart, S., Baldermann, S., Hanschen, F. S., Klopsch, R., Wiesner-Reinhold, M. and Schreiner, M. 2018. The intrinsic quality of brassicaceous vegetables: How secondary plant metabolites are affected by genetic, environmental, and agronomic factors. *Sci. Hortic.* 233: 460–478.

Nguyen, V. P., Stewart, J. D., Ioannou, I. and Allais, F. 2021. Sinapic acid and sinapate esters in brassica: Innate accumulation, biosynthesis, accessibility via chemical synthesis or recovery from biomass, and biological activities. *Front. Chem.* 9: 664602.

Nugrahedi, P. Y., Dekker, M. and Verkerk, R. 2016. Processing and preparation of *brassica* vegetables and the fate of glucosinolates. In: *Glucosinolates. Reference Series in Phytochemistry*, Mérillon, J. M. and Ramawat, K. (Eds). Springer, Cham.

Oliveira, S. M., Ramos, I. N., Brandão, T. R. S. and Silva, C. L. M. 2015. Effect of air-drying temperature on the quality and bioactive characteristics of dried galega kale (*Brassica oleracea* L. var. *Acephala*). *J. Food Process. Preserv.* 39(6): 2485–2496.

Oliviero, T., Verkerk, R. and Dekker, M. 2018. Isothiocyanates from *brassica* vegetables-effects of processing, cooking, mastication, and digestion. *Mol. Nutr. Food Res.* 62(18): e1701069.

Owis, A. I. 2015. Broccoli; The Green beauty: A review. *J. Pharm. Sci. Res.* 7(9): 696–703.

Pareek, S., Sagar, N. A., Sharma, S., Kumar, V., Agarwal, T., González-Aguilar, G. A. and Yahia, E. M. 2017. Chlorophylls: Chemistry and biological functions. In: *Fruit and Vegetable Phytochemicals*, Yahia, E. M. (Ed). John Wiley & Sons, Ltd, Hoboken, NJ; Chichester.

Pham, N. A., Jacobberger, J. W., Schimmer, A. D., Cao, P., Gronda, M. and Hedley, D. W. 2004. The dietary isothiocyanate sulforaphane targets pathways of apoptosis, cell cycle arrest, and oxidative stress in human pancreatic cancer cells and inhibits tumor growth in severe combined immunodeficient mice. *Mol. Cancer Ther.* 3(10): 1239–1248.

Pinheiro-Sant'ana, H. M., Guinazi, M., Oliveira, D. S., Della Lucia, C. M., Reis, B. L. and Brandão, S. C. 2011. Method for simultaneous analysis of eight vitamin E isomers in various foods by high performance liquid chromatography and fluorescence detection. *J. Chromatogr. A* 1218(47): 8496–8502.

Plaszkó, T., Szűcs, Z., Vasas, G. and Gonda, S. 2021. Effects of glucosinolate-derived isothiocyanates on fungi: A comprehensive review on direct effects, mechanisms, structure-activity relationship data and possible agricultural applications. *J. Fungi (Basel)* 7(7): 539.

Podsedek, A. 2007. Natural antioxidants and antioxidant capacity of Brassica vegetables: A review. *LWT Food Sci. Technol.* 40(1): 1–11.

Prieto, M. A., López, C. J. and Simal-Gandara, J. 2019. Glucosinolates: Molecular structure, breakdown, genetic, bioavailability, properties and healthy and adverse effects. *Adv. Food Nutr. Res.* 90: 305–350.

Ramirez, D., Abellán-Victorio, A., Beretta, V., Camargo, A. and Moreno, D. A. 2020. Functional ingredients from Brassicaceae species: Overview and perspectives. *Int. J. Mol. Sci.* 21(6): 1998.

Ramos, S. J., Yuan, Y., Faquin, V., Guilherme, L. R. G. and Li, L. 2011. Evaluation of genotypic variation of broccoli (*Brassica oleracea* var. *italica*) in response to selenium treatment. *J. Agric. Food Chem.* 59(8): 3657–3665.

Ramos García, M., Hernández López, M., Barrera Necha, L. L., Bautista Baños, S., Troncoso Rojas, R. and Bosquez Molina, E. 2012. In vitro response of *Fusarium oxysporum* isolates to isothiocyanates application. *Rev. Mex. Fitopatol.* 30(1): 1–10.

Rechner, A. R., Smith, M. A., Kuhnle, G., Gibson, G. R., Debnam, E. S., Srai, S. K. S., Moore, K. P. and Rice-Evans, C. A. 2004. Colonic metabolism of dietary polyphenol: Influence of structure on microbial fermentation products. *Free Radic. Biol. Med.* 36(2): 212–225.

Riso, P., Martini, D., Møller, P., Loft, S., Bonacina, G., Moro, M. and Porrini, M. 2010. DNA damage and repair activity after broccoli intake in young healthy smokers. *Mutagenesis* 25(6): 595–602.

Ritz, S. A., Wan, J. and Diaz-Sanchez, D. 2007. Sulforaphane-stimulated phase II enzyme induction inhibits cytokine production by airway epithelial cells stimulated with diesel extract. *Am. J. Physiol. Lung Cell. Mol. Physiol.* 292(1): L33–L39.

Rosa, E. A. S. 1997. Daily variation in glucosinolate concentrations in the leaves and roots of cabbage seedlings in two constant temperature regimes. *J. Sci. Food Agric.* 73(3): 364–368.

Roy, M. K., Juneja, L. R., Isobe, S. and Tsushida, T. 2009. Steam processed broccoli (*Brassica oleracea*) has higher antioxidant activity in chemical and cellular assay systems. *Food Chem.* 114(1): 263–269.

Sachdeva, V., Roy, A. and Bharadvaja, N. 2020. Current prospects of nutraceuticals: A review. *Curr. Pharm. Biotechnol.* 21(10): 884–896.

Sanlier, N. and Guler, S. M. 2018. The benefits of brassica vegetables on human health. *J. Hum. Health Res* 1(1): 104.

Sarris, J., Byrne, G. J., Stough, C., Bousman, C., Mischoulon, D., Murphy, J., Macdonald, P., Adams, L., Nazareth, S., Oliver, G., Cribb, L., Savage, K., Menon, R., Chamoli, S., Berk, M. and Ng, C. H. 2019. Nutraceuticals for major depressive disorder- more is not merrier: An 8-week double-blind, randomised, controlled trial. *J. Affect. Disord.* 245: 1007–1015.

Scalbert, A. and Williamson, G. 2000. Dietary intake and bioavailability of polyphenols. *J. Nutr.* 130(8S Suppl): 2073S–2085S.

Schneider, C. 2005. Chemistry and biology of vitamin E. *Mol. Nutr. Food Res.* 49(1): 7–30.

Schreiner, M., Peters, P. and Krumbein, A. 2007. Changes of glucosinolates in mixed fresh-cut broccoli and cauliflower florets in modified atmosphere packaging. *J. Food Sci.* 72(8): S585–S589.

Shao, Y., Jiang, J., Ran, L., Lu, C., Wei, C. and Wang, Y. 2014. Analysis of flavonoids and hydroxycinnamic acid derivatives in rapeseeds (*Brassica napus* L. var. *napus*) by HPLC-PDA-ESI(−)-MS^n/HRMS. *J. Agric. Food Chem.* 62(13): 2935–2945.

Shapiro, T. A., Fahey, J. W., Wade, K. L., Stephenson, K. K. and Talalay, P. 2001. Chemoprotective glucosinolates and isothiocyanates of broccoli sprouts: Metabolism and excretion in humans. *Cancer Epidemiol. Biomarkers Prev.* 10(5): 501–508.

Silberberg, M., Morand, C., Mathevon, T., Besson, C., Manach, C., Scalbert, A. and Remesy, C. 2006. The bioavailability of polyphenols is highly governed by the capacity of the intestine and of the liver to secrete conjugated metabolites. *Eur. J. Nutr.* 45(2): 88–96.

Skandrani, I., Limem, I., Neffati, A., Boubaker, J., Ben Sghaier, M., Bhouri, W., Bouhlel, I., Kilani, S., Ghedira, K. and Chekir-Ghedira, L. 2010. Assessment of phenolic content, free-radical-scavenging capacity genotoxic and anti-genotoxic effect of aqueous extract prepared from *Moricandia arvensis* leaves. *Food Chem. Toxicol.* 48(2): 710–715.

Smith, T. K., Lund, E. K., Parker, M. L., Clarke, R. G. and Johnson, I. T. 2004. Allyl-isothiocyanate causes mitotic block, loss of cell adhesion and disrupted cytoskeletal structure in HT29 cells. *Carcinogenesis* 25(8): 1409–1415.

Song, B., Xu, H., Chen, L., Fan, X., Jing, Z., Chen, S. and Xu, Z. 2020. Study of the relationship between leaf color formation and anthocyanin metabolism among different purple Pakchoi lines. *Molecules* 25(20): 4809.

Song, L., Morrison, J. J., Botting, N. P. and Thornalley, P. J. 2005. Analysis of glucosinolates, isothiocyanates, and amine degradation products in vegetable extracts and blood plasma by LC-MS/MS. *Anal. Biochem.* 347(2): 234-243.

Song, L. and Thornalley, P. J. 2007. Effect of storage, processing and cooking on glucosinolate content of Brassica vegetables. *Food Chem. Toxicol.* 45(2): 216-224.

Sosnowska, B., Penson, P. and Banach, M. 2017. The role of nutraceuticals in the prevention of cardiovascular disease. *Cardiovasc. Diagn. Ther.* 7 (Suppl. 1): S21-S31.

Sotelo, T., Lema, M., Soengas, P., Cartea, M. E. and Velasco, P. 2015. In vitro activity of glucosinolates and their degradation products against brassica-pathogenic bacteria and fungi. *Appl. Environ. Microbiol.* 81(1): 432-440.

Sun, J., Wang, Y., Pang, X., Tian, S., Hu, Q., Li, X., Liu, J., Wang, J. and Lu, Y. 2021. The effect of processing and cooking on glucoraphanin and sulforaphane in brassica vegetables. *Food Chem.* 360: 130007.

Swallah, M. S., Fu, H., Sun, H., Affoh, R. and Yu, H. 2020. The impact of polyphenol on general nutrient metabolism in the monogastric gastrointestinal tract. *J. Food Qual.* 2020: 1-12:5952834.

Tabela da Composição dos Alimentos (TCA) 2019. *Instituto Nacional de Saúde Doutor Ricardo Jorge*. Portuguesa. Available at: http://portfir.insa.pt/foodcomp/introduction (accessed 9 Octorber 2021).

Tang, L., Zhang, Y., Jobson, H. E., Li, J., Stephenson, K. K., Wade, K. L. and Fahey, J. W. 2006. Potent activationof mitochondria-mediated apoptosis and arrest in S and M phases of cancer cells by a broccoli sprout extract. *Mol. Cancer Ther.* 5(4): 935-944.

Tarko, T., Duda-Chodak, A. and Zajac, N. 2013. Digestion and absorption of phenolic compounds assessed by in vitro simulation methods. A review. *Rocz. Panstw. Zakl. Hig.* 64(2): 79-84.

Thakur, N., Raigond, P., Singh, Y., Mishra, T., Singh, B., Kumar Lal, M. K. and Dutt, S. 2020. Recent updates on bioaccessibility of phytonutrients. *Trends Food Sci. Technol.* 97: 366-380.

Thejass, P. and Kuttan, G. 2007. Immunomodulatory activity of sulforaphane, a naturally occurring isothiocyanate from broccoli (*Brassica oleracea*). *Phytomedicine* 14(7-8): 538-545.

Tilg, H. 2012. Diet and intestinal immunity. *N. Engl. J. Med.* 366(2): 181-183.

Tomás-Barberán, F. A., Gil-Izquierdo, A. and Moreno, D. A. 2009. 8-Bioavailability and metabolism of phenolic compounds and glucosinolates. In: *Woodhead Publishing Series in Food Science, Technology and Nutrition, Designing Functional Foods* (vol. 2009), McClements, D. J. and Decker, E. A. (Eds). Woodhead Publishing, Cambridge, pp. 194-229. ISBN: 9781845694326.

Truong, T., Baron-Dubourdieu, D., Rougier, Y. and Guénel, P. 2010. Role of dietary iodine and cruciferous vegetables in thyroid cancer: A countrywide case-control study in New Caledonia. *Cancer Causes Control* 21(8): 1183-1192.

Vallejo, F., Garcia-Viguera, C. and Tomás-Barberán, F. A. 2003. Changes in broccoli (*Brassica oleracea* L. var. *italica*) health-promoting compounds with inflorescence development. *J. Agric. Food Chem.* 51(13): 3776-3782.

Vasanthi, H. R., ShriShriMal, N. and Das, D. K. 2012. Retraction Notice: Phytochemicals from plants to combat cardiovascular disease. *Curr. Med. Chem.* 19(14): 2242–2251.

Vazquez, M. A. and Miatello, R. M. 2010. Organosulfur compounds and cardiovascular diseases. *Mol. Aspects Med.* 31(6): 540–545.

Verkerk, R., Schreiner, M., Krumbein, A., Ciska, E., Holst, B., Rowland, I., de Schrijver, R., Hansen, M., Gerhäuser, C., Mithen, R. and Dekker, M. 2009. Glucosinolates in Brassica vegetables: The influence of the food supply chain on intake, bioavailability and human health. *Mol. Nutr. Food Res.* 53 (Suppl. 2): S219.

Wang, H., Khor, T. O., Shu, L., Su, Z. Y., Fuentes, F., Lee, J. H. and Kong, A. N. 2012. Plants vs. Cancer: A review on Natural Phytochemicals in Preventing and Treating Cancers and Their druggability. *Anti Cancer Agents Med. Chem.* 12(10): 1281–1305.

Wei, J., Miao, H. Y. and Wang, Q. M. 2011. Effect of glucose on glucosinolates, antioxidants and metabolic enzymes in *Brassica* sprouts. *Sci. Hort.* 129(4): 535–540.

Wiktorska, K., Misiewicz-Krzeminska, I., Rafal, S., Lubelska, K. and Kasprzycka-Guttman, T. 2009. Sulforaphane and its analogues inhibit CYP1A1 and CYP1A2 activity induced by benzo[a]pyrene. *J. Biochem. Mol. Toxicol.* 23(1): 18–28.

Williams, D. J., Critchley, C., Pun, S., Chaliha, M. and O'Hare, T. J. 2009. Differing MEChA-nisms of simple nitrile formation on glucosinolate degradation in *Lepidium sativum* and *Nasturtium officinale* seeds. *Phytochemistry* 70(11–12): 1401–1409.

Winter, J., Popoff, M. R., Grimont, P. and Bokkenheuser, V. D. 1991. *Clostridium orbiscindens* sp. Nov., a human intestinal bacterium capable of cleaving the flavonoid C-ring. *Int. J. Syst. Bacteriol.* 41(3): 355–357.

Xu, L., Nagata, N. and Ota, T. 2018. Glucoraphanin: A broccoli sprout extract that ameliorates obesity-induced inflammation and insulin resistance. *Adipocyte* 7(3): 218–225.

Yilmaz, M. S., Şakiyan, Ö., Barutcu Mazi, I. and Mazi, B. G. 2019. Phenolic content and some physical properties of dried broccoli as affected by drying method. *Food Sci. Technol. Int.* 25(1): 76–88.

Yoon, G. A., Yenum, K. J., Cho, Y. S., Chen, C. Y., Tang, G., Blumberg, J. B., Russell, R. M., Yoon, S. and Lee-Kim, Y. C. 2012. Carotenoids and total phenolic contents in plant food commonly consumed in Korea. *Nutr. Res. Pract.* 6(6): 481–490.

Zhang, D. and Hamauzu, Y. 2004. Phenolics, ascorbic acid, carotenoids and antioxidant activity of broccoli and their changes during conventional and microwave cooking. *Food Chem.* 88(4): 503–509.

Zhang, Y., Munday, R., Jobson, H. E., Munday, C. M., Lister, C., Wilson, P., Fahey, J. W. and Mhawech-Fauceglia, P. 2006. Induction of GST and NQO1 in cultured bladder cells and in the urinarybladders of rats by an extract of broccoli (*Brassica oleracea italica*) sprouts. *J. Agric. Food Chem.* 54(25): 9370–9376.

Zhao, Z. H. and Moghadasian, M. H. 2010. Bioavailability of hydroxycinnamates: A brief review of in vivo and in vitro studies. *Phytochem. Rev.* 9(1): 133–145.

Zhu, M., Li, W., Dong, X., Chen, Y., Lu, Y., Lin, B., Guo, J. and Li, M. 2017. Benzyl-isothiocyanate induces apoptosis and inhibits migration and invasion of hepatocellular carcinoma cells in vitro. *J. Cancer* 8(2): 240–248.

Index

Lightning Source UK Ltd.
Milton Keynes UK
UKHW020747020922
408225UK00001B/24

9 781786 768506